Anticarcinogenesis and Radiation Protection

Anticarcinogenesis and Radiation Protection

Edited by

Peter A. Cerutti
Swiss Institute for Experimental Cancer Research
Lausanne, Switzerland

Oddvar F. Nygaard
Case Western Reserve University
Cleveland, Ohio

and

Michael G. Simic
National Bureau of Standards
Gaithersburg, Maryland

Plenum Press • New York and London

Library of Congress Cataloging in Publication Data

International Conference on Anticarcinogenesis and Radiation Protection (2nd: 1987: Gaithersburg, Md.)
Anticarcinogenesis and radiation protection.

"Proceedings of the Second International Conference on Anticarcinogenesis and Radiation Protection, held March 8-12, 1987, in Gaithersburg, Maryland"—T.p. verso.
Includes bibliographies and index.
1. Tumors, Radiation-induced—Prevention—Congresses. 2. Radiation-protective agents—Congresses. I. Cerutti, Peter A. II. Nygaard, Oddvar F. III. Simic, Michael G. IV. Title. [DNLM: 1. Antineoplastic Agents—congresses. 2. Carcinogens—congresses. 3. Radiation Protection—congresses. W3 IN1682 2nd 1987a / QV 269 I602 1987a]
RC268.55.I57 1987 616.99'4 87-36133
ISBN 978-1-4615-6464-5 ISBN 978-1-4615-6462-1(eBook)
DOI 10.1007/978-1-4615-6462-1

Proceedings of the Second International Conference on Anticarcinogenesis and Radiation Protection, held March 8-12, 1987, in Gaithersburg, Maryland

Softcover reprint of the hardcover 1st edition 1987
A Division of Plenum Publishing Corporation
233 Spring Street, New York, N.Y. 10013

PREFACE

This book is based on the invited and contributed papers presented at the 2nd International Conference on Anticarcinogenesis and Radiation Protection held at the National Bureau of Standards, Gaithersburg, Maryland, USA, on March 8-12, 1987. The conference documented developments that have taken place in areas that were addressed during the first conference in 1982. A number of new topics, such as biological response modifiers, were included because of their emerging relevance to anticarcinogenesis and radiation protection. The organization of the material in this book does not follow the conference program; rather, we have attempted to provide a different sequence for didactic reasons.

The aim of the conference, which is reflected in this book, was to promote further development of mechanistic approaches to cancer prevention and treatment based on recent progress in molecular biology and free radical chemistry. At the basis of carcinogenesis lie changes in the dynamics of growth and differentiation of specific cell subpopulations in the target tissue. These changes are brought about by selective toxicity and modulation of gene expression that are induced by xenobiotic carcinogens and affected by physiological and genetic factors. The book deals with oxidative stress and molecular damage caused by radiation and chemical pro-oxidants and their role in carcinogenesis, and it discusses mechanisms of deregulation of the expression of oncogenes and other genes involved in carcinogenic initiation and promotion. Attention is given to the mechanisms of chemoprevention, including the effects of some dietary constituents, and to approaches that amplify the cellular and organismic defenses against the initiation and growth of tumors.

The organizers of the conference believe that the publication of these proceedings will stimulate further the interaction and cross-fertilization of ideas among chemists, biochemists, and molecular biologists, as well as clinicians concerned about the causes and treatment of cancer.

P. Cerutti
O.F. Nygaard
M.G. Simic

ACKNOWLEDGMENTS

The Chairman and the Co-Chairmen of the 2nd International Conference on Anticarcinogenesis and Radiation Protection wish to acknowledge financial support and/or sponsorship by The National Bureau of Standards, The Environmental Protection Agency, The National Cancer Institute, and The National Council for Radiation Protection and Measurements.

A great debt of gratitude is owed to the members of the Organizing Committee: Donald C. Borg, Randall S. Caswell, Bernard D. Goldstein, Eric J. Hall, John B. Little, David G. Longfellow, Dan L. Longo, and to the Session Chairmen: William A. Pryor, Lawrence J. Marnett, Stuart H. Yuspa, Curtis C. Harris, Arthur B. Pardee, Isaiah J. Fidler, James E. Talmadge, and Charles H. Hennekens, all of whom assisted in the organization of the program.

The continued support and encouragement by Ernest Ambler, Director of The National Bureau of Standards; Raymond G. Kammer, Deputy Director of the National Bureau of Standards; and Richard H. Adamson, Director of The Division of Cancer Etiology, National Cancer Institute, have been greatly appreciated.

Finally, we wish to express our gratitude to The National Bureau of Standards for providing facilities and management assistance for the Conference. Special thanks go to Janet N. Hauber, Conference Coordinator; to the NBS Arrangement Committee: Sara R. Torrence, Kathy C. Stang and Mary E. Conneran; to other NBS staff: Gloria Wiersma and Cassandra D. Beck; and to Technical Writer Karen A. Taylor of the International Life Sciences Institute.

CONTENTS

FREE RADICAL MECHANISMS

OXYGEN SPECIES IN CARCINOGENESIS

ENZYMATIC REPAIR OF DNA DAMAGE

TUMOR PROMOTION AND PROGRESSION

CHEMOPREVENTION

RADIATION DAMAGE AND PROTECTION

THIOLS AS ANTICARCINOGENS AND RADIOPROTECTORS

CANCER THERAPY

BIOLOGICAL RESPONSE MODIFIERS

THE INVOLVEMENT OF FREE RADICALS IN CHEMICAL CARCINOGENESIS

William A. Pryor

Biodynamics Institute, Louisiana State University
711 Choppin
Baton Rouge, Louisiana 70803

INTRODUCTION

Cancer is a multi-step process, involving three stages that can be operationally distinguished: initiation, promotion and progression. Initiation involves an irreversible alteration of the cellular DNA that permits the transformation of the cell to a non-malignant state. Promotion produces conditions that allow the initiated cell to become malignant, and progression is the growth of the malignant cell to a tumor.

Promotion, which occurs over a fairly long period and is reversible, involves a complex cascade of cellular changes, not all of which are understood. However, superoxide and other oxy-radicals are known to be involved (1-4). Since promotion is reversible, there continues to be hope that the use of antioxidant and other strategies that control free radical reactions can protect initiated cells against promotion and, thus, prevent the ultimate development of a tumor (5,6).

The involvement of radicals in promotion has been reviewed by a number of authors (2,7-10). In this article, I wish to turn instead to a more controversial and even less understood area, the involvement of free radicals in the initiation of tumor development.

MECHANISMS BY WHICH RADICALS MIGHT BE INVOLVED IN TUMOR INITIATION

Table 1 summarizes five possible mechanisms for the activation and binding of carcinogens to DNA based on whether radicals are involved in the activation or the binding steps. Class I, a group that does not involve radical reactions, has become familiar from the studies of benzo[a]pyrene (BaP) activated by cytochrome P450. Most workers believe that P450 catalyzes the oxidation of BaP to the dihydrodiol-epoxide (PBDE) by reactions that do not involve <u>free</u> radicals (11). [Radicaloid species, such as ferryl species, may be involved, however (12).] Thus, the entry in the first column in Table 1 is "no," since the activation of BaP by P450 does not appear to involve free radical intermediates. Furthermore, the BPDE intermediate binds to DNA by a mechanism that involves attack on the epoxide ring by nucleophilic centers in the DNA. Thus, binding of this carcinogen to DNA involves even-electron processes, and not free radicals, so mechanism I also gets a "no" in column 2 of Table 1.

Table 1. Possible roles of free radical reactions in chemical carcinogenesis.

Mechanism number	Are radical reactions involved in the activation of the procarcinogen to its carcinogenic form?	Are radicals involved in the binding of the carcinogen to DNA?	Possible examples (see text)
I	No	No	BaP/oxidation by P450
II	Yes	No	BaP/oxidation by PGS
III	No	Yes	?
IV	Yes	Yes	?
V	Oxy radicals produce DNA damage (e.g., nicks & thymine glycol) but binding of the carcinogens does not occur.		Radiation, bleomycin, cigarette tar

Mechanism II also is a well-established pathway. One of the best-known examples is the oxidation of the dihydrodiol of BaP to the BPDE intermediate by radicals produced during the activity of the prostaglandin synthetase (PGS) enzyme system (13). Thus, the same BPDE intermediate is produced from BaP by this route as by the P450 route, but in this case the PGS enzyme system is responsible and radicals are involved in the activation step, so mechanism II gets a "yes" in the first column of Table 1. However, the dihydrodiolepoxide again binds to DNA by a process that involves nucleophilic attack on the epoxide ring, not free radicals, so mechanism II has a "no" in column 2 of Table 1.

Another possibility for mechanism II is the oxidation of a PAH to hydroquinones and quinones. The oxidation of the PAH to the quinone could involve radical reactions (14), but the binding of the quinone or the hydroquinone to DNA could involve nucleophilic addition of DNA to the quinone. Thus, the activation of the procarcinogen could involve a radical but the binding is non-radical, as in mechanism II.

Mechanisms III and IV, which involve the binding of a radical species to DNA, appear to be quite rare. There are remarkably few examples of large organic radicals binding to DNA by free radical mechanisms, and, as will be seen below, those examples that are known appear to involve simplified in vitro systems that employ high radical concentrations. In contrast, a large number of examples are known in which positively charged ionic species or uncharged electrophilic species attack and bind to DNA by two-electron mechanisms; in fact, for these electrophilic cases, both regio- and stereoselectivities have been worked out in great detail. Oxy-radicals of the types that generally are of interest in free radical biology are electrophilic species (15,16). However, as will be discussed below, they are not sufficiently electrophilic to add to the double bonds of DNA bases in competition with abstracting hydrogen atoms from the deoxyribose portion of DNA.

Another possibility exists for mechanisms III and IV. Some quinones can be reduced to the hydroquinone by DT-diaphorase enzymes in a two-electron process that does not involve free radicals. The resulting hydroquinone, however, can spontaneously oxidize to the semiquinone. Thus, although the enzymatic conversion of the quinone to the hydroquinone may not involve radicals, semiquinone radicals may be formed, and these could bind to DNA by a radical pathway, in theory at least. This, therefore, is a mechanism with some features of type III, although, as we will discuss below, most semiquinones do not bind to DNA by radical mechanisms, but instead reduce oxygen to superoxide.

Despite these possibilities for mechanisms III and IV to rationalize radical involvement in chemical carcinogenesis, it appears that mechanism V, the production of hydroxyl and related oxy-radicals, is the most common way in which free radicals produce damage to DNA. Ionizing radiation, bleomycin, adriamycin, and cigarette tar all produce superoxide and a cascade of activated oxygen species that ultimately lead to the hydroxyl radical (21-25,29,30). The hydroxyl radical then is able to both abstract hydrogen atoms from DNA (producing nicks resulting from attack on the ribose or converting thymine to hydroxymethyluracil) and add to DNA (producing thymine glycol). In the case of these very different oxidative threats to the DNA, therefore, large organic free radicals do not bind to DNA and all of the damage is mediated through small oxy-radicals such as the hydroxyl radical.

IN VITRO EXAMPLES OF THE ADDITION OF RADICALS TO DNA

Where radicals have been found to bind to DNA, the systems appear to involve high radical concentrations. For example, ketyl radicals, produced by the abstraction of the hydrogen atom from the alpha-carbon of an alcohol, adds to both purines and pyrimidines (17). These addition reactions can be induced by ionizing radiation, by light, or by the thermal decomposition of peroxidic compounds. For example, a mixture of thymidine 5'-monophosphate (TMP) and adenosine 5'-monophosphate (AMP) adds isopropanol to give 38% of the TMP and 36% of the AMP addition product under photochemical conditions. With isopropanol for example, TMP forms the 5,6-dihydro-6-hydroxyalkyl adduct; the mechanism involves hydrogen abstraction from the alpha-carbon of isopropanol to give the ketyl radical, $(CH_3)_2C\text{-}OH$, which adds to the 6 position to give a carbon-centered radical that then abstracts hydrogen from another isopropanol molecule to produce the TMP derivative and another ketyl radical that can continue the chain. [The reaction mixtures were deoxygenated before irradiation, so the carbon-centered radical adds to the 5,6-double bond of thymine rather than reacting with dioxygen to produce a peroxyl radical.] The AMP product is produced by the addition of the ketyl radical to the 8-position, ultimately to give an 8-hydroxylalkyladenine derivative. Similar reactions appear to occur with amino acids (18).

It at first seems surprising that radical additions to DNA bases is not a common observation, since most heterocylic compounds add radicals under favorable circumstances (19,20). For the reactions of electrophilic oxy-radicals with DNA, however, one must consider not only whether or not the radical has the thermokinetic energy to add to a base but also the relative rate of that process versus hydrogen-atom abstraction reactions, for example from the sugar portion of the DNA. Of course, the deoxyribose is an H-atom donor toward oxy-radicals; in fact, alcohols often are used as hydroxyl radical scavengers since they possess alpha-hydrogen atoms that are particularly reactive toward electrophilic oxy-radicals. In fact, hydrogen-atom abstraction from the deoxyribose appears to be the type of reaction that is observed with most oxy-radical generating systems, leading to DNA strand nicking (21-27).

Table 2. The ratio of the rate constant for addition to double bonds or to rings over that for the abstraction of hydrogen atoms for oxy-radicals and carbon-centered radicals.

Radical	k_{ad}/k_H	Reference
tert-$C_4H_9O\cdot$	ca 0.03	Houk (30)
ROO·	ca 1	Houk (30)
HO·	> 10	Houk (30)
Methyl	ca 10	Pryor (47)
Phenyl	ca 1	Pryor (48)

(a) Note that these ratios are approximate, since the ratios for different types of substrates (including both olefins and aromatic systems for addition) and H-atom donors of differing reactivity for hydrogen-atom abstraction have been combined to give an overall profile.

THE HYDROXYL RADICAL AS AN UNUSUAL CASE: AN EXTREMELY ELECTROPHIC RADICAL WITH HIGH REACTIVITY

Although most oxy-radicals do not add to DNA, the hydroxyl radical does, and apparently does so very readily in vivo (28,29). The hydroxyl radical also abstracts hydrogen atoms, both from the deoxyribose and from allylic positions in the DNA bases such as the thymine methyl group. Why is this?

The unique reactivity of the hydroxyl radical results in part because it is the most electrophilic radical to which DNA is normally exposed. The ratio of the rate constant for addition to double bonds divided by the rate of H-atom abstraction from good H-atom donors by a free radical, k_{ad}/k_H, for a series of oxy-radicals correlates with the electrophilicity of the radical (30). Table 2 shows the ratio of rate constants for their addition to double bonds divided by the rate constant for H-abstraction by a series of oxy- and carbon-centered radicals for which data are available. Note that the hydroxyl radical is the only radical that appears to undergo both addition and H-abstraction reactions and that it is by far the most electrophilic radical among the oxy-radicals.

Carbon-centered radicals such as the methyl or phenyl radical also add to rings and double bonds. Why is the addition of such radicals not observed <u>in vivo?</u> Probably because there is no mechanism to generate carbon-centered radicals near cellular DNA; for the hydroxyl radical, species are known that can bind to DNA and then produce hydroxyl radicals. For a radical such as the methyl radical, this is not the case.

Thus, it takes a radical of very high electrophilicity and high reactivity (so the radical does not desplay high selectivity in its reactions) to penetrate the shield of easily abstracted H-atoms in the sugar moieties and reach the DNA bases. The hydroxyl radical meets these requirements, but slightly less reactive and less electrophilic radicals, such as alkoxyl or peroxyl, or nucleophilic radicals such as alkyl radicals (15-16), never reach the bases; their reduced electrophilicity allows them to abstract hydrogen atoms in the periphery of the DNA helix instead.

ASSOCIATION OF HYDROXYL RADICAL GENERATING SYSTEMS WITH DNA

Most workers believe that hydroxyl radical reactions with DNA are preceeded by the association of a species [e.g., the cigarette tar radical (21,22)] with DNA with some regio-specificity, followed by the production of the hydroxyl radical at that particular site, a type of reaction that has come to be called site specific damage to DNA (31,32). Thus, the nature of the binding of the species to DNA and the stereo- and regioselectivity of production of the radical in the vicinity of DNA may control the types of reactions that are observed. For example, the hydroxyl radical, which is so reactive that it can only diffuse 5-10 molecular diameters before it reacts (33), may be formed in a particular site on the DNA chain and then react close to that site (31,32).

IS THE INCORPORATION OF A HYDROXYL SUBSTITUENT INTO A DNA BASE A MARKER OF FREE RADICAL ATTACK?

Recently, 8-hydroxyguanine (8-HO-Gua) residues have been detected in DNA; these residues could be taken to be a marker of radical damage, resulting from the addition of a hydroxyl radical to guanine. However, in the specific case of 4-nitroquinoline 1-oxide (4NQO), a mechanism for production of 8-HO-Gua has been suggested that involves only even-electron reactions and not the hydroxyl radical (34). The mechanism involves the conversion of the nitro group to an amino group, which is acylated and leads to a nitronium ion that adds to Gua. The net reaction that occurs is the addition of RNH-OH across the 7,8-double bond of Gua, followed by the elimination or RNH-H, leaving a hydroxyl substituent at the 8-position. Thus, in this case, the incorporation of a hydroxyl group into Gua cannot be taken as evidence for the involvement of the hydroxyl radical.

Mitomycin C is another interesting case. This antitumor drug forms a stable semiquinone radical. However, the mechanism of binding of the drug to DNA involves the nucleophilic attack by a DNA amino group on an azocyclopropane ring of the mitomycin C, followed by a Michael addition of an amino group from the complementary DNA strand to a conjugated diene functionality (35). Thus, all of these processes are even-electron reactions of DNA nucleophiles with electrophilic centers in the drug. The mitomycin C semiquinone is present during these transformations, but it apparently plays no direct role in the binding.

SUMMARY AND CONCLUSIONS

Cells appear to be bathed in a continuous flux of oxy-radicals, produced by the oxidative stress that is commonplace to aerobically metabolizing cells. Oxygenated derivatives of DNA bases are continually excreted by animals at extremely high levels (23). Thus, radical damage to DNA is an ongoing fact of life that the cell must deal with (36-45). However, only a narrow band of electrophilicity, reactivity, and also a mode of production near DNA results in a radical that can add to a DNA base and produce a point mutation by direct chemical means. Of the oxy-radicals, only the hydroxyl radical appears able to meet these requirements.

It may be critically important that most radicals do not bind to DNA. Instead large organic radicals, such as semiquinones, produce a reasonably similar pattern of damage that results from the reaction of hydroxyl radicals produced by redox cycling. This may have the result in producing a limited variety of damage to DNA that can be enzymatically repaired in a limited set of processes (29,46).

ACKNOWLEDGEMENT

Research reported in this manuscript was supported in part by a grant from the National Institutes of Health and a contract from the National Foundation for Cancer Research. I also want to thank my coworkers for suggestions and ideas throughout the course of this research, particularly Drs. E. T. Borish and Professor D. F. Church.

REFERENCES

1. P. A. Cerutti, Prooxidant states and tumor promotion. Science **227**, 375 (1985).

2. T. J. Slaga, A. J. P. Klein-Szanto, L. L. Triplett, L. P. Yotti, and J. E. Trosko, Skin-tumor-promoting activity of benzoyl peroxide, a widely used free radical-generating compound. Science **213**, 1023-1025 (1981).

3. T. W. Kensler and B. G. Taffe, Free radicals in tumor promotion. Adv. Free Radical Biol. Med. 2, 347-388 (1986).

4. H. C. Birnboim, DNA strand breakage in human leukocytes exposed to a tumor promoter, phorbol myristate acetate. Science **215**, 1247 (1982).

5. R. J. Shamberger, F. F. Baughman, S. L. Kalchert, C. E. Willis, and G. C. Hoffman, Carcinogen-induced chromosomal breakage decreased by antioxidants. Proc. Natl. Acad. Sci. **70**, 1461-1468 (1973).

6. D. M. Shankel, P. F. Hartman, T. Kada, and H. Hollaender, (Eds.) Mechanisms of Antimutagenesis and Anticarcinogenesis, Plenum Press, New York (1986).

7. E. Heckler, N. E. Fusenig, W. Kunz, F. Marks, and H. W. Thielmann, (Eds.) Cocarcinogenesis and Biological Effects of Tumor Promoters, Raven Press, New York (1982).

8. T. J. Slaga, Overview of Tumor Protection in Animals. Environ. Health Perspect. **50**, 3-12 (1983).

9. C. V. Smith, H. Hughes, B. G. Lauterberg, and J. R. Mitchell, Chemical nature of reactive metabolites determines their biological interactions with glutathione. In Functions of Glutathione: Biochemical, Physiological, Toxicological, and Clinical Aspects (A. Larsson, Ed.) p. 125. Raven Press, New York (1983).

10. W. Troll, G. Witz, B. Goldstein, D. Stone, and T. Sugimura, The role of free oxygen radicals in tumor promotion and carcinogenesis. In Carcinogenesis A Comprehensive Survey (E. Hecker, N. E. Fusenig, W. Kunz, F. Marks, and H. W. Thielmann, Eds.) p. 593. Raven Press, New York (1982).

11. R. G. Harvey, (Ed.) Polycyclic Hydrocarbons and Carcinogenesis, American Chemical Society, ACS Symposium Series No. 283, Washington (1985).

12. P. R. Ortiz De Montellano, O. Augusto, F. Viola, and K. L. Kunze, Carbon radicals in the metabolism of alkyl hydrazines. J. Biol. Chem. **258**, 8623-8629 (1983).

13. L. J. Marnett, Hydroperoxide-dependent oxidation during prostaglandin

biosynthesis. In Free Radicals in Biology, Volume VI (W. A. Pryor, Ed.) pp. 64-95. Academic Press, New York (1984).

14. P. O. P. Ts'O, W. J. Caspary, and R. J. Lorentzen, The involvement of free radicals in chemical carcinogenesis. In Free Radicals in Biology, Volume III (W. A. Pryor, Ed.) pp. 251-303. Academic Press, New York (1977).

15. W. A. Pryor, F. Y. Tang, R. H. Tang, and D. F. Church, Polar character, rho value for the reaction with toluene, and the effect of radical polarity on the ratio of benzylic hydrogen abstraction to addition to aromatic rings. J. Am. Chem. Soc. **104**, 2885-2891 (1982).

16. W. A. Pryor, D. F. Church, F.Y. Tang, and R.H. Tang, The role of polar effects and bond dissociation energies in radical reactivities. In Frontiers of Free Radical Chemistry (W. A. Pryor Ed.) pp. 355-379. Academic Press, New York (1980).

17. A. A. Frimer, A. Havron, D. Leonov, J. Sperling, and D. Elad, Ultraviolet and gamma ray induced radical reactions of nucleic acid constituents. J. Am. Chem. Soc. **98**, 6026-6033 (1976).

18. J. E. Byfield, Y. C. Lee, and L. R. Bennett, Bonding of small molecules to DNA following izonizing radiation. Nature **225**, 859-860 (1970).

19. M. F. Zady and J. L. Wong, Kinetics and mechanism of carbon-8 methylation of purine bases and nucleopsides by methyl radical. J. Am. Chem. Soc. **99**, 5096-6002 (1977).

20. A. Sugimori, T. Yamada, H. Ishida, M. Nose, K. Terashima, and N. Oohata, Radiation induced alkylation of quinoline derivatives with alcohol. Bull. Chem. Soc. Japan **59**, 3905-3909 (1986).

21. E. T. Borish, S. Venugopal, W. A. Pryor, and W. A. Deutsch, DNA synthesis is blocked at cigarette tar-induced DNA single strand breaks (in press).

22. W. A. Pryor, K. Uehara, and D. F. Church, The chemistry and biochemistry of the radicals in cigarette smoke: ESR evidence for the binding of the tar radical to DNA and polynucleotides. In Oxygen Radicals in Chemistry and Biology (W. Bors, M. Saran and D. Tait, Eds.) pp. 193-201. Walter de Gruyter & Co., Berlin (1984).

23. E. T. Borish, J. P. Cosgrove, D. F. Church, W. A. Deutsch, and W. A. Pryor, Cigarette tar causes single strand breaks in DNA. Biochem. Biophys. Res. Commun. 133, 780 (1985).

24. E. T. Borish, J. P. Cosgrove, D. F. Church, W. A. Deutsch, and W. A. Pryor, Cigarette smoke, free radicals, and biological damage. In Superoxide and Superoxide Dismutase in Chemistry, Biology and Medicine (G. Rotilio, Ed.) p. 467. Elsevier, New York (1986).

25. D. F. Church and W. A. Pryor, The free radical chemistry of cigarette smoke and its toxicological implications. Environ. Health Perspect. **64**, 111 (1985).

26. J. P. Cosgrove, E. T. Borish, D. F. Church, and W. A. Pryor, The metal-mediated formation of hydroxyl radical by aqueous extracts of cigarette tar. Biochem. Biophys. Res. Commun. 132, 390 (1985).

27. L. Rabow, J. Stubbe, J. W. Kozarich, and J. A. Gerlt, Identification of the alkaline labile product accompanying cytosine relase during bleomycin mediated degradation of d(CGCGCG). J. Am. Chem. Soc. **108**, 7130-7131 (1987).

28. B. N. Ames, R. L. Saul, E. Schwiers, R. Adelman, and R. Cathcart, Oxidative DNA damage as related to cancer and aging: the assay of thymine glycol, thymidine glycol, and hydroxymethyluracil in human and rat urine. In Molecular Biology of Aging: Gene Stability and Gene Expression. Raven Press, New York (1984).

29. B. Demple and J. Halbrook, Inducible repair of oxidative DNA damage in Escherichia coli. Nature **304**, 466 (1983).

30. K. Houk, Molecular distortions and organic reactivity; additions, cycloadditions, and free radical reactions. In Frontiers of Free Radical Chemistry (W. A. Pryor, Ed.) pp. 43-71. Academic Press, New York (1980).

31. D. S. Sigman, Nuclease activity of 1,10-phenantroline-copper ion. Acc. Chem. Res. **19**, 180-186 (1986).

32. S. M. Hecht, The chemistry of activated bleomycin. Acc. Chem. Res. **19**, 383-391 (1986).

33. W. A. Pryor, Oxy-radicals and related species: their formation, lifetimes, and reactions. Annu. Rev. Physiol. **48**, 657-663 (1986).

34. D. Kohda, M. Tada, H. Kasai, S. Nishimura, and Y. Kawazoe, Formation of 8-hydroxylguanine residues in cellular DNA exposed to the carcinogen 4-nitroquinoline 1-oxide. Biochem. Biophis. Res. Commun. **139**, 626-632 (1986).

35. M. Tomaz, R. Lipman, D. Chowdary, J. Pawlak, G. L. Verdine, and K. Nakanishi, Isolation and structure of a covalent crosslink adduct between mitomycin C and DNA. Science **235**, 1204-1208 (1987).

36. E. L. Cavalieri and E. G. Rogan, One-electron and two-electron oxidation in aromatic hydrocarbon carcinogenesis. In Free Radicals in Biology, Volume VI (W.A. Pryor, Ed.) pp. 324-370. Academic Press, New York (1984).

37. E. L. Cavalieri and E. G. Rogan, One-electron oxidation in aromatic hydrocarbon carcinogenesis. In Polycyclic Hydrocarbons & Carcinogenesis (R. G. Harvey, Ed.) p. 289. American Chemical Society, Washington, DC (1985).

38. R. G. Cutler, Antioxidants, aging, and longevity. In Free Radicals in Biology, Volume VI (W. A. Pryor, Ed.) pp. 371-429. Academic Press, New York (1984).

39. W. A. Pryor, The role of free radical reactions in biological systems. In Free Radicals in Biology, Volume I (W. A. Pryor, Ed.) pp. 1-49. Academic Press, New York (1976).

40. W. A. Pryor, Free radicals in biology. The involvement of radical reactions in aging and carcinogenesis. In Medicinal Chemistry V (J. Mathieu, Ed.) pp. 331-345. Elsevier, Amsterdam (1977).

41. W. A. Pryor, The formation of free radicals and the consequences of their reactions in vivo. Photochem. Photobiol. **28**, 787-796 (1978).

42. W. A. Pryor, Free radical biology: xenobiotics, cancer, and aging. Ann. N. Y. Acad. Sci. **393**, 1-30 (1982).

43. W. A. Pryor, Free radicals in autoxidation and in aging. Part I. Kinetics of the autoxidation of linoleic acid in SDS micelles: calculations of radical concentrations, kinetic chain lengths, and the effects of vitamin E. Part II. The role of radicals in chronic human diseases and in aging. In Molecular Biology, Aging and Disease (D. Armstrong, R. S. Sohal, R. G. Cutler, and T. F. Slater, Eds.) pp. 13-41. Academic Press, New York (1984).

44. S. A. Lesko, R. J. Lorentzen and P. O. P. Ts'O, Role of superoxide in deoxyribonucleic acid strand scission. Biochemistry **19**, 305 (1980).

45. W. A. Pryor, The role of free radicals in chemical carcinogenesis. In Mechanisms of Antimutagenesis and Anticarcinogenesis (D. Shankel, P. Hartman, T. Kada, and A. Hollaender, Eds.) pp. 45-60. Plenum Press, New York (1986).

46. B. Demple and S. Linn, 5,6-saturated thymine lesions in DNA: production by ultraviolet light or hydrogen peroxide. J. Bacteriol. **153**, 1079-1086 (1982).

47. W. A. Pryor, D. L. Fuller, and J. P. Stanley, Reactivity patterns of the methyl radical. J. Am. Chem. Soc. **94**, 1632-1638 (1972).

48. W. A. Pryor, J. T. Echols, and K. Smith, Rates of the reactions of substituted phenyl radicals with hydrogen donors. J. Am. Chem. Soc. **88**, 1189-1199 (1966).

AUTOXIDATION AND ENZYMATIC OXIDATION OF UNSATURATED LIPIDS

Ned A. Porter

Department of Chemistry
Paul M. Gross Chemical Laboratory
Duke University
Durham, NC 27706 USA

AUTOXIDATION

Oxygen may act not only as an oxidant to initiate free radical reactions, but it can also act as a substrate for the propagation of these reactions. The spontaneous reaction of molecular oxygen with radicals is commonly referred to as autoxidation. Autoxidation is responsible for the deterioration of many manufactured plastics and rubber goods. Rancidity and spoilage of foodstuffs is a direct result of the autoxidation of fats, which are most susceptible to air oxidation and present, to a large extent, in virtually all foods.

The autoxidative process is commonly represented as consisting of chain initiation, propagation, and termination steps:

Initiation: $$In\cdot + RH \longrightarrow R\cdot + InH \qquad \text{(Eq. I-1)}$$

Propagation: $$R\cdot + O_2 \longrightarrow ROO\cdot \qquad \text{(Eq. I-2)}$$

$$ROO\cdot + RH \xrightarrow{k_p} R\cdot + ROOH \qquad \text{(Eq. I-3)}$$

Termination: $$2ROO\cdot \longrightarrow [ROOOOR] \longrightarrow \overset{\text{non radical}}{\text{products}}\ O_2 \qquad \text{(Eq. I-4)}$$

The key event in initiation is the formation of R•. There are many sources of radical species which may serve to abstract a hydrogen atom from RH. These include nitric oxide (NO), nitrogen dioxide (NO_2), and ozone, all of which are common environmental pollutants and have been shown to initiate autoxidation via hydrogen atom abstraction (1). Another method of generating R• is via thermal or photochemical induced homolytic scission.

$$RH \xrightarrow[\text{UV radiation}]{\text{heat or}} R\cdot + H\cdot$$

In the initial propagation step of autoxidation (Eq. I-2), molecular oxygen adds to R•. At partial oxygen pressures above 100 mm Hg, this addition approaches the diffusion controlled limit (ca. 10^9 1/mol-s) (2). This means that the major radical in solution is peroxy radical ROO• and not R•. As a result, it is unlikely that any termination reactions involving R• will take place, provided that oxygen is present in sufficient concentration. In the second propagation reaction (Eq. I-3), ROO• abstracts hydrogen atom from RH at rate k_p to generate more R• in what is the rate-determining step. For each hydroperoxide product formed, another radical R• is generated. This process could proceed ad infinitum, save for termination reactions that interrupt propagation. One potential termination step shown is that of peroxy radical coupling to ultimately give oxygen and non-radical organic products (Eq. I-4) (3,4). Another possible terminating route is that of radical disproportionation.

The rate constant k_p for hydrogen atom abstraction depends primarily on the activation energy of the reaction, i.e., the strength of the C-H bond being broken. Since ROO• is strongly resonance stabilized and a comparatively unreactive radical (2), it is quite selective in abstraction from hydrocarbons and prefers the most weakly bound hydrogen atom. The selectivity of peroxy radicals has been demonstrated in experiments where hydrogen is abstracted in marked preference to deuterium, resulting in a large kinetic isotope effect (5).

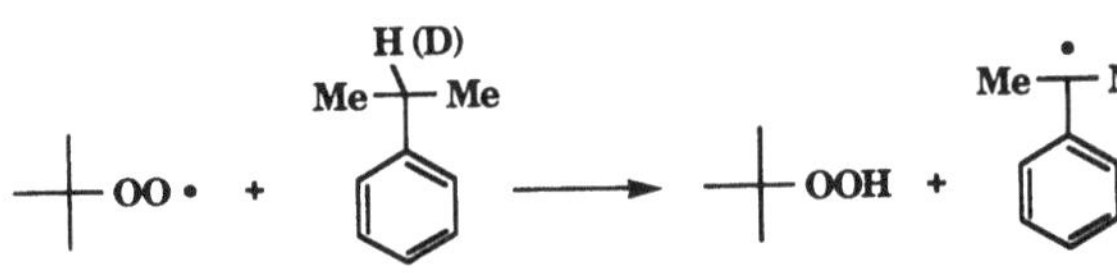

Another route of peroxy radical reaction is that of addition to a double bond to give the more stable β-peroxy alkyl radical. Addition will tend to be the favored route when the

double bond is in conjugation with stabilizing groups, such as vinyl, carbonyl, or aromatic. Intramolecular peroxy radical addition reactions result in cyclic peroxides, provided that orientation and ring size are favorable for such a reaction to occur.

DIENE AUTOXIDATION

The competition of the propagation step illustrated in equation 1 and peroxyl radical β-fragmentation has one important consequence in fatty acid or ester autoxidation; the stereochemistry of products formed in autoxidation is dependent on these competitive pathways. As an example of this, consider the autoxidation of a simple diene fatty acid, linoleic acid (6-10). Four conjugated diene hydroperoxide products are formed, **1-4**. Two of these products, **1** and **2**, have conjugated diene stereochemistry with the double bond nearest the hydroperoxide being trans and the remote double bond being cis. The two other products have trans/trans diene stereochemistry. Two of the products are substituted at the 9 position, and two at the 13 position of the eighteen carbon chain.

The mechanism for the formation of these four products has been discussed in detail elsewhere (11) and will be described here in outline form. Briefly stated, the mechanism proceeds by H-atom removal from C11 to generate a pentadienyl radical, **5**, with stereochemistry as shown in Scheme I (12).

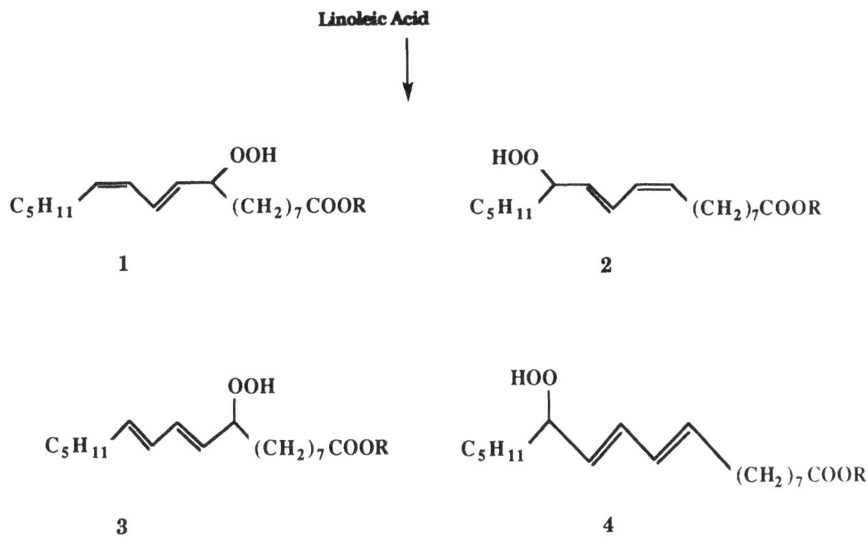

To simplify the discussion, consider only oxygen coupling at C9 (recognizing that similar chemistry occurs at C13), which gives the peroxyl radical **6** a radical with trans/cis conjugated diene stereochemistry. Entrapment of this radical before β-fragmentation leads to **1**, the kinetic product, the trans/cis compound. If good H-atom donors are not present (R-H or antioxidants, such as tocopherol) then β-fragmentation leading to radicals **5** and **7** occurs and oxygen coupling at C13 of the carbon radical **7** leads ultimately to the thermodynamic products, the 13 trans/trans product **4**. The competition between H-atom transfer to give kinetic product and β-fragmentation to give thermodynamic product thus accounts for the four major products from linoleate autoxidation.

The mechanism outlined in Scheme I is consistent with several experimental observations (8,13). First, higher concentrations of linoleate leads to more trans/cis products. This is because at higher [R-H], the reaction is driven toward kinetic control, since linoleate is an H-atom donor. Second, addition of other good H-atom donors, such as 1,4-cyclohexadiene, causes the product mixture to consist of primarily trans/cis products (kinetic control). Third, product distributions are independent of oxygen pressure between 10 and 1000 mm Hg. Oxygen addition to carbon radicals is not rate limiting at these pressures and β-fragmentation is independent of oxygen concentration.

In summary, the mechanisms of autoxidation propagation of diene lipids consist of three steps: (1) H-atom transfer from substrate to peroxy radical; (2) oxygen coupling to intermediate carbon radicals; and (3) β-fragmentation of intermediate peroxyl radical to remove the peroxyl. The kinetics and product distributions of autoxidation can be understood by this fundamental mechanism.

ENZYME OXIDATION OF PUFA'S

It is know that virtually all plants have the ability to oxidize PUFA's to conjugated fatty acid hydroperoxides (FAHP's). Enzymes which catalyze this transformation are referred to as lipoxygenases. For example, maize lipoxygenase yields exclusively the 9-d-hydroperoxylinoleate from linoleic acid and the 5-d-hydroperoxyarachidonate from arachidonic acid. In a similar manner, soybean lipoxygenase gives rise to the 13-l-hydroperoxylinoleate and 15-l-hydroperxoyarachidonate from linoleic and arachidonic

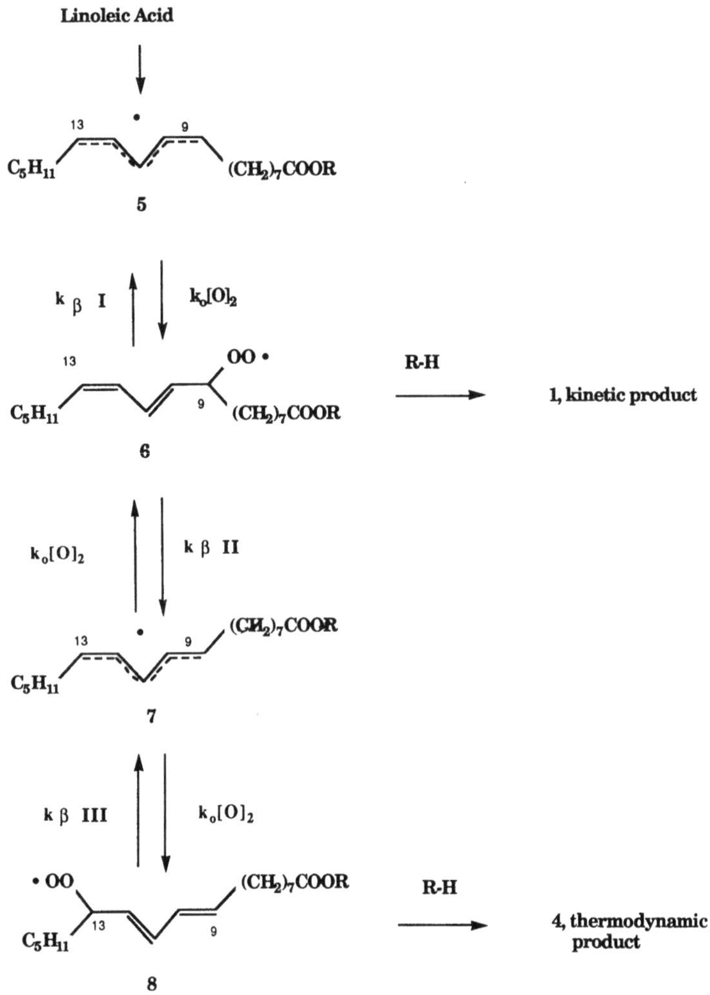

Scheme 1

acids, respectively. The overall conversion to a conjugated diene hydroperoxide involves bisallylic hydrogen atom removal and subsequent addition of molecular oxygen. Both of these steps are thought to take place stereospecifically. It is known that addition of molecular oxygen is anti to hydrogen atom removal and the resultant diene stereochemistry is always trans/cis relative to the hydroperoxide functionality. Further studies have established that hydrogen atom removal is rate-determining in the overall conversion.

Although lipoxygenase enzymes are universal in the plant kingdom, no biological function has ever been assigned directly to FAHP's resulting from lipoxygenase-mediated oxidation in plants. In sharp contrast to this observation is the fact that FAHP's resulting from lipoxygenase action in physiological systems are of considerable biological importance. For instance, in human platelets, lipoxygenase enzymes are able to stereospecifically oxygenate arachidonic acid to afford hydroperoxyeicostatetraenoic acids (HPETE's) which, following enzymatically controlled dehydration, lead to an important class of bioregulators know as leukotrienes. Further reactions of the LTA_4's with nucleophiles, such as cysteine and water, lead to still more leukotrienes - all of which have unique and far-ranging biological effects. It is known, for example, that LTB_4 is a neutrophil activator causing chemotactic movement of leukocytes (15); leukotrienes C_4, D_4, and E_4, all cysteinyl-containing leukotrienes and often referred to a slow-reacting substances of anaphylaxis (SRS-As), are potent bronchoconstrictors and have pro-inflammatory effects in humans (16).

In addition to lipoxygenase, there are other enzymes which also convert arachidonic acid to biochemically important oxygenated compounds, all of which form an integral part of the Arachidonic Acid Cascade (17). In a process catalyzed by cyclooxygenase enzymes, arachidonic acid is stereospecifically oxygenated at the 11 position and the resultant 11 HPETE cyclized to give endoperoxide PGG_2. Chemical models mimicking this enzymatic conversion suggest that the reaction probably follows a free radical pathway (18-20). This route includes two consecutive 5-exo radical cyclizations to form the 2,3-dioxabicyclo-[2.2.1]heptane skeleton and subsequent entrapment of molecular oxygen and hydrogen atom to yield the final prostaglandin product.

Not stored in the body, but biosynthesized upon demand, PGG_2 is the precursor to PGH_2, which in turn is the prostaglandin from which all other PG's ultimately originate.

REFERENCES

1. A. P. Autor, Oxygen toxicity in eukaryotes. In: The Biology and Chemistry of Active Oxygen (J. V. Bannister, W. H. Bannister, Eds), pp. 139-145. Elsevier, New York, 1984.

2. K. U. Ingold, Peroxy radicals. Acc. Chem. Res., 2, 1-9 (1969).

3. J. A. Howard, Free radicals (J. K. Kochi, Ed.), pp. 3-62. Wiley, New York, 1973.

4. G. A. Russell, Deuterium-isotope effects in the autoxidation of aralkyl hydrocarbons. Mechanism of the interaction of peroxy radicals. J. Am. Chem. Soc., 79, 3871-3877 (1957).

5. J. A. Howard, K. U. Ingold, and M. Symonds, Absolute rate constants for hydrocarbon oxidation. VIII. The reactions of cumylperoxy radicals. Can J. Chem., 45, 1017-1022 (1968).

6. F. Haslbeck, and W. Grosch, HPLC analysis of all positional isomers of the monohydroperoxides formed by soybean lipoxygenase during oxidation of linoleic acid. J. Food Biochem., 9, 1-3 (1985).

7. H. W. S. Chan, V. K. Newby, and G. Levett, Metal ion-catalyzed oxidation of linoleic acid. Lipoxygenase-like regioselectivity of oxygenation. J. Chem. Soc., Chem. Commun., 82-83 (1978).

8. N. A. Porter, B. A. Weber, H. Weenen, and J. A. Khan, The autoxidation of polyunsaturated lipids: factors controlling the stereochemistry of product hydroperoxides. J. Am. Chem. Soc., 102, 5597-5601 (1980).

9. Yamamoto, S. Haga, E. Niki, and Y. Kamiya, Oxidation of lipids. V. Oxidation of methyl linoleate in aqueous dispersion. Bull. Chem. Soc., Jpn, 57, 1260-1264 (1984).

10. Y. Yamamoto, E. Niki, and Y. Kamiya, Oxidation of lipids. III. Oxidation of methyl linoleate in solution. Lipids, 17, 870-877 (1982).

11. N. A. Porter, Mechanisms for the autoxidation of polyunsaturated lipids. Acc. Chem. Res., 19, 262-268 (1986).

12. E. Bascetta, F. D. Gunstone, and J. C. Walton, An electron spin resonance study of fatty acids and esters. Part I. Hydrogen abstraction from olefinic and acetylenic long-chain esters. *J. Chem. Soc., Perkin Trans. II*, 603-613 (1983).

13. N. A. Porter, L. S. Lehman, B. A. Weber, and K. J. Smith, A unified mechanism for polyunsaturated fatty acid autoxidation: competition of peroxy radical hydrogen atom abstract, β-scission and cyclization. J. Am. Chem. Soc., 103, 6447-6455 (1981).

14. M. R. Egmond, J. F. G. Vliegenthart, and J. Boldingh, Stereospecifity of the hydrogen abstraction at carbon atom n-8 in the oxygenation of linoleic acid by lipoxygenases from corn germs and soya beans. Biochem. Biophys. Res. Commun., 48, 1055-1060 (1972).

15. J. K. Beckman, J. C. Gray, A. R. Brash, J. N. Lukens, and J. A. Oates, Differential effects of lipoxygenase products on FMLP and LTB_4 evoked neutrophil aggregation. Lipids, 20, 357-360 (1985).

16. B. Samuelsson, Leukotrienes: mediators of immediate hypersensitivity reactions and inflammation. Science, 220, 568-575 (1983).

17. J. E. Pike, and D. R. Morton, Jr., Eds., Advances in prostaglandin, thromboxane, and leukotriene research, Vol. 14, pp. 156-157. Raven Press, New York, 1985.

18. N. A. Nelson, R. C. Kelly, and R. A. Johnson, Prostaglandins and the arachidonic acid cascade. Chem. Eng. News, 60 (33) 30-44 (1982).

19. E. J. Corey, J. Simoji, and C. Shih, Synthesis of prostaglandins via a 2,3-dioxa-bicyclo[2.2.1]heptane (endoperoxide) intermediate. Stereochemical divergence of enzymatic and biomimetic chemical cyclization reactions. J. Am. Chem. Soc., 106, 6425-6427 (1984).

20. N. A. Porter, P. J. Zuraw, and J. A. Sullivan, Peroxymercuration-dermercuration of lipid hydroperoxides. Tetrahedron Lett., 25, 807-810 (1984).

ELECTRON vs. H-ATOM TRANSFER IN CHEMICAL REPAIR

Michael G. Simic, Edward P.L. Hunter and
Slobodan V. Jovanovic*

Center for Radiation Research, National Bureau of Standards
Gaithersburg, MD 20899, U.S.A., and *Laboratory of Solid
State Physics and Radiation Chemistry, Boris Kidric Institute
of Nuclear Sciences 11001 Beograd, Yugoslavia

INTRODUCTION

Free radicals are generated by ionizing radiations (e.g., X rays, γ rays, electrons, neutrons, and Rn α-particles),[1] and the biological effects they cause are in some ways a consequence of their reactions.[2] Free radicals may also be generated by other processes and agents and numerous chemicals generate them under physiological conditions. Cancer promoters[3] and antineoplastic drugs[4] cause DNA strand breaks via free radical reactions. Metabolic processes are also perceived as possible sources of free radicals.[5-7] The repair of free radicals and the elimination (or reduction) of their biological effects by radioprotectors (mainly sulfhydryls) is a well-established defense mechanism in biological systems[8]. Although the interaction of anticarcinogens (primarily antioxidants) with free radicals is feasible, this is a poorly understood process.[8,9]

In this paper, the mechanisms of inactivation of different types of free radicals by representatives of classes of radioprotectors and anticarcinogens in model systems are discussed. The described models provide an insight into the mechanisms of the protective action of these agents against radiation and chemical carcinogens.

SULFHYDRYLS AS RADIOPROTECTORS

Historically, sulfhydryls have been considered to be H-atom donors.[10] Radiation generated radicals, •Bio, from biomolecules, BioH,

$$\mathrm{BioH} \rightsquigarrow \cdot\mathrm{Bio} + \mathrm{H}\ (\text{or } e^- + \mathrm{H}^+) \qquad (1)$$

can be restituted by H-atom donation,

$$\cdot\mathrm{Bio} + \mathrm{RSH} \rightarrow \mathrm{BioH} + \mathrm{RS}\cdot \qquad (2)$$

Reaction (2) is considered to be in simple competition with "oxygen fixation" of the damaged sites,[11]

$$\cdot\mathrm{Bio} + \mathrm{O_2} \rightarrow \mathrm{Bio\text{-}OO}\cdot \qquad (3)$$

This model, however, does not discriminate between different types of radicals or the actual kinetics of these processes. It also does not consider the possible repair of peroxy radicals, Bio-OO•.

GENERATION OF FREE RADICALS

In DNA exposed to ionizing radiations, free radicals are generated by direct ionization of its components (direct action) or by the reactions of OH radicals and electrons (indirect action). The OH radicals are formed by the radiation-induced decomposition of water,

$$H_2O \rightsquigarrow \cdot OH,\ e^-,\ H^+ \qquad (4)$$

The electrons, e^-, are formed by both the direct ionization of biomaterial and by the ionization of water, as shown in reaction (4). The reactions of electrons are not considered to be damaging to biosystems. Hence, our considerations are restricted only to radicals resulting from the other two sources.

TYPES OF BIORADICALS

The direct ionization of DNA leads to a cascade or charge migration which ends at guanine.[12] The positive ion of guanine loses a proton under normal biological conditions, becoming a highly resonant guanyl radical,[13]

$$dR\text{-}G \rightarrow dR\text{-}\dot{G}^+ \rightleftarrows dR\text{-}\dot{G}(-H) + H^+ \qquad (5)$$

$$pK_a < 3$$

Conversely, an •OH radical reacts with all four bases and deoxyribose. In the reaction of an •OH radical with guanine, two types of radicals are generated. In the initial step, the •OH radical adds to the purine ring,

$$dR\text{-}G + \cdot OH \rightarrow dR\text{-}\dot{G}\text{-}OH \qquad (6)$$

Some of the OH-adducts (50%) then lose water,[13]

$$dR\text{-}\dot{G}\text{-}OH \rightarrow dR\text{-}\dot{G}(-H) + H_2O \qquad (7)$$

Hence, •OH radical leads in part to the formation of the guanyl radical.

Similarly, pyrimidines also yield OH-adducts when they react with •OH radicals. A smaller fraction of •OH (10% in the reaction with thymine) abstracts hydrogen from the methyl group of thymine. The three types of thymine radicals are shown in reaction (8).

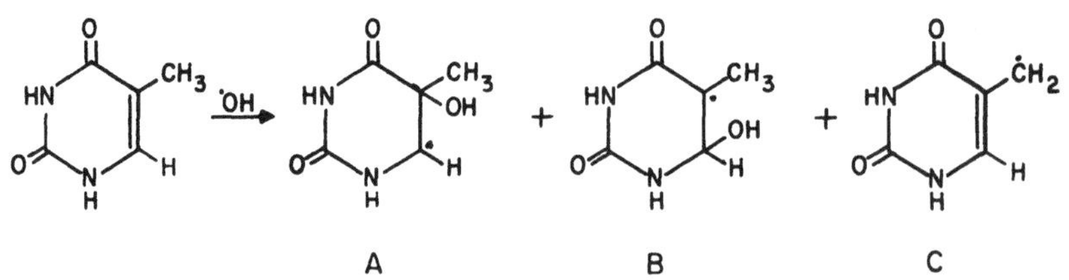

(8)

The reaction of •OH radical with deoxyribose leads to abstraction of H from positions C1 to C5 and to the formation of five different radicals,[15]

$$\cdot OH + dR \rightarrow d\dot{R}(-H) + H_2O \qquad (9)$$

Some of these $d\dot{R}(-H)$ radicals are responsible for DNA strand breaks.

Radicals A and C of thymine (see equation (8)) are C-centered radicals, as are deoxyribose radicals. Radical B of thymine is a resonant radical with the unpaired electron residing on the carbonyl group,[16]

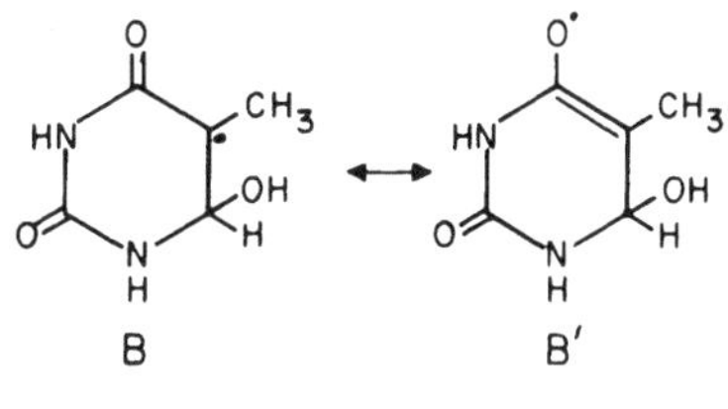

The C-centered radicals, •R, react with oxygen at very high rates,

$$\cdot R + O_2 \rightarrow ROO\cdot \qquad (10)$$

$$k \sim 10^9\ M^{-1}\ s^{-1}$$

The guanyl radical, $dR\text{-}\dot{G}(-H)$, does not react with oxygen because of the highly delocalized unpaired electron. Other resonant radicals may or may not react with oxygen depending on their resonance energy. Hence, in the presence of oxygen, some bioradicals will be converted into peroxy radicals.

We conclude, therefore, that there are three major types of DNA radicals: (a) C-centered, (b) resonant, and (c) peroxy radicals. We shall briefly discuss their reactivities with repair agents (radioprotectors and antioxidants).

C-CENTERED RADICALS

The rate of repair of C-centered radicals by sulfhydryls depends on the type of free radical and of sulfhydryl. For example glutathione and para-methoxythiophenol differ by an order of magnitude, in reactivity with a simple α-hydroxy radical such as hydroxy-methyl radical, which can be easily generated in a reaction of •OH radicals with methanol,

$$\cdot CH_2OH + GSH \rightarrow CH_3OH + GS\cdot \qquad (11)$$

$$k = 5.9 \times 10^7\ M^{-1}\ s^{-1} \text{ (Ref. 17)}$$

$$\cdot CH_2OH + p\text{-}CH_3O\phi SH \rightarrow CH_3OH + p\text{-}CH_3O\phi S\cdot \qquad (12)$$

$$k = 4 \times 10^8\ M^{-1}\ s^{-1}$$

The •OH radical adducts of thymine react with sulfhydryls similarly. Here again the aromatic sulfhydryl is more reactive,

$$\dot{T}\text{-}OH + GSH \rightarrow TH\text{-}OH + GS\cdot \qquad (13)$$

$$k \sim 10^6\ M^{-1}\ s^{-1}$$

$$\dot{T}\text{-}OH + p\text{-}CH_3O\phi SH \rightarrow TH\text{-}OH + p\text{-}CH_3O\phi S\cdot \qquad (14)$$

$$k = 4.6 \times 10^7\ M^{-1}\ s^{-1}$$

The sulfhydryls discussed here are in the following acid-base equilibria,

$$GSH \rightleftarrows GS^- + H^+ \tag{15}$$

$$pK_a = 8.66 \text{ (Ref. 18)}$$

p-Methoxythiophenol being a much stronger acid than glutathion has a lower pK_a value,

$$p\text{-}CH_3O\phi SH \rightleftarrows p\text{-}CH_3O\phi S^- + H^+ \tag{16}$$

$$pK_a = 6.7 \quad \text{(Ref. 19)}$$

Although the RSH form is an excellent H-atom donor, as demonstrated by examples given in equations (11)-(14), it can act also as an electron donor. The RS^- form by definition can be only an electron donor. The behavior of those two forms of sulfhydryls towards a C-centered radical $(\cdot CH_2(CH_3)_2COH)$ is shown in Figure 1.

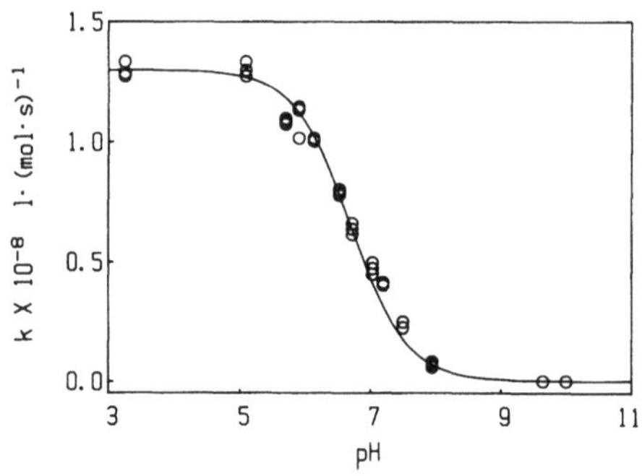

Figure 1. pH dependence of reactivity of tert-butanol radical with p-methoxythiophenol in aqueous solutions, as measured by pulse radiolysis.[20] Conditions: N_2O-saturated aqueous solution, 20°C, dose/pulse = 2.7 Gy, [tert-butanol] = 0.25 M, 10 mM phosphate buffer.

As indicated in equilibrium (16), at low pH the predominant form is -SH, whereas the predominant form is $-S^-$ at high pH.

It is clear from Figure 1 that C-centered radicals react very poorly, if at all, with electron donors,

$$R\dot{C}H_2 + p\text{-}CH_3O\phi S^- \rightarrow \text{no reaction observed} \tag{17}$$

Similarly, phenolic antioxidants, such as para-methoxyphenol, which are good electron donors and poor H- atom donors, are unreactive with C-centered radicals,

$$R\dot{C}H_2 + p\text{-}CH_3O\phi OH \rightarrow \text{no reaction observed} \tag{18}$$

In both cases $k < 10^4\ M^{-1}\ s^{-1}$.

RESONANT RADICALS

Resonant bio-radicals have not been studied extensively because their distinctive properties have been recognized only very recently.[13] No reaction is observed with the protonated sulfhydryl form,

$$dR\text{-}\dot{G}(\text{-}H) + GSH \rightarrow \text{no reaction observed}. \tag{19}$$

In contrast, electron donors are excellent repairing agents of the guanyl radical because guanyl radical is a strong oxidant (E_7 = 0.91 V),[21]

$$dR\text{-}\dot{G}(\text{-}H) + p\text{-}CH_3O\phi SH \rightarrow dR\text{-}G + p\text{-}CH_3O\phi S\cdot \qquad (20)$$

$$k = 1.9 \times 10^9\ M^{-1}\ s^{-1}$$

Consequently, antioxidants should be capable of restitution of DNA damaged by the direct action of radiation.

PEROXY RADICALS

Peroxy radicals have a high electron affinity as do other oxy radicals (•OH and RO•), and, as a consequence, they are inactivated by a variety of electron donors. The rate constants increase with decreasing redox potentials of antioxidants and with increasing electron affinity of peroxy radicals. For example, BHA has a lower redox potential than BHT, but a higher redox potential than α-tocopherol. Hydrocarbon peroxy radicals, ROO•, have a lower redox potential than α-halogenated peroxy radicals. Hence, the highest rate is expected for α-tocopherol + $Cl_3COO\cdot$, and the lowest rate for BHT + ROO•. The measured rate constants of these reactions are summarized in Table 1.

TABLE 1. Rate Constants, k (in $M^{-1}\ s^{-1}$), for the Reaction of Antioxidants with Peroxy Radicals[22]

	BHT	BHA	α-tocopherol
ROO•	$\sim 10^4$	3.4×10^6	7.9×10^6
$Cl_3COO\cdot$	6.1×10^5	3.9×10^7	1.8×10^8

Peroxy radicals react very slowly with H-atom donors. Bisallylic H in linoleic acid, LH_2, has one of the weakest bond energies, which should qualify linoleic acid as one of the best H-atom donors. Reaction with peroxy radicals, however, is still very slow,

$$ROO\cdot + LH_2 \rightarrow ROOH + \cdot LH \qquad (21)$$

$$k = 60\ M^{-1}\ s^{-1}\ \text{(Ref. 23)}.$$

Such low rates can not be measured by pulse radiolysis. On the other hand using the more electrophilic trichloromethyl peroxy radicals and aromatic sulfhydryls the rate constant was considerably higher and consequently measurable,

$$Cl_3COO\cdot + p\text{-}CH_3O\phi SH \rightarrow Cl_3COOH + p\text{-}CH_3O\phi S\cdot \qquad (22)$$

$$k = 5 \times 10^7\ M^{-1}\ s^{-1}$$

From k_{22}, however, one still could not decide whether the above process involved an H or electron transfer, since the redox potential of p-methoxythiophenol is fairly low (E_7 = 0.47 V). Proof that reaction (22) is in fact an electron transfer process was provided by the pH dependence of k_{22}, Figure 2.

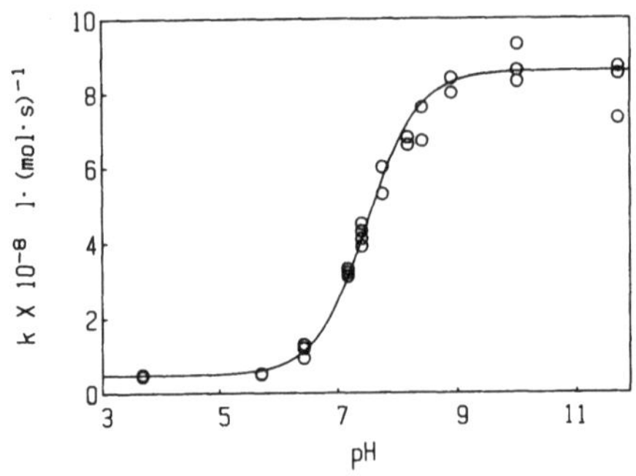

Figure 2. pH dependence of the reaction rate constant for the reaction between $Cl_3COO\cdot$ and p-methoxythiophenol.

At pH above 9, where the $-S^-$ form predominates the rate constant is much higher than at pH 3 where -SH form is present. Hence in alkaline solutions,

$$Cl_3COO\cdot + p\text{-}CH_3O\phi S^- \rightarrow Cl_3COO^- + p\text{-}CH_3O\phi S\cdot \quad (23)$$

$$k = 8.7 \times 10^8\ M^{-1}\ s^{-1}$$

If reaction (22) were an H-atom transfer, the rate constant would have decreased as in Figure 1. Since k_{23} was found to be higher than k_{22}, it may be concluded that the reaction of peroxy radicals with sulfhydryls is an electron transfer process.

These experimental findings[22] can be summarized by a general form,

$$Z\text{-}AH + X\text{-}ROO\cdot \rightarrow Z\text{-}\dot{A}H^+ + X\text{-}ROO^- \quad (24)$$

in which AH is an antioxidant; ROO•, peroxy radical; Z, electron-donating substituent; and X, electron-withdrawing substituent. As Z becomes a better electron donor (e.g., the CH_3O^- group) and as the electron affinity of X increases, a faster reaction rate is observed for inactivation of peroxy radicals.

MECHANISMS OF CHEMICAL PROTECTORS

Protection mechanisms from free radical damage to biosystems consists of three components:

(1) Prevention of damage by inactivating potentially damaging free radicals before they interact with biomolecules.

(2) Repair of free radicals generated on "target", i.e., relevant biomolecules in which the original molecules are restituted.

(3) Prevention of further (consecutive) damage (i.e., creation of multiple-damage sites) by inactivation of free radicals on biomolecules.

The type of protective agents in class (1) would depend on the type of free radicals and the nature of the target biomolecule.

Complete restitution of damaged biomolecules before the repair enzymes are set in motion is also possible. In this class fall, e.g., sugar radicals which can be repaired by H-atom donors, and guanyl radical which can be repaired by electron donors.

The class (3) type of damage has not been fully recognized as such and its consequences have not been considered in cells. This type of damage may be induced by DNA base or DNA peroxy radicals which may generate another radical in close proximity by H-atom abstraction, as observed in poly-U[23] and poly-dT[24]. Under oxyc conditions, inhibition of propagation of damage by peroxy radicals would be most efficient by electron donors which are good inactivators of peroxy radicals, provided these agents are capable of reaching the damaged sites.

CONCLUSIONS

Understanding whether the repair mechanisms involve H atom or electron transfer, or both may provide a base for a design of specific repair agents. Furthermore, utilization of these highly specific repair agents in chemical repair of free radical damage may allow new insights into a damage-consequence causality. We should understand whether the H atom or electron donating ability of a protector is crucial in the repair of damage that contributes to mutation, transformation, and lethality.

REFERENCES

1. M. Z. Hoffman, Ed. Radiation Chemistry. J. Chem. Ed. 58, 83 (1981).
2. E. J. Hall, Radiobiology for the Radiobiologist. Harper & Row, New York, 1978.
3. H. C. Birnboim, Importance of DNA strand-break damage in tumor promotion. In: Radioprotectors and Anticarcinogens (O. F. Nygaard and M. G. Simic, Eds.) pp 539-56. Academic Press, New York, 1983.
4. S. M. Hecht, DNA strand scission by activated bleomycin group antibiotics. This volume.
5. J. M. McCord and I. Fridovich, The reduction of cytochrome-c by milk xanthine oxidase. J. Biol. Chem. 243, 5753 (1968).
6. R. G. Cutler, Antioxidants and Longevity. In: Free Radicals in Molecular Biology, Aging, and Disease (D. Armstrong, R. S. Sohal, R. G. Cutler, and T. F. Slater, Eds.) pp. 371-428 Raven Press, New York, 1984.
7. B. N. Ames, Dietary Carcinogens and Anticarcinogens. Science 221, 1256 (1983).
8. O. F. Nygaard and M. G. Simic, Eds., Radioprotectors and Anticarcinogens. Academic Press, New York, 1983.
9. L. W. Wattenberg, Protective effects of 2(3)-tert-butyl-4-hydroxyanisole on chemical carcinogenesis. Food Chem. Toxicol. 24, 1099 (1986).
10. P. Alexander and A. Charlesby, Physico-chemical methods of protection against ionizing radiations. In: Radiobiology Symposium (Z. M. Bacq and P. Alexander, Eds.) pp.49-59. Academic Press, New York, 1955.
11. T. Alper and P. Howard-Flanders, Role of oxygen in modifying the radiosensitivity of E. coli B, Nature 178, 978 (1956).
12. P. M. Cullis and M.C.R. Symmons, Electron spin resonance studies of the mechanisms of radiation damage to DNA. In: Mechanisms of DNA Damage and Repair (M. G. Simic, L. Grossman, and A. C. Upton, Eds.) pp. 29-37. Plenum Press, New York, 1986.
13. M. G. Simic and S. V. Jovanovic, Free radical mechanisms of DNA base damage, ibid, p. 39-49.
14. S. V. Jovanovic and M. G. Simic, Mechanisms of OH radical reactions with thymine and uracil derivatives. J. Am. Chem. Soc. 118, 5968 (1986).
15. C. von Sonntag, The Chemical Basis of Radiation Biology. Taylor and Francis, London, 1987.
16. S. Fujita and S. Steenken, Pattern of OH radical addition to uracil and methyl- and carboxyl substituted uracils. Electron transfer of OH adducts with N,N,N',N'-tetramethyl-p-phenylenediamine and tetranitromethane. J. Am. Chem. Soc. 97, 2277 (1981).

17. R. L. Willson, Free radical repair mechanisms, in: "Radioprotectors and Anticarcinogens", (O. F. Nygaard and M. G. Simic, Eds.) pp. 1-22. Academic Press, New York, 1983.
18. The Merck Index, 10th ed. Merck & Co., Inc., Rahway, 1983.
19. M. G. Simic and E.P.L. Hunter, Reaction mechanisms of peroxy and C-centered radicals with sulfhydryls. J. Free Radial. Biol. Med. 2, 227 (1986).
20. E.P.L. Hunter, B. D. Michael, and M. G. Simic, Use of an optical multichannel analyzer for recording absorption spectra of short-lived transients. Rev. Sci. Instrum. 56, 2199 (1985).
21. M. G. Simic and S. V. Jovanovic, Acid-base and redox properties of the guanyl radical. J. Phys. Chem., submitted for publication.
22. M.G. Simic and E.P.L. Hunter, Interaction of free radicals and antioxidants. In: Radioprotectors and Anticarcinogens (O. F. Nygaard and M. G. Simic, Eds.) pp. 449-460. Academic Press, New York, 1983.
23. D. Shulte-Frohlinde, G. Behrens, and A. Onal, Lifetime of peroxy radicals of poly(U), poly(A), and single- and double-stranded DNA and the rate of their reaction with thiols. Int. J. Radiat. Biol. 50, 103 (1986).
24. L. Karam, M. G. Simic, and M. Dizdaroglu, Free radical-induced cross-linking of polydeoxythymidylic acid in deoxygenated aqueous solution. Int. J. Radiat. Biol. 49, 67 (1986).

RADIATION INDUCED REACTIONS OF GLUTATHIONE WITH OXYGEN AND THEIR POSSIBLE ROLE IN BIOLOGICAL SYSTEMS

M. Tamba, G. Simone and M. Quintiliani[a]

Istituto di Fotochimica e Radiazioni di Alta Energia (C.N.R.)
Via de' Castagnoli, 1 - 40126 Bologna, Italy and [a]Istituto
T.B.M. (C.N.R.), Via Morgagni 30/E - 00161 Roma, Italy

INTRODUCTION

The reactivity of some thiyl radicals with molecular oxygen were measured and reported much before the intracellular thiols began to arouse the wave of interest that has characterized the last decade or so (1,2). Cellular thiols, almost entirely consisting of glutathione (GSH), are considered at the present time as one of the most important system capable of protecting cells against free radicals formed during oxidative metabolism or from exposure to drugs or ionizing radiation.

In schemes proposed to visualize the interactions leading to the above effects, reactions of thiol radicals with oxygen are usually disregarded or, at most, indicated as precursor of the corresponding sulphinic acid, whose production is usually low at physiological pH (3).

This seems to be hardly consistent with the yield of both thiyl and thiol peroxyl radicals that can be expected in irradiated cells in view of: 1. the intracellular GSH concentration; 2. the rate of reaction of thiyl radicals with oxygen; 3. the possibility, indicated by radiation chemical experiments, that this process might involve a short chain reaction. This is why, in our opinion, the intracellular production of GSOO˙ radicals should be far from negligible. It should be also noted, at this point, that the final product of GSOO˙ decay is almost exclusively the corresponding disulfide as it would be in the absence of oxygen (2).

THE RADIOLYSIS OF THIOLS IN OXYGEN-CONTAINING SOLUTIONS

Radiation chemical data indicate that the process can be described by the following series of reactions:

(1) $RSH \rightleftharpoons RS^- + H^+$

(2) $RSH(RS^-) + OH \rightarrow RS\cdot + H_2O(OH^-)$

(3) $RS^{\cdot} + RS^{-} \rightleftharpoons RSSR^{\dot{-}}$

(4) $RS^{\cdot} + O_2 \rightleftharpoons RSOO^{\cdot}$

(5) $RSSR^{\dot{-}} + O_2 \longrightarrow RSSR + O_2^{\dot{-}}$

(6) $RSOO^{\cdot} + RSH \longrightarrow RSOOH + RS^{\cdot}$

(7) $O_2^{\dot{-}} + RSH \xrightarrow{H^+} RS^{\cdot} + H_2O_2$

(8) $RS^{\cdot} + RS^{\cdot} \longrightarrow RSSR$

The reaction of molecular oxygen with thiyl radical (reaction 4) can be demonstrated in the pulse radiolysis of oxygen-containing solutions of thiol compounds, either with an indirect method or with a direct one.

The indirect method is based on the competition between the complexation of the thiyl radical with the ionized thiol and the oxygen addition reaction: reaction 3 and 4. Under these conditions the classical competition equation can be applied, namely:

$$[I] \qquad OD_o/OD = 1 + \frac{k_4[O_2]}{k_3[RS^-]}$$

where the ratio of the optical densities in the absence and in the presence of oxygen is equal to 1, plus the ratio of the rate constant for reaction 4 times the oxygen concentration to the rate constant for reaction 3 times the concentration of the ionized thiol. This treatment has been applied to the experimental data obtained with solution of GSH and penicillamine (PnSH) with the result shown in Figure 1.

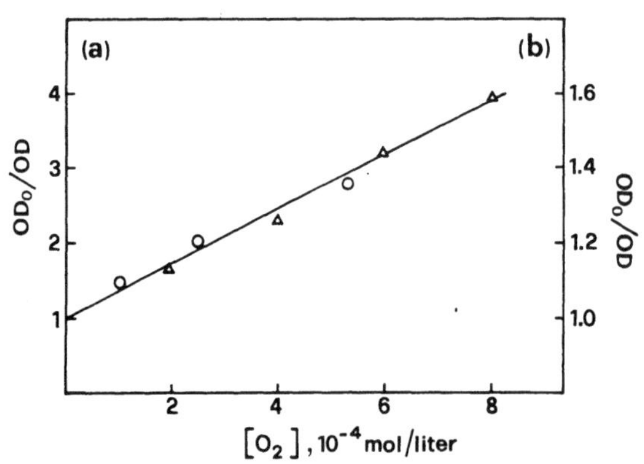

Fig. 1. Indirect metnod for the determination of $k(RS^{\cdot} + O_2)$. Plot of OD_o/OD (from $RSSR^{\dot{-}}$ measured at 420 nm) against $[O_2]$ from N_2O/O_2 saturated solutions of 1 mmol dm^{-3} glutathione (pH 9.2) and penicillamine (pH 8.1). (A),o, GSH, Dose= 12 Gy; (B),Δ, PnSH, Dose = 10 Gy.

Table 1. Rate constants (k_4) for the reaction: $RS^{\cdot} + O_2$, as determined with different methods.

Compounds	Indirect[a]	Direct[b]
Glutathione (GSH)	16	20
Penicillamine (PnSH)	14	17

k values are in unit of $10^8\ dm^3\ mol^{-1}\ s^{-1}$;

[a]Determined using $k_{RS.+RS^-}$ values of 8.0 (8) and 36 (this work) for GSH and PnSH, respectively;

[b]Determined according to eqn. II.

From the slope of the line and from the value of the rate constant for reaction 3, that is k= 8.0 and 36 ($x10^8 dm^3 mol^{-1} s^{-1}$) for GSH and PnSH, respectively, the rate constant values reported in Table 1 column a (Indirect) can be calculated.

The direct method is based, instead, on the kinetic analysis of the weak transient absorption at 540 nm, produced on pulse irradiation of oxygenated solutions of thiols and attributed to the corrispondent peroxyl radicals (4). Figure 2 illustrates this finding for GSH. The same pattern, however, can be obtained on the irradiation of cysteine (CySH) and of PnSH under the same experimental conditions. It has been shown that the yield of the 540 nm absorption increases with increasing oxygen concentration even at levels where thiyl radicals would be expected to be completely converted into peroxyl radicals. This suggests that reaction 4 is indeed an equilibrium reaction, and the analysis of the absorption build-up supports this suggestion (4). Such an analysis, in fact, shows that peroxyl radical is formed exponentially and that the observed rate constant for its formation is experimentally related to the oxygen concentration according to eq. [II]:

$$[II] \qquad k_{obs} = k_4[O_2] + k_{-4}$$

The application of eq. II to the experimental data relative to pulse irradiation of GSH and PnSH in oxygen-containing solutions generates the curves reported in Figure 3. From these curves the rate constant values reported in Table 1 column b (Direct) can be calculated. Theoretically, the same rate constants could be also evaluated from the decay kinetics of the thiyl radical absorption at 320-330 nm as a function of oxygen concentration. However, in this spectral region the interference from the tail of the absorption due to the peroxyl radical often makes this kinetic treatment difficult.

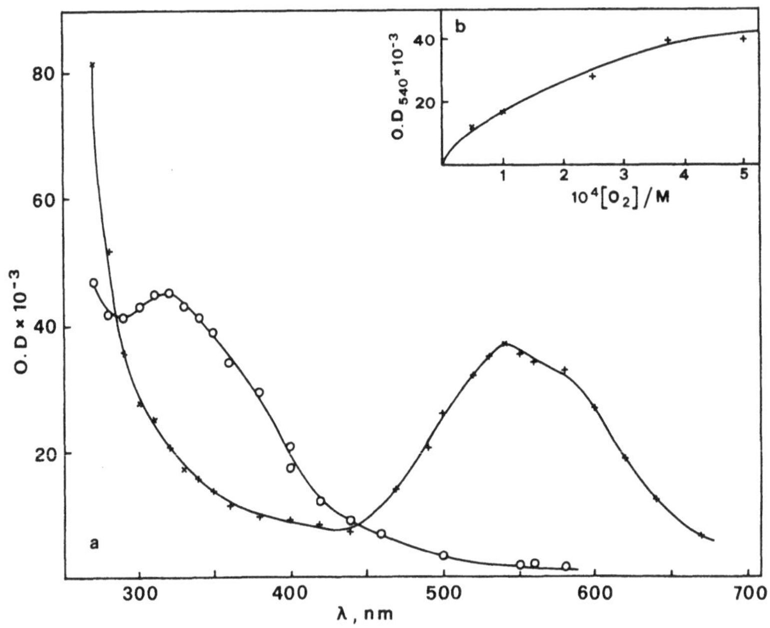

Fig. 2. Transient spectra of 1 mmol dm^{-3} GSH at pH 5.5 (a), o, N_2O saturated solutions, ≅ 4 μs after the pulse; x, N_2O/O_2 (60 : 40 v/v)-saturated solutions, where $[O_2]$ = 0.5 mmol dm^{-3}, ≅ 5 μs after the pulse. Dose ≅ 21 Gy, path lenght = 0.5 dm. (b) Optical absorption at 540 nm as a function of oxygen concentration, Dose = 26.5 Gy.

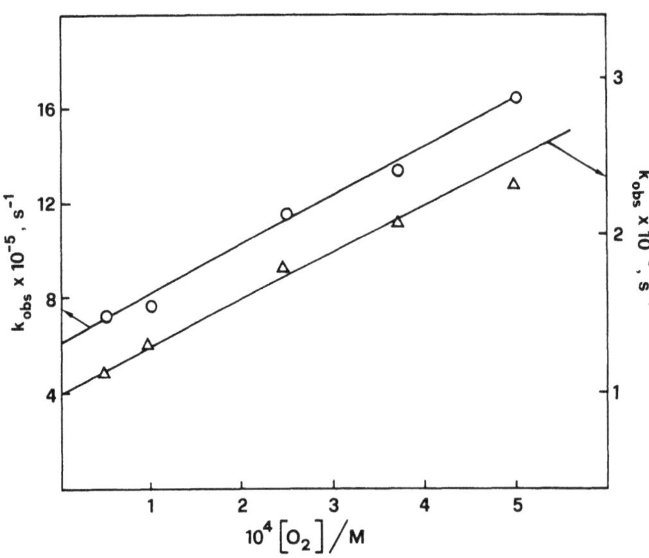

Fig. 3. Effect of oxygen concentration on the first-order build-up of the 540 nm absorption in pulse irradiated 1 mmol GSH (o) and PnSH (Δ) at pH 5.5 Dose = 24 Gy.

The agreement between the indirect and the direct estimates of the rate constants reported in Table 1 seems to resolve the uncertainties linked to the experimental method used in the determination of an unexpected high rate constant for the reaction under investigation. These values are, in fact, even larger than for the addition of dioxygen to several radicals, including simpler RS· radicals and carbon-centered-species.

As mentioned in the introductory section, the radiolytic processes of several thiols in aqueous oxygenated solutions appears to involve some short chain reactions. This can be shown by the rate of the thiol disappearance under irradiation (2,3) as well as by the rate of the oxygen consumption (Simic, personal communication). Reactions 6 and 7 can be propagating steps in the chain.

$$(6) \qquad RSOO\cdot + RSH \longrightarrow RSOOH + RS\cdot$$

$$(7) \qquad O_2^{\bar{\cdot}} + RSH \xrightarrow{H^+} RS\cdot + H_2O_2$$

Finally, both thiyl and thiol-peroxyl radicals appear to be susceptible to reaction with reducing agents. Such an interaction is observed for thiyl and peroxyl radicals derived from GSH with several natural or man-made antioxidants, such as ascorbic acid and 2,2'-azino-di(3-ethylbenzothiazoline-6-sulphonate (ABTS):

$$(9) \qquad GS\cdot + D \longrightarrow GS^- + D^{\dot{+}}$$

$$(10) \qquad GSOO\cdot + D \longrightarrow GSOO^- + D^{\dot{+}}$$

where D indicates the antioxidant agent. The k_9 and k_{10} valeus have been determined either by measuring the rate of formation of the absorbing radical $D^{\dot{+}}$ or by the decay of GSOO· at 540 nm. These values for ascorbic acid and ABTS are of the order of $10^8 dm^3 mol^{-1} s^{-1}$, reflecting comparable redox properties for both GS· and GSOO· radicals.

IN VITRO BIOLOGICAL ACTIVITY OF GSOO·

Radiation induced formation of GSOO·, or more generally of thiol-peroxyl radicals RSOO·, rises the question of their roles in living systems. Peroxyl radicals are, in fact, injurious to cells. In model experiments, peroxyl radicals have been shown to inactivate phage particles (5, 6) and to cause extensive damage to biologically active macromolecules (7). With regard to thiols, it has been known for many years that, while the inactivation of enzymes is not much affected by the presence or the absence of oxygen, in thiol-containing solutions a greater inactivation occurs in the presence of oxygen than in its absence. This finding has usually been interpreted as a typical expression of the competition between the fixation of damage by oxygen and its repair by hydrogen donation. On the other hand, the results of experiments carried out in the authors' laboratory on two enzymes, lysozyme and trypsin (8), support the interpretation that the greater inactivation in the presence of oxygen should be attributed to the inactivating effect of thiol peroxyl radicals.

SPECULATIONS ON THE ACTIVITY OF GSH AND ITS RADICALS IN VIVO

Current ideas still rest on the radical repair model proposed by Howard-Flanders in 1960 (9). The model is represented in the following scheme:

$$TH \longrightarrow T^{\cdot} \begin{cases} \xrightarrow{O_2} TO_2^{\cdot} \quad \text{(Damage Fixation)} \\ \xrightarrow{RSH} TH + RS^{\cdot} \quad \text{(Damage repair)} \end{cases}$$

postulating that thiols compete with oxygen for transiently damaged biological targets: the two competing mechanisms being the damage repair by hydrogen donation from sulphydryls and the damage fixation by peroxidation from molecular oxygen.

The assumption that the hydrogen donation mechanism can operate in vivo is supported by very convincing experimental evidence obtained in in vitro studies (10, 11), as well as by the relatively high intracellular GSH concentration. A finding that is difficult to reconcile with the radical repair model is that in many cell lines GSH depletion increases the radiation sensitivity of the cells, not only in the absence but also in the presence of oxygen (12, 13).

With regard to the results reported above, if one speculate that part of the oxygen enhancement ratio can be attributed to GSH peroxyl radicals, one would encounter the same difficulty. GSH depletion should, in fact, decrease the amount of GSOO$^{\cdot}$ produced during oxic irradiation and the radiosensitivity of cells should not be increased.

In order to explain the increased radiosensitivity of GSH depleted cells in the presence of oxygen, Biaglow et al. (14) have extended the radical repair model postulating that GSH can repair, by hydrogen donation, not only the radical sites produced in anoxia on biologically important molecules, but also the corresponding peroxyl radicals. According to this hypothesis, peroxyl radicals are first reduced to hydroperoxides and then to harmless alcohols through a reaction catalysed by glutathione peroxidase. The process can be represented as follows:

(11) $$TOO^{\cdot} + GSH \longrightarrow TOOH + GS^{\cdot}$$

(12) $$TOOH + 2GSH \xrightarrow{enz} TOH + H_2O + GSSG$$

The pentose cycle will then take care of providing the reducing equivalents to rigenerate GSH from GSSG. The whole process appears quite possible, however the implication of an enzymatic step rises the question as to whether it might be fast enough in comparison with the time-scale of the oxygen effect in irradiated cells (15).

In our opinion reaction 4, and possibly also reaction 5, might be involved in the process of radiation damage fixation by oxygen acting in opposition to it. There is little doubt, in fact, that if reaction 4 and 5 occur in oxically irradiated cells they would be in competition with damage fixation reaction or with any other harmful reaction due to molecular oxygen. According to this point of view, glutathione intermediates, by reacting with oxygen, might reduce the instantaneous concentration of this species available for some specific damaging reaction, producing instead species such GSOO· or $O_2^{\cdot -}$ (16, 17) that are somewhat less harmful to biological structures. Moreover, the observation that reaction 4 is indeed an equilibrium, might be favorable in order to insuring that, even in the presence of oxygen, the major product of GSH radiolysis is GSSG.

These ideas clearly need more extensive experimental substantiation.

ACKNOWLEDGMENT

This work has been supported in part by a grant from the Italian National Research Council, Special Project "Oncology".

REFERENCES

1. J.P. Barton and J. E. Packer, The radiolysis of oxygenated cysteine solutions at neutral pH. The role of $RSSR^{\cdot -}$ and $O_2^{\cdot -}$. Int. J. Radiat. Phys. Chem. 2, 159-166 (1970).
2. M. Quintiliani, R. Badiello, M. Tamba, A. Esfandi, and G. Gorin, Radiolysis of glutathione in oxygen-containing solutions of pH 7. Int. J. Radiat. Biol. 50, 195-202 (1977).
3. M. Lal, ^{60}Co-γ radiolysis of reduced glutathione in aerated solutions at pH values between 1-7.0. Can. J. Chem. 54, 1092-1097 (1976).
4. M. Tamba, M. Quintiliani, and G. Simone, Interactions of thiyl free radicals with oxygen: a pulse radiolysis study. Int. J. Radiat. Biol. 50, 595-600 (1986).
5. J. Blok and W. S. D. Verhey, The attack of free radicals on biologically active DNA in irradiated aqueous solutions. Radiat. Res. 34, 689-703 (1968).
6. J. De Jong, H. Loman and J. Blok, Inactivation of biologically active DNA by radiation-induced phenylalanine radicals. Int. J. Radiat. Biol. 22, 11-21 (1972).
7. R. L. Willson, Organic peroxy free radicals as ultimate agents in oxygen toxicity. In Oxidative Stress (H. Sies, Ed.) pp. 41-72, Academic Press, London, 1985.
8. G. Simone, J. C. M. Bremner, M. Tamba, B. R. Guerra, and M. Quintiliani, Thiol peroxyl radicals and radiation-induced enzyme inactivation. Int. J. Radiat. Biol. 49, 739 (1986) [Abstract].
9. P. Howards-Flanders, Effect of oxygen on the radiosensitivity of bacteriophage in the presence of sulphydryl compounds. Nature 186 485-487 (1960).
10. K.D. Held, H. A. Harrop, and B. D. Michael, Effects of oxygen and sulphydryl-containing compounds on irradiated transforming DNA. Part II. Glutathione, cysteine and cysteamine. Int. J. Radiat. Biol. 40, 615-626 (1984).

11. M. Tamba and M. Quintiliani, Kinetic studies of reactions involved in hydrogen transfer from glutathione to carbohydrate radicals. Radiat. Phys. Chem. 23, 259-263 (1984).
12. R. J. Hodgkiss and R. W. Middleton, Enhancement of misonidazole radiosensitization by an inhibitor of glutathione biosynthesis. Int. J. Radiat. Biol. 43, 179-183 (1983).
13. J. K. Dethmers and A. Meister, Glutathione export by human lymphoid cells: depletion of glutathione by inhibition of its synthesis decreases export and increases sensitivity to irradiation. Proc. Natl. Acad. Sci. USA, 78, 7492-7496 (1981).
14. J. E. Biaglow, M. E. Varnes, E. P. Clark and E. P. Epp, The role of thiols in cellular response to radiation and drugs. Radiat. Res. 95, 437-455 (1983).
15. J. E. Biaglow, M. E. Varnes, E. P. Epp, E. P. Clark and M. Astor, Redox enzymes and thiol radicals. In Oxygen and Sulfur Radicals in Chemistry and Medicine (A. Breccia, M. A. J. Rodgers and G. Semerano, Eds.) pp. 89-102, Edizioni Scientifiche "Lo Scarabeo", Bologna, Italy, 1986.
16. D. Ewing, T. M. Koval, and H. L. Walton, Radiation sensitization by oxygen of in vitro mammalian cells: is O_2^- involved? Radiat.Res. 106, 356-365 (1986).
17. H. Durchshlag and P. Zipper, Post-irradiation inactivation, protection, and repair of the sulfhydryl enzyme malate synthase. Effects of formate, superoxide dismutase, catalase and dithiothreitol. Radiat. Environ. Biophys. 24, 99-111 (1985).

POTENTIAL LIMITATION TO HYDROGEN ATOM DONATION AS A MECHANISM OF REPAIR IN CHEMICAL MODELS OF RADIATION DAMAGE

J.A. Raleigh, A.F. Fuciarelli and C.R. Kulatunga

Radiobiology, Cross Cancer Institute
11560 University Avenue
Edmonton, Alberta, Canada T6G 1Z2

Ionizing radiation can be lethal to mammalian cells. For those cells which are irradiated and survive, irradiation can also be mutagenic or, in the case of cells in animal tissues, carcinogenic. The biological effects of ionizing radiation are generally believed to originate in free radical reactions. In particular, a radical competition model has been proposed to account for the "oxygen effect" on radiation lethality - the so-called "oxygen fixation" hypothesis (Figure 1) (1-5).

Central to the model is the target radical, T•, which is formed indirectly by hydroxyl radical attack or directly by ionization of the target, TH. The fate of T• is believed to determine the biological outcome - either "repair" by hydrogen atom or electron donation or "fixation" by oxidation or oxygenation. The model is kinetically feasible (6) and can, in general, account for the effects of chemical modifiers such as hydroxyl radical scavengers (7-9) and hypoxic cell radiosensitizers (10,11) on the lethal effects of ionizing radiation. Whether such a model can serve to direct effective chemical interventions in the process of radical-induced mutation is not yet known but some fundamental limitations of the model in this regard are apparent.

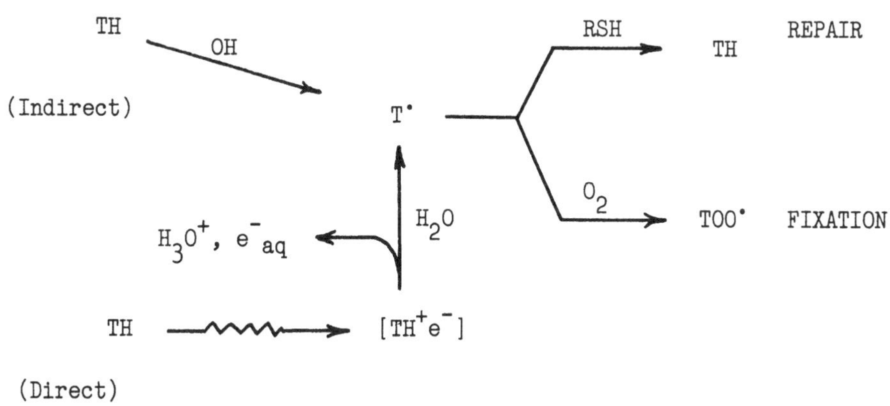

Figure 1. Radical competition model.

Holian and Garrison (12) pointed out many years ago that hydrogen atom donation to a target radical formed by OH addition to pyrimidine bases would not restore the original target (Figure 2). This would also be true of OH adducts of purine bases (13,14). In both cases hydrogen atom donation represents misrepair. In principle, such misrepair is not limited to base damage. Three out of the five carbon atoms in the sugar phosphate backbone of DNA constitute chiral centres. Two of the chiral centres - the C-1' and C-4' positions in the deoxyribose ring - are considered to be particularly sensitive to OH attack (15). Hydrogen atom abstraction produces intermediate carbon radicals which could assume a planar configuration. Hydrogen atom donation to the intermediate radical can occur in one of two ways which are not biologically equivalent. Hydrogen atom addition from below the plane of the deoxyribose ring regenerates the target to the natural β-configuration. Hydrogen atom addition from above the plane of the ring does not regenerate the original configuration at C-1' but the unnatural α configuration (Figure 3).

This possibility of hydrogen atom donation leading to chemical misrepair at the sugar phosphate backbone in DNA or, indeed, at chiral centres in proteins, phospholipids and carbohydrates has received little attention in the literature. Cadet, Berger and Shaw have observed that irradiation of frozen aqueous solutions of deoxyadenosine produces α-deoxyadenosine and that thiols did not appear to affect the yield of the unnatural α-isomer (16). Kochetkov and colleagues had previously reported that irradiation of monosaccharides in the frozen aqueous state caused epimerization at chiral centres (17).

Perhaps the most thorough studies of radiation-induced epimerization are those of Gold and his colleagues who followed the process in organic molecules dissolved in tritiated water (18 and references therein). These authors have proposed a mechanism for epimerization which, in the microcosm of a spur, is an analogue of the "repair reaction" in the oxygen fixation hypothesis (Figure 1). Gold and coworkers found that hydroxyl radicals alone are not sufficient to bring about hydrogen atom exchange and attendant epimerization at chiral centres. This, combined with the observation that electron scavengers such as silver ions inhibit the reaction, led to the proposal that hydrogen atom exchange reactions involved in epimerization at chiral centres occurred

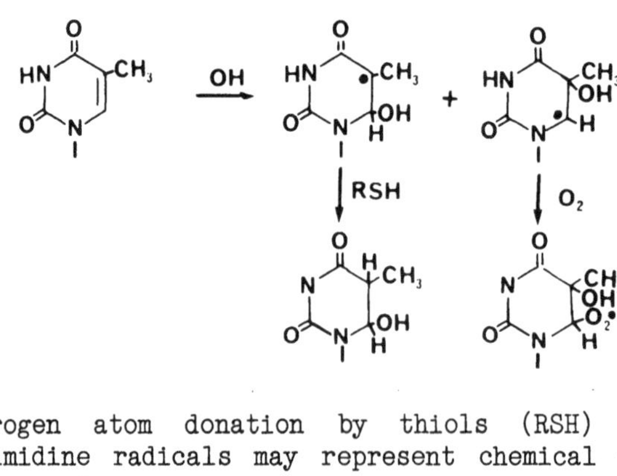

Figure 2. Hydrogen atom donation by thiols (RSH) to intermediate pyrimidine radicals may represent chemical misrepair rather than repair.

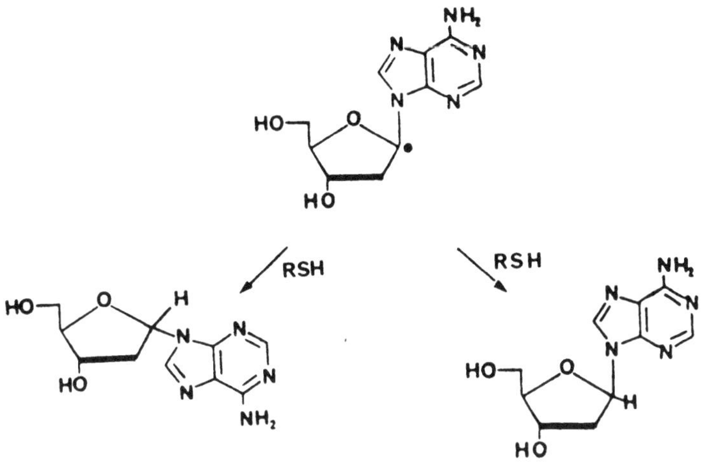

Figure 3. Possible mechanism of chemical misrepair by hydrogen atom donation to radicals on the deoxyribose moiety - epimerization at C-1'.

within spurs and required the sequential attack of both OH and hydrated electrons. This is illustrated for cyclohexane-1,2-diol in Figure 4.

Although Kochetkov and co-workers claimed that the epimerization reaction occurred only in frozen aqueous solutions and Cadet, Berger and Shaw also carried out their studies with frozen aqueous solutions of 2'-deoxyadenosine, we have found that the epimerization reaction at C-1' occurs in aqueous solutions of both 5'-AMP (Figure 5) and poly A (Figure 6) (19). Lesiak and Wheeler have reported recently that the epimerization reaction at C-1' also occurs in irradiated solutions of poly dA (20).

Cadet, Berger and Shaw (16) proposed a mechanism for epimerization that involved hydrogen atom abstraction at C-4' by OH followed by opening and resealing of the deoxyribose ring in a way which permitted epimerization at C-1'. However, comparison of the epimerization process

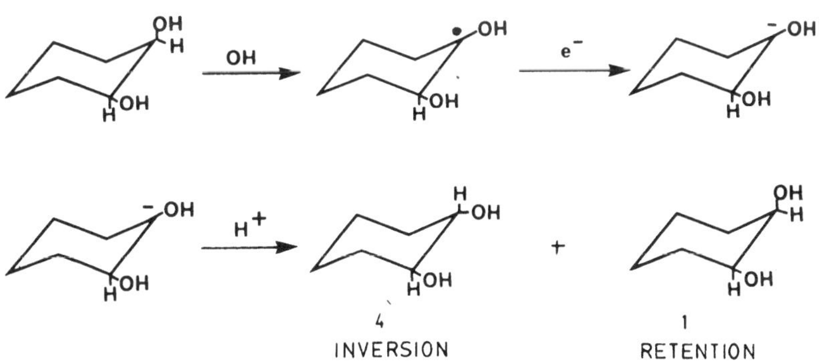

Figure 4. Mechanism proposed for epimerization of cyclohexane-1,2-diol in [^{3}H]-H_2O. Sequential OH and e^-_{aq} attack at the same carbon atom produces a planar carbanion which, upon protonation, leads to inversion of configuration at the chiral centre (18).

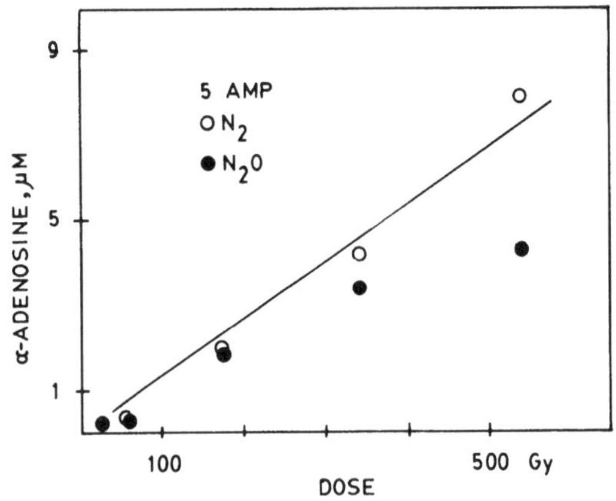

Figure 5. Dose-yield response for radiation-induced epimerization in 5'-AMP irradiated in N_2 (O) or N_2O (●) saturated solutions of 5'-AMP (10^{-3} M). Gammacell 220 ^{60}Co-γ irradiation at 61.5 Gy/min. Data points are averages of two separate experiments. Analysis by HPLC of α-adenosine following hydrolysis of irradiated 5'-AMP by alkaline phosphatase (19).

in solutions saturated with nitrous oxide or nitrogen showed that increasing the yield of hydroxyl radicals ($N_2O + e^-_{aq} \xrightarrow{H^+} OH + N_2$) (21) does not lead to an increase in α-adenosine formation in either 5'-AMP or poly A (Figures 5 and 6). It seems unlikely that the epimerization process is occurring in spurs in the case of 5'-AMP or poly A irradiated with ^{60}Co-γ radiation in dilute aqueous solution. Alternative mechanisms include the possibility that the initial hydrogen atom abstraction at C-1' is followed by the addition of a freely diffusing hydrated electron to the intermediate radical at C-1' and subsequent protonation of the intermediate carbanion to produce α-adenosine (Figure 7). Radical disproportionation involving C-1'

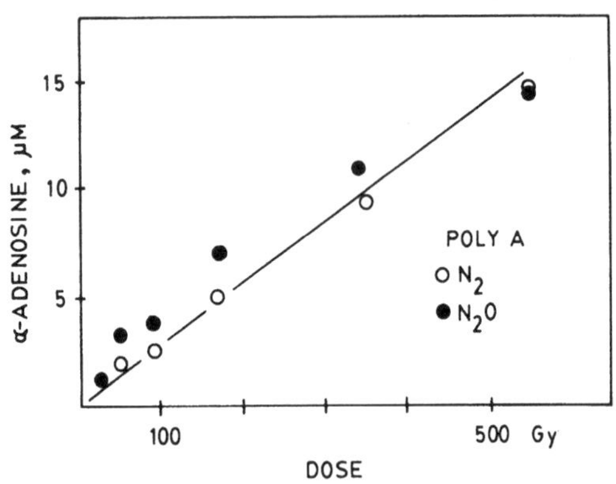

Figure 6. Dose-yield response for radiation-induced epimerzation in poly A irradiated in N_2 (O) or N_2O (●) saturated solutions of poly A (10^{-3} M in mononucleotide). Irradiation conditions and analysis as in Figure 5 except that nuclease P1 hydrolysis of irradiated poly A to the mononucleotide level preceded alkaline phosphatase hydrolysis (19).

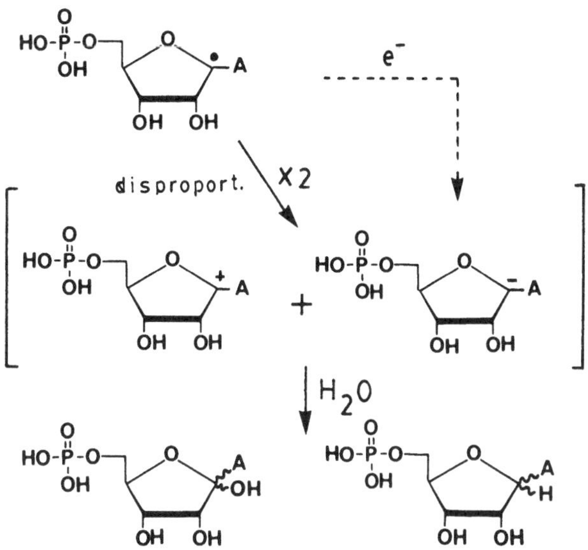

Figure 7. Possible mechanisms of radiation-induced epimerization at C-1' in 5'-AMP. Radical disproportionation seems less likely for poly A in which case electron addition to an intermediate C-1' radical may be involved (dashed line).

radicals could also produce an intermediate carbanion at C-1' and this mechanism must be considered possible for 5'-AMP although it would be less likely for radicals in polymeric poly A. The epimerization process in nitrogen-saturated solution is relatively efficient in both 5'-AMP and poly A giving G values of 0.12 and 0.3, respectively.

While hydrogen atom or electron donation to carbon radicals has limitations as a mechanism of chemical repair of nucleic acids irradiated in the absence of oxygen, this is not the case for hydrogen atom donation to peroxy radicals formed during irradiation in the presence of oxygen. Peroxy radicals are clearly implicated in radiation-induced damage in membranes (22) and there is evidence that peroxy radicals in nucleic acids may represent more than "fixation" of damage as defined in Figure 1. This opens up the possibility that quenching of intermediate peroxy radicals by hydrogen atom donation ($ROO^\bullet + R'SH \longrightarrow ROOH + R'S^\bullet$) can in fact protect nucleic acids from more extensive damage. Schulte-Frohlinde and Bothe have proposed (23) that peroxy radicals ($ROO^\bullet$) which are estimated to be $\sim 10^7$ times less reactive than OH radicals (see 22) can, nevertheless, be involved in reactions which enhance damage to nucleic acids. Hydrogen donating antioxidants could intercept these processes by scavenging intermediate peroxy radicals.

In summary, hydrogen atom or electron donation is limited as a chemical mechanism of repair in the absence of oxygen. Epimerization reactions may, in fact, be a major class of damage along with intramolecular (24) and intermolecular (25) cross-linking reactions for irradiations in the absence of oxygen. On the other hand, quenching of peroxy radicals by hydrogen atom donation, whether they occur in membrane constituents or nucleic acids, may be an important mechanism of protection against radiation-induced lethality under aerobic conditions. The extent to which this holds for radiation-induced mutation and carcinogenesis is not yet clear.

ACKNOWLEDGEMENT

The authors thank the Alberta Cancer Board and the Alberta Heritage Savings Trust Fund for financial assistance and Ms. Gina Kennedy for assistance in preparing the manuscript.

REFERENCES

1. P. Alexander and A. Charlesby, Physico-chemical methods of protection against ionizing radiations. In Radiobiology Symposium 1954 (Z.M. Bacq and P. Alexander, Eds.) pp. 49-59. Butterworth, London.
2. P. Howard-Flanders and T. Alper, The sensitivity of microorganisms to irradiation under controlled gas conditions. Radiat. Res. 1, 518-540 (1957).
3. P. Howard-Flanders, The effect of oxygen on the radiosensitivity of bacteriophage in the presence of sulphydryl compounds. Nature (London) 186, 485-487 (1960).
4. F. Hutchison, Sulfhydryl groups and the oxygen effect on irradiated dilute solutions of enzymes and nucleic acids. Radiat. Res. 14, 721-731 (1961).
5. For recent review see numerous contributors in "Radioprotectors and Anticarcinogens, 1983 (O.F. Nygaard and M.G. Simic, Eds.). Academic Press, New York.
6. B.D. Michael, K.D. Held and H.A. Harrop, Biological aspects of DNA radioprotection. In Radioprotectors and Anticarcinogens, 1983 (O.F. Nygaard and M.G. Simic, Eds.) pp. 325-338. Academic Press, New York.
7. T. Sanner and A. Pihl, Significance and mechanism of the indirect effect in bacterial cells. The relative protective effect of added compounds in Escherichia coli B irradiated in liquid and in frozen suspension. Radiat. Res. 37, 216-227 (1969).
8. R. Roots and S. Okada, Protection of DNA molecules of cultured mammalian cells from radiation-induced single-strand scissions by various alcohols and SH compounds. Int. J. Radiat. Biol. 21, 329-342 (1972).
9. J.D. Chapman, A.P. Reuvers, J. Borsa and C.L. Greenstock, Chemical radioprotection and radiosensitization of mammalian cells growing in vitro. Radiat. Res. 56, 291-306 (1973).
10. G.E. Adams, Molecular mechanisms of cellular radiosensitization and protection. In Radiation Protection and Sensitization, 1970 (H.L. Moroson and M. Quintiliani, Eds.) pp. 3-14. Taylor and Francis, London.
11. R.L. Willson and P.T. Emmerson, Reaction of triacetoneamine-N-oxyl with radiation-induced radicals from DNA and from deoxyribonucleotides in aqueous solution. In Radiaiton Protection and Sensitization, 1970 (H.L. Moroson and M. Quintiliani, Eds.) pp. 73-79. Taylor and Francis, London.
12. J. Holian and W.M. Garrison, Reconstitution mechanisms in the radiolysis of aqueous biochemical systems: inhibitive effects of thiols. Nature (London) 221, 57 (1969).
13. M.G. Simic and S.V. Jovanovic, Free radical mechanisms of DNA base damage. In Mechanisms of DNA Damage and Repair, 1986 (M.G. Simic, L. Grossman and A.C. Upton, Eds.) pp. 39-49. Plenum Press, New York.
14. P. O'Neill, Pulse radiolytic study of the interaction of thiols and ascorbate with OH adducts of dGMP and dG: implications for DNA repair processes. Radiat. Res. 96, 198-210 (1983).
15. C. von Sonntag, U. Hagen, A. Schon-Bopp and D. Schulte-Frohlinde, Radiation-induced strand breaks in DNA: chemical and enzymatic

analysis of end groups and mechanistic aspects. Adv. Radiat. Biol. 9, 109-142 (1981).
16. J. Cadet, M. Berger and A. Shaw, The radiation chemistry of the purine bases within DNA and related model compounds. In Mechanisms of DNA Damage and Repair, 1986 (M.G. Simic, L. Grossman and A.C. Upton, Eds.) pp.69-74. Plenum Press, New York.
17. N.K. Kochetkov, L.I. Kudrjashov, M.A. Chlenov and T. Ya. Livertovskaya, The epimerization of monosaccharides by γ-irradiation in frozen and aqueous solutions. Carbohydr. Res. 28, 86-88 (1973).
18. V. Gold and M.E. McAdam, Radiation-induced organic hydrogen isotope exchange reactions in aqueous solution. Acc. Chem. Res. 11, 36-43 (1978).
19. A.J. Alexander, A.F. Fuciarelli, P. Kebarle and J.A. Raleigh, Characterization of radiation-induced damage to polyadenylic acid using high-performance liquid chromatography - tandem mass spectrometry (HPLC-MS-MS). Anal. Chem., 1987 (Submitted).
20. K.B. Lesiak and K.T. Wheeler, Radiation damage to deoxyadenosine in solutions of double stranded poly (dA.T). Abstract (Ei-2) Radiation Research Society Thirty-Fifth Annual Meeting, Atlanta, Georgia, February 21-26, 1987.
21. F.S. Dainton and D.B. Peterson, Forms of H and OH produced in the radiolysis of aqueous systems. Proc. R. Soc. London, Ser. A. 267, 443-463 (1962).
22. J.A. Raleigh and F.Y. Shum, Radioprotection in model lipid membranes by hydroxyl radical scavengers: supplemental role for α-tocopherol in scavenging secondary peroxy radicals. In Radioprotectors and Anticarcinogens, 1983 (O.F. Nygaard and M.G. Simic, Eds.) pp. 87-102. Academic Press, New York.
23. D. Schulte-Frohlinde and E. Bothe, Identification of a major pathway of strand break formation in poly U induced by OH radicals in the presence of oxygen. Z. Naturforsch. Teil. C 39: 315-319, 1984.
24. A.F. Fuciarelli, G.G. Miller and J.A. Raleigh, An immunochemical probe for 8,5'-cyclodeoxyadenosine-5'-monophosphate and its deoxy analog in irradiated nucleic acids. Radiat. Res. 104, 272-283 (1985).
25. For recent review see, N.L. Oleinick, S-M. Chiu, L.R. Friedman, L-Y. Xue and N. Ramakrishnan, DNA-protein cross-links: new insights in their formation and repair in irradiated mammalian cells. In Mechanisms of DNA Damage and Repair, 1986 (M.G. Simic, L. Grossman and A.C. Upton, Eds.) pp. 181-192. Plenum Press, New York.

MECHANISMS OF INACTIVATION OF OXYGEN SPECIES BY CAROTENOIDS

Norman I. Krinsky

Department of Biochemistry and Pharmacology
Tufts University School of Medicine
Boston MA 02111, USA

INTRODUCTION

Although carotenoid pigments have been implicated as anti-carcinogenic compounds for several years, based on both epidemiological evidence (1) as well as experiments in animals (2,3), the exact mechanism whereby this widely distributed class of componds functions is still poorly understood. What appears to be important however, is the fact that many of the effects of carotenoids in vivo and in vitro can be observed with pigments that do not function as precursors of vitamin A (retinol). For example, beta-carotene may exert its biological effects merely by functioning as a precursor of retinal and retinol. On the other hand, there are carotenoid pigments, such as canthaxanthin (4,4'-diketo-beta-carotene) which also exhibit anti-carcinogenic properties and cannot be converted to retinol (Figure 1). Under these circumstances, we must look at the properties of the intact molecules in order to understand their functions.

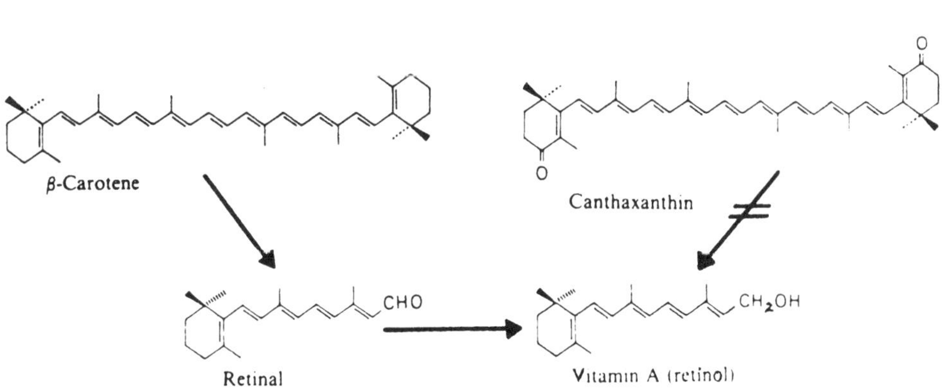

Fig. 1. The intestinal metabolism of beta-carotene to retinal, followed by a reduction to retinol. Canthaxanthin cannot be converted to either of these compounds.

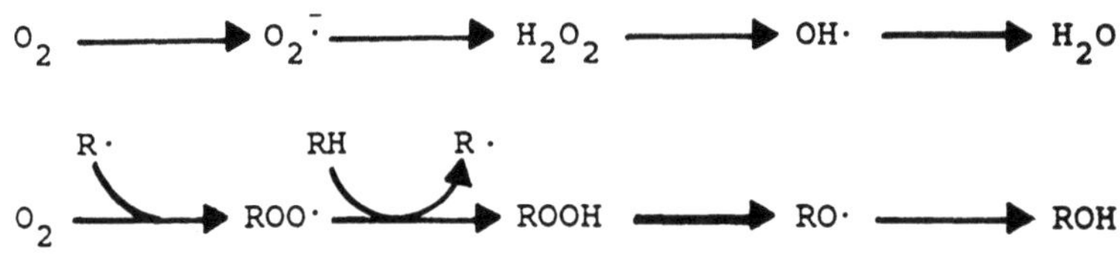

Fig. 2. The reduction products of oxygen initiated by either the addition of an electron (A) or an alkyl radical (B).

ACTIVE OXYGEN SPECIES

There are classes of active oxygen species that can have wide-ranging biological effects. Singlet excited oxygen ($^1\Delta_g$), an electronically excited form of oxygen, plays a role in biological systems after photochemical sensitization has occurred (4). Oxygen-centered radicals, such as superoxide ($O_2^-\cdot$), or the hydroxyl radical (OH·) are formed during normal metabolism or from activated phagocytes such as polymorphonuclear leukocytes (PMNL) (5,6). These intermediates, which arise from the 1-electron reduction of molecular oxygen, have analogous reactions that occur when oxygen reacts with an alkyl radical to give a peroxyl and consequently an alkoxy radical (Figure 2).

Singlet Oxygen Reactions

With respect to an interaction of carotenoid pigments with active oxygen species, the clearest demonstration and best understood process involves the physical quenching of singlet excited oxygen by a wide variety of carotenoids. This reaction was first demonstrated in homogenous solution by Foote and Denny (7) and has also been reported in artificial membrane systems by Anderson and Krinsky (8). In fact, the ubiquitous distribution of carotenoid pigments among all photosynthetic organisms can best be understood in terms of their role as protective agents against photosensitized oxidations initiated by singlet oxygen (9). It is through photosensitized reactions that singlet oxygen plays a role in biological systems, although there are a number of reports indicating that singlet oxygen might arise in biological systems through non-photochemical means (10,11). The mechanism of carotenoid quenching involves a direct energy transfer to form the carotenoid triplet, with subsequent loss of the excess energy to the medium by way of rotational and vibrational interactions with the regeneration of ground-state carotenoid, as seen below:

$$\begin{aligned} &1.\quad {}^1O_2 + \text{Carotenoid} \dashrightarrow {}^3O_2 + {}^3\text{Carotenoid} \\ &2.\quad {}^3\text{Carotenoid} \dashrightarrow \text{Carotenoid} + \text{heat} \\ &\text{Sum:}\quad {}^1O_2 + \text{Carotenoid} \dashrightarrow {}^3O_2 + \text{Carotenoid} + \text{heat} \end{aligned}$$

Oxygen-centered Radical Reactions

The interaction of carotenoid pigments with oxygen-centered radicals has been demonstrated in many systems but we still lack a clear understanding of the mechanisms involved in these processes. One demonstration of carotenoid interaction with active oxygen species is the observation of bleaching of the carotenoid molecule, indicating a chemical reaction which has led to the interruption of the conjugated chromophore of these pigments. This bleaching has been used as a measure of interaction of carotenoids with radical species and used successfully by Bors et al. (12) to determine rate constants for a series of antioxidants which can compete for the alkoxyl radical bleaching of the digentiobioside carotenoid, crocin, as well as canthaxanthin. In addition, this same assay has been used to indicate that $O_2^-\cdot$ does not bleach crocin, whereas OH· readily reacts with crocin (13). Klein et al. (14) have recently reviewed lipid peroxidations catalyzed by plant lipoxygenases which have been followed by measuring carotenoid bleaching. Packer et al. (15) have reported that peroxyl radicals generated by pulse radiolysis are quenched by carotenoids as demonstrated by the bleaching of the carotenoid pigment.

CAROTENOID FUNCTIONS IN VITRO

In vitro, carotenoids have been demonstrated to prevent lipid peroxidation induced by a variety of factors. In homogenous systems, carotenoid pigments have been demonstrated to inhibit lipid peroxidation induced by active oxygen species generated by the use of xanthine oxidase and acetaldehyde (16). As seen in Figure 3, carotenoids have also been demonstrated to inhibit iron-induced lipid peroxidation in liposomes (17,18), or in microsomes using NADPH to initiate lipid peroxidation (19). Burton and Ingold (20) have reported that beta-carotene is an excellent inhibitor of lipid peroxidation at low oxygen tensions. They suggested that the mechanism for this effect might involve the inhibition of the propagating peroxyl radical, ROO via the formation of a resonance-stabilized carbon-centered radical by addition of the peroxyl radical to the carotenoid (20).

In addition, hypochlorous acid has also been demonstrated to result in the rapid bleaching of beta-carotene (21) and this may explain some of the observations implicating carotenoid pigments as protective agents in reactions initiated by reactive oxygen species generated by polymorphonuclear leukocytes (PMNL).

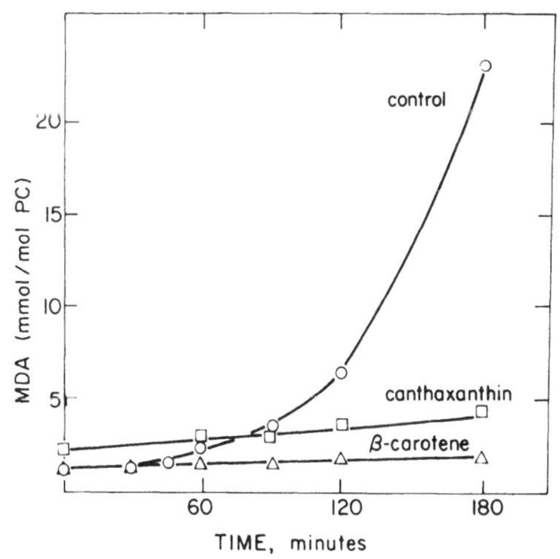

Fig. 3. Carotenoid protection of Fe-induced lipid peroxidation in egg phosphatidyl choline liposomes (17,18).

CAROTENOID FUNCTIONS IN VIVO

As mentioned earlier, one of the best known functions of carotenoids in vivo is to act as quenchers of photochemically-derived singlet oxygen (9). Mathews-Roth (22) has followed the bleaching of 1,3-diphenylisobenzofuran (DPBP) in mouse epidermis of light-sensitive porphyric mice that received either beta-carotene or canthaxanthin in their diets. When DPBF was added to the skin of these mice, the extent of bleaching was markedly decreased in those animals that received carotenoids in their diets.

In a different type of experiment, Krinsky (23) demonstrated that the presence of colored carotenoids in wild-type Sarcina lutea rendered this strain more resistant to killing by human PMNL than a mutant carotenoidless strain. Subsequent experiments have confirmed these observations (24). Weitberg et al. (25) have studied the effects of antioxidants on preventing the oxidant damage induced by activated PMNL. They reported that a number of antioxidants, including beta-carotene, could significantly reduce the appearance of sister chromatid exchanges in cultured mammalian cells exposed to activated PMNL. We have recently demonstrated that the addition of carotenoid pigments to PMNL results in a decreased generation of superoxide when these PMNL are stimulated by soluble activating agents such as the phorbol esters (26).

MECHANISMS

We have many examples of carotenoid pigments functioning as antioxidants, presumably by interacting with and quenching active oxygen species. However, with the exception of singlet oxygen, the mechanism whereby this process takes place is still poorly understood and requires additional experimentation in order to formulate a rational basis for the use of carotenoid pigments as anticarcinogenic compounds. The following diagram represents our present level of understanding of how carotenoids interact with active oxygen species.

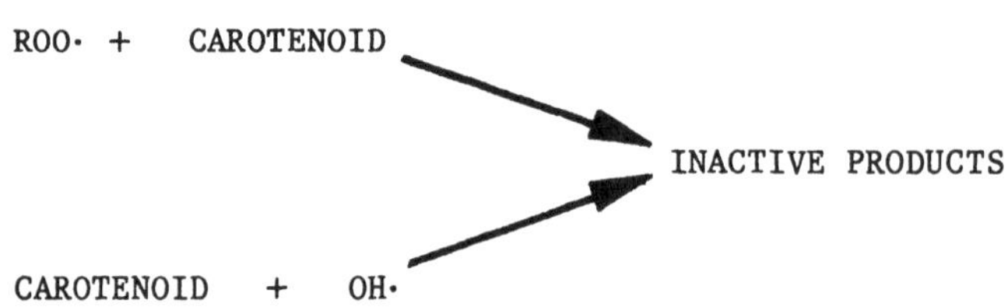

REFERENCES

1. R. Peto, R. Doll, J.D. Buckley, and M.B. Sporn. Can dietary beta-carotene materially reduce human cancer rates? Nature 290, 201-208 (1981).
2. M.M. Mathews-Roth. Carotenoids and cancer prevention -- experimental and epidemiological studies. Pure Appl. Chem. 57, 717-722 (1985).
3. L. Santamaria, L. Bianchi, A. Arnaboldi, L. Andreoni, and P. Bermond. Dietary carotenoids block photocarcinogenic enhancement by benzo(a)pyrene and inhibit its carcinogenesis in the dark. Experientia 39, 1043-1045 (1983).
4. N.I. Krinsky, Biology and Photobiology of Singlet Oxygen. in Oxygen Radicals In Chemistry and Biology (W. Bors, M. Saran and D. Tait, Eds.) pp. 453-464. Walter de Gruyter, Berlin, 1984.

5. I. Fridovich. Biological effects of the superoxide radical. Arch. Biochem. Biophys. 247, 1-11 (1986).
6. A.I. Tauber and B.M. Babior. Neutrophil oxygen reduction: The enzymes and the products. Adv. Free Radical Biol. Med. 1, 265-307 (1985).
7. C.S. Foote, and R.W. Denny. Chemistry of Singlet Oxygen. VII. Quenching by beta-carotene. J. Am. Chem. Soc. 90, 6233-6235 (1968).
8. S.M. Anderson, and N.I. Krinsky. Protective action of carotenoid pigments against photodynamic damage to liposomes. Photochem. Photobiol. 18, 403-408 (1973).
9. N.I. Krinsky. Carotenoid protection against oxidation. Pure Appl. Chem. 51, 649-660 (1979).
10. J.R. Kanovsky and B. Axelrod. Singlet oxygen production by soybean lipoxygenase isozymes. J. Biol. Chem. 261, 1099-1104 (1986).
11. E. Cadenas. Oxidative stress and formation of excited species. in Oxidative Stress (H. Sies, Ed.) pp. 311-330. Academic Press, New York, 1985.
12. W. Bors, C. Michel and M. Saran. Inhibition of the bleaching of the carotenoid crocin. A rapid test for quantifying antioxidant activity. Biochim. Biophys. Acta 796, 312-319 (1984).
13. W. Bors, M. Saran, and C. Michel. Radical intermediates involved in the bleaching of the carotenoid crocin. Hydroxyl radicals, superoxide anions and hydrated electrons. Int. J. Radiat. Biol. 41, 493-501 (1982).
14. B.P. Klein, D. King, and S. Grossman. Cooxidation reactions of lipoxygenase in plant systems. Adv. Free Radical Biol. Med. 1, 308-343 (1985).
15. J.E. Packer, J.S. Mahood, V.O. Mora-Arellano, T.F. Slater, R.L. Willson, and B.S. Wolfenden. Free radicals and singlet oxygen scavengers: reaction of a peroxy-radical with beta-carotene, diphenylfuran and 1,4-diazobicyclo(2,2,2)-octane. Biochem. Biophys. Res. Commun. 98, 901-906 (1981).
16. E.W. Kellogg, III and I. Fridovich. Superoxide, hydrogen peroxide, and singlet oxygen in lipid peroxidation by a xanthine oxidase system. J. Biol. Chem. 250, 8812-8817 (1975).
17. N.I. Krinsky. Carotenoid pigments: multiple mechanisms for coping with the stress of photosensitized oxidations. in Strategies of Microbial Life in Extreme Environments (M. Shilo, ed.) pp. 163-177, Dahlem Konferenzen, Berlin, 1979.
18. N.I. Krinsky, and S.M. Deneke. Interaction of oxygen and oxy-radicals with carotenoids. JNCI 69, 205-210 (1982).
19. R. Dixit, H. Mukhtar, and D.R. Bickers. Studies on the role of reactive oxygen species in mediating lipid peroxide formation in epidermal microsomes of rat skin. J. Invest. Dermatol. 81, 369-375 (1983).
20. G.W. Burton, and K.W. Ingold. Beta-Carotene: an unusual type of lipid antioxidant. Science 224, 569-573 (1984).
21. J.M. Albrich, C.A. McCarthy, and J.K. Hurst. Biological reactivity of hypochlorous acid. Implications for microbicidal mechanisms of leukocyte myeloperoxidase. Proc. Natl. Acad. Sci. U.S.A. 78, 210-214 (1981).
22. M. M. Mathews-Roth, Porphyrin photosensitiztion and carotenoid protection in mice; in vitro and in vivo studies. Photochem. Photobiol. 40, 63-67 (1984).
23. N.I. Krinsky. Singlet excited oxygen as a mediator of the antibacterial action of leukocytes. Science 186,363-365 (1974).
24. H. Rosen, and S.J. Klebanoff. Bactericidal activity of a superoxide anion-generating system. A model for the polymorphonuclear leukocyte. J. Exp. Med. 149, 27-39 (1979).

25. A.B. Weitberg, S.A. Weitzman, E.P. Clark, and T.P. Stossel. Effects of antioxidants on oxidant-induced sister chromatid exchange formation. J. Clin. Invest. 75, 1835-1841 (1985).
26. P.A. Rubin, S. Welankiwar, and N.I. Krinsky. Unpublished observations (1987).

(Supported by NIH 5U01 CA42587-02)

NIK:IV:8
6/17/87/vo

INHIBITION OF AUTOXIDATION BY VITAMIN E AND BILIRUBIN

Mohammed Al-Sheikhly and Michael G. Simic

Center for Radiation Research
National Bureau of Standards
Gaithersburg, MD 20899

INTRODUCTION

Antioxidants are known to have an anticarcinogenic effect[1,2]. Although they are capable of reducing the incidence of tumors induced by chemical carcinogens, their effect on radiation-induced tumors has not been clearly demonstrated[3,4]. The mechanism of the anticarcinogenic action of antioxidants is poorly understood, and despite numerous mechanistic studies of autoxidation processes and antioxidants in model systems, detailed reaction mechanisms under physiological conditions are still elusive. However, inhibition of free radical processes may be an important part of this activity.

The processes associated with membranes and micelles are specially interesting because the oxidation of fatty acids has been implicated in the generation of clastogenic factors[5]. In this paper we present our findings on free radical mechanisms associated with the chain-breaking action of Vitamin E, a well-known physiological antioxidant, and bilirubin, a metabolic product of porphyrin, in rod-shaped micelles of linoleic acid. The similarities and differences between Vitamin E and bilirubin are discussed, especially with regard to their roles as chain-breaking antioxidants in fatty acid aggregates.

Experimental Approach

Rod-shaped micelles,[6] consisting of about 300 linoleic acid (LH_2) molecules, in aqueous solutions were used as a simplified membrane model[7]. The overall LH_2 concentration was 20 mM, which corresponds to a rod concentration of about 7×10^{-5} M.

Radiation-generated OH radicals were used as free radical initiators. In $N_2O + O_2$ (4:1) solutions, hydrated electrons are converted (> 95%) to OH radicals. Hence, under these conditions OH is the predominant reactant, G(OH) = 5.6, with only small amounts of H atoms, G(H) = 0.6.

Chain autoxidation reactions were monitored by oxygen uptake measurements [OU, i.e., $G(-O_2)$ value] using a standard Clark electrode[7]. Typical OU measurements are shown in Fig. 1. The kinetics of free radical reactions were measured by pulse radiolysis using the NBS pulse radiolysis assembly[8].

The yield of products and the consumption of stable constituents are expressed as G-values (number of molecules and/or radicals produced or consumed per 100 eV of absorbed energy)[9].

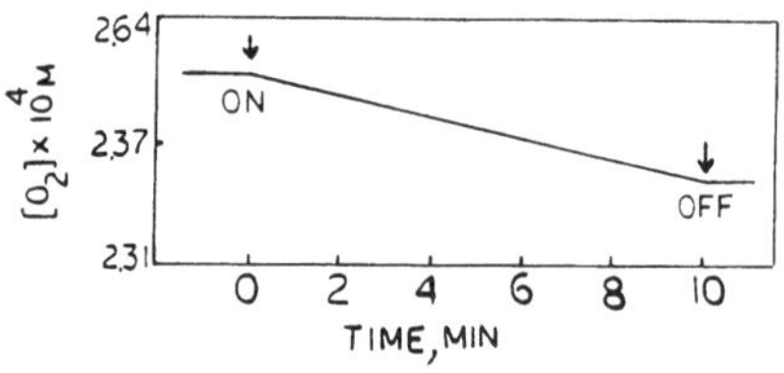

Fig. 1. Oxygen uptake in irradiated aqueous solutions of linoleic acid.

Autoxidation Reactions

In the first reaction step, three types of linoleic acid radicals were formed at a diffusion controlled rate by •OH radicals. OH radicals preferentially add to a double bond of linoleic acid to form $\cdot LH_2$-OH. They also abstract hydrogen from the bisallylic positions to form •LH and from other positions within LH_2 to form $\cdot LH_2(-H)$ radicals. Hence,

$$LH_2 + OH \rightarrow \cdot LH_2\text{-}OH, \quad \cdot LH, \quad \cdot LH_2(-H) \qquad (1)$$

$$k \sim 1.2\ 10^{10}\ M^{-1}\ s^{-1}$$

All three types of linoleic acid radicals (designated here as •L) react rapidly with oxygen to give corresponding peroxy radicals, as shown in the following reaction,

$$\cdot L + O_2 \rightarrow LOO\cdot \qquad (2)$$

$$k = 1.3 \times 10^8\ M^{-1}\ s^{-1}$$

In the next step, the peroxy radicals react preferentially (>99%), with the bisallylic hydrogen,

$$LOO\cdot + LH_2 \rightarrow LOOH + \cdot LH \qquad (3)$$

$$k = 60\ M^{-1}\ s^{-1}\ \text{(ref. 10)}$$

The •LH radical reacts with oxygen and propagates the chain[10]. Consequently, an initial •OH radical leads to more than one oxidation product of linoleic acid (hydroperoxides) and a higher oxygen consumption.

The chain autoxidation is terminated by a radical-radical reaction,

$$LHOO\cdot + LHOO\cdot \rightarrow \text{products} \qquad (4)$$

In homogeneous systems, k_4 is of the order of $2.4 \times 10^7\ M^{-1}\ s^{-1}$. In micellar systems the k_4 is much slower for two reasons: (1) there is never more than one peroxy radical in a rod, and (2) the rods have multiple negative charges that inhibit rapid inter-rod reactions.

Inhibition of Chain Autoxidation Reactions

In the absence of antioxidants, the autoxidation chain reaction was dependent on dose rate. At low dose rates the chain was fairly long. At a dose rate of 1.6×10^{-2} Gy•s^{-1}, $G(-O_2) = 117$. In the presence of vitamin E (d-α-tocopherol) and bilirubin, OU was reduced and was dependent on the concentration of the antioxidant (Fig. 2). OU reaches the plateau value $G(-O_2) = 10$ at about $(6\text{-}7) \times 10^{-5}$ M antioxidant. Because this concentration of the antioxidant should correspond to the concentration of the rods, it appears that, on average, one antioxidant molecule per rod is sufficient to stop the autoxidation chain reaction. However, the OU value does not drop to $G(-O_2) = 3$, which is the OU value in the absence of a chain reaction. The difference between the values G = 10 and G = 3 may be due to processes occurring outside the rod (e.g., in a number of small aggregations of LH_2 that are in an equilibrium with the rods).

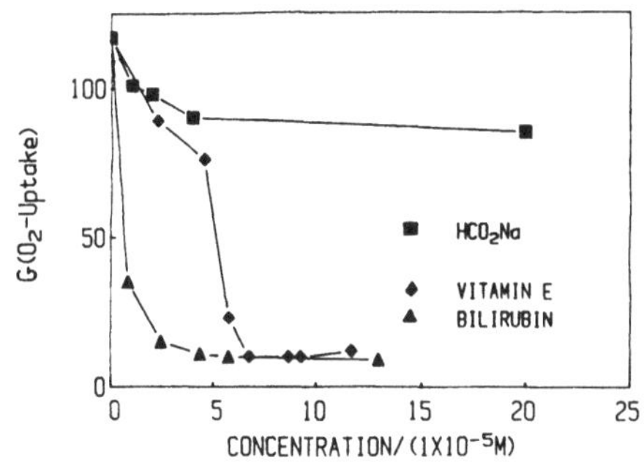

Fig. 2. Inhibition of chain autoxidation of linoleic acid molecules in rod-shaped micelles by vitamin E and bilirubin.

Both bilirubin and vitamin E are hydrophobic molecules that are not soluble in water. At high pH, bilirubin has limited solubility. Under the experimental conditions described, neither of these antioxidants is present in the aqueous phase. The decreased OU due to these antioxidations is not a consequence of OH radical scavenging. The classical OH radical scavenger formate, HCO_2^-, which does not induce a chain reaction,[7] has a very small effect on the OU when added to the system at concentrations equivalent to those of bilirubin and vitamin E. This fact clearly indicates that neither bilirubin nor vitamin E react with •OH under these experimental conditions, and that inhibition of OU must be a consequence of their reactions with peroxy radicals. This is in agreement with the recently found antioxidant activity of bilirubin in liposomes.[12]

The kinetics of the reactions of vitamin E and bilirubin[13] with oxidizing species were determined by pulse radiolysis. In both cases direct use of •OH is unsuitable for oxidative studies because it adds to the benzene ring of vitamin E and to the double bonds of bilirubin.

The reactivity of bilirubin (BH_2) with oxidative species was studied using carbonate radicals in aqueous solution at pH 11,

$$BH_2 + \bullet CO_3^- \rightarrow \bullet BH + CO_3^{2-} + H^+ \qquad (5)$$

$$k = 5 \times 10^8 \ M^{-1} \ s^{-1}$$

The carbonate radicals were generated by,

$$CO_3^{2-} + \bullet OH \rightarrow \bullet CO_3^- + OH^- \qquad (6)$$

Since the carbonate radical $\bullet CO_3^-$ is, strictly speaking, an electron acceptor and not an H-atom acceptor[14], it appears that bilirubin is an electron donor rather than an H atom-donor.

CONCLUSIONS

The chain-breaking antioxidant activity of bilirubin is considerably higher than that of vitamin E (Fig. 2). The chain autoxidation reaction in the rod and in the membrane would occur at the level of the bisallylic hydrogen (C11). The relatively hydrophilic chromanol head of vitamin E would be expected to be found at the surface of a micelle, a rod, or a phospholipid membrane. At present, it is not clear how the linoleic peroxy radicals, which are generated at the C11 level, are inactivated by the chromanol head. In contrast, bilirubin is considerably more hydrophobic than vitamin E and should easily reach the linoleic acid peroxy radical, an assumption that is supported by its more efficient chain-breaking antioxidant activity.

Our studies appear to indicate that the best defense of organisms against the deleterious effects of free radicals and their products in membranes should be a combination of antioxidants. For example, vitamin E and bilirubin would be a good combination for protecting against oxidants coming from the aqueous phase, chain autoxidation reactions within membranes, and autoxidation processes in micellar systems in the gut.

REFERENCES

1. L. Wattenberg and L.K.T. Lan, Phenolic antioxidants as protective agents in chemical carcinogenesis. Radioprotectors and Anticarcinogenesis, 1983 (O. F. Nygaard and M. G. Simic, Eds.) pp. 461-470, Academic Press, New York.
2. M. G. Simic and E.P.L. Hunter, Interaction of free radicals and antioxidants. Ibid, pp 449-460.
3. P. A. Cerutti and I. Emerit, Tumor promotor phorbol-myristate-acetate induces membrane-mediated chromorsomal damage. Ibid, pp 527-538.
4. W. Bors, M. Saran and D. Tait, Eds. Oxygen Radicals in Chemistry and Biology (1984), Walter Gruyter and Co., Berlin, New York.
5. P. A. Cerutti, Prooxidant states and tumor promotion. Science, 227, 375-381 (1985).
6. J. K. Thomas, "The Chemistry of Excitation at Interfaces" (1984), American Chemical Society, Washington, D.C.
7. M. Al-Sheikhly and M.G. Simic, J. Phys. Chem. submitted.
8. E.P.L. Hunter, M. G. Simic and B. D. Michael, Use of an optical multichannel analyzer for recording absorption spectra of short-lived transients. Rev. Sci. Instru. 56, 2199-2204 (1985).
9. A. J. Swallow, "Radiation Chemistry" (1973), John Wiley & Sons, London.
10. J. A. Howard and K. U. Ingold, Absolute rate constants for hydrocarbon autoxidation V. The hydroperoxy radical chain propagation and termination. Can. J. Chem. 45, 785-792 (1966).
11. N. A. Isildar, M. N. Schuchmann, D. Schulte-Frohlinde, C. von Sonntag, Oxygen uptake in the radiolysis of aqueous solutions of nucleic acids and their constituents. Int. J. Radiat. Biol. 41, 525-533 (1982).
12. R. Stocker, Y. Yamamoto, A. F. McDonagh, A. N. Glazer, and B. N. Ames, Bilirubin is an antioxidant of possible physiological in importance. Science 235, 1043-1045 (1987).
13. M. G. Simic, M. Z. Hoffman, and I. A. Taub, Association for Radiation Research, University of Newcastle upon Tyne, 6th-8th January 1976 Book of Abstracts.
14. M. G. Simic and E.P.L. Hunter, The reactivity of organic oxygen (oxy) radicals. Oxygen Radicals in Chemistry 1984 (W. Bors, M. Saran and D. Tait, Eds.) pp 109-121, Walter de Gruyter and Co., Berlin. New York.

THE ROLE OF ACTIVE OXYGEN AND THE METABOLISM OF ARACHIDONIC ACID IN THE FORMATION OF CLASTOGENIC FACTOR BY HUMAN MONOCYTES

Walter Kozumbo, *Dominique Muehlematter, *Takafumi Ochi and *Peter Cerutti
Center for Environmental Medicine, University of North Carolina, Chapel Hill, North Carolina. *Department of Carcinogenesis, Swiss Institute for Experimental Cancer Research, 1066 Epalinges/Lausanne, Switzerland

INTRODUCTION

The mouse skin tumor promoter phorbol-12-myristate-13-acetate (PMA) is a potent inflammatory agent that not only elicits the infiltration of neutrophils, monocytes and lymphocytes to the site of application (1) but also induces human neutrophils and monocytes, but not lymphocytes, to release clastogenic factor(s) (CF) (2). CF are characterized as lipophilic (ethylacetate extractable) substances of relatively low molecular weight (<10 kDa) whose formation is mediated by superoxide anions and whose activity is inhibited by CuZn superoxide dismutase (SOD), free-radical scavengers and antagonists of arachidonic acid (AA) metabolism (3). A correlation between CF formation and lipid peroxidation has been demonstrated (4). These properties of CF are suggestive of free-radical chain reactions which can give rise to lipid peroxides and possibly account for the formation and/or action of CF.

The prooxidant activity of PMA may in part be due to the stimulation of both cyclooxygenase and lipoxygenase pathways, resulting in the production of intermediate peroxy-metabolites (endo- and hydroperoxides) of AA (5). These peroxides may play a role in CF formation and/or activity as evidenced by the anticlastogenic action of several antagonists of the AA cascade (6). The formation of AA peroxides, however, may also occur independently of oxygenase activity. That is, oxy radicals produced by PMA-stimulated monocytes (7) may initiate free-radical chain reactions leading to the non-enzymatic peroxidation of AA and, ultimately, the formation of CF. It seems reasonable to suspect the formation of a clastogenic AA product that results from a reaction between superoxide anion (or derivative) and AA or its metabolites. An alternative possibility, however, is that superoxide radical or a derived product alters the activity of enzymes involved in the release and/or metabolism of AA so as to generate a clastogenic metabolite. To investigate these possibilities and to elucidate the clastogenic mechanisms, our strategy is to identify metabolites of AA or other AA-derived clastogenic lipids (such as 4-hydroxynonenal and platelet activating factor) in ultrafiltrates from PMA-stimulated monocytes in order that we may examine them individually for their capacities to induce DNA and chromosomal damage in test cell systems.

RESULTS

Composition of Lipids in CF-containing Medium from PMA-stimulated Monocytes

Human monocytes were equilibrium-labeled with tritiated AA for 20 h, washed and treated with PMA (30 ng/ml) for 1 h in medium 199, as was previously demonstrated to generate CF (2). Chloroform:methanol (2:1) extractions of the medium were separated by thin-layer chromatography (TLC) (8) to determine the composition of radioactively labeled lipids. Medium from unstimulated control monocytes contained only 3.4% of the total radiolabeled cellular lipids in comparison to 13.8% for the clastogenic medium of PMA-stimulated monocytes, a 4-fold increase in release of AA-associated lipids. Nearly 88% of the radiolabeled lipids in the clastogenic medium was composed of free AA plus metabolites while the remaining 12% consisted of 7% phospholipids, 4% mono- and diacylglycerols, and 1% triacylglycerols. In control medium a similar distribution was found, i.e., 82% AA and metabolites, 13% phospholipids, 6% mono- and diacylglycerols, and 1% triacylglycerols. However, in contrast to the unstimulated cells the PMA-stimulated monocytes released greater amounts (determined by released cpm as a percentage of initial total incorporated cpm) of each class of radiolabeled lipid. PMA stimulation increased the release of free AA (including metabolites), phospholipids, mono-di- and triacylglycerols by 4.3, 2.3, 3.0 and 1.3 fold, respectively. Thus, in the clastogenic medium free AA plus metabolites composed nearly all (88%) of the radiolabeled lipids and, in comparison to the AA-esterified lipids, displayed the largest increased release factor (4.3 fold) in response to PMA stimulation.

AA Metabolites Found in Clastogenic Medium from PMA-stimulated Monocytes

High-performance liquid chromatographic (HPLC) analysis established by Eling and coworkers (9) was used to separate both cyclooxygenase and lipoxygenase products found in control and clastogenic media. Figure 1 shows representative elution profiles of control (panel B) and clastogenic (panel C) media. Results from a number of experiments showed that clastogenic medium contained the same AA metabolites as control medium but at levels from 3- to 4-fold higher. These metabolites consisted of the cyclooxygenase products prostaglandins $F_{2\alpha}$ and E_2 ($PGF_{2\alpha}$ and PGE_2), thromboxane B_2 (TxB_2), 12-hydroxy-5,8,10-heptadecatrienoic acid (HHT) and the lipoxygenase-type products 5-, 11- and 15-monohydroxy-eicosatetraenoic acids (HETEs), AA itself and 3 unidentified products. The 3- to 4-fold increase of most metabolites was proportional to the increase in release of total free AA (including metabolites). This was further indicated by the fact that the sum of all metabolites when represented as a fraction of total released AA was the same in control and clastogenic media, i.e., representing 10.5 and 10.0%, respectively, of the total released AA.

Since SOD inhibits the formation of CF by PMA-stimulated monocytes (2), we examined the effects of anticlastogenic concentrations of SOD on the production of AA metabolites. Except for occasional insignificant decreases in the amounts of HETEs, SOD (10-25 µg/ml) failed in five experiments to alter qualitatively or quantitatively the release of AA and its metabolites from PMA-stimulated monocytes. Catalase at 50 µg/ml was also without effect.

Additional analyses by HPLC showed that leukotrienes were not present in the clastogenic or control media. Both an ilium contraction bioassay for slow-reacting substance and a radioimmune assay for leukotriene LTC_4 also produced negative results for both control and PMA-stimulated monocytes. A radioimmune assay for LTB_4 indicated that 1.00 ng was released into 1 ml of medium by 1 million unstimulated monocytes. LTB_4 production, however, apparently is not related to CF formation because the slightly elevated levels found in medium from PMA-stimulated monocytes of 1.31 ng/ml were un-

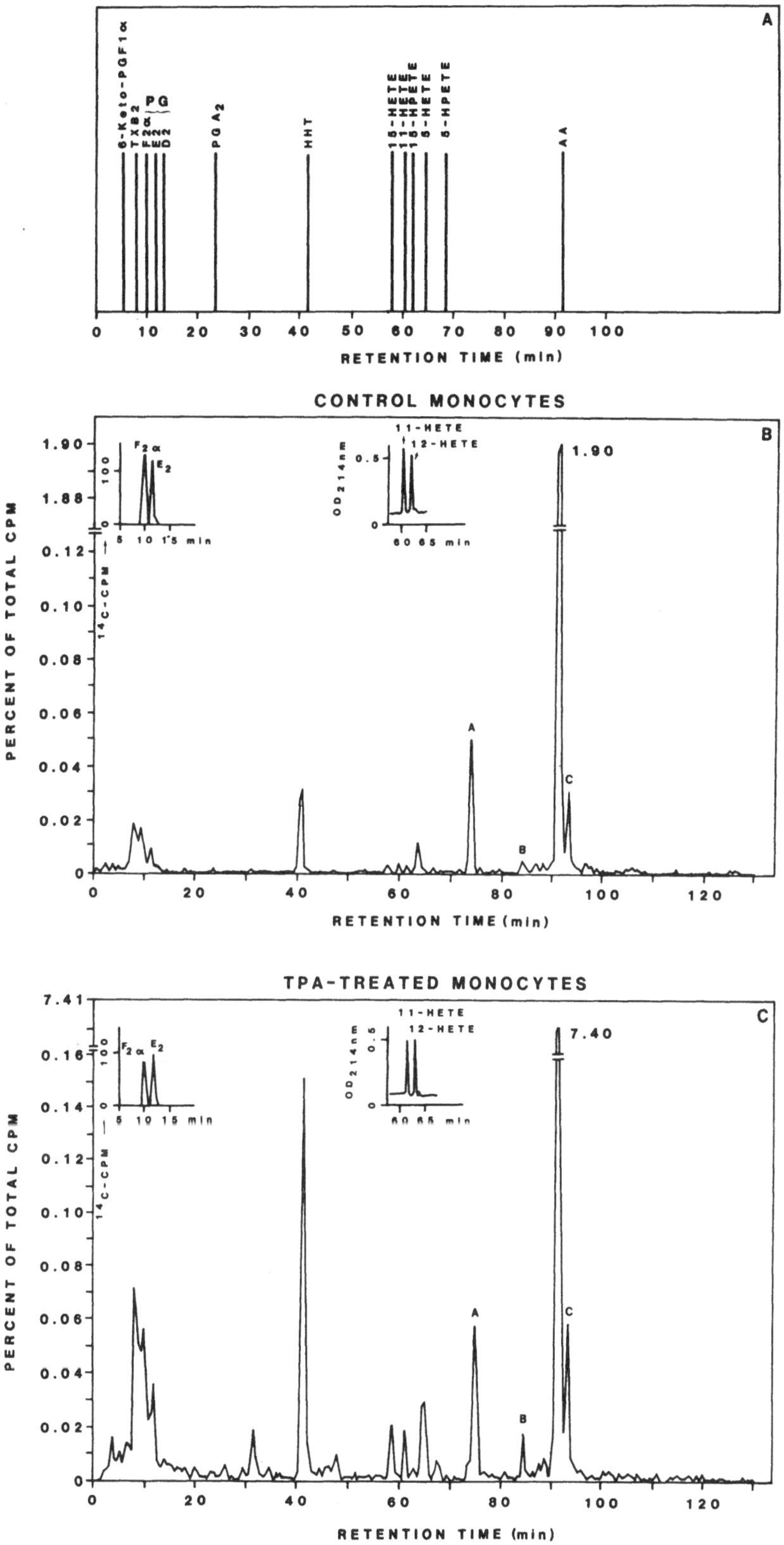

Fig.1. HPLC of media extracts of control (B) & PMA(TPA)-treated (C) monocytes (from ref. 14).

changed when monocytes were stimulated with PMA in the presence of concentrations of SOD (10 μg/ml) that had inhibited the formation of CF.

4-Hydroxynonenal, a Cytotoxic Aldehydic Degradation Product of AA, Was Not Found in Clastogenic Media

The possibility was considered that peroxidation products of AA degrade into a class of highly toxic aldehydes identified as 4-hydroxyalkenals (10) and that these aldehydes may possess clastogenic action. Since 4-hydroxynonenal quantitatively represents the most abundant aldehyde of this class (10), HPLC analysis was used to separate it from the ^{3}H-AA-radiolabeled media components of PMA-stimulated monocytes. UV detection of a non-radiolabeled 4-hydroxynonenal standard never coincided with a detectable radioactive peak, indicating that the breakdown of peroxidized AA into 4-hydroxynonenal did not occur in clastogenic media analyzed after monocytes were stimulated for 1 h with PMA.

Platelet-activating Factor (PAF) Was Not Detected in Clastogenic Media

Because PAF is formed only by CF-forming leukocytes (monocytes and neutrophils but not lymphocytes, 11) and had been reported to stimulate both an oxidative burst and AA metabolism in phagocytes (12,13), we investigated whether PMA induced the formation and release of PAF by human monocytes. Human monocytes were treated for 1 h in sodium acetate-depleted medium 199 containing 2.5 mg/ml lipid-free BSA as carrier for PAF and ^{3}H-sodium acetate (200 μCi/ml) as substratefor the transacetylation reaction. TLC analysis (8) of media from control and treated cells showed that the second stage tumor promoter A23187 but not PMA caused an increase in the release of PAF in comparison to controls, indicating that PAF most probably did not account for the clastogenic action found in ultrafiltrates from PMA-stimulated monocytes. An original account of the results summarized above has been given by Kozumbo and coworkers (14).

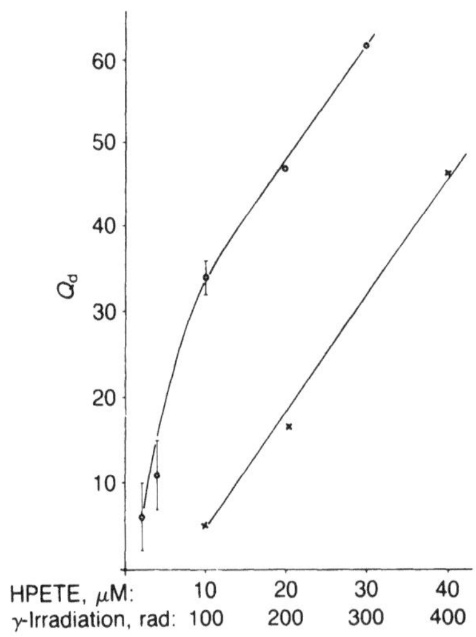

Fig. 2. Dose-response curve for the formation of DNA strand breakage and alkali-labile sites by 15-HPETE and γ-rays under aerobic conditions measured by alkaline elution (from ref. 15) O, 15-HPETE; X, aerobic γ-rays.

Clastogenic Effect of Monocyte-derived Hydroperoxy-5,8,11,13-eicosatetraenoic Acids (HPETEs) on Mouse Embryo Cells C3H/10T½

In order to test whether hydroperoxy-precursors of the HETEs from PMA-stimulated monocytes are clastogenic, we incubated various concentrations of 15-HPETE for 15 min with pre-labeled (^{14}C-thymidine, 20 h) mouse fibroblasts in PBS and then examined the cells for DNA single-strand breaks and/or alkali-labile sites using the alkaline elution assay as described previously by Ochi and Cerutti (15). As shown in Figure 2, 15-HPETE possessed a dose-dependent clastogenic activity that was inhibited by chelators of iron (desferrioxamine) and calcium (EGTA plus quin-2). Other hydroperoxy metabolites of AA (12-, 9- and 8 HPETE) were also clastogenic. Glutathione peroxidase, but neither SOD, catalase nor the free-radical scavenger, butylated hydroxyanisole (BHT), inhibited the clastogenic action of HPETEs in this test cell system.

DISCUSSION

The inhibitory effect of SOD on the formation of lipophilic CF by PMA-stimulated monocytes indicated that the formation of CF requires an interaction between superoxide anions and lipids present in the medium. This interaction is further suggested by observations that non-CF-producing lymphocytes are transformed into CF-producing cells when exposed to a non-cellular source of superoxide and that the appearance of lipid peroxidation products are correlated with CF formation (4). The anticlastogenic activities of several AA antagonists further suggest that lipid metabolites of AA are involved in the clastogenic action of PMA-stimulated monocytes (5). Herein we have shown that during production of CF by PMA-stimulated monocytes there was not only a 4-fold increased release of AA (representing nearly 90% of all released radiolabeled constituents) but also a corresponding increase in both cyclooxygenase and lipoxygenase pathways, resulting in a rapid production of those labile and reactive hydroperoxy-intermediates (HPETEs) of AA which are reduced eventually to their hydroxy-counterparts. Extraction and chromatographic analysis of media from PMA-stimulated monocytes demonstrate the presence of 5-, 11- and 15-HPETEs, suggesting that their corresponding hydroperoxy-precursors existed in clastogenic medium. We have also shown that several of these HPETEs are in fact clastogenic to mouse embryo fibroblasts C3H/10T½ and that ionized iron and calcium apparently mediate this clastogenic action since chelation of either cation sharply reduced HPETE-induced clastogenicity (Fig. 2, ref. 15). In addition to this HPETE-induced clastogenicity, hydrogen peroxide has also been reported to damage cellular DNA (16) and in the presence of iron was clastogenic as well (17). As a result of these findings, we propose that the CF of ultrafiltrates from PMA-stimulated monocytes consists of a mixture of hydrogen peroxide, AA-OOH and superoxide reduced iron (Fe^{++}). Once chemically reduced by superoxide, iron may readily participate in Fenton-type reactions with AA-OOH, generating a cascade of free radicals involved in the clastogenic action of monocytes (see Fig. 3). In further support of this scheme is the recent observation that chelation of iron by desferrioxamine inhibited the clastogenic action of preformed CF (I. Emerit, personal communication). The fact that EGTA and quin-2, but neither SOD, catalase nor the radical scavenger BHT, could inhibit the clastogenic action of the monocyte product 15-HPETE suggests that unbound calcium may act at a late, post free-radical stage in the clastogenic pathway.

Lewis et al. (16) have reported that a combination of both active oxygen and lipoxygenase products produced by macrophages was substantially more effective in damaging DNA of cocultured fibroblasts than was either of these products alone. Such DNA damage may be mediated by a mechanism that is similar to our proposed pathway (Fig. 3) of clastogenic action. The capacity of inflammatory cells to release DNA-damaging factors indicates a possible relationship between chronic inflammation and the process of carcinogenesis.

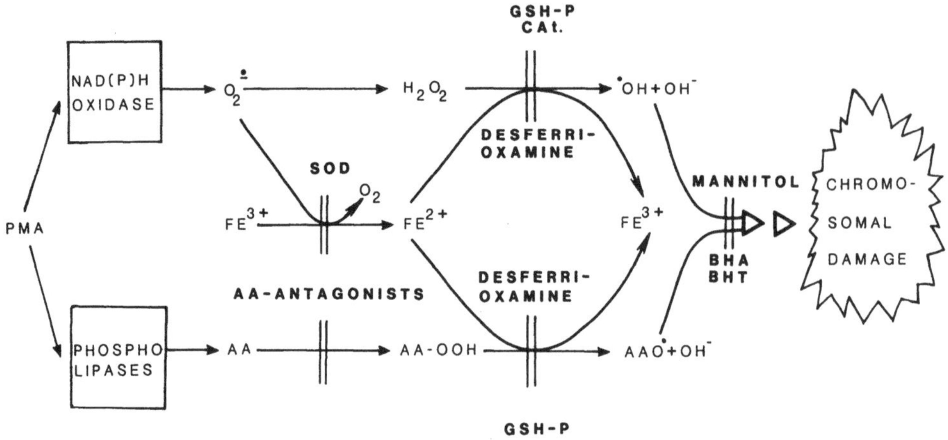

Fig. 3. Clastogenic action of PMA-stimulated human monocytes (from ref. 14).

REFERENCES

1. A. Viaje, T. Slaga, M. Wigler and I. Weinstein, Effect of anti-inflammatory agents on mouse skin tumor promotion, epidermal DNA synthesis, phorbol ester-induced cellular proliferation and production of plasminogen activator. Cancer Res. 37, 1530-1536 (1977).
2. I. Emerit and P. Cerutti, Clastogenic action of tumor promoter phorbol-12-myristate-13-acetate in mixed human leukocyte cultures. Carcinogenesis 4, 1313-1316 (1983).
3. I. Emerit and P. Cerutti, Tumor promoter phorbol-12-myristate-13-acetate induces a clastogenic factor in human lymphocytes. Proc. Natl. Acad. Sci. U.S.A. 79, 7509-7513 (1982).
4. I. Emerit, S. Khan and P. Cerutti, Treatment of lymphocyte cultures with a hypoxanthine-xanthine oxidase system induces the formation of transferable clastogenic material. J. Free Radical Biol. Med. 1, 51-57 (1985).
5. P. Cerutti, Prooxidant states and tumor promotion. Science 22, 375-381 (1985).
6. I. Emerit, A. Levy and P. Cerutti, Suppression of tumor promoter phorbol myristate acetate-induced chromosome breakage by antioxidants and inhibitors of arachidonic acid metabolism. Mutat. Res. 110, 327-335 (1983).
7. Y. Keisari, I. Geva, E. Flescher, H. Golden and G. Lavie, Stimulation of human monocyte oxidative burst and related cytotoxicity by tumor-promoting and non-tumor-promoting diterpene esters, indole alkaloids and polyacetate-type agents. Int. J. Cancer 36, 467-472 (1985).
8. F. Chilton, J. Ellis, S. Olson and R. Wykle, 1-O-alkyl-2-arachidonyl-sn-glycerol-3-phosphocholine. J.Biol. Chem. 259, 12014-12019 (1984).
9. T. Eling, B. Tainer, A. Ally and R. Warnock, Separation of archidonic acid metabolites by high pressure liquid chromatography. Methods Enzymol. 86, 511-517 (1982).
10. A. Benedetti, R. Fulceri and M. Comporti, Inhibition of calcium sequestration activity of liver microsomes by 4-hydroxyalkenals originating from the peroxidation of liver microsomal lipids. Biochim. Biophys. Acta 793, 489-493 (1984).
11. G. Camussi, M. Aglietta, R. Coda, F. Bussolino, W. Piacibello and C. Tetta, Release of platelet-activating factor (PAF) and histamine. Immunology 42, 191-199 (1982).
12. T. Yasaka, L. Boxer and R. Baehner, Monocyte aggregation and superoxide anion release in response to formyl-methionyl-leucyl-phenylalanine (FMLP) and platelet-activating factor (PAF). J. Immunol. 128, 1939-1944 (1982).

13. F. Chilton, J. O'Flaherty, C. Walsh, M. Thomas, R. Wykler, L. de Chatelet and B. Waite, Platelet-activating factor. J.Biol. Chem. 257, 5402-5407 (1982).
14. W. Kozumbo, D. Muehlematter, A. Jorg, I. Emerit and P. Cerutti, Phorbol ester-induced formation of clastogenic factor from human monocytes. Carcinogenesis 8, 521-526 (1987).
15. T. Ochi and P. Cerutti, Clastogenic action of hydroperoxy-5,8,11,13-icosatetraenoic acids on the mouse embryo fibroblasts C3H/10T½. Proc. Natl. Acad. Sci. U.S.A. 84, 990-994 (1987).
16. J. Lewis, T. Hamilton and D. Adams, The effect of macrophage development on the release of reactive oxygen intermediates and lipid oxidation products, and their ability to induce oxidative DNA damage in mammalian cells. Carcinogenesis 7, 813-818 (1986).
17. S. Lesko, R. Lorentzen and P. Ts'o, Role of superoxide in deoxyribonucleic acid strand scission. Biochemistry 19, 3023-3028 (1980).

CLASTOGENIC FACTORS, A LINK BETWEEN

CHRONIC INFLAMMATION AND CARCINOGENESIS

Ingrid Emerit

Institute Biomedical des Cordeliers
Universite Paris VI and CNRS
Paris, France 75006

INTRODUCTION

Clastogenic factors (CF), also called chromosome breakage factors, were first described by radiobiologists in 1968 (for review see 1). It was noted that not only do therapeutically or accidentally irradiated persons show an increased frequency of chromosome breaks and rearrangements in their own cells, but the plasma of these persons contains chromosome damaging material able to induce chromosome damage after transfer into cell cultures. These CF were circulating in the blood stream of the irradiated persons even years after the irradiation event (2). Plasma from irradiated animals had a tumorigenic effect in rats (3), suggesting a role for CF in carcinogenesis.

Work of our laboratory demonstrated the presence of similar transferable clastogenic material in plasma of patients with chronic inflammatory diseases, explaining the so-called "spontaneous" chromosome instability regularly observed in blood or fibroblast cultures from these patients. CF were not only found in the plasma and in the supernatants of cell cultures from patients with collagen diseases such as progressive systemic sclerosis (4), lupus erythematosus (5), and rheumatoid arthritis, but also in diseases of the gastrointestinal tract (6) and the nervous system (7). (For review see 8). For systemic sclerosis and Crohn's disease, an association with cancer is confirmed by epidemiological studies (9, 10).

The clastogenic material described as CF in these diseases and in irradiated persons is of low molecular weight. CF pass through filters with a cut-off at 10,000 daltons, but are retained by filters with a cut-off at 1000. The clastogenic activity of CF is rapidly destroyed at room temperature or overnight in a refrigerator, but conserved over long periods at -20°. The CF are not destroyed by proteolytic enzymes and have no optical density at a wavelength of 260-280 nm. Their UV and fluorescent spectra are not characteristic; the activity is distributed over more than one fraction after HPLC separation.

The observation that radiation-induced and so-called "spontaneous" chromosome damage is prevented by the enzyme superoxide dismutase (SOD) was a first indication for a role of the superoxide anion O_2^- in the breakage phenomenon (8). Also, the clastogenic effect of patients' serum in the test culture system was inhibited by SOD. Further work showed that CF may be generated in vitro, either by ionizing radiation or by exposure of cells to an oxyradical-generating system. The tumor promoter TPA, known to stimulate a respiratory burst and to initiate the arachidonic acid cascade, induces a CF in PHA-stimulated lymphocyte cultures and in Chinese hamster lung cells. Exposure of rat mesothelial cell cultures to asbestos fibers (chrysotile), known to elicit phagocytosis and to activate phospholipase A_2, results in chromosome damage and sister chromatid exchanges. The ultrafiltrates of these culture media, which are free of asbestos, are clastogenic in a human lymphocyte test culture. These experiments show that not only cells capable of a respiratory burst, but also other cell types respond by CF formation. Treatment of culture media in absence of cells does not result in CF generation. When cells are exposed in presence of SOD, no clastogenic activity is detectable. Also, exposure of cells to O_2^- generated by a xanthine oxidase reaction, results in release of CF. The reaction generates O_2^- during 2-3 hours during which CF formation can be prevented by SOD. Added later, SOD is no longer protective (11, 12).

As suggested by the anticlastogenic effect of SOD in CF-treated test cultures, O_2^- is also involved in CF action. This could be confirmed by direct measurement of O_2^- with the cytochrome-C assay in supernatants of neutrophils and monocytes exposed to increasing doses of CF. Cytochrome-C reduction increased with exposure time and as a function of cell number; no cytochrome-C reduction was observed in the presence of SOD (Khan and Emerit, unpublished data).

Despite the protector effect of SOD, O_2^- itself may not be the ultimate DNA-damaging radical species. The powerfully oxidizing hydroxyl radical, $OH^{\bullet}$, considered as responsible for radiation-induced cell killing, may be formed by interaction with H_2O_2 in a metal-catalyzed Haber-Weiss reaction. Catalase was protective against the clastogenic effect of CF in test cultures, but did not prevent CF formation.

In addition, not only direct radical attack of DNA has to be considered. CF preparations may contain more long-lived, non-radical clastogenic compounds that may be derived from lipid peroxidation of membranes. An increased level of free arachidonic acid (AA) was found in CF preparations obtained by exposure of cells to TPA, asbestos and xanthine oxidase. This may be a consequence of phospholipase A_2 activation. Inhibitors of this enzyme, such as bromophenacyl bromide and fluocinolone, indeed were anticlastogenic. According to Fridovich and Porter (13), exposure to a xanthine oxidase reaction results in autocatalytic autoxidation of AA with accumulation of conjugated dienes. The minimum delay of 18 hours necessary for CF production in the culture system, which was observed independent of the cell type used, is in favor of an autoxidative process rather than of AA oxidation via the cyclo- and the lipoxygenase pathways. No increases in the products of these pathways (PGE_2, PGF_2 alpha, thromboxane, 5, 12 and 15 HETES) were observed by gas chromatography in samples collected after 48 hours of cultivation, when clastogenic activity has reached its maximum (Esterbauer, personal communication).

An aldehydic breakdown product of AA, 4-hydroxynonenal, was detected at micromolar concentrations by HPLC, using a wavelength of 222 nm and 42% acetonitrile in water. This highly clastogenic product was present in samples collected at 48 and 72 hours after the start of the cultures, but not in samples collected as early as after 24 hours. Hydroxynonenal reacts with nonprotein and protein SH groups (14). This may lead to glutathione depletion and to inactivation of the functional SH group of DNA polymerase. Both mechanisms can be indirectly responsible for the chromosomal instability observed as a consequence of CF. A continuous overload of the DNA repair system may ultimately lead to cancer. This is documented by the increased risk of cancer for most of the situations discussed above.

New Zealand Black (NZB) mice represent a good animal model for this system. These mice, used for many years as a model for the study of lupus erythematosus, exhibit an increase in chromosome breaks and rearrangements, as is observed also in the human autoimmune disease. By selective matings based on chromosome breakage frequencies in bone marrow cells, we could develop two NZB sublines, a high breakage (HB) and a low breakage (LB) line, which differ in the presence of CF in the plasma. Resident peritoneal macrophages of HB mice produce "spontaneously" O_2^- and H_2O_2 after attachment to Falcon dishes. This is not the case for macrophages for the non-autoimmune strains, NZW and Swiss mice, under identical conditions. Macrophages from LB mice, in which the disease develops later or not at all (15), produce only small quantities of active oxygen species. Under conditions of low "spontaneous" production NZB mouse macrophages respond to TPA stimulation with considerable oxyradical production. They therefore behave like immunologically elicited macrophages. Resident macrophages, on the other hand, are known to respond poorly to a chemical or particulate stimulant. The increase in O_2^- and H_2O_2 production of NZB mice was positively correlated with age and with the severety of the disease state. Older NZB mice develop lymphoreticular as well as epithelial neoplasia. The frequency of tumors in mice aged 18 months is 4 times higher in HB than in LB animals (15). It is not known yet whether the incidence of tumors can be influenced by SOD, but the observation that intraperitoneal injection of bovine Cu-Zn SOD to HB mice reduces the chromosomal instability could be an indication for a possible anticarcinogenic effect of this enzyme via inhibition of CF-induced oxidative stress (16).

REFERENCES

1. C. B. Faguet, S. M. Reichard and D. A. Welter, Radiation-induced clastogenic factors. Cancer Genet. Cytogenet. 12:73-83 (1984).
2. D. Scott, The effect of irradiated plasma on normal human chromosomes and its relevance to the long-lived lymphocyte hypothesis. Cell Tissue Kinet. 2:295-305 (1969).
3. J. Souto, Tumour development in the rat induced by blood of irradiated animals. Nature 195:1317-1318 (1962).
4. I. Emerit, A. Levy and E. Housset, Breakage factor in systemic sclerosis and protector effect of L-cysteine. Humangenetik 25:221-225 (1974).
5. I. Emerit and A. M. Michelson, Chromosome instability in human and murine autoimmune diseases. Anticlastogenic effect of superoxide dismutase, Acta Physiol. Scand. 492 (Suppl.) 59-65 (1980).
6. I. Emerit and R. Marteau, Chromosome studies in 14 patients with disseminated sclerosis. Humangenetik 13:25-29 (1971).
7. I. Emerit and A. M. Michelson, Chromosomal breakage in Crohn's disease. In: Recent Advances in Crohn's Disease (A. S. Pena, Ed.), pp. 225-230 Martinus Nighoff Publishers, The Hague, 1981.
8. I. Emerit, Chromosome breakage factors: origin and possible significance. Progr. Mutat. Res. 4:61-74 (1982).
9. A. D. Roumm and T. A. Medsger, Cancer and systemic sclerosis. An epidemiologic study. Arthritis Rheum. 28:1336-1340 (1985).
10. D. D. Weedon, R. G. Shorter, D. M. Ilstrup, K. A. Huizenga, and W. F. Taylor, Crohn's disease and cancer. N. Engl. J. Med. 289:1099-1102 (1973).

11. I. Emerit and P. Cerutti, Tumor promoter phorbol-12-myristate-13-acetate induces chromosomal damage via indirect action. Nature (London) 293:144-146 (1981).
12. I. Emerit, S. H. Kahn and P. Cerutti, Treatment of lymphocyte cultures with a hypoxanthine-xanthine oxidase system induces the formation of transferable clastogenic material. J. Free Radicals Biol. Med. 1:51-57 (1985).
13. I. E. Fridovich and N. A. Porter, Oxidation of arachidonic acid in micelles by superoxide and hydrogen peroxide. J. Biol. Chem. 256:260-265 (1981).
14. H. Esterbauer, Aldehydic products of lipid peroxidation. In: Free Radicals, Lipid Peroxidation and Cancer (D. C. McBrien and T. F. Slater, Eds.), pp. 101-128. Academic Press, London, 1982.
15. I. Emerit, J. Feingold, A. Levy, E. Martin and E. Housset, Tumor incidence and development of autoimmune hemolytic anemia in two breeding lines of the NZB mouse strain that differ in chromosome breakage. J. Natl. Cancer Inst. 64:510-520 (1980).

ROLE OF EICOSANOIDS IN TUMOR CELL-PLATELET-

ENDOTHELIAL CELL INTERACTIONS

Bruce W. Steinert[1], Bonnie F. Sloane[2], David G. Menter[3,5], John D. Crissman[4], John D. Taylor[3] and Kenneth V. Honn[1,3]

Departments of Radiation Oncology[1], Pharmacology[2] Biological Sciences[3] and Pathology[4], Wayne State University, Detroit, MI

METASTATIC CASCADE

Metastases represent perhaps the single most important hindrance to improved cancer patient survival. The formation of metastases is the culmination of a complex series of tumor cell-host interactions called the metastatic cascade (Figure 1; 1,2). During tumor progression, tumor cells become separated from the primary tumor mass and invade into the surrounding tissue. Occasionally, tumor cells will invade blood vessels or lymphatics (intravasation) and are readily disseminated throughout the host. A relatively small percentage of the circulating tumor cells will arrest at the vessel wall and invade into the surrounding tissue (extravasation) to establish a secondary tumor site (3-5). The interaction of the circulating tumor cells with host immune cells may result in destruction of the tumor cells, whereas interaction with host platelets and/or blood coagulation components may enhance the metastatic process. Evidence for the involvement of platelets in metastasis has been demonstrated by several investigators (for review see 6,7). Enhancement of tumor cell arrest and adhesion to the vessel wall by platelets has been postulated as the mechanism for this phenomenon, but its exact nature is undetermined. Because of the potent effects of arachidonic acid metabolites (i.e., prostacyclin and thromboxane A_2) on platelet function, we have hypothesized that tumor cells shift the balance between these metabolites in favor of platelet aggregation (8). We have previously proposed the use of modifiers of arachidonic acid metabolism as antimetastatic agents, although the efficacy of antimetastatic therapy with these modifiers has been inconsistent. Prostacyclin (PGI_2) has been recently shown to reduce the incidence of pulmonary

[5]Present address: M.D. Anderson Hospital and Tumor Inst., Houston, TX.

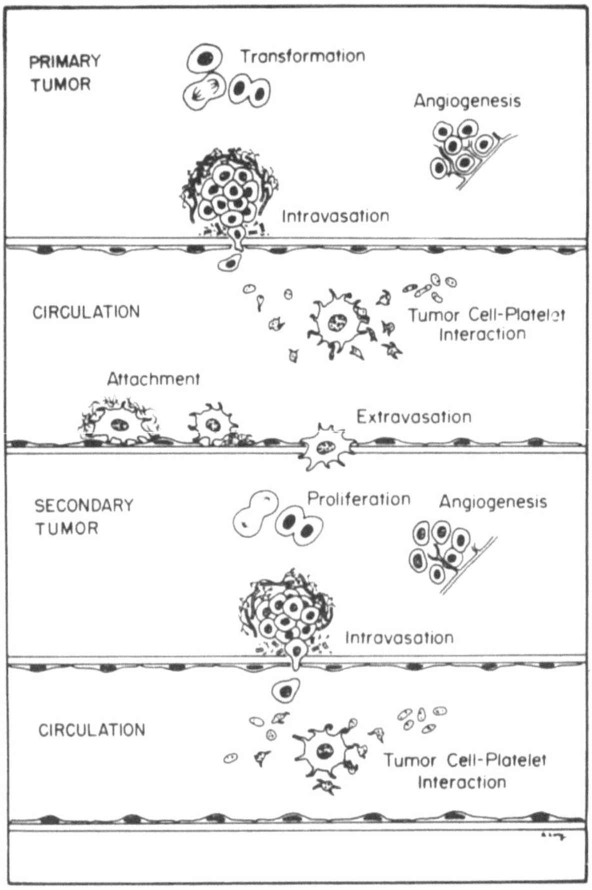

Fig. 1. Metastatic cascade. The formation of metastases involves a complex process series of events which include tumor cell migration, intravasation and hematogenous dissemination of the tumor cells, arrest and adhesion of tumor cells on the vessel wall and extravasation into the surrounding normal tissue. (Reprinted with permission; ref 9.)

tumor colonies in experimental and spontaneous metastasis models in vivo and inhibit tumor cell-induced platelet aggregation (TCIPA) in vivo (9-14). Compounds which inhibit PGI_2 production (e.g., indomethacin) increase the incidence of pulmonary colony formation and do not inhibit TCIPA. PGI_2-enhancing agents (e.g., Nafazatrom), while reducing the incidence of pulmonary tumor colony formation, fail to inhibit TCIPA in vitro (15). Similarly, inhibitors of thromboxane A_2 (TXA_2) synthesis reduce the incidence of pulmonary tumor colonies in vivo, although they fail to inhibit TCIPA in vivo (16). The failure of PGI_2-enhancing agents and thromboxane synthetase inhibitors to inhibit TCIPA in vivo indicates that the present understanding of tumor cell-platelet interactions is incomplete. A better understanding of the mechanism of TCIPA could resolve the controversy surrounding the use of antiplatelet and antithrombotic compounds as antimetastatic agents.

ARACHIDONIC ACID METABOLISM AND TUMOR CELL-INDUCED PLATELET AGGREGATION

The arrest and adhesion of circulating tumor cells to vessel walls is critical to the metastatic process, as extravasation of tumor cells at the secondary site would not be possible without adhesion. As mentioned above, the control of platelet aggregation is highly dependent on the metabolism of arachidonic acid (PGI_2 inhibits; TXA_2 enhances). PGI_2 is the most potent antithrombotic agent known (17), whereas thromboxane A_2 (TXA_2) strongly enhances the aggregation response. Tumor cells and/or their shed vesicles may shift the PGI_2/TXA_2 balance to favor tumor cell adhesion and platelet aggregation. Menter _et al_ (14) have shown that Walker 256 carcinosarcoma (W256), B16 amelanotic melanoma, and 15091A adenocarcinoma cells and the spontaneously shed vesicles of 15091A cells induce platelet aggregation in vitro. TCIPA is completely inhibited in vitro by exogenous PGI_2 (14), but not by thromboxane synthetase inhibitors, yet both classes of compounds reduce the incidence of pulmonary tumor colonies _in vivo_. Recent investigations in our laboratory have shed some light on this controversy and have led us to conclude that TXA_2 alone is not sufficient to induce TCIPA, but rather enhances the aggregation response initiated by another stimulus. We also conclude that the antimetastatic potential of thromboxane synthetase inhibitors may be dependent on sensitization of the platelets to the action of endogenous PGI_2 in addition to the reduction of TXA_2 synthesis (14).

We have examined the roles of agonist strength and the cyclooxygenase (COX) and lipoxygenase (LOX) pathways of arachidonic acid metabolism in TCIPA in vitro. Preliminary studies indicated that both the degree of platelet aggregation and the ability of the COX or LOX inhibitors to affect TCIPA were dependent on the strength of the inducing agonist (i.e., tumor cell number). The addition of 8.75×10^4 W256 cells (low agonist concentration) to rat platelet-rich plasma resulted in minimal aggregation and production of the platelet-derived arachidonic acid metabolites (i.e., TXA_2 and 12-HETE), whereas the addition of 5×10^5 W256 cells (high agonist concentration) results in a significantly greater aggregation response and metabolite production (Figure 2, Table 1). At high agonist concentration, COX inhibitors alone failed to inhibit TCIPA (Figure 2), whereas TCIPA by a low agonist concentration could be completely inhibited by COX inhibitors (e.g., aspirin) alone. When aggregation was induced by a high agonist concentration, inhibition of both COX and LOX pathways was required to inhibit TCIPA (data not shown).

ARACHIDONIC ACID METABOLISM AND PLATELET-ENHANCED TUMOR CELL ADHESION

The adhesion of circulating tumor cells to the vessel endothelium or subendothelial matrix must occur prior to

TABLE 1. Effect of Aspirin and Tumor Cell Concentration on Platelet Eicosanoid Metabolism

Agonist concentration	TXA_2 (ng/ml/15 min) (measured as TXB_2)	12-HETE (ng/ml/15 min)
5 x 10^5 cells	16.5 +/- 2.5	4,600 +/- 700
5 x 10^5 cells + 250 µM aspirin	1.5 +/- 0.2	5,900 +/- 900
5 x 10^4 cells	9.0 +/- 1.0	2,100 +/- 300
5 x 10^4 cells + 250 µM aspirin	1.7 +/- 0.5	1,600 +/- 200

extravasation. Tumor cell adhesion to plastic or nylon fibers can be enhanced by tumor promoters (e.g., phorbol myristate acetate), chemotactic agents (e.g., f-met-leu-phe; 18) and platelets and blood coagulation factors (9,19).

The examination of biochemical and morphological aspects of TCIPA and platelet-enhanced tumor cell adhesion have been restricted due to the lack of an appropriate in vitro model. Previous models have studied only a limited number of the cell types (e.g., tumor cells and platelets or tumor cells and endothelial cells) which would interact during adhesion and extravasation of the tumor cells; or involved the use of static conditions (e.g., tumor cell adhesion to cultured endothelial cell monolayers). Recent studies have clearly shown that hemodynamic shear forces, which mimic the forces present in vivo, can alter the morphology and biochemistry of endothelial cells in vitro (20-22). Static models, or models using a limited number of cell types, therefore are not an adequate representation of in vivo events. We have described a new aggregometry model, using rat aortic endothelial cells grown on microcarrier beads (MEC; Ventregel, Ventrex Bioventures Group, Portland, ME), which permits the examination of tumor cell-platelet-endothelial cell interactions under dynamic conditions (23). We have examined the adherence of Walker 256 carcinosarcoma cells to MEC in suspension culture in the presence or absence of homologous platelets. This study has demonstrated that the presence of MEC (possibly acting through the production of endogenous PGI_2) significantly inhibited tumor cell adhesion to the endothelial cells covering the beads. Following approximately 1 h of coincubation, however, tumor cell adhesion was observed to increase dramatically even in the presence of MEC (data not shown). The adhesion of tumor cells to MEC or microcarrier beads alone was minimal in the absence of platelets regardless of the time of coincubation. These results suggest that inhibition of platelet-enhanced tumor cell adhesion by endothelial cells may be limited under conditions of prolonged exposure to tumor cells. The mechanism of this loss of inhibitory activity (e.g.,

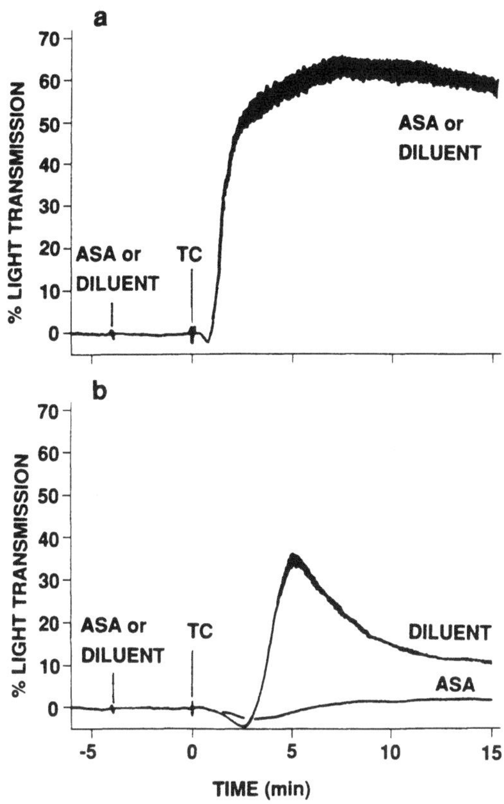

Fig. 2 Effect of agonist concentration and aspirin (ASA) on tumor cell-induced platelet aggregation. The ability of W256 cells to induce platelet aggregation varied with cell concentration in a dose dependent manner. Inhibition of TCIPA by ASA (250 µM) alone was also dependent on W256 cell concentration. ASA failed to significantly inhibit TCIPA induced by 5×10^5 W256 cells (a), whereas ASA significantly inhibited TCIPA at lower concentrations of W256 cells compared to solvent treated control (b). TCIPA was not induced by 2.5×10^4 W256 cells either in the presence or absence of ASA (not shown).

depletion of arachidonic acid, inactivation of metabolic enzymes such as COX, alteration of receptor affinity) have yet to be determined.

CONCLUSIONS

Metastasis represents one of the most important hindrances to improved cancer patient survival. We believe arachidonic acid metabolism by platelets and endothelial cells to be intimately coupled to the formation of metastases, although the exact mechanism remains to be determined. Numerous arachidonic acid-modulating agents (e.g., PGI_2, PGI_2-enhancing agents, etc.) have been shown to reduce the incidence of pulmonary tumor colonies in experimental metastasis models and inhibit TCIPA and platelet-enhanced tumor cell adhesion in vitro. Investigations with a newly developed aggregometry model have shown that endothelial cells grown on microcarrier beads can completely inhibit TCIPA and platelet-enhanced tumor cell adhesion in vitro, but this inhibition decreases substantially with prolonged coincubation of endothelial cells, platelets and tumor cells. Further investigations are currently in progress to determine the roles of endothelial cells and platelets in metastasis under dynamic conditions.

ACKNOWLEDGEMENTS

This work was supported by Public Health Service Grants CA-29997 and CA-29405 awarded to Dr. Honn by the National Cancer Institute, Department of Health and Human Services. Dr. Sloane is the recipient of Research Career Development Award CA-00921 from the National Cancer Institute, Department of Health and Human Services.

REFERENCES

1. I.J. Fidler, D.M. Gersten and I.R. Hart The biology of cancer invasion and metastasis. Adv. Cancer Res. 28:149-250 (1985).
2. G.L. Nicolson Cancer metastasis. Organ colonization and the cell surface properties of malignant cells. Biochem. Biophys. Acta 695:113-176 (1982).
3. L. Weiss, Cancer cell traffic from the lungs to the liver: an example of metastatic inefficiency. Int. J. Cancer 25:385-392 (1980).
4. L. Weiss, Metastatic inefficiency. In: "Tumor Invasion and Metastasis" (L.A. Liotta and I.R. Hart, Eds), pp. 81-98. Martinus-Nijhoff, The Hague, 1982.
5. L. Weiss, Overview of the metastatic cascade. In: "Hemostatic Mechanisms and Metastasis" (K.V. Honn and B.F. Sloane, Eds), pp. 15-38. Martinus-Nijhoff, The Hague, 1983.
6. K.V. Honn and B.F. Sloane (Eds.), "Hemostatic Mechanisms and Metastasis". Martinus Nijhoff, the Hague, 1983.
7. G.A. Jamieson (Ed.), "Interaction of Platelets and Tumor Cells". Alan R. Liss, New York, 1982.
8. K.V. Honn, W.D. Busse and B.F. Sloane, Commentary. Prostacyclin and thomboxanes. Implications for their role in tumor cell metastasis. Biochem. Pharmacol. 32:1-11, (1983).
9. K.V. Honn, D.G. Menter, J.M. Onoda, J.D. Taylor and B.F. Sloane. Role of Prostacyclin as a natural deterrent to hematogenous tumor metastasis. In: "Cancer Invasion and Metastasis: Biologic and Therapeutic Aspects" (G.L. Nicolson and L. Milas, Eds), pp. 361-388. Raven Press, New York, 1984.
10. K.V. Honn, B. Cicone and A. Skoff, Prostacyclin: a potent antimetastatic agent. Science 212:1270-1272 (1981).

11. K.V. Honn and J.R. Dunn, Nafazatrom (Bay g 6575) inhibition of tumor lipoxygenase activity and cellular proliferation. FEBS Lett. 139:65-68 (1982).
12. K.V. Honn, Inhibition of tumor cell metastasis by modulation of the vascular prostacyclin/thromboxane A_2 system. Clin. Exp. Metastases 1:103-114 (1983).
13. K.V. Honn, D. Menter, C. Moilanen, P.G. Cavanaugh, J.D. Taylor and B.F. Sloane, Role of prostacyclin and thromboxanes in tumor cell-platelet-vessel wall interactions. In: "Protective Agents in Cancer (D.C.H. McBrian and T.F. Slater, Eds.), pp. 57-79. Academic Press, New York, 1983.
14. D.G. Menter, J.M. Onoda, J.D. Taylor and K.V. Honn. Effects of prostacyclin on tumor cell-induced platelet aggregation. Cancer Res. 44:450-456, 1984.
15. K.V. Honn and B.F. Sloane, Prostacyclin, thromboxanes and hematogenous metastasis. Adv. Prostaglandin Thromboxane Leukotriene Res. 12:313-318 (1983).
16. K.V. Honn, D. Menter, P.G. Cavanaugh, G. Neagos, D. Moilanen, J.D. Taylor and B. F. Sloane, A review of prostaglandins and the treatment of tumor metastasis. Acta Clin. Belg. 38:53-67 (1983).
17. S. Moncada and J.R. Vane, The discovery of prostacyclin (PGX): a fresh insight into arachidonic acid metabolism. In: "Biochemical Aspects of Prostaglandins and Thromboxanes" (N. Kharasch and J. Fried, Eds.), pp. 155-177. Academic Press, New York, 1977.
18. J. Fantone, S. Kunkel and J. Varani, Inhibition of tumor cell adherence by prostaglandins. In: "Prostaglandins and Cancer" (T.J. Powles, R.S. Bockman, K.V. Honn and P. Ramwell, Eds.), pp. 673-677. Alan R. Liss, New York, 1982.
19. Y. Hara, M. Steiner and M.G. Baldini. Characterization of the platelet aggregating activity of tumor cells. Cancer Res. 40:1217-1222 (1980).
20. J.A. Frangos, L.V. McIntire, S.G. Eskin and C.L. Ives, Flow effects on prostacyclin production by cultured human endothelial cells. Science 227:1477-1479 (1985).
21. C.F. Dewey, Jr., Effect of fluid flow on living vascular cells. In: Proceedings of Poster Communications, 7th International Symposium on Athersclerosis, p. 190. Int. Atherosclerosis Soc., 1984.
22. W.E. Stehbens, Hemodynamics and atherosclerosis. Biorheology 19:95-101 (1983).
23. D.G. Menter, B.W. Steinert, J.M. Onoda, J.D. Taylor and K.V. Honn. Development of a new aggregometry model using microcarrier grown endothelial cells to examine tumor cell-platelet-endothelial cell interactions. J. Cell Biol. 103:471a (1986).

THE INVOLVEMENT OF PEROXYL FREE RADICALS IN TUMOR INITIATION AND PROMOTION

Lawrence J. Marnett

Department of Chemistry
Wayne State University
Detroit, MI 48202

INTRODUCTION

In recent years, there has been considerable interest in the involvement of free radicals in carcinogenesis (1-4). Much of the research emphasis has focused on intermediates of oxygen reduction such as superoxide anion and hydroxyl radical (2). Organic analogs of superoxide anion and its protonated form, perhydroxyl radical, have received less attention but may be just as important biologically as reduced oxygen intermediates. Our laboratory is interested in the role of hydroperoxide-dependent oxidations in carcinogenesis and much of our research is concerned with the chemistry and biology of peroxyl radicals generated by metal-catalyzed fragmentations of unsaturated fatty acid hydroperoxides. We believe evidence exists to support the hypothesis that peroxyl free radicals contribute to metabolic activation of chemical carcinogens (e.g., polycyclic hydrocarbons) in the initiation phase of tumorigenesis; circumstantial evidence implicates peroxyl radicals in tumor promotion as well.

PEROXYL FREE RADICALS

The principal pathway of peroxyl radical formation in biological systems is spontaneous or enzyme-catalyzed autoxidation of unsaturated fatty acids (e.g., lipid peroxidation and prostaglandin biosynthesis, respectively.) (5). Compared to other oxygen-centered free radicals, peroxyl radicals are stable species capable of diffusing to remote cellular locations. Pryor has estimated that the half-lives of peroxyl radicals are on the order of seconds (6). Changing the organic group attached to the oxygen does not significantly alter the reactivity of peroxyl radicals but it can have a dramatic effect on the physical properties of the radicals. The two general types of reaction that oxygen radicals participate in are H abstraction and addition to double bonds (eqs 1 and 2). Both reactions produce carbon-centered radicals as intermediates. The predominant fate of carbon-centered radicals is coupling with O_2 to produce peroxyl radicals. The fact that O_2 is itself a triplet diradical greatly facilitates coupling, which occurs at a diffusion-controlled rate (7). The implication of the sequence of reactions in eqs 1 and 2 is that formation of oxygen-centered free radicals in a cell, either by O_2 reduction or autoxidation, eventually results in the generation of peroxyl radicals. It is likely that peroxyl radicals are the major oxygen-centered free radicals generated as a result of oxygen activation and their stability indicates that they can diffuse to cellular loci remote from the site of their generation. Thus, they should be considered prime mediators of radical-induced pathology.

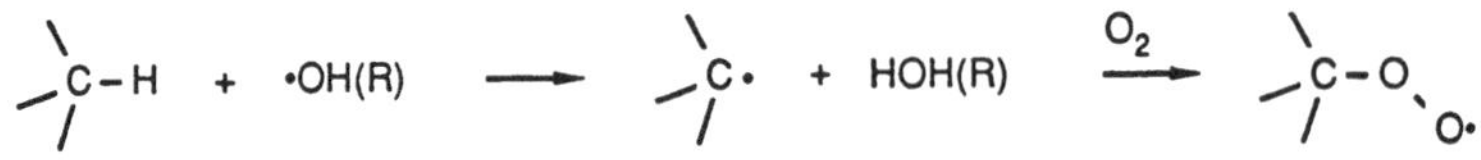

(eq 1)

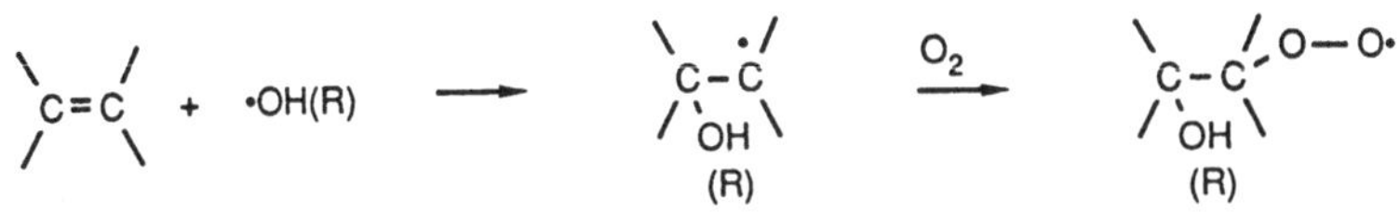

(eq 2)

One possibility relevant to carcinogenesis is that peroxyl radicals react with DNA. Although a good deal is known about the reaction of hydroxyl radical with DNA, the kinetics and products of reaction of peroxyl radicals with DNA have not been investigated so it is difficult to speculate on the likelihood of this possibility. Because of their stability, peroxyl radicals will be much more selective than hydroxyl radical in their reactions with DNA and other macromolecules. An alternate possibility is that peroxyl radicals react with other cellular components converting them to derivatives that react with DNA. Peroxyl radicals react with polycyclic hydrocarbons to form quinones. For example, benzo[a]pyrene reacts with peroxyl radical generating systems to form a mixture of 1,6-, 3,6-, and 6,12-quinones (8). When dihydroaromatic metabolites of polycyclic hydrocarbons react with peroxyl radicals they introduce an oxygen atom into the non-aromatic double bond resulting in epoxidation. In the case of 7,8-dihydroxy-7,8-dihydrobenzo[a]pyrene (BP-7,8-diol), the major product is a dihydrodiolepoxide that represents the ultimate carcinogenic form of benzo[a]pyrene (eq 3). This reaction provides a potential link between peroxyl radical formation and metabolic activation of polycyclic hydrocarbons or their metabolites to tumor-initiating derivatives.

(eq 3)

STEREOCHEMISTRY OF PEROXYL RADICAL-DEPENDENT EPOXIDATION OF DIHYDRODIOLS

The stereoselectivity of epoxidation of BP-7,8-diol by peroxyl radicals is distinct from that exhibited by cytochromes P-450 (9). Oxygen introduction by peroxyl radicals occurs on the same side of the molecule as the 8-hydroxyl group, whereas oxygen introduction by cytochromes P-450 is determined primarily by the orientation of the pyrene moiety with respect to the active site of the enzyme (9,10). Figure 1 outlines the stereochemistry of epoxidation of both enantiomers of BP-7,8-diol by peroxyl radicals and cytochromes P-450. (-)-BP-7,8-diol - the enantiomer generated *in vivo* from BP - is epoxidized by both pathways to the (+)-enantiomer of the *anti*-dihydrodiolepoxide. (+)-BP-7,8-diol is epoxidized by peroxyl radicals mainly to the (-)-enantiomer of the *anti*-diolepoxide but by cytochromes P-450 to the (+)-enantiomer of the *syn*-diolepoxide. The distinctive stereochemistry of epoxidation of (+)-BP-7,8-diol by peroxyl radicals and cytochromes P-450 provides a useful method for differentiating the two pathways of epoxidation in vitro and in vivo

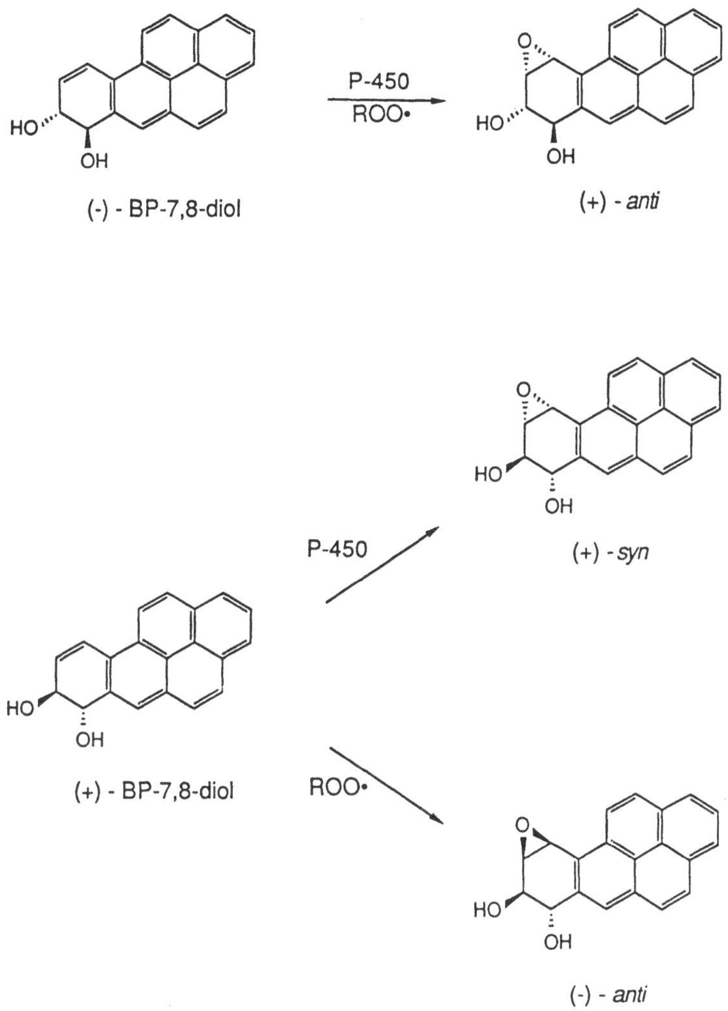

Figure 1. Stereochemistry of Epoxidation of BP-7,8-diol by Cytochromes P-450 and by Peroxyl Radicals

The reactivity of different peroxyl radicals is similar because the organic group is two atoms removed from the unpaired electron and does not sterically hinder its reaction with other molecules. One expects, therefore, that any system that generates peroxyl radicals is capable of epoxidizing BP-7,8-diol. BP-7,8-diol is epoxidized by the following biochemical systems that generate peroxyl radicals - ascorbic acid- and NAPDH-dependent lipid peroxidation (11), cumene hydroperoxide plus rat liver microsomes (12), phenylbutazone peroxidation by peroxidases (13), and sulfite peroxidation or autoxidation (14). In each case, the major epoxidation product is the anti-dihydrodiolepoxide. Dihydrodiol metabolites of benz[a]-anthracene, and chrysene are also oxidized to mutagenic products by peroxyl radicals and in the case of 3,4-dihydroxy-3,4-dihydrobenz[a]-anthracene the principal oxidation product is the anti-dihydrodiol-epoxide (15.16). The extent of epoxidation of BP-7,8-diol is approximately three times greater than that of BA-3,4-diol (16). The high reactivity of BP-7,8-diol to peroxyl radicals and the unique stereochemistry of epoxidation suggests that BP-7,8-diol and other dihydrodiol metabolites of polycyclic hydrocarbons are excellent diagnostic probes for the detection of peroxyl radicals in chemical or biological systems.

PEROXYL RADICALS AND TUMOR INITIATION

It is important to determine if peroxyl radical-dependent epoxidation plays a role in metabolic activation of BP or its dihydrodiol metabolites. This requires a characterization of the pathways of metabolism and DNA binding in target tissues for polycyclic hydrocarbon carcinogenesis. A collaborative effort between Thomas Eling's laboratory at the NIEHS and my laboratory at Wayne State University approached this problem using freshly-isolated epidermal cells from newborn mouse skin, a target tissue for BP carcinogenesis (17). Cells were isolated by a trypsin flotation method and immediately incubated in a HEPES-PBS solution containing 5 mM glucose. The cell preparation was a mixture of approximately 20% basal cells and 80% keratinocytes. Cell viability was initially greater than 82%; no decrease in viability, judged by vital dye exclusion and lactate dehydrogenase release, was observed during the time course of incubation (up to 90 min).

Figure 2 displays the profiles of radioactive metabolites eluting from reverse-phase HPLC columns following incubation of (+)-BP-7,8-diol with 7 x 10^7 cells prepared from animals pretreated with acetone or ß-naphthoflavone (17). Cells from acetone-treated animals exhibited the same pattern of metabolism as those from untreated animals and were used as the control preparation. Cells from ß-naphthoflavone-treated animals represented a preparation with high levels of the isoenzyme of cytochrome P-450 that exhibits the highest specific activity of polycyclic hydrocarbon oxidation of any of the cytochromes P-450 (18,19). A striking contrast exists between the profiles of oxidation products of (+)-BP-7,8-diol by the control and induced cell preparations. The major oxidation product detected in control cells was the anti-dihydrodiolepoxide whereas the syn-dihydrodiolepoxide was the major product in ß-naphthoflavone-induced cells. Oxidation by control cells increased linearly up to 90 min whereas oxidation by ß-naphthoflavone-induced cells plateaued after 30 min. The phenolic antioxidant butylated hydroxyanisole was a potent inhibitor of (+)-BP-7,8-diol epoxidation by control cells (I_{50} = 1 µM) but had no effect on epoxidation by ß-naphthoflavone-induced cells. Investigations of the metabolism of (+)-BP-7,8-diol by epidermal homogenates of untreated and ß-naphthoflavone-treated mice indicate that anti-dihydrodiolepoxide formation is NADPH-independent not stimulated by inducers of cytochrome P-450, and potently inhibited by phenolic antioxidants. In contrast, syn-dihydrodiolepoxide formation is NADPH-dependent, strongly stimulated by cytochrome P-450 inducers, and unaffected by phenolic antioxidants (20). All the observations suggest that two distinct pathways of epoxidation operate in cellular and subcellular epidermal preparations and that they can be modulated independently (Figure 3). Peroxyl free radicals appear to carry out epoxidation in control cells whereas cytochrome P-450 effects epoxidation in ß-naphthoflavone-induced cells. The capacity for epoxidation appears greater in induced cells because the yield of syn-dihydrodiolepoxide in ß-naphthoflavone-induced cells is approximately 10 times higher than the yield of anti-dihydrodiolepoxide in control cells.

If these results can be extrapolated to mouse skin in vivo they suggest that peroxyl free radicals play an important role in oxidative activation of carcinogens. Peroxyl radicals will be more important for activation of molecules that do not induce cytochrome P-450 (assuming that the molecule in question is oxidized by peroxyl radicals). One such molecule appears to be BP-7,8-diol. It is a potent tumor initiator when applied topically to Sencar or CD-1 mice, followed by phorbol ester tumor promotion (100% tumor incidence at 150 nmol (±)-BP-7,8-diol per animal) (21). Topical applicat of this dose of BP-7,8-diol to mice induces only a very slight increase in AHH activity (approximately 30%) whereas the same dose of ß-naphthoflavone increases AHH activity tenfold (17). This means that the pathways responsible for metabolic activation of BP-7,8-diol in mouse skin are those operative at the time of its administration and not any pathways induced as a

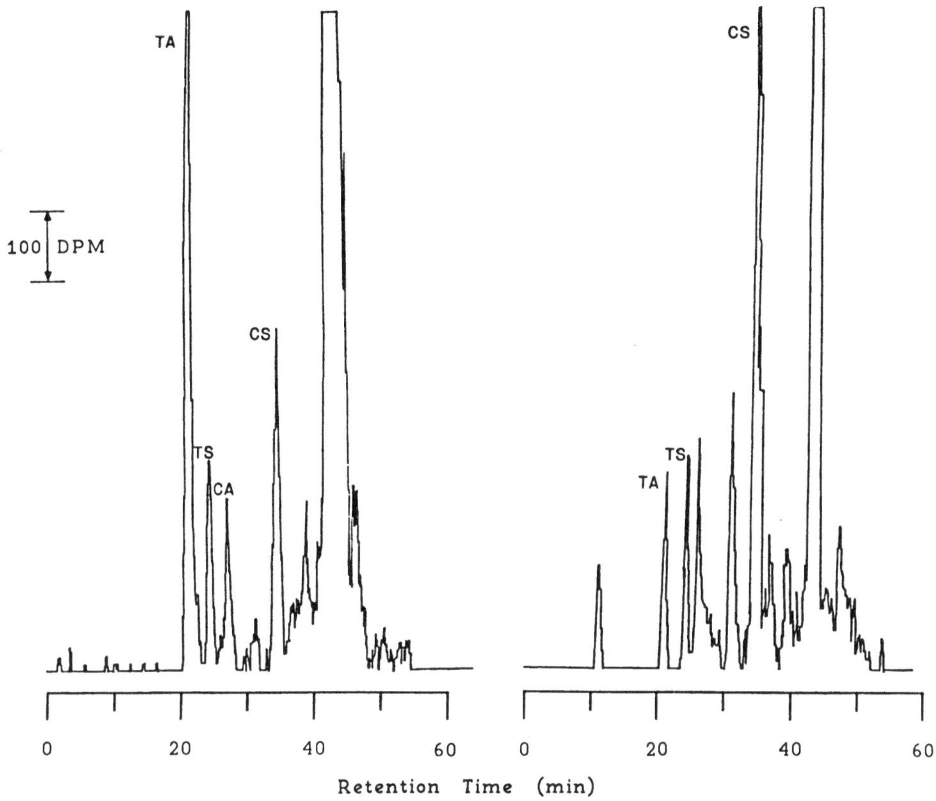

Figure 2. HPLC Profiles of Oxidation Products Formed by Incubation of BP-7,8-diol with Freshly Isolated Epidermal Cells from Control or β-Napthoflavone-Pretreated Animals.

Control (left) and Induced (right). Details are provided in reference (17). TA and CA represents the trans and cis tetraol hydrolysis products of the anti-dihydrodiolepoxide. TS and CS represent the trans and cis tetraol hydrolysis products of *syn*-dihydrodiolepoxide.

result of the treatment. The relative contribution of peroxyl radicals to BP-7,8-diol epoxidation appears to be greater than the contribution of cytochrome P-450 in untreated animals.

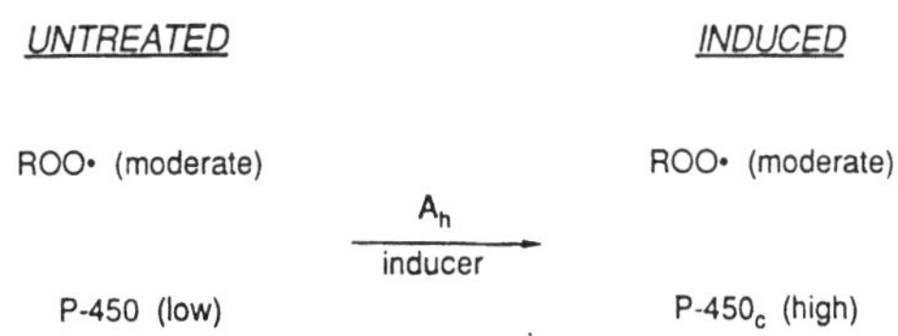

Figure 3. Pathways of Metabolic Activation in Skin of Uninduced and A_h-Induced Mice.

PEROXYL RADICALS AND TUMOR PROMOTION

The tumor-promoting activity of peroxides that are chemical sources of free radicals and the ability of antioxidants to inhibit promotion constitutes some of the major supportive evidence for the hypothesis that free radicals are involved in tumor promotion (22,23). So far, it has not been possible to directly determine whether compounds like benzoyl peroxide actually enhance free radical formation in vivo under conditions where they act as promoters. All techniques for studying free radical involvement in promotion are by nature indirect and are based on chemical trapping experiments or spectroscopic experiments to provide evidence for the existence of unstable radical intermediates. For example, Kensler and Taffe have recently reported trapping methyl radicals with the spin trap 5,5-dimethyl-1-pyrroline-N-oxide following addition of cumene hydroperoxide to isolated mouse epidermal cells (4).

The discussion in the previous sections suggests that BP-7,8-diol should be extremely useful as a probe for the generation of peroxyl radicals in intact cells or perhaps even in vivo. The high rate and stereoselectivity of the reaction of BP-7,8-diol with peroxyl radicals means that one should be able to quantitate the oxidation of radiolabeled material (preferably the (+)-enantiomer, which is commercially available) by cell or tissue preparations to tetrahydro-tetraols that are diagnostic of the intermediacy of peroxyl radicals in the oxidations. If the levels of oxygen-centered free radicals increase following administration of tumor promoters the increases should be reflected in the levels of peroxyl radicals trapped by BP-7,8-diol. Conversely, agents that inhibit free radical formation or scavenge them should decrease the peroxyl radical-mediated epoxidation of BP-7,8-diol. Thus, the experimental strategy developed to test the role of peroxyl radicals in tumor initiation should be equally useful for testing their involvement in promotion.

The anti-promoting activity of antioxidants is hypothesized to result from radical scavenging. Antioxidant compounds usually contain a phenol or aromatic amine functionality that is the radical trapping moiety. However, some compounds that are potent inhibitors of tumor promotion, but are not considered classic antioxidants, also may act by altering the steady-state levels of free radicals in initiated epidermal cells. Retinoids (e.g. all-trans- or 13-*cis*-retinoic acid) are potent inhibitors of epidermal promotion by phorbol esters and are generally believed to act by affecting the state of differentiation of the cells or interrupting message transduction by the promoter (24-27). Consistent with the latter view, retinoids prevent induction of ornithine decarboxylase in mouse skin following administration of phorbol esters and they inhibit protein kinase C activity *in vitro* (28,29).

The chemical structures of retinoids suggest they should be quite reactive with free radicals at a number of sites in the extended polyene chain. In fact, we have found that incubation of 13-*cis*-retinoic acid with peroxidases and hydroperoxides or with peroxyl radical-generating systems leads to rapid oxygen uptake, retinoid oxidation, and formation of oxygenated products (30). The major product of reaction of 13-*cis*-retinoic acid with peroxyl radicals is 5,6-epoxy-13-*cis*-retinoic acid and the rate of its formation is somewhat higher than the rate of formation of the anti-dihydrodiol-epoxide from BP-7,8-diol. Reaction with peroxidases and hydroperoxides produces a mixture of 4-hydroxy-13-*cis*-retinoic acid and 5,6-epoxy-13-*cis*-retinoic acid (31). Mechanistic studies indicate that peroxidase higher oxidation states remove the H-atom from the 4 position generating a carbon-centered radical that couples with O_2 to generate a retinoid peroxyl radical (eq 4). Addition of the peroxyl radical to the 5,6-double bond produces 5,6-epoxy-13-*cis*-retinoic acid and a retinoid alkoxyl radical that is reduced to 4-hydroxy-13-cis-retinoic acid (Figure 4).

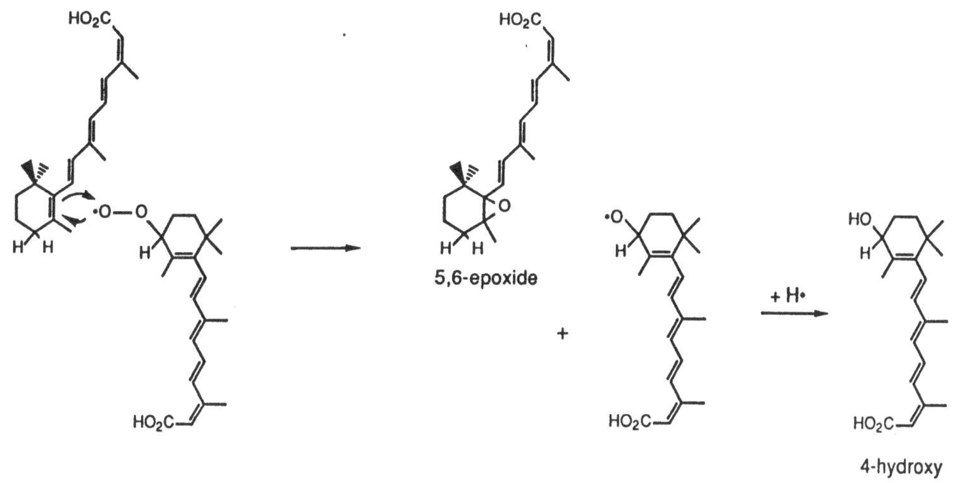

(eq 4)

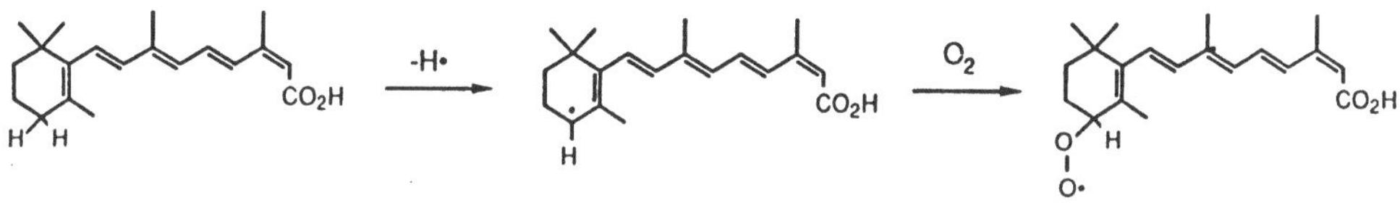

Figure 4. Mechanism of Epoxidation of 13-cis-Retinoic Acid by Retinoid-Derived Peroxyl Radicals.

The complex nature of the reaction of retinoids with peroxyl radicals and peroxidases suggests that they are not only efficient at trapping reactive oxidants , thereby lowering the steady-state oxidant concentration, but that under certain conditions they can actually enhance peroxyl and alkoxyl free radical formation. This may be relevant to occasional reports that retinoids enhance rather than inhibit tumor promotion and highlights another set of metabolic experiments that seem worthwhile to perform (32-34). Despite the potent chemopreventive activity of retinoids and related compounds and despite the fact that they possess chemical functionalities that are extremely reactive toward oxidizing agents, the fate of these molecules on mouse skin has never been determined. This may be related to observations that oxidized retinoic acid derivatives exhibit somewhat lower biological activities in cell culture than do the parent molecules (35). The opinion may have arisen that retinoids do not require metabolic activation and although this may be correct for hormonal activities or even chemopreventive activities in internal organs, it may not be possible to extrapolate this conclusion to mouse skin, a tissue exposed to high oxygen tensions and possessing active oxygenase-dependent metabolic pathways. In fact, 13-cis-retinoic acid is the most reactive molecule toward peroxyl radicals that we have worked with. Nearly complete oxidation in vitro is detected at initial retinoid concentrations of 1 μM, a concentration that approaches the level attained in tumor promotion experiments. If BP-7,8-diol is epoxidized in mouse skin cells by peroxyl radicals, it is virtually certain that 13-cis-retinoic acid will be epoxidized under similar conditions because it is more reactive to peroxyl radicals. The ease of oxidation of certain retinoids by transient oxidants likely to be generated in promotion-sensitive cells may provide a new basis for testing the biochemical mechanisms of their antipromoter or promoter action.

ACKNOWLEDGMENTS

This work was supported by research grants from the American Cancer Society (BC 244I) and the National Institutes of Health (CA 43209 and GM 23642). L. J. M. is a recipient of a Faculty Research Award of the American Cancer Society (FRA 243).

REFERENCES

1. Copeland, E. S. (1983) Free radicals in promotion - A chemical pathology study section report. Cancer Res. 43, 5631-5637.

2. Troll, W. and Weisner, R. (1985) The role of oxygen radicals as a possible mechanism of tumor promotion. Ann. Rev. Pharmacol. Toxicol. 25, 509-528.

3. Cerutti, P. (1985) Prooxidant states and tumor promotion. Science 227, 375-381.

4. Kensler, T. W. and Taffe, B. G. (1986) Free radicals in tumor promotion. Adv. in Free Radical Biology and Medicine 2, 347-387.

5. Porter, N. A. (1986) Mechanisms for the autoxidation of polyunsaturated lipids. Acc. Chem. Res. 19, 262-268.

6. Pryor, W. A. (1986) Oxy-radicals and related species: their formation, lifetimes, and reactions. Ann. Rev. Physiol. 48, 657-667.

7. Walling, C. (1957) In Walling, C. (ed) Free Radicals in Solution, John Wiley and Sons, New York, pp 418-427.

8. Marnett, L. J. (1981) Polycyclic hydrocarbon oxidation during prostaglandin biosynthesis. Life Sci., 29, 531-546.

9. Panthananickal, A. and Marnett, L. J. (1981) Arachidonic acid-dependent metabolism of 7,8-dihydroxy-7,8-dihydrobenzo[a]pyrene to polyguanylic acid-binding derivatives. Chem. Biol. Interactions 33, 239-252.

10. Jerina, D. M., Michaud, D. P., Feldman, R. J., Armstrong, R. N., Vyas, K. P., Thakker, D. R., Yagi, H., Thomas, P. E., Ryan, D. E., and Levin, W. (1982) Stereochemical modeling of the catalytic site of cytochrome P-450c. In Sato, R. and Kato, R. (eds) Microsomes, Drug Oxidations, and Drug Toxicity Wiley-Interscience, New York, pp 195-201.

11. Dix, T. A. and Marnett, L. J. (1983) Metabolism of polycyclic aromatic hydrocarbon derivatives to ultimate carcinogens during lipid peroxidation. Science 221, 77-79.

12. Battista, J. R., Dix, T. A., and Marnett, L.J. (1984) The mechanism of hydroperoxide-dependent epoxidation of 7,8-dihydroxy-7,8-dihydrobenzo[a]pyrene by rat liver microsomes. Proc. Amer. Assoc. Cancer Res. 25, 114.

13. Reed, G. A., Brooks, E. A., and Eling, T. E. (1984) Phenylbutazone-dependent epoxidation of 7,8-dihydroxy-7,8-dihydrobenzo[a]pyrene. J. Biol. Chem. 259, 5591-5595.

14. Reed, G. A., Curtis, J. F., Mottley, C., Eling, T. E., and Mason, R. P. (1986) Epoxidation of (±)-7,8-dihydroxy-7,8-dihydrobenzo[a]pyrene durin (bi)sulfite autoxidation: activation of a procarcinogen by a cocarcinogen. Proc. Natl. Acad. Sci. USA 83, 7499-7502.

15. Guthrie, J., Robertson, I. G. C., Zeiger, E., Boyd, J. A., and Eling, T. E. (1982) Selective activation of dihydrodiols of several polycyclic

aromatic hydrocarbons to mutagenic products by prostaglandin synthetase. Cancer Res. 42, 1620-1623.

16. Dix, T. A., Buck, J. R., and Marnett, L. J. (1986) Hydroperoxide-dependent epoxidation of of 3,4-dihydroxy-3,4-dihydrobenzo[a]anthracene by ram seminal vesicles and by hematin. Biochem. Biophys. Res. Comm. 140, 181-187.

17. Eling, T. E., Curtis, J., Battista, J. R., and Marnett, L. J. (1986) Oxidation of (+)-7,8-dihydroxy-7,8-dihydrobenzo[a]pyrene by mouse keratinocytes: evidence for peroxyl radical- and monooxygenase-dependent metabolism. Carcinogenesis 7, 1957-1963.

18. Deutsch, J., Leutz, J. C., Yang, S. K., Gelboin, H. V., Chang, Y. L., Vatsis, K. P., and Coon, M. J. (1978) Regio- and stereoselectivity of various forms of cytochrome P-450 in the metabolism of benzo[a]pyrene and (±)-trans-7,8-dihydroxy-7,8-dihydrobenzo[a]pyrene as shown by product formation and binding to DNA. Proc. Natl. Acad. Sci. USA 75, 3123-3127.

19. Thakker, D. R., Yagi, H., Akagi, H., Koreeda, M., Lu, A. Y. H., Levin, W., Wood, A. W., Conney, A. H., and Jerina, D. M. (1977) Stereoselective metabolism of benzo[a]pyrene and benzo[a]pyrene-7,8-dihydrodiol to diolepoxides. Chem. Biol. Interactions 16, 281-300.

20. Preuss-Schwartz, D., Nimesheim, A., and Marnett, L.J., submitted for publication.

21. Levin, W., Wood, A. W., Chang, R. L., Slaga, T. J., Yagi, H., Jerina, D. M., and Conney, A. H. (1977) Marked differences in the tumor-initiating activity of optically pure (+)- and (-)-trans-7,8-dihydroxy-7,8-dihydrobenzo[a]pyrene on mouse skin. Cancer Res. 37, 2721-2725.

22. Slaga, T. J., Klein-Szanto, A. J. P., Triplett, L. L., Yotti, L. P., and Trosko, J. E. (1981) Skin tumor promoting activity of benzoyl peroxide, a widely used free radical generating compound. Science 213, 1023-1025.

23. Slaga, T. J., Solanki, V., and Logani, M. (1983) Studies on the mechanism of action of antitumor promoting agents: suggestive evidence for the involvement of free radicals in promotion. In Nygaard, O. and Simic, M. G. (eds) Radioprotectors and Anticarcinogens Academic Press, New York, pp 471-485.

24. Bollag, W. (1972) Prophylaxis of chemically induced benign and malignant epithelial tumors by vitamin A acid (retinoic acid).Eur. J. Cancer 8, 689-693.

25. Verma, A. K., Rice, H. M., Shapas, B. G., and Boutwell, R. K. (1978) Cancer Res. 38, 793-801.

26. Verma, A. K., Shapas, B. G., Rice, H. M., and Boutwell, R. K. (1979) Cancer Res. 39, 419-425.

27. Roberts, A. B. and Sporn, M. B. (1984) Cellular biology and biochemistry of the retinoids. In Sporn, M. B., Roberts, A. B., and Goodman, D. S. (eds) The Retinoids Vol. 2 Academic Press, New York, pp 210-286.

28. Verma, A.K. and Boutwell, R. K. (1977) Vitamin A acid (retinoic acid), a potent inhibitor of 12-O-tetradecanoyl-phorbol-13-acetate-induced ornithine decarboxylase activity in mouse epidermis. Cancer Res. 37, 2196-2201.

29. Cope. F. O., Howard. B. D., and Boutwell, R. K. (1986) The in vitro characterization of the inhibition of mouse brain protein kinase-C by retinoids and their receptors. Experientia 42, 1023-1027.

30. Samokyszyn. V. M., Sloane, B. F., Honn, K. V., and Marnett, L. J. (1984) Cooxidation of 13-cis retinoic acid by prostaglandin H synthase. Biochem. Biophys. Res. Comm. 124, 430-436.

31. Samokyszyn, V.M. and Marnett, L.J. (1987) J. Biol. Chem. in press.

32. Fischer, S. M., Klein-Szanto, A. J. P., Adams, L. M., and Slaga, T. J. (1985) Carcinogenesis 6, 575-578.

33. Forbes, P. D., Urbach, F., and Davies, R. E. (1979) Enhancement of experimental photocarcinogenesis by topical retinoic acid. Cancer Letts. 7, 85-90.

34. Hennings, H., Wenk, M. L., and Donahoe, R. (1982) Retinoic acid promotion of papilloma formation in mouse skin. Cancer Letts.16, 1-5.

35. Sporn, M. B. and Roberts, A. B. (1984) Biological methods of analysis and assay of retinoids - relationships between structure and activity. In Sporn, M. B., Roberts, A. B., and Goodman, D. S. (eds) The Retinoids Vol. 2 Academic Press, New York, pp 236-279.

AMPLIFICATION OF TISSUE PEROXIDES IN DISEASE

William E.M. Lands and Robert B. Pendleton

Dept. Biological Chemistry, Univ. Illinois at Chicago
Chicago, IL 60612, USA

INTRACELLULAR SIGNALS

The fatty acid oxygenases, cyclooxygenase and lipoxygenase, require hydroperoxides to achieve maximal rates of reaction, and their reaction products are hydroperoxides that can stimulate a faster oxygenation reaction. The cyclooxygenase can be stimulated to half-maximal velocity by 20 picomoles of lipid hydroperoxide per ml (1), while it can generate nanomoles of the hydroperoxide, prostaglandin G_2 (PGG_2). Soybean lipoxygenase, and all other lipoxygenases, also react faster as the level of lipid hydroperoxide is amplified. Thus, intracellular amplification of lipid hydroperoxides by the fatty acid oxygenases may be closely related to high rates of formation of eicosanoids (prostaglandins and leukotrienes). This relationship has broader importance when overproduction of eicosanoids is recognized to be a common occurrence in many major pathophysiological events (2). The intracellular amplification by the fatty acid oxygenases of "triggering" levels of hydroperoxide (3) to higher levels that permit pathological rates of eicosanoid synthesis needs careful study. Intracellular levels of 10^{-9} to 10^{-7} M lipid hydroperoxide seem likely to be involved in this amplification.

The amplification process is particularly important in interpreting the efficacy of certain therapeutic agents that antagonize the hydroperoxide activation of the cycloxygenase (4). The phenolic type of inhibitors (such as acetamidophenol) antagonize the hydroperoxide activation (5) but they are effective antagonists only when the ambient peroxide tone is kept low (6). The fact that acetamidophenol is more effective in vivo as an analgesic agent than as an antiinflammatory agent led to the concept that inflammatory conditions are associated with amplified intracellular levels of cellular hydroperoxides whereas hyperalgesia may involve less highly amplified amounts of hydroperoxide (4). Apparently, the phenolic type of inhibitor can effectively interfere with early triggering events before the amplification has proceeded to high levels.

EXTRACELLULAR SIGNALS

Although lipid hydroperoxides are several hundred-fold more effective than hydrogen peroxide in stimulating the cyclooxygenase activity of prostaglandin H (PGH) synthase to generate more prostaglandins and hydroperoxide, hydrogen peroxide is often generated at very high extracellular concentrations by inflammatory cells such as neutrophils and macrophages. The hydrogen peroxide may be formed directly as a product of flavoprotein oxidase action (e.g., NADPH oxidase) or indirectly from the conversion of superoxide to H_2O_2. Extracellular levels of 20 to 30 nanomoles H_2O_2 per ml can be generated by 10^6 neutrophils/ml (7), and local abundances of 2 to 7 x 10^6 neutrophils/ml may occur in an inflamed region. Rapid diffusion of H_2O_2 across cellular membranes can then provide a stimulatory signal to the oxygenases in adjacent cells and trigger oxygenase activity with a resultant intracellular amplification of hydroperoxides in those cells (7). This process provides an extracellular form of signalling that transforms the initial stimulation of a cell into a H_2O_2-mediated extracellular signal that provokes other cells to give an eicosanoid response. Extracellular levels of 10^{-6} to 10^{-5} M hydrogen peroxide seem likely to be involved in this mode of signal amplification. Such stimulation of eicosanoid biosynthesis may be an important way in which commonly designated "oxy radicals" promote pathological events by extracellular signalling with highly selective, receptor-mediated actions rather than putative non-selective, destructive events that seem to require more than 10^{-4} M hydrogen peroxide (8). The formation of extracellular lipid hydroperoxides may be secondary to the generation of H_2O_2, and experiments showed a requirement for the presence of iron as well as added unsaturated fatty acid (7). Presumably traces of extracellular transition metals could occur at sites of tissue trauma and react with O_2^{-} or H_2O_2 to generate the extracellular lipid hydroperoxides.

Because eicosanoids can promote chemotactic responses, they promote a type of extracellular amplification by recruiting more peroxide-forming cells to a given region. These newly arrived cells can increase the extracellular levels of lipid hydroperoxides to 10^{-6} to 10^{-5} M and increase the probability of stimulating intracellular hydroperoxide formation by the oxygenases (3). Factors influencing the intracellular and extracellular amplification of hydroperoxides may represent useful points for intervention in disease processes. The interstitial and plasma levels of hydroperoxides may indicate the severity of the extracellular amplification associated with tissue injury.

PLASMA HYDROPEROXIDES IN HUMANS

We are currently employing the accelerative, amplifying action of the cyclooxygenase to assay picomole amounts of lipid hydroperoxide in biological systems (9). The assay permitted a recognition that the oxygenase activator found in normal human plasma was primarily lipid hydroperoxides and not H_2O_2 (10).

Further study showed that patients had elevated hydroperoxide levels in their plasma during sepsis. Peripheral arterial blood was collected from 11 healthy male volunteers and 18 patients either with or without septic shock. Blood was drawn into Vacutainer tubes containing buffered sodium citrate as anticoagulant, and plasma was prepared by centrifugation at 1,500 x g for 30 min at 4°C. Aliquots of the supernatant plasma were either treated immediately with ethanol (1:1 by volume) to precipitate proteins or were stored at -70°C and then thawed on ice immediately before the precipitation procedure. The ethanol-treated plasma samples were centrifuged at 1,500 x g for 30 min at 4°C, and 100µl aliquots of the resulting ethanolic supernatant were assayed by the cyclooxygenase activation method.

Samples were obtained from septic patients on two different occasions. Fig. 1A shows that the average plasma peroxide concentration in septic patients in 1983 was 3.0 fold greater than that of the non-septic patients. The confidence interval for this observation was greater than 99% ($p < 0.01$) using Student's t test for normally distributed populations with unknown and unequal variances. Also, the average levels of plasma hydroperoxides were 2.8 fold ($p < 0.01$; Fig. 1A, 1983) and 4.0 fold ($p < 0.1$; Fig. 1B, 1986) greater in septic patients than in healthy volunteers. Differences in the absolute values of peroxide between 1983 and 1986 may be due to several changes in the assay procedures and materials. It is interesting to note that the patients in the Surgical Intensive Care Unit who were not septic had significantly lower plasma peroxides than had the volunteers ($p < 0.05$; Fig. 1B, 1983).

The results indicate that significant amounts of extracellular lipid hydroperoxide may accumulate in conditions of systemic inflammation. The possibility of using this direct measure of the actual hydroperoxide content at a given time makes it feasible to monitor the severity of such disorders and also monitor the degree of success of the therapeutic interventions being used.

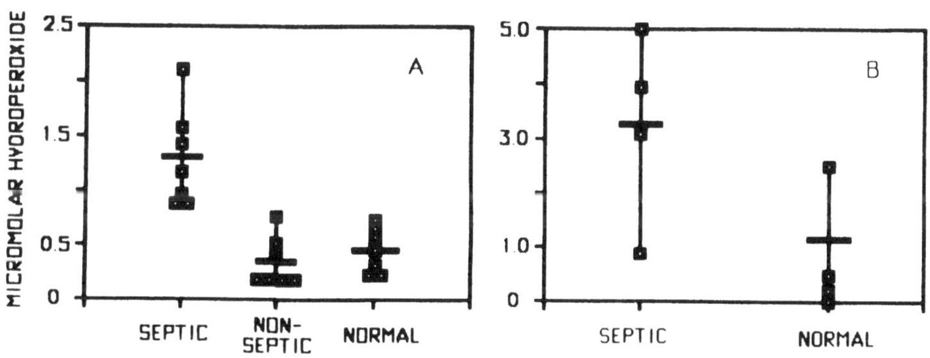

Fig. 1. Elevated plasma hydroperoxides in sepsis. (A) Samples were collected in 1983 by M.A. Warso, M.D. from patients in the Surgical Intensive Care Unit, and were analyzed according to the method in reference 1. (B) Samples were collected in 1986 by D. Lange, M.D. from patients in the Trauma Intensive Care Unit, and were assayed at 25°C in 90 mM potassium phosphate, pH 7.2, in the presence of 10 mM NaCN. Bars = average values.

REFERENCES

1. R.J. Kulmacz and W.E.M. Lands, Requirements for hydroperoxide by the cyclooxygenase and peroxidase activities of prostaglandin H synthase. Prostaglandins 25, 531-540 (1983).
2. W.E.M. Lands, Interactions of lipid hydroperoxides with eicosanoid biosynthesis. J. Free Radicals Biol. Med. 1, 97-101 (1985).
3. W.E.M. Lands and P.J. Marshall, Peroxide stimulators of prostaglandin biosynthesis. IN: Rheumatology 85, (P.M. Brooks and J.R. York, Eds.), pp. 47-50, Elsevier, New York, 1985.
4. W.E.M. Lands and A.M. Hanel, Phenolic anticyclooxygenase agents in antiinflammatory and analgesic therapy. Prostaglandins 24, 271-277 (1982).
5. W.E.M. Lands, H.W. Cook and L.H. Rome, Prostaglandin biosynthesis: Consequences of oxygenase mechanism upon in vitro assays of drug effectiveness. Adv. Prost. Thromb. Res., 1, 7-17 (1976).
6. A.M. Hanel and W.E.M. Lands, Modification of anti-inflammatory drug effectiveness by ambient lipid peroxides. Biochem. Pharmacol. 31, 3307-3311 (1982).
7. P.J. Marshall and W.E.M. Lands, In vitro formation of activators for prostaglandin synthesis by neutrophills and macrophages from humans and guinea pigs. J. Lab. Clin. Med. 108, 525-534 (1986).
8. B. Chance, H. Sies and A. Boveris, Hydroperoxide metabolism in mammalian organs. Physiol. Rev. 59, 527-605 (1979).
9. P.J. Marshall, M.A. Warso and W.E.M. Lands, Selective microdetermination of lipid hydroperoxides. Anal. Biochem. 145, 192-199 (1985).
10. M.A. Warso and W.E.M. Lands, Presence of lipid hydroperoxide in human plasma. J. Clin. Invest. 75, 667-671 (1985).

PEROXISOME PROLIFERATION-RELATED OXIDATIVE STRESS AND HEPATOCARCINOGENESIS

Janardan K. Reddy and M. Sambasiva Rao

Department of Pathology
Northwestern University Medical School
303 East Chicago Avenue
Chicago, Illinois 60611

INTRODUCTION

Investigation of the phenomenon of xenobiotic-induced peroxisome proliferation in liver parenchymal cells continues to provide fascinating details about the differential regulation of peroxisomal enzymes, their role in the production of intrahepatic oxidative stress and the possible relationship between sustained oxidative stress and the eventual development of hepatocellular carcinomas (1-3). Hepatic peroxisome proliferation was first noted nearly twenty years ago in the livers of rats treated with the hypolipidemic drug clofibrate (4,5). Subsequently, several structurally diverse chemicals have been identified as hepatic peroxisome proliferators (1,2,6). Long-term feeding studies demonstrated the hepatocarcinogenicity of several of these agents (2,7-9), although they are not genotoxic in short-term tests (10-13). On the basis of these observations it was proposed that peroxisome proliferators form a novel class of chemical carcinogens (8). An understanding of the mechanism of induction of peroxisome proliferation and the metabolic events that accompany persistent increase in the number of these organelles is necessary to resolve the question whether these nongenotoxic hepatocarcinogenic peroxisome proliferators pose the same carcinogenic risk to humans as those that are genotoxic.

PEROXISOME PROLIFERATORS AS NONGENOTOXIC CARCINOGENS

In 1976, we reported that chronic dietary administration of nafenopin, a peroxisome proliferator, results in the development of hepatocellular carcinomas (7). During the past ten years we have demonstrated that six other peroxisome proliferators, including the widely used hypolipidemic drug clofibrate (Fig.1), induce liver tumors in rats and/or mice (7-9). Since none of these compounds exerted genotoxicity in short-term tests, it was postulated that the hepatocarcinogenicity is related to peroxisome proliferation (8). Subsequently, several other hypolipidemic agents and the widely used phthalate ester plasticizer, di(2-ethylhexyl)phthalate (DEHP), were identified as hepatocarcinogenic agents (2,14). All these agents induce peroxisome proliferation and are nongenotoxic in short-term tests. The lack of mutagenicity of these compounds in Salmonella/microsome mutagenesis assay (15) has been well documented (10,11). No adducts were noted in studies using ^{14}C or ^{3}H-

labeled peroxisome proliferators such as clofibrate, fenofibrate, Wy-14643, nafenopin and DEHP (16,17). The results of the ^{32}P-postlabeling study provide convincing evidence that clofibrate, ciprofibrate, Wy-14643 and DEHP do not form adducts with hepatocyte DNA under *in vivo* and *in vitro* conditions (13). Evidence also suggests that peroxisome proliferators are largely ineffective in inducing unscheduled DNA synthesis in

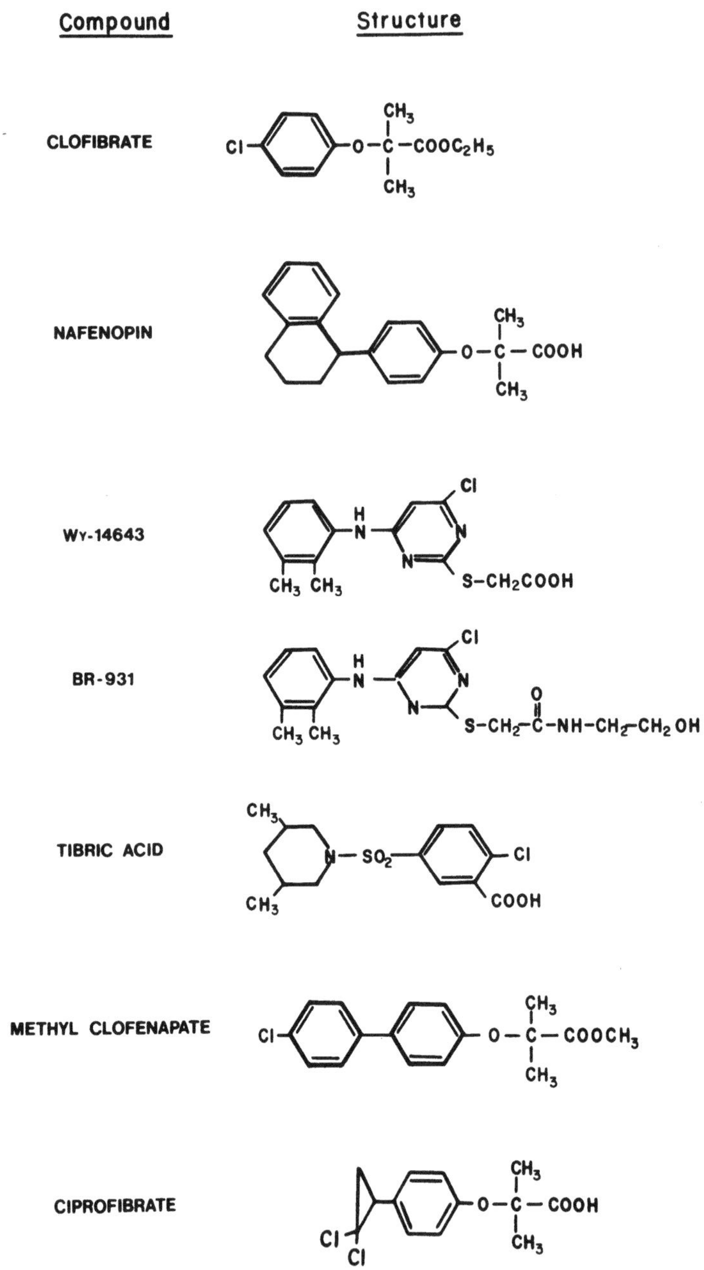

Fig.1.

Chemical structures of some of the peroxisome proliferators identified as hepatocarcinogens: Clofibrate; Nafenopin; Wy-14643; BR-931; tibric acid; methyl clofenapate; and ciprofibrate.

primary hepatocyte culture assay (11). Thus, it is reasonable to conclude that carcinogenicity of peroxisome proliferators is attributable to factors other than the direct effect of these chemicals or their metabolites on DNA.

TISSUE SPECIFICITY OF PEROXISOME PROLIFERATOR PLEIOTROPIC RESPONSE

Peroxisome proliferators, as a group, produce a predictably similar pleiotropic response, which includes hepatomegaly, an increase in the number of peroxisomes in liver parenchymal cells, and increase in the activity of several enzymes especially those associated with peroxisomes (1,2). Peroxisomes contain catalase and several other enzymes involved in the synthesis and catabolism of lipids (16-18). It is now firmly established that the xenobiotic-induced increase in peroxisome number is associated with less than two-fold increase in the activity, content and synthesis of the peroxisomal marker enzyme, catalase (1,2), whereas the enzymes of the peroxisomal β-oxidation system are increased by over twenty-five fold (1,2). Peroxisome proliferators, DEHP, clofibrate, ciprofibrate and Wy-14643 were shown to increase the translatable mRNA levels of the β-oxidation enzymes in rat liver; these observations were further confirmed by blot hybridization measurements using the cDNA probes (19). The increase in mRNA levels is due to rapid and coordinately transcriptional activation of the β-oxidation genes (19). The catalase mRNA increases less than two-fold and there appears no change in its transcription rate (19). The coordinated transcriptional regulation of the genes for peroxisomal fatty acid β-oxidation enzyme by structurally diverse chemicals implies the existence of a common control mechanism (2,3). We proposed that peroxisome proliferators exert their effect by a ligand-receptor-mediated mechanism (20-22). Tissue specificity of the

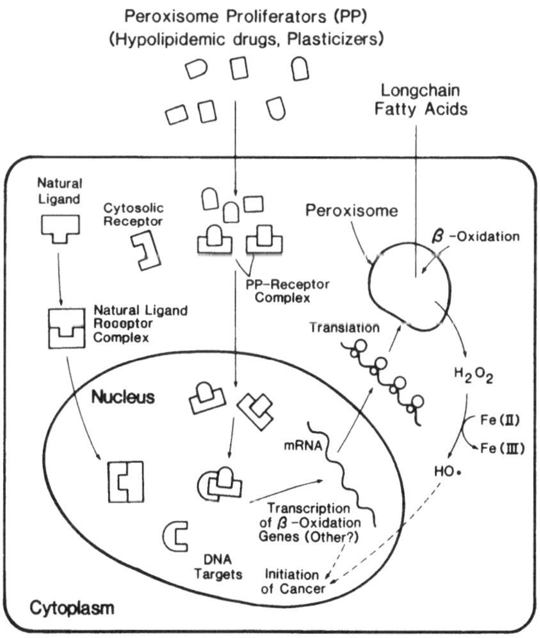

Fig.2.

Proposed model for the induction of peroxisome proliferation and liver tumors by peroxisome proliferators.

pleiotropic response, response of extrahepatic hepatocytes to the inductive effects of peroxisome proliferators, inducibility of peroxisome proliferation in primary cultures of liver cells, rapid increase in the transcriptional rate, and the presence of specific binding protein in liver cytosol for clofibric acid, nafenopin and ciprofibrate, are considered as evidence supporting the existence of peroxisome proliferator receptor(s) in rat liver (Fig.2).

OXIDATIVE STRESS ASSOCIATED WITH HEPATIC PEROXISOME PROLIFERATION

All peroxisome proliferators tested so far in long-term studies in rodents have been found to induce liver tumors (2). Accordingly, liver carcinogenicity is considered to be a delayed component of the pleiotropic response resulting from the prolonged exposure to these agents. The lack of genotoxicity of all peroxisome proliferators in prokaryotic and eukaryotic test systems (10-13) has generated considerable interest in the mechanism by which these agents induce hepatocellular carcinomas. We have proposed that the carcinogenicity of this novel class of hepatocarcinogens is related to the generation of increased levels of H_2O_2 by peroxisomal oxidases (2,8). The lack of transcriptional activation of catalase gene by peroxisome proliferators also contributes to the increased H_2O_2 levels in the liver. Increases in the activities of peroxisomal β-oxidation enzymes and in the steady-state levels of H_2O_2 in the livers appear to correlate well with the carcinogenic potential of these xenobiotics (23,24). Support for the intrahepatic oxidative stress comes from the observation that prolonged administration of peroxisome proliferators is associated with increased accumulation of lipofuscin (25) which is generally considered as an indicator of membrane damage by lipid peroxidation. Furthermore, the hepatocarcinogenicity of the potent peroxisome proliferator, ciprofibrate, was markedly reduced in rats fed antioxidants ethoxyquin and butylated hydroxyanisole (9), suggesting that reactive oxygen species may be involved in liver tumor induction by peroxisome proliferators. Goel et al. (26) and Lake et al. (27) have provided evidence for the enhanced hepatic lipid peroxidation in the livers of rats chronically treated with peroxisome proliferators ciprofibrate, Wy-14643 and DHEP. In these studies increased levels of conjugated dienes and fluorescent products were noted. Additional evidence for increased intrahepatic oxidative stress was obtained recently by Elliott and coworkers (28). They noted the rate of hydroxyl radical generation was increased in the livers of rats fed DEHP, clofibrate or methyl clofenapate. It is thus clear that increased levels of H_2O_2-generating peroxisomal β-oxidation enzyme system in the livers of rats treated chronically with peroxisome proliferators lead to increased lipid peroxidation and production of hydroxyl radicals. The reductions in catalase, glutathione transferase and glutathione peroxidase activities in these livers also led to increased intrahepatic oxidative stress (27,29,30). It is now generally accepted that oxidative stress contributes to initiation and promotion of neoplasia (31) and aging (32). Despite this assumption, the precise role of oxidative damage in the causation of cancer is not clear. H_2O_2 and hydroxyl radicals generated by peroxisomal oxidases may induce DNA damage directly (Fig.3) or cause alterations in the expression of specific oncogenes or growth factor genes. Additional studies are necessary to examine the role of intrahepatic oxidative stress induced by persistent increase in the number of peroxisomes in liver carcinogenesis.

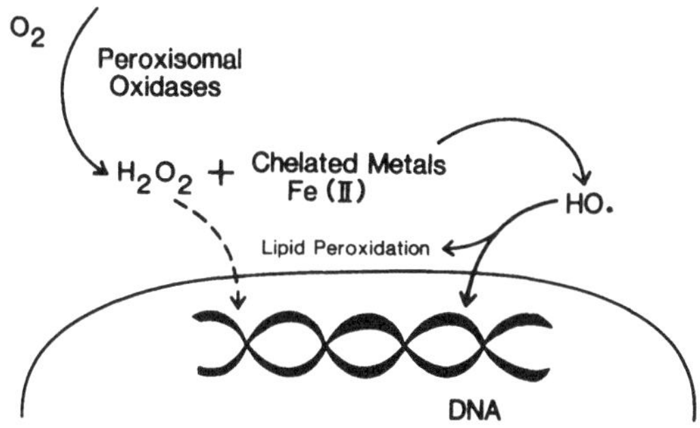

Fig.3.

Peroxisomal oxidases and the production of intrahepatic oxidative stress. Peroxisome proliferators increase the activity of peroxisomal fatty acyl-CoA oxidase. This flavoprotein utilizes oxygen and generates H_2O_2. DNA damage, if necessary for the initiation of neoplasia, may be caused by H_2O_2 directly or as a result of hydroxyl radical (HO) generation. It is also possible that lipid peroxidation, H_2O_2 and HO may lead to activation of oncogene(s).

ACKNOWLEDGMENTS

This work was supported by NIH Grants GM 23750 and CA 32504.

REFERENCES

1. J.K. Reddy, J.R. Warren, M.K. Reddy, and N.D. Lalwani, Hepatic and renal effects of peroxisome proliferators: Biological implications. Ann.N.Y.Acad.Sci. 386, 81-110 (1982).
2. J.K. Reddy and N.D. Lalwani, Carcinogenesis by hepatic peroxisome proliferators: Evaluation of the risk of hypolipidemic drugs and industrial plasticizers to humans. Crit.Rev.Toxicol. 12, 1-58 (1983).
3. J.K. Reddy and M.S. Rao, Peroxisome proliferators and cancer: Mechanisms and implications. Trends Pharmacol.Sci. 7, 438-443 (1986).
4. R. Hess, W. Staubli, and W. Riess, Nature of the hepatomegalic effect produced by ethylchlorophenoxyisobutyrate in the rat. Nature (London) 208, 856-858 (1985).
5. D.J. Svoboda and D.L. Azarnoff, Response of hepatic microbodies to a hypolipidemic agent, ethylchlorophenoxyisobutyrate (CPIB). J.Cell Biol. 30, 442-450 (1966).
6. J.K. Reddy and T.P. Krishnakantha, Hepatic peroxisome proliferation: Induction by two novel compounds structurally unrelated to clofibrate. Science 190, 787-789 (1975).
7. J.K. Reddy, M.S. Rao, and D.E. Moody, Hepatocellular carcinomas in acatalasemic mice treated with nafenopin, a hypolipidemic peroxisome proliferator. Cancer Res. 36, 1211-1217 (1976).
8. J.K. Reddy, D.L. Azarnoff, and C.E. Hignite, Hypolipidemic hepatic peroxisome proliferators form a novel class of chemical carcinogens. Nature (London) 283, 397-398 (1980).
9. M.S. Rao, N.D. Lalwani, T.K. Watanabe, and J.K. Reddy, Inhibitory effect of antioxidants ethoxyquin and 2(3)-tert-butyl-4-hydroxyanisole on hepatic tumorigenesis in rats fed ciprofibrate, a peroxisome proliferator. Cancer Res. 44, 1072-1076 (1984).
10. J.R. Warren, V.F. Simmon, and J.K. Reddy, Properties of hypolipidemic peroxisome proliferators in the lymphocyte (^{3}H)thymidine and Salmonella mutagenesis assays. Cancer Res. 40, 36-41 (1980).

11. H.P. Glauert, J.K. Reddy, Y.S. Kennan, G.L. Sattler, V. Subbarao, and H.C. Pitot, Effect of hypolipidemic peroxisome proliferators on unscheduled DNA synthesis in cultured hepatocytes and on mutagenesis in Salmonella. Cancer Lett. 24, 147-156 (1984).
12. A. Von Daniken, W.K. Lutz, R. Jaekl, and C. Schlatter, Investigation of the potential for binding of di(2-ethylhexyl)phthalate (DEHP) and di(2-ethylhexyl)adipate (DEHA) to liver DNA in vivo. Toxicol.Appl. Pharmacol. 73, 373-387 (1984).
13. R.C. Gupta, S.K. Goel, K. Earley, B. Singh, and J.K. Reddy, ^{32}P-postlabeling analysis of peroxisome proliferator-DNA adduct formation in rat liver in vivo and hepatocytes in vitro. Carcinogenesis 6,933-936 (1985).
14. W.M. Kluwe, J.K. Haseman, J.F. Douglas, and J.E. Huff, The carcinogenicity of dietary di(2-ethylhexyl)phthalate (DEHP) in Fischer 344 rats and B6C3F1 mice. J.Toxicol.Environ.Health 10, 797-815 (1982).
15. B.N. Ames, Identifying environmental chemicals causing mutations and cancer. Science 204, 587-593 (1979).
16. P.B. Lazarow and C. de Duve, A fatty acyl-CoA oxidizing system in rat liver peroxisomes; Enhancement by clofibrate, a hypolipidemic drug. Proc.Natl.Acad.Sci.U.S.A. 73, 2043-2046 (1976).
17. N.E. Tolbert, Metabolic pathways in peroxisomes and glyoxysomes. Annu.Rev.Biochem. 50, 133-157 (1981).
18. A.K. Hajra and J.E. Bishop, Glycerolipid biosynthesis in peroxisomes via the acyl dihydroxyacetone phosphate pathway. Ann.N.Y.Acad.Sci. 386, 170-182 (1982).
19. J.K. Reddy, S.K. Goel, M.R. Nemali, J.J. Carrino, T.G. Laffler, M.K. Reddy, S.J. Sperbeck, T. Osumi, T. Hashimoto, N.D. Lalwani, and M.S. Rao, Transcriptional regulation of peroxisomal fatty acyl-CoA oxidase and enoyl-CoA hydratase/3-hydroxyacyl-CoA dehydrogenase in rat liver by peroxisome proliferators. Proc.Natl.Acad.Sci.U.S.A. 83, 1747-1751 (1986).
20. N.D. Lalwani, W.E. Fahl, and J.K. Reddy, Detection of a nafenopin-binding protein in rat liver cytosol associated with the induction of peroxisome proliferation by hypolipidemic compounds. Biochem.Biophys. Res.Commun. 116, 383-393 (1983).
21. N.D. Lalwani, K. Alvares, M.K. Reddy, M.N. Reddy, I. Parikh, and J.K. Reddy, Peroxisome proliferator binding protein: Identification and partial characterization of nafenopin, clofibric acid and ciprofibrate binding proteins from rat liver. Proc.Natl.Acad.Sci.U.S.A. (in press).
22. M.S. Rao and J.K. Reddy, Peroxisome proliferation and hepatocarcinogenesis. Carcinogenesis 8, 631-636 (1987).
23. N.D. Lalwani, M.K. Reddy, S.A. Qureshi, and J.K. Reddy, Development of hepatocellular carcinomas and increased peroxisomal fatty acid β-oxidation in rats fed [4-chloro-6-(2,3-xylidino)-2-pyrimidinylthio] acetic acid (Wy-14,643) in the semipurified diet. Carcinogenesis 7, 645-650 (1981).
24. J.K. Reddy, M.K. Reddy, M.I. Usman, N.D. Lalwani, and M.S. Rao, Comparison of hepatic peroxisome proliferative effect and its implication for hepatocarcinogenicity of phthalate esters, di(2-ethylhexyl) phthalate, and di(2-ethylhexyl)adipate with a hypolipidemic drug. Environ.Health Perspect. 65, 317-327 (1986).
25. J.K. Reddy, N.D. Lalwani, M.K. Reddy, and S.A. Qureshi, Excessive accumulation autofluorescent lipofuscin in the liver during hepatocarcinogenesis by methyl clofenapate and other hypolipidemic peroxisome proliferators. Cancer Res. 42, 259-266 (1982).
26. S.K. Goel, N.D. Lalwani, and J.K. Reddy, Peroxisome proliferation and lipid peroxidation in rat liver. Cancer Res. 46, 1324-1330 (1986).
27. B.G. Lake, S.L. Kozlen, J.G. Evans, T.J.B. Gray, P.J. Young, and S.D. Gangoli, Effect of prolonged administration of clofibric acid

and di(2-ethylhexyl)phthalate on hepatic enzyme activities and lipid peroxidation in the rat. Toxicology 44, 213-228 (1987).
28. B.M. Elliott, N.J.F. Dodd, and C.R. Elcombe, Increased hydroxyl radical production in liver peroxisomal fractions from rats treated with peroxisome proliferators. Carcinogenesis 7, 795-799 (1986).
29. Y.C. Awasthi, S.V. Singh, S.K. Goel, and J.K. Reddy, Irreversible inhibition of hepatic glutathione s-transferase by ciprofibrate, a peroxisome proliferator. Biochem.Biophys.Res.Commun. 123, 1012-1018 (1984).
30. K. Furukawa, S. Numoto, K. Furuya, N.T. Furukawa, and G.M. Williams, Effects of the hepatocarcinogen nafenopin, a peroxisome proliferator, on the activities of rat liver glutathione-requiring enzymes and catalase in comparison to the action of phenobarbital. Cancer Res. 45, 5011-5019 (1985).
31. P.A. Cerutti, Prooxidant states and tumor promotion. Science 228, 375-381 (1985).
32. B.N. Ames, Dietary carcinogens and anticarcinogens. Science 221, 1256-1264 (1983).

HYDROXYL FREE RADICAL MEDIATED STRAND BREAKS IN PBR322 PLASMID DNA

E. Schneider, J.J. Watson, and R.A. Floyd

Molecular Toxicology Research Group
Oklahoma Medical Research Foundation
Oklahoma City, Oklahoma 731040

INTRODUCTION

Oxygen free radicals react with and cause damage to DNA. Hydroxyl free radicals (·OH) add to DNA to yield several products, one of which is 8-hydroxydeoxyguanine (1). Digestion of DNA to the nucleoside level yields 8-hydroxydeoxyguanosine (8-OHdG) which can be very sensitively quantitated (to 20 femtomoles) by HPLC-electrochemical detection (2,3). The potential significance of the 8-OHdG adduct in living systems has been demonstrated by its presence in the DNA of human granulocytes following TPA-stimulated oxidative damage (3). The amount of 8-OHdG present in the granulocyte DNA was decreased by the addition of superoxide dismutase and/or catalase, thus emphasizing that oxygen free radical intermediates are the responsible agents (3).

Characteristics of ·OH damage with regard to its preferred form of DNA, types of strand breakage, and specificity of target sites on the DNA are important questions. We have investigated structural changes in supercoiled pBR322 resulting from exposure to an Fe(III)/ascorbate system which causes the formation of hydroxyl free radicals. Changes in the rate of migration of the plasmid DNA in agarose gels allows supercoiled, single-strand nicked and double-strand cut forms of the plasmid to be separated and quantitated.

RESULTS AND DISCUSSION

Figure 1 shows that nicking of supercoiled plasmid (as evidenced by the increased levels of open circular plasmid forms) by the ascorbate-iron system progresses with time of incubation. After incubation for 180 minutes, the linear form of the plasmid appears, indicating the presence of double strand breaks. Double strand breaks are expected to result from nicking of both complementary strands within about a 5 base pair region at a particular location on the plasmid. Double strand breaks by separate, possibly concerted less frequent events, however, cannot be ruled out.

Lanes 1,3,5,7 and 9 (without ascorbate) show that single strand nicking is ascorbate dependent and, since catalytic levels of iron is present in the buffer, it is highly likely that hydroxyl free radicals

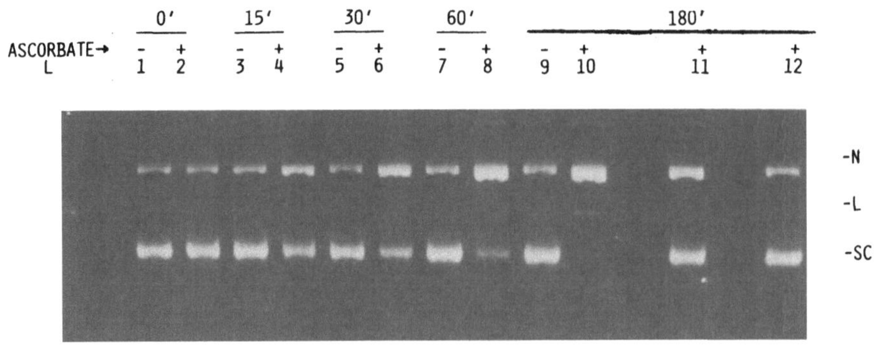

Figure 1. Iron/ascorbate mediated formation of single stranded and double strand breaks in plasmid pBR322 supercoiled DNA. Each reaction mixture of 50µl contained 5µg of DNA, 100mM sodium phosphate buffer, pH 7.0, and 100µM ascorbate. Iron was estimated to be present at a level of 20µM as a contaminant of the sodium phosphate (1). Reactions were begun by ascorbate addition. Incubations were at 30°C, and were terminated at the indicated time by the addition of an equal volume of agarose gel loading solution (30% glyceral, 0.01% bromphenol blue) and cooling to 4°C until loading onto a 1.0% agarose gel in TAE (2). Gels were electrophoresed 15 hours at 25 volts, stained with 0.5µg/ml ethidium bromide and photographed with Polaroid type 55 film through an orange filter during the exposure of the gel to 300nm U.V. light. Lane designations: L) linear plasmid produced by digestion with EcoR1 restriction enzyme; plus (+) and (-) indicates the presence or absence of 100 µM adscorbate in the reaction; 11 and 12 contained 100 µM mannitol and DTPA, respectively, in the nicking reaction. Label abbreviations: N, open circle (representing single strand nicked plasmids); L, linear (representing double strand breaks) and sc, supercoiled plasmid.

are generated. Nicking is iron-dependent since it is inhibited by the addition of the iron chelator, diethylenetriamine pentaacetic acid (DTPA) (lane 12). Iron is introduced into this system as a component of the sodium phosphate buffer at an approximate level of 20 µM (4). In results not shown, nicking is also inhibited by Desferal (desferrioxamine), a specific iron chelator. Lane 11 shows that 100 µM mannitol, a scavenger of hydroxyl free radicals inhibits the nicking activity of the iron/ascorbate system, but not as completely as does DTPA (lane 12). Table 1 shows that mannitol is, however, an effective inhibitor of the nicking reaction, since its presence in the nicking reaction allows only 5-10 percent nicking during a 180 minute incubation period (reaction No. 2) while reaction No. 6, without mannitol for the same duration, causes nearly complete loss of the supercoiled form of the plasmid. Iron dependent nicking in the presence of 100 µM mannitol demonstrates the great potential for DNA damage residing in iron-DNA complexes. In addition, although the detection of single strand breaks in supercoiled DNA can be very sensitive, a great deal of damage, such as base modifications and apurinic/apyrimidinic sites can be formed in DNA that do not directly result in strand scissions, and which may occur much more frequently than do strand breaks. Measurements of the 8-hydroxydeoxyguanosine content of the DNA corresponding to the reaction conditions in Table 1 are in progress. Since the mannitol carbons may have a similar reactivity with ·OH as do deoxyribose carbons involved in strand scission, a preference with regard to which ·OH targets are protected may be seen at low levels of ·OH. These ideas are in the process of further testing with regard to the effects of DNA concentration, DNA conformation, and other scavengers. Since low levels of oxidative damage in DNA may escape repair processes in the cell, we plan to further explore how DNA function is impaired by low levels of oxidative damage (e.g. before strand scission occurs) and by residual damage following repair activities. This research was supported in part by NIH Grant No. CA42854.

Table 1. The Effect of Mannitol on Ascorbate/Iron-Induced Strand Nicking.

Reaction	100μM Ascorbate	Time of Mannitol Addition	Time of Incubation at 30°C	Nicking
1	none	180 min	180 min	0
2	+	0 "	180 "	5-10
3	+	15 "	180 "	10-20
4	+	30 "	180 "	20-30
5	+	60 "	180 "	40-60
6	+	180 "	180 "	90-95

*Reaction mixture contained, in 500μl, 100mM sodium phosphate, pH 7.0, 30μg pBR322 DNA from Applied Genetics, 100μM ascorbate, 100μM Mannitol (added at the times indicated). The hydroxyl free radical generation was begun with the addition of ascorbate and by placing the reaction mixture in a 30°C bath. It was terminated and prepared for 8-hydroxyguanosine content determination by cooling it to 4°C and by rapid precipitation of the DNA by the cetylpyridinium bromide procedure (5). The precipitated DNA was rinsed with 80% ethanol in 0.1M sodium acetate and resuspended in 150μl H_2O. 100μl of each sample was digested to the nucleoside level and the amount of 8-hydroxydeoxyguanosine determined by HPLC-ED (2) (data not shown). Estimation of nicking was by visual inspection of the photograph of an aliquot of the DNA before digestion evaluated by gel analysis as described in the legend to Figure 1. Control experiments (results not shown) showed that the precipitation procedure did not alter the nicking seen as compared to gel analysis without precipitation.

REFERENCES

1. H. Kasai and S. Nishimura, Hydroxylation of deoxyguanosine at the C-8 position by ascorbic acid and other reducing agents. Nucleic Acids Res. 12, 2137-2145 (1984).
2. R.A. Floyd, J.J. Vatson, P.K. Wong, D.H. Altmiller, R.C. Rickard, Hydroxyl free radical adduct of deoxyguanosine: sensitive detection and mechanisms of formation. Free Radical Res. Commun. 1, 163-172 (1986).
3. R.A. Floyd, J.J. Watson, J. Harris, M. West, and P.K. Wong, Formation of 8-hydroxyguanosine, hydroxyl free radical adduct of DNA in granulocytes exposed to the tumor promoter, tetradeconylphoboloutate. Biochem. Biophys. Res. Commun. 137, 841-846 (1986).
4. K. Brawn and I. Fridovich, DNA strand scission by enzymatically generated oxygen radicals. Arch. Biochem. Biophys. 206, 414-419 (1981).
5. P. Geck and I. Nasz, Concentrated, digestible DNA after hydroxylapetite chromatography with cetylpyridinium bromide precipitation. Anal. Biochem. 135, 264-268 (1983).

RADIATION-LIKE MODIFICATION OF DNA AND H_2O_2 FORMATION BY ACTIVATED HUMAN POLYMORPHONUCLEAR LEUKOCYTES (PMNs)

Krystyna Frenkel[1,2] and Kazimierz Chrzan[1]

Departments of Environmental Medicine[1] and Pathology[2]
New York University Medical Center
New York, NY 10016

Active oxygen species generated by stimulated PMNs were found to be mutagenic and carcinogenic (1,2). We (3-5) and others (6,7) have shown that oxygen species produced by phagocytic cells are capable of modifying bases in DNA exposed to them. Those identified to date are 5-hydroxymethyl uracil (HMU) (3-5), thymine glycol (TG) (4-6) and 8-hydroxyguanine (8OHG) (7). All of them are known to be formed by the action of ionizing radiation (8-11). HMU and TG are excreted in the urine by humans and rats which points to these thymine derivatives as products of the normal biological processes (12). That these oxidized DNA base derivatives are potentially harmful to the well-being of humans and animals is shown by the existence of the repair enzymes HMU- and TG-glycosylases in mammalian cells (13-16). There is also evidence that 8OHG might be repaired as well (11). However, when removal of abnormal bases from DNA is not complete or timely, they have the potential to exert their deleterious effects. For example, TG is thought to provide a replication block (17,18), whereas, HMU and particularly its deoxyribonucleoside HMdU are mutagenic (19). HMdU, which is known to be incorporated into cellular DNA, is also cytotoxic and cytostatic to a number of mammalian cells, and acts as a radiomimetic agent as it causes diarrhea and leukopenia, symptoms of acute radiation sickness (20-23). Thus PMNs, whose role is to protect against invading bacteria or other opsonized particles, are capable of imparting a radiation-like damage, certain types of which might be heritable.

Phagocytes can also be activated by other stimuli such as some lectins and also tumor promoters (24). Stimulation of PMNs results in formation of active oxygen species which include superoxide anion radicals, H_2O_2, hypochlorite and hydroxyl radicals (24). Since H_2O_2 is a precursor of the actual DNA-damaging species, its formation provides a measure of a potential for causing heritable damage. We have found recently that generation of H_2O_2 correlates with the first-stage promoting activity of tumor promoters that were used to stimulate PMNs, and with the formation of HMU and TG in DNA exposed to these cells (5). Tumor promoters used for stimulation of PMNs were: 12-O-tetradecanoyl-phorbol-13-acetate (TPA), a complete tumor promoter with a potent first-stage activity; 12-O-retinoyl-phorbol-13-acetate (RPA), a promoter having weak first-stage activity in most animal models; and mezerein (Mez), with an intermediate first-stage activity (25). Assuming formation of H_2O_2, HMU and TG by TPA-activated PMNs to be 100%, Mez- or RPA-activated cells formed 75% or 30% H_2O_2, 70% or 25% HMU, and 45% or 5% TG, respectively (Table 1) (5). In the presence of autologous plasma there was 70% H_2O_2, 50% HMU and 40% TG formed by Mez-induced cells, whereas there was 30% H_2O_2, 30% HMU and 20% TG formed by RPA-induced PMNs (Table 1, unpublished data) (26). These results show that formation of H_2O_2 may indeed provide a predictive value for the DNA damaging potential, as measured by the appearance of oxidized bases.

Formation of modified bases in DNA by active oxygen species generated by stimulated PMNs requires a presence of exogenously-supplied Fe ions and is potentiated by autologous plasma in their absence (4). This latter finding was surprising because human plasma is known to contain substantial amounts of antioxidants and free radical scavengers (24). However, in addition to the detoxifying agents, iron-containing proteins such as transferrin and ferritin are also present. Lactoferrin, another iron-containing protein, is released from granules of PMNs during phagocytosis (24,27). In fact, we found that Fe-loaded transferrin also catalyzes formation of HMU in DNA exposed to TPA-stimulated PMNs in the absence of plasma (4). The activity of iron complexed with transferrin was approximately 25-fold higher than that of Fe chelated by EDTA. Iron ions catalyze formation of hydroxyl radicals (•OHs) in the Haber-Weiss sequence of reactions (28). Superoxide anion radicals generated by stimulated PMNs dismutate both enzymatically and spontaneously to H_2O_2 and they also reduce Fe(III) to Fe(II). It is Fe(II) which catalyzes formation of •OHs from H_2O_2. These •OHs cause formation of both HMU and TG, but in addition, they also catalyze decomposition of the TG ring, which leads to the formation of other products in TG-containing DNA (4). In contrast, the HMU moiety is not susceptible to further interaction with •OHs under conditions causing total TG degradation. In fact, with an increased flux of •OHs, larger amounts of HMU are formed in DNA, whereas levels of intact TG decrease to the non-detectable. Using G = 0.002, obtained for formation of HMU in γ-irradiated DNA (10), the calculated radiation equivalency values for formation of HMU in DNA exposed to TPA-activated PMNs varied between 0.2 to 65 krads, depending upon whether Fe/EDTA or autologous plasma was used during incubation of DNA with TPA-activated PMNs (4). Recently, Fe(II) ions were reported to be strongly mutagenic in the ØX system with the effect abolished by desferioxamine, an Fe-chelator, or by catalase and mannitol, which remove H_2O_2 and •OHs, respectively (29). These findings strengthen the hypothesis that Fe ions are involved in the active oxygen species-induced damage to the genetic material.

Table 1. Percent formation of H_2O_2, HMdU and dTG by human PMNs activated with Mez or RPA, in comparison to TPA. The analyses of H_2O_2 were carried out as in Ref. 5, and of HMdU and dTG as in Refs. 4 and 5.

TUMOR PROMOTER	H_2O_2	HMdU	dTG
	NO PLASMA		
TPA	100	100	100
Mez	75	70	45
RPA	30	25	5
	10% AUTOLOGOUS PLASMA		
TPA	100	100	100
Mez	70	50	40
RPA	30	30	20

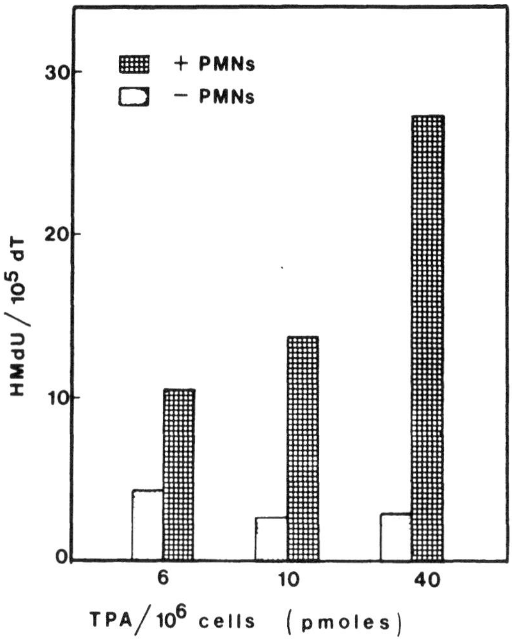

Fig. 1. Formation of 5-hydroxymethyl-2'-deoxyuridine (HMdU) in DNA of HeLa cells (prelabeled with [6-^{3}H]thymidine; 8) through the action of TPA-activated PMNs. Each sample (+PMNs) and control (-PMNs) was incubated at the same time at 37°C for 2 hr. DNA was isolated, enzymatically digested to 2'-deoxyribonucleosides, and analyzed by HPLC and microderivatization (4,10).

Of the active oxygen species generated by TPA-stimulated PMNs only H_2O_2 can easily cross cellular membranes (24,30). To test whether exogenously-generated H_2O_2 is capable of damaging the DNA of neighboring cells, we decided to look for the formation of modified bases in the DNA of HeLa cells co-incubated with TPA-activated PMNs. As our preliminary data depicted in Fig. 1 (unpublished results) show, increasing amounts of TPA caused increased formation of HMU (measured as HMdU) in HeLa cell DNA exposed to TPA-stimulated PMNs in the presence of 10% autologous plasma (26). Moreover, there was some HMU formed in the DNA of HeLa cells when they were treated with TPA in the absence of PMNs. As Fig. 2 (unpublished results) shows, increased concentrations of TPA caused increased formation of HMU in the DNA of HeLa cells but with an order of magnitude lower yield than in the presence of PMNs. At the toxic concentrations of TPA (as determined by trypan blue exclusion), HMU formation drastically decreased. The yield of HMU in DNA exposed to 25nM TPA-activated PMNs in the presence of 10% autologous plasma varied between 65 and 1800 residues per 10^5 thymine residues (4). In the case of HeLa cells, the same concentration of TPA caused formation of approximately 20 HMU residues per 10^5 thymines in the presence of activated PMNs and plasma, whereas about 2 HMU per 10^5 thymines were formed through the action of TPA in the absence of PMNs (unpublished results). These data show that tumor promoters such as TPA are capable of "oxidative activation" of cells even in the absence of phagocytes. This activation results in modification of DNA bases, the formation of which can be potentiated by the action of stimulated phagocytic cells.

We found that it is indeed H_2O_2 which is necessary for the formation of HMU and TG in DNA exposed to TPA-activated PMNs, as catalase abolished their formation (unpublished data). Since H_2O_2 mediates DNA modification, a decrease in its production might lead to a decrease in that damage. We have found that μmolar concentrations of chymotrypsin-specific protease inhibitors such as potato inhibitor 1 are capable of inhibiting H_2O_2 generation by TPA-stimulated PMNs, whereas trypsin-specific inhibitors such as the Kunitz inhibitor are virtually inactive (31). Bifunctional protease inhibitors, i.e. those which are capable of binding both trypsin and chymotrypsin such as potato inhibitor 2, Bowman-Birk inhibitor and chicken ovoinhibitor, also decreased formation of H_2O_2 but with the intermediate values between those of potato 1 and Kunitz inhibitors. Recently, evidence was presented for the existence of chymotrypsin-like proteases (not necessarily the same) in cellular membranes and that they seem to be responsible for cellular transformation (32,33). These conclusions were based on the inhibition of X-ray-induced transformation by Bowman-Birk inhibitor and potato inhibitor 1.

Wounding and inflammation cause vasodilation which leads to infiltration of PMNs and diffusion of serum into affected areas (24). Although blood serum contains antioxidants and free radical scavengers, it potentiates formation of modified thymines in DNA exposed to TPA-activated PMNs (4). Thus, it could be that the co-carcinogenic effect of wounding (34) is due to the formation of at least one of the modified bases in the DNA of neighboring cells. Since chymotrypsin-like protease seems to be involved

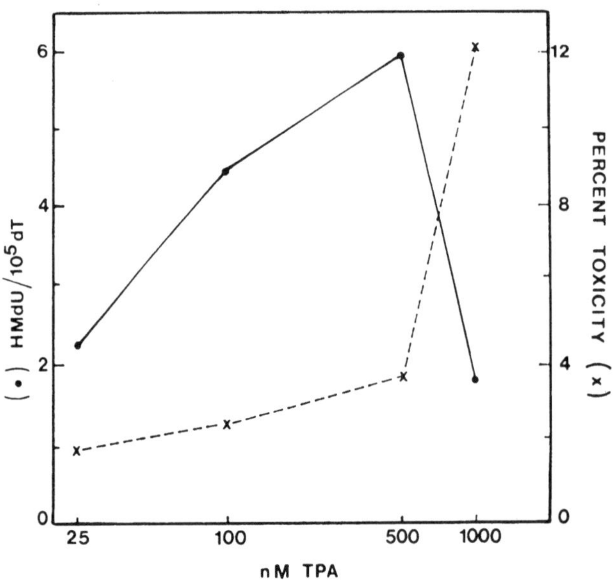

Fig. 2. Formation of HMdU in DNA of HeLa cells (1.8×10^6/ml) incubated with TPA at 37°C for 2 hr in the presence of 10% human plasma. DNA was analyzed as described in the legend to Fig. 1.

in the process(es) leading to the production of H_2O_2 by PMNs, it is possible that treatment with chymotrypsin-specific protease inhibitors will lead to a decrease in inflammation, DNA damage and, consequently, to a decrease in carcinogenic effects.

This work was supported in part by NIH Grant CA 37858 and Center Grants ES 00260 and CA 13343.

REFERENCES

1. M. Barak, S. Ulitzur and D. Merzbach, Phagocytosis-induced mutagenesis in bacteria. Mutat. Res. 121, 7-16 (1983).
2. S. A. Weitzman, A. B. Weitberg, E. P. Clark and T. P. Stossel, Phagocytes as carcinogens: malignant transformation produced by human neutrophils. Science, 227, 1231-1233 (1985).
3. W. Troll, K. Frenkel and G. W. Teebor, Free oxygen radicals: Necessary contributors to tumor promotion and cocarcinogenesis. In Cellular Interactions by Environmental Tumor Promoters 1984 (H. Fujiki et al. Eds.) pp. 207-218. Japan Scientific Societies Press, Tokyo.
4. K. Frenkel, K. Chrzan, W. Troll, G. W. Teebor and J. J. Steinberg, Radiation-like modification of bases in DNA exposed to tumor promoter-activated polymorphonuclear leukocytes. Cancer Res. 46, 5533-5540 (1986).
5. K. Frenkel and K. Chrzan, H_2O_2 formation and DNA base modification by tumor promoter-activated polymorphonuclear leukocytes. Carcinogenesis, 8, 455-460 (1987).
6. J. G. Lewis and D. O. Adams, Induction of 5,6-ring saturated thymine bases in NIH-3T3 cells by phorbol ester-stimulated macrophages: role of reactive oxygen intermediates. Cancer Res. 45, 1270-1275 (1985).
7. R. A. Floyd, J. J. Watson, J. Harris, M. West and P. K. Wong, Formation of 8-hydroxydeoxyguanosine, hydroxyl free radical adduct of DNA in granulocytes exposed to the tumor promoter, tetradecanoylphorbolacetate. Biochem. Biophys. Res. Comm. 37, 841-846 (1986).
8. P. A. Cerutti, Base damage induced by ionizing radiation. Photochem. Photobiol. Nucleic Acids. 2, 375-401 (1976).
9. K. Frenkel, M. S. Goldstein and G. W. Teebor, Identification of the cis-thymine glycol moiety in chemically oxidized and γ-irradiated deoxyribinucleic acid by high-pressure liquid chromatography. Biochemistry, 20, 7566-7571 (1981).
10. K. Frenkel, A. Cummings, J. Solomon, J. Cadet, J. J. Steinberg and G. W. Teebor, Quantitative determination of the 5-(hydroxymethyl)-uracil moiety in the DNA of γ-irradiated cells. Biochemistry, 24, 4527-4533 (1985).
11. H. Kasai, P. F. Crain, Y. Kochino, S. Nishimura, A. Ootsuyama, and H. Tanooka, Formation of 8-hydroxyguanine moiety in cellular DNA by agents producing oxygen radicals and evidence for its repair. Carcinogenesis, 7, 1849-1851 (1986).
12. R. Cathcart, E. Schwiers, R. L. Saul and B. Ames, Thymine glycol and thymidine glycol in human and rat urine: a possible assay for oxidative DNA damage. Proc. Natl. Acad. Sci. USA, 81, 5633-5637 (1984).
13. M. C. Hollstein, P. Brooks, S. Linn and B. Ames, Hydroxymethyluracil DNA glycosylase in mammalian cells. Proc. Natl. Acad. Sci. USA, 81, 4003-4007 (1984).
14. R. J. Boorstein, D. D. Levy and G. W. Teebor, 5-Hydroxymethyluracil-DNA glycosylase activity may be a differentiated mammalian function. Mutat. Res. 183, 257-263 (1987).
15. P. W. Doetsch, D. E. Helland and W. A. Haseltine, Mechanism of action of a mammalian DNA repair endonuclease. Biochemistry, 25, 2212-2220 (1986).
16. S. A. Higgins, K. Frenkel, A. Cummings and G. W. Teebor, Definitive

characterization of human thymine glycol N-glycosylase activity. Biochemistry, 26, 1683-1688 (1987).
17. P. Rouet and J. M. Essigman, Possible role for thymine glycol in the selective inhibition of DNA synthesis on oxidized DNA templates. Cancer Res. 45, 6113-6118 (1985).
18. H. Ide, Y. W. Kow and S. S. Wallace, Thymine glycol and urea residues in M13 template constitute replicative blocks in vitro. Nucleic Acids Res. 13, 8035-8052 (1985).
19. L. Shirnamé-Moré, T. G. Rossman, W. Troll, G. W. Teebor and K. Frenkel, Genetic effects of 5-hydroxymethyl-2'-deoxyuridine, a product of ionizing radiation. Mutation Res. 178, 177-186 (1987).
20. S. Waschke, J. Reefschlager, D. Barwolff and P. Langen, 5-Hydroxymethyl-2'-deoxyuridine, a normal DNA constituent in certain Bacillus subtilis phages is cytostatic for mammalian cells. Nature (London), 225, 629-630 (1975).
21. J. B. Meldrum, V. S. Gupta, N. R. Lowes and A. R. P. Paterson, Toxicologic and antitumor studies on 5-hydroxymethyldeoxyuridine. Toxicol. Appl. Pharmacol. 79, 423-435 (1985).
22. L. I. Kahilainen, D. E. Bergstrom and J. A. Vilpo, 5-Hydroxymethyl-2'-deoxyuridine, cytotoxicity and DNA incorporation studied by using a novel [2-^{14}C] derivative with normal and leukemic human hematopoietic cells. Acta Chem. Scand. B, 39, 477-484 (1985).
23. E. R. Kaufman, Biochemical analysis of toxic effects of 5-hydroxymethyl-2'-deoxyuridine in mammalian cells. Somat. Cell Mol. Genet. 12, 501-512 (1986).
24. J. C. Fantone and P. A. Word, Role of oxygen-derived free radicals and metabolites in leukocyte-dependent inflammatory reactions. Am. J. Pathol. 107, 397-418 (1982).
25. T. J. Slaga, S. M. Fisher, C. E. Weeks, K. Nelson, M. Mamrack and A. J. P. Klein-Szanto, Specificity and mechanism(s) of promoter inhibitors in multistage promotion. In Carcinogenesis, Vol. 7 1982 (E. Hecker, W. Kuntz, N. E. Fusenig, F. Marks and H. W. Thielmann, Eds.) pp. 19-34. Raven Press, New York.
26. K. Frenkel, and K. Chrzan, Correlation between formation of H_2O_2 by tumor promoter-activated human polymorphonuclear leukocytes (PMNs), modification of bases in DNA exposed to them, and first stage tumor promotion. Proc. Am. Assoc. Cancer Res. 28, 160 (1987).
27. D. R. Ambruso and R. B. Johnston, Lactoferrin enhances hydroxyl radical production by human neutrophils, neutrophil particulate fractions, and an enzymatic generating system. J. Clin. Invest. 67, 352-360 (1981).
28. I. Fridovich, Superoxide radical: an endogenous toxicant. Ann. Rev. Pharmacol. Toxicol. 23, 239-257 (1983).
29. L. A. Loeb, E. A. James, A. M. Waltersdorph, K. Stokes and S. Klebanoff. Proc. Am. Assoc. Cancer Res. 28, 162 (1987).
30. A. C. de Mello-Filho and R. Meneghini, Protection of mammalian cells by o-phenanthroline from lethal and DNA-damaging effects produced by active oxygen species. Biochem. Biophys. Acta, 847, 82-89 (1985).
31. K. Frenkel, K. Chrzan, C. A. Ryan, R. Wiesner, and W. Troll, Chymotrypsin-specific protease inhibitors decrease H_2O_2 formation by activated human polymorphonuclear leukocytes. Carcinogenesis, in press (1987).
32. J. Yavelow, M. Caggana and K. A. Beck, Proteases occurring in the cell membrane: a possible cell receptor for the Bowman-Birk type of protease inhibitors. Cancer Res. 47, 1598-1601 (1987).
33. P. C. Billings, W. St. Clair, C. A. Ryan and A. R. Kennedy, Inhibition of radiation-induced transformation of C3H/10T½ cells by chymotrypsin inhibitor 1 from potatoes. Carcinogenesis, in press (1987).
34. L. J. Dunham, Cancer in man at a site of prior benign lesion of skin or mucous membrane: a review. Cancer Res. 32, 1359-1374 (1972).

MITOCHONDRIAL DNA DAMAGE DURING MITOCHONDRIAL LIPID PEROXIDATION

Andrew M. Hruszkewycz

Department of Pathology
George Washington University Medical Center
Washington, D.C. 20037

INTRODUCTION

It is well known that the carcinogenic potency of radiation and chemical carcinogens is closely correlated with the induction of DNA damage (1-3); however, the ultimate mechanisms involved are far from clear. In the case of high energy radiation, generation of hydroxyl radicals (OH·) probably plays a key role, either by direct attack of OH· on DNA, or indirectly via initiation of lipid peroxidation (4,5). Initiation of lipid peroxidation may also be a consequence of the generation of oxygen centered radicals in the oxidative stress associated with metabolism of some chemical carcinogens(6). Thus lipid peroxidation may be one of the mechanisms linking both radiation and metabolism of chemical carcinogens to DNA damage. This view is supported by reports (7-12) that products of lipid peroxidation can cause DNA damage in model systems and mutations in prokaryotes. However, actual data showing DNA damage to eukaryotic genes by lipid peroxidation is limited and indirect (5,13). In this communication, evidence is reported showing extensive damage to DNA of liver mitochondria occurring concomitantly with mitochondrial lipid peroxidation. Mitochondrial rather than nuclear DNA was studied since the former exists as a discrete circular molecule of single molecular weight, which facilitates detection of small degrees of DNA damage. The study of the relationship between lipid peroxidation and DNA damage in mitochondria is important in the assessment of the carcinogenic potency of oxidative stress, especially since many strong carcinogenic agents preferentially attack mitochondrial DNA (14).

MATERIALS AND METHODS

Animals. Animals used in this study were male rats of the Sprague Dawley strain obtained from Hilltop Laboratories, Inc. The animals were fed Purina rat chow and had water available *ad libitum*. The animals were fasted for 18 to 24 hours before sacrifice. Rats were killed by decapitation, and allowed to exsanguinate. Most of the rats used were within a weight range of 300 to 450 gm body weight, although occasionally lighter or heavier rats were used. No effects of differences in rat body weights were noted with respect to any of the parameters which were monitored.

Preparation of Liver Mitochondrial Fraction. Following decapitation and exsanguination, the liver was removed, chilled in ice-cold homogenization medium, and then a 10 per cent whole liver homogenate was prepared. The homogenization medium was 0.25 M sucrose which contained 5 mM Hepes (pH 7.4). Homogenization was carried out with a glass Potter-Elvijhem homogenizer equipped with a Teflon pestle. The homogenate was centrifuged at 600 g for 12 minutes in a SS-34 rotor of a SS-3 Sorval centrifuge, at 2-4°C. The supernatant fraction was then centrifuged at 9,000 g for 12 minutes. The "fluffy layer" was removed from the centrifuged pellet, and the remaining pellet containing sedimented mitochondria was washed three times with 0.15 M Tris-HCl (pH 7.4) and finally resuspended in the same buffer.

Incubation Procedures: Overall Experimental Design. The essential experimental design consisted of promoting lipid peroxidation of rat liver mitochondria in vitro with NADPH and ferrous chloride. Lipid peroxidation was followed by measuring the evolution of malonic dialdehyde (MDA). Changes in mitochondrial DNA were monitored by evaluating patterns of electrophoretic migration.

The reaction mixtures contained: 150 mM Tris-HCl (pH 7.4), 1.0 mM KH_2PO_4 and 1.6 mg of mitochondrial protein per ml in a final volume of 40 ml. Additions of 0.4 mM NADPH and 0.1 mM $FeCl_2$ were made as indicated in Table 1. To prepare anaerobic controls, reaction mixtures contained within 125 ml flasks were flushed with 10 liters of nitrogen free of oxygen. Experimental and control systems were incubated and shaken in a water bath at 37°C for 20 minutes. Immediately after incubation all test systems were cooled on ice and made 10 mM with respect to EDTA. Aliquots (1.2 ml) were then removed for MDA assay and the remaining mitochondria were processed for evaluation of mitochondrial DNA integrity.

Malonic dialdehyde was assayed according to Ghoshal and Recknagel (15). Mitochondrial protein content was measured by the method of Lowry et al.(16).

Preparation of Mitochondrial DNA. Mitochondrial DNA was isolated by a modification of the method of Palva and Palva (17). The mitochondria from each reaction mixture were concentrated by centrifugation (15,000 g for 15 minutes) and then resuspended to a final volume of 1.5 ml in ice cold buffer containing 50 mM of glucose, 1.5 mM EDTA and 25 mM Tris-HCl (pH 7.4). To lyse mitochondria and denature any contaminating nuclear DNA, 4.0 ml of alkaline SDS solution was mixed into the suspension. Two ml of 3.0 M potassium acetate were then added and the resultant supernatant was then mixed with an equal volume of isopropanol and kept at -20°C for 12 hours, after which the DNA was collected by centrifugation (12,000 g for 30 minutes). Each DNA sample was resuspended in 1.5 ml of a solution containing Tris-HCl (pH 7.8), 5.0 mM EDTA and 0.5% SDS and treated with 6 units of Proteinase K for 30 minutes at 37°C. The solution was then phenol extracted two times and the residual phenol was removed with chloroform. After reprecipitation with ethanol, the DNA was dried and resuspended in 50 ul of water. Quantitative estimation of DNA isolation was performed by spectrophotometric analysis. Restriction analysis was achieved with EcoR I endonuclease. DNA electrophoresis was done with 1.0% agarose slab gels, followed by ethidium bromide staining.

RESULTS

Table 1 shows malonic dialdehyde (MDA) evolution is a variety of experimental conditions. Aerobic incubation of mitochondria with the

Table 1. Effect of NADPH and ferrous chloride on MDA production in mitochondrial suspensions, incubated under aerobic and anaerobic conditions.

Additions	Malonic dialdehyde production (nMole/mg protein)
None (aerobic)	0.64 ± 0.15
NADPH (aerobic)	5.80 ± 4.55
Fe (aerobic)	4.17 ± 0.68
Fe + NADPH (aerobic)	17.89 ± 0.58
Fe + NADPH (anaerobic)	0.53 ± 0.20

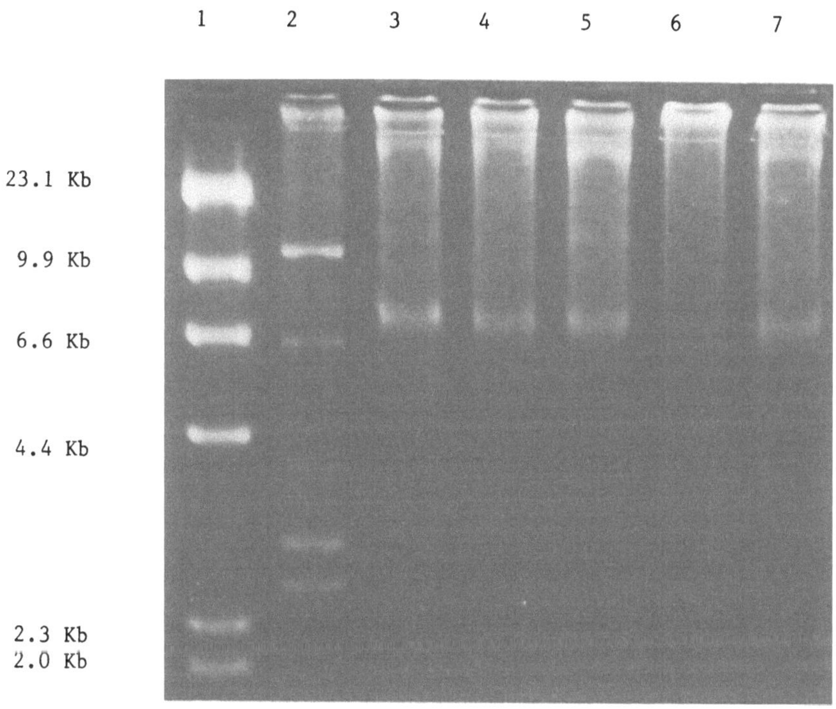

Figure 1. Change in electrophoretic mobility of mitochondrial DNA during mitochondrial lipid peroxidation.

Lane 1. Hind III digested Lambda phage DNA.
Lane 2. EcoR I digest of DNA from untreated mitochondria.
Lane 3. Undigested DNA from untreated mitochondria.
Lane 4. Undigested DNA from mitochondria incubated aerobically with NADPH alone.
Lane 5. Undigested DNA from mitochondria incubated aerobically with Fe alone.
Lane 6. Undigested DNA from mitochondria incubated aerobically with Fe and NADPH.
Lane 7. Undigested DNA from mitochondria incubated anaerobically with Fe and NADPH.

addition of NADPH and ferrous chloride led to marked peroxidation as evidenced by the evolution of MDA. In the absence of oxygen or any of the additives, MDA production was markedly diminished.

Agarose gel electrophoresis of the DNA samples extracted from the controls showed the typical banding pattern of mitochondrial DNA (18). The exact nature of the DNA was verified by electrophoresis of the EcoR I endonuclease digest (Figure 1, lane 2), which shows the banding pattern typical of mitochondrial DNA from Sprague Dawley rats (19). All DNA isolated from mitochondria which had undergone little or no peroxidation showed prominent DNA bands of similar intensity (lanes 3, 4, 5 and 7). However, when there was marked evolution of MDA, there was complete obliteration of the banding pattern of intact mitochondrial DNA. Densitometric analysis of agarose gels from these experiments showed that most of the nucleic acid from peroxidized mitochondria was unable to migrate into the agarose and accumulated at the surfaces of the sample wells.

DISCUSSION

These experiments show severe mitochondrial DNA damage during NADPH stimulated lipid peroxidation. In more recent experiments, vitamin E concomitantly prevented both mitochondrial lipid peroxidation and mitochondrial DNA damage (20). This latter observation supports the view that the DNA damage reported here may be due to lipid peroxidation.

Although these experiments do not permit formulation of a precise mechanism by which lipid peroxidation may affect DNA, one obvious possibility is that lipid peroxy radicals (LOO·) or alkoxy radicals (LO·) attack DNA directly. Such a radical attack would be similar to the well known postulated attack on DNA by hydroxyl radicals (OH·), for example generated by radiation. Attack on DNA by oxidized polyenoic fatty acids is known to occur and is known to cause strand scission in plasmid DNA (21). The attack on DNA by toxic products of fatty acid oxidation may yield cross-linked DNA aggregates similar to those produced when carcinogen metabolites interact with plasmid DNA (22). The results presented here are consistent with the observations by Emerit et al.(13) that suggest that clastogenic eicosanoids generated by cultured lymphocytes treated by a superoxide generating system are composed of hydroperoxides of arachidonic acid and their breakdown products.

The data presented provide evidence that oxidative free radical processes can destroy mitochondrial DNA, and that lipid peroxidation may be important in the mechanism of DNA damage. As the main site of cellular oxygen activation, mitochondrial membranes are vulnerable to peroxidation. Since mitochondrial DNA is free of histones and does not have a DNA repair system comparable to that of nuclear DNA (23), it may be readily vulnerable to clastogenic substances that may be produced at the mitochondrial membrane due to oxidative stress. Mitochondrial DNA undergoes spontaneous mutation at a higher frequency than nuclear DNA (24). This work shows that mitochondria can provide a test system for the study of DNA damage induced by free radicals and lipid peroxidation.

ACKNOWLEDGEMENT

The technical assistance of Vinay Bararia is gratefully acknowledged. This work was supported by Grant CA 07955 National Cancer Institute, United States Public Health Service.

REFERENCES

1. D. Grunberger and R.M. Santella, Conformational changes in DNA induced by chemical carcinogens. In Genes and Proteins in Oncogenesis (I. B. Weinstein and H. J. Vogel, Eds.) pp. 13-40. Academic Press, New York (1983).
2. I. Emerit and P. Cerutti, Icosanoids and chromosome damage. In Icosanoids and Cancer (H. Thaler-Dao, A. Crastes de Paulet and R. Paoletti, Eds.) pp. 127-138. Raven Press, New York (1984).
3. R. L. Willson, Iron and hydroxyl free radicals in enzyme inactivation and cancer. In Free Radicals, Lipid Peroxidation and Cancer (D.C.H. McBrien and T.F. Slater, Eds.) pp. 275-303. Academic Press, New York (1982).
4. M.G. Simic and S.V. Jovanovic, Free radical mechanisms of DNA base damage. In Mechanisms of DNA Damage and Repair (M.G. Simic, L. Grossman and A.C. Upton, Eds.) pp. 39-49. Plenum Press, New York (1986).
5. B.N. Ames and R.L. Saul, Oxidative DNA damage as related to cancer and aging. Prog. Clin. Biol. Res. 209A, 11-26 (1986).
6. P.A. Cerutti, Prooxidant states and tumor promotion. Science 227, 375-381 (1985).
7. U. Reiss and A.L. Tappel, Fluorescent product formation and changes in structure of DNA reacted with peroxidizing arachidonic acid. Lipids 8, 199-202 (1973).
8. S. Yonei and H. Furui, Lethal and mutagenic effects of malondialdehyde, a decomposition product of peroxidized lipids, on Escherichia coli with different DNA-repair capacities. Mutation Res. 88, 23-32 (1981).
9. F. H. Mukai and B.D. Goldstein, Mutagenicity of malonaldehyde, a decomposition product of peroxidized polyunsaturated fatty acids. Science 191, 868-869 (1976).
10. U.M. Marinari, M. Ferro, L. Sciaba, R. Finollo, A.M. Bassi and G. Brambilla, DNA-damaging activity of biotic and xenobiotic aldehydes in chinese hamster ovary cells. Cell Biochem. and Func. 2, 243-248 (1984).
11. L. J. Marnet, H.K. Hurd, M.C. Hollstein, D.E. Levin, H. Esterbauer and B.N. Ames, Naturally occurring carbonyl compounds are mutagens in Salmonella tester strain TA 104. Mutation Res. 148, 25-34 (1985).
12. S. Alaska, Inactivation of transforming activity of plasmid DNA by lipid peroxidation. Biochim. Biophys. Acta, 867, 201-208 (1986).
13. I. Emerit, S.H. Khan and P.A. Cerutti, Treatment of lymphocyte cultures with a hypoxanthine-xanthine oxidase system induces the formation of transferable clastogenic material. J. Free Rad. Biol. Med. 1, 51-57 (1985).
14. B.G. Niranjan, N.K. Bhat, and N.G. Avadhani, Preferential attack of mitochondrial DNA by aflatoxin B_1 during hepatocarcinogenesis. Science 215, 73-75 (1985).
15. A.K. Ghoshal and R.O. Recknagel, Positive evidence of acceleration of lipoperoxidation in rat liver by carbon tetrachloride: In vitro experiments. Life Sci. 4, 1521-1530 (1965).
16. O.H. Lowry, N.J. Rosebrough, A.L. Farr and R.J. Randall, Protein measurement with the folin phenol reagent. J. of Biol. Chem. 193, 265-275 (1951).
17. T.K. Palva and E.J. Palva, Rapid isolation of animal mitochondrial DNA by alkaline extraction. FEBS Letters 192, 267-170 (1985).
18. R. Wilkinson, A. Hawks, and A.E. Peggy, Methylation of rat liver mitochondrial deoxyribonucleic acid by chemical carcinogens and associated alterations in physical properties. Chem.-Biol.

Interactions 9, 157-167 (1975).
19. J. F. Francisco. F.F. Vissering and M.V. Simpson, Two aspects of mitochondrial DNA structure: The occurrence of two types of mitochondrial DNA in rat liver and the isolation from rat liver of DNA complexes of high buoyant density. In Mitochondria 1977 (W. Bendlow, R.J. Schueyen, K. Wolf and K. Kardewitz, Eds.) pp. 25-37. Walter de Gruyter, New York.
20. A.M. Hruszkewycz (unpublished results).
21. S. Inouye, Site-specific cleavage of double-stranded DNA by hydroperoxide of linoleic acid. FEBS Letters 172, 231-234 (1984).
22. R.M. Fourney, P.J. O'Brien and W.S. Davidson, Peroxidase catalyzed aggregation of plasmid pBR322 DNA by benzidine metabolites in vitro. Carcinogenesis 7, 1535-1542 (1985).
23. W.M. Brown, Mechanisms of evolution in animal mitochondrial DNA. Annals New York Acad. Sci. 361, 119-134 (1980).
24. W.M. Brown, M. George, Jr. and A.C. Wilson, Rapid evolution of animal mitochondrial DNA. Proc. Nat. Acad. Sci. 76, 1967-1971 (1979).

MITIGATION OF SINGLET OXYGEN ($^1\Delta_gO_2$) DAMAGE IN BIOLOGICAL SYSTEMS

Thomas A. Dahl[1], Philip E. Hartman[1] and W. Robert Midden[2*]

[1]Department of Biology and [2]Department of Environmental Health Sciences, The Johns Hopkins University, Baltimore, MD 21218

Active oxygen species have been implicated in a wide variety of environmental and health effects, including aging, heart disease, and induction of cancers. One of these reactive species, singlet oxygen ($^1\Delta_gO_2$), the lowest energy electronically excited state of molecular oxygen, has evoked particular interest, because of its potential to react with a variety of biologically important substrates.and because it may be generated in living systems by the decomposition or interconversion of other active oxygen species, photosensitization reactions involving endogenous sensitizers, electron transport systems, and some enzyme systems (references cited in 1).

Singlet oxygen has been reported to oxidize free guanine and deoxyguanosine in solution (2-5) and, far more slowly, guanine residues in single-stranded DNA, while duplex DNA is essentially unreactive toward singlet oxygen (6,7). Some amino acids (e.g. cys, his, met, trp, tyr) react readily with singlet oxygen. Small peptides are oxidized at about the same rate as their constituent amino acids, while the same amino acid residues in proteins are generally oxidized more slowly (8). Photosensitized oxidation of unsaturated fatty acids also may involve singlet oxygen. A recent review of photosensitized oxidations of biomolecules can be found in (9).

Singlet oxygen has been invoked as a toxic intermediate in bacteriophage ΦX174 inactivation (2), killing of yeast (10), and destruction of mouse mammary carcinoma cells (11). We have demonstrated directly the potent cytotoxicity of pure singlet oxygen (at least 10^4times more potent than H_2O_2) for several strains of bacteria (12).

In addition to direct toxic effects, singlet oxygen has been implicated in the mutagenic activation of polycyclic aromatic hydrocarbons (13,14), as a base substitution mutagen in bacteria (15), and in induction of genetic changes in yeast (10). Pure exogenous singlet oxygen, however, does not appear to produce significant genetic damage in *Salmonella* (16).

Participation of singlet oxygen in reactions is almost always ambiguous in complex biological systems because of the presence of oxygen radicals and other excited state molecules. The separated-surface-sensitizer (S-S-S) described by Midden and Wang (4) provides a simple means of excluding non-singlet oxygen reactions from photosensitized oxidations, thus eliminating any ambiguities regarding the identification of the reactive species.

---

* Present address: Center for Photochemical Sciences, Chemistry Department, Bowling Green State University, Bowling Green, Ohio 43403

In order to mitigate the adverse effects of singlet oxygen, living systems may employ at least two defense strategies; these include interception of singlet oxygen either by physical or chemical quenchers (17-22) and prevention of singlet oxygen formation by direct quenching of excited state sensitizers (17,18,20,22). In this study we investigate the abilities of several prevalent biomolecules at physiologically relevant concentrations and neutral pH to protect the target substrate 2,5-bis(hydroxymethyl)-furan (BHMF) against oxidation by singlet oxygen, using the S-S-S singlet oxygen-generating system as well as illuminated solubilized sensitizer.

For studies involving sensitizer in solution, reaction mixtures contained 10^{-5} *M* rose bengal and 10^{-4}*M* BHMF in 0.1 *M* sodium phosphate buffer, pH 7.0, with or without putative protectors. Solutions were placed in the shallow depressions of a Boerner slide. The slide was then covered with a glass plate, and the system illuminated with a 150 W tungsten filament lamp. Temperature was maintained at 10°C with a cooling plate and an IR filter.

For pure singlet oxygen exposures, solutions containing the reference substrate BHMF at 10^{-4} *M* in 0.1 *M* sodium phosphate buffer, pH 7.0, with or without putative protectors were placed in the shallow depressions of a Boerner slide. A glass plate to which rose bengal-coated glass beads (4) were attached with double-stick tape were placed sensitizer-side down over the wells, after removing loose sensitizer with a gentle stream of compressed air. Illuminations were carried out as above.

To determine the extent of BHMF oxidation, samples were illuminated for various lengths of time and duplicate aliquots analyzed by HPLC using a C18 reverse phase column. Samples were eluted with 7% MeOH in H_2O, except for solutions containing homocarnosine, which were eluted with 5% MeOH in H_2O. The samples were monitored at 220 nm. The amount of BHMF was then compared to that in aliquots of unilluminated sample. Illumination without sensitizer had no effect on BHMF levels. The amount of BHMF after illumination divided by the amount in the unilluminated sample is reported as fraction BHMF remaining, averaged for at least three duplicate experiments.

L-histidine is one of the most reactive quenchers of singlet oxygen among biomolecules (19). Except for the muscles of some fish, however, free histidine is not commonly found in high concentrations in nature (23). Carnosine (β-alanyl-L-histidine), on the other hand, is present in striated muscles of many organisms, most notably mammals, in concentrations up to 40 m*M* (24). Carnosine is a better quencher of singlet oxygen than is equimolar histidine at neutral pH, both in solubilized sensitizer studies (Table 1) and in the pure singlet oxygen system (Table 2). In the system involving pure singlet oxygen, 1 m*M* carnosine quenches as well as 3 m*M* histidine alone, and equally well as a mixture of 3 m*M* histidine and 3 m*M* β-alanine (Table 2) which is surprising since β-alanine also protects to some extent. One role of carnosine in muscle tissue therefore may be to reduce damage caused by singlet oxygen, and possibly other active oxygen species, as suggested by others (25). Homocarnosine (γ-aminobutyryl-L-histidine) is present in the brain and cerebrospinal fluid of many mammals, at concentrations in the low micromolar range (26,27). The physiological role of homocarnosine within the central nervous system is yet unknown. Our studies indicate that homocarnosine and histidine quench singlet oxygen at similar rates (Table 2), but the concentration of homocarnosine required for significant quenching is much higher than is found in nature. Thus, homocarnosine is probaably not acting as a protective agent against singlet oxygen damage in the mode suggested for carnosine.

Urate, as an antioxidant, has been proposed to be responsible for increases in longevity of species and for reduction of age-specific cancer rates in humans compared to shorter-lived species. Although uric acid has been reported to quench singlet oxygen in photosensitized studies (21), our results with pure singlet oxygen indicate that 1 m*M* urate (i.e., nearly saturated solution) fails to protect against singlet oxygen oxidation of the target substrate (Table 2). Entirely consistent with previous reports, however, are our results with illuminated sensitizer in solution, in which 0.5 m*M* urate protected about as well as the traditional singlet oxygen quencher azide at 1 m*M* (Table 1). Since oxidation of BHMF in this system proceeds almost exclusively by reaction with singlet oxygen (5), we conclude that urate cannot quench singlet oxygen once formed but can efficiently prevent its formation by excited state sensitizer molecules.

Table 1. Relative Protection Against BHMF Oxidation By Illuminated Rose Bengal In Solution.

sample	fraction BHMF remaining after 40 min illumination*+
BHMF alone	.24 ±.01 (4)
+ 1 m*M* azide	.58 ±.03 (3)
+ 1 m*M* histidine	.27 ±.01 (3)
+ 1 m*M* carnosine	.36 ±.03 (3)
+ 1 m*M* ergothioneine	.76 ±.02 (3)
+ 0.5 m*M* urate	.53 ±.02 (3)

* Plots of log(BHMF Remaining) versus time yield linear curves for this range of BHMF concentrations.
+ Values are expressed as mean ± standard error of the mean. Numbers in parentheses are numbers of duplicate exposures.

L-ergothioneine (2-thiolhistidine betaine) is synthesized by fungi and assimilated by higher plants and many animals. Ergothioneine is highly conserved in nature, and is found in egg white, milk, semen, liver, red blood cells, bone marrow, and the central nervous system, often in millimolar amounts (28 and references therein). One millimolar ergothioneine, like uric acid fails to protect the target substrate from oxidation by pure singlet oxygen in the S-S-S system (Table 2), yet protects the substrate well in sensitizer solution studies (Table 1). Again we conclude that protection is due to prevention of singlet oxygen formation by excited sensitizer. Ergothioneine also protects bacteria from mutagenesis induced by organic hydroperoxides (28) and protects phages T4 and P22 against inactivation by low level γ-irradiation (29). These modes of defense may be important in fungi. In extension of this line of reasoning, we conjecture that ergothioneine may be sequestered for similar purposes in the tissues of organisms that ingest but do not synthesize it, and provides an effective adjunct defense against oxidative damage.

Table 2. Relative Protection Against BHMF Oxidation by Singlet Oxygen In the S-S-S System

sample	fraction BHMF remaining after exposure to singlet oxygen*+
BHMF alone	.31 ± .01 (14)
+ 1 m*M* azide	.45 ± .01 (3)
+ 1 m*M* histidine	.33 ± .01 (4)
+ 2 m*M* histidine	.38 ± .01 (3)
+ 3 m*M* histidine	.42 ± .03 (3)
+ 1 m*M* carnosine	.41 ± .01 (9)
+ 1 m*M* homocarnosine	.30 ± .02 (3)
+ 3 m*M* homocarnosine	.42 ± .01 (3)
+ 5 m*M* β-alanine	.38 ± .01 (3)
+ 3 m*M* histidine, 3 mM β-alanine	.42 ± .01 (3)
+ 1 m*M* ergothioneine	.28 ± .02 (3)
+ 1 m*M* urate	.32 ± .01 (4)

* Approximately 10^{-4} moles singlet oxygen reaching the surface of the solution, as determined from dosimetry in (12).
+ Values are expressed as mean ± standard error of the mean. Numbers in parentheses are numbers of duplicate exposures.

We conclude that these prevalent compounds may represent at least two distinct biologically relevant mechanisms by which living systems protect themselves against the harmful, particularly toxic, effects of singlet oxygen. Some compounds (ergothioneine, urate) prevent the formation of singlet oxygen, by excited state sensitizers while others (e.g. carnosine) intercept singlet oxygen after it is formed.

REFERENCES

(1) A Naqui, B Chance and E Cadenas (1986) Reactive oxygen intermediates in biochemistry. Annu. Rev. Biochem. 55:137-166.

(2) N Houba-Herin, CM Calberg-Bacq, J Piette and A Van de Vorst (1982) Mechanisms for dye-mediated photodynamic action: singlet oxygen production, deoxyguanosine oxidation and phage inactivating efficiencies. Photochem. Photobiol. 36:297-306.

(3) J Cadet, C Decarroz, SY Wang and WR Midden (1983) Mechanisms and products of photosensitized degradation of nucleic acids and related model compounds. Isr. J. Chem. 23:420-429.

(4) WR Midden and SY Wang (1983) Singlet oxygen generation for solution kinetics: clean and simple. J. Am. Chem. Soc. 105:4129-4135.

(5) SY Wang and WR Midden (1983) Photodynamic action of nucleic acids and their components. Stud. Biophys. 94:7-12.

(6) S Kawanishi, S Inoue, S Sano H Aiba (1986) Photodynamic guanine modification by hematoporphyrin is specific for single-stranded DNA with singlet oxygen as mediator. J. Biol. Chem. 261: 6090-6095.

(7) E Hildebrand, WR Midden and BL Murr (1987) DNA damage induced by singlet oxygen. Photochem. Photobiol. 45S:8S.

(8) JD Spikes and ML MacKnight (1970) Dye-sensitized photooxidation of proteins. Ann. NY Acad. Sci. 171:149-162.

(9) RC Straight and JD Spikes, Photosensitized oxidation of biomolecules. In: Singlet Oxygen, Vol.IV, (AA Frimer, Ed.), Ch.2, CRC Press, Boca Raton, FL.(1985)

(10) K Kobayashi and T Ito (1976) Further *in vivo* studies on the participation of singlet oxygen in the photodynamic inactivation and induction of genetic changes in *S. cerevisiae*. Photochem. Photobiol. 23:21-28.

(11) KR Weishaupt, CJ Gomer and TJ Dougherty (1976) Identification of singlet oxygen as the cytotoxic agent in photoinactivation of a murine tumor. Cancer Res. 36: 2326-2329.

(12) TA Dahl, WR Midden and PE Hartman (1987) Pure singlet oxygen cytotoxicity for bacteria. Photochem. Photobiol 46: in press.

(13) CL Greenstock and RH Wiebe (1978) Photochem. Photobiol. 28: 863-867.

(14) EC McCoy and HS Rosenkranz (1980) Activation of polycyclic aromatic hydrocarbons to mutagens by singlet oxygen: an enhancing effect of atmospheric pollutants? Cancer Lett. 9:35-42.

(15) B Gutter, WT Speck and HS Rosenkranz (1977) A study of the photoinduced mutagenicity of methylene blue. Mutat. Res. 44:177-182.

(16) TA Dahl, PE Hartman and WR Midden (1987) No mutagenicity observed in *Salmonella* with extracellularly generated singlet oxygen. Photochem. Photobiol. 45S: 9S.

(17) RS Davidson and KR Tretheway (1975) The mechanism of dye-sensitized photo-oxygenation of amines. J. Chem. Soc. Chem. Commun.: 674-675.

(18) RS Davidson and KR Tretheway (1976) The role of excited singlet state dyes in dye sensitized photooxygenation reactions. J. Am. Chem. Soc. 98:4008-4009.

(19) D Bellus (1979) Physical quenchers of singlet molecular oxygen. Adv. Photochem. 11:105-205.

(20) CS Foote, Quenching of singlet oxygen. In: Singlet Oxygen, (HH Wasserman and RW Murray, Eds.), pp.139-171, Academic Press, New York (1979).

(21) BN Ames, R Cathcart, E Schwiers and P Hochstein (1981) Uric acid provides an antioxidant defense in humans against oxidant- and redical-caused aging and cancer: a hypothesis. Proc. Natl. Acad. Sci. USA 78:6858-6862.

(22) MP Percival and AD Dodge (1983) Photodynamic damage to chloroplast membranes, photosensitized oxidation of chloroplast acyl lipid. Plant Sci. Lett. 29:255-264.

(23) H Abe (1983) Distribution of free L-histidine and related compounds in marine fishes. Bull. Jpn. Soc. Sci. Fish. 49:1683-1687.

(24) KG Crush (1970) Carnosine and related substances in animal tissues. Comp. Biochem. Physiol. 34:3-30.

(25) AM Dupin, AA Boldyrev, YV Arkhipenko and VE Kagan (1984) Protection of Ca^{++} transport by carnosine against disturbances induced by lipid peroxidation. Bull. Exp. Biol. Med. 98:1071-1073.
(26) P Bohlen, G Tell, PJ Schechter, J Koch-Weser, Y Agid, G Coquillot, G Chazot and C Fischer (1980) Cerebrospinal fluid homocarnosine in Huntington's disease. Life Sci. 26: 1009-1012.
(27) TL Perry, SJ Kish, O Sjaastad, LR Gjessing, R Nessbakken, H Schrader and AC Loken (1979) Homocarnosinosis: increased content of homocarnosine and deficiency of homocarnosinase in brain. J. Neurochem. 32: 1637-1640.
(28) Z Hartman and PE Hartman (1987) Interception of some direct-acting mutagens by ergothioneine. Environ. Molec. Mutagen. 10:in press.
(29) PE Hartman and M Citardi (1986) Protection afforded by carnosine and by ergothioneine against bacteriophage P22 inactivation by γ-irradiation. Environ. Mutagen. 8(suppl 6):35.

REACTIVE OXYGEN SPECIES ARE MUTAGENIC TO MAMMALIAN CELLS

Abraham W. Hsie

The University of Texas Medical Branch
Galveston, Texas 77550

INTRODUCTION

Radiation biologists have proposed that various reactive oxygen species (referred to hereafter as reactive oxygens) which are produced during reactions between radiation and the biological system, are largely responsible for the adverse biological effects of radiations (1). Recently, reactive oxygens have been implicated in the toxic action of numerous chemicals (2).

Since 1974, we have used Chinese hamster ovary (CHO) cell clone K1-BH4 to study mutations induced by various chemical and physical agents at the hypoxanthine-guanine phosphoribosyltransferase (hgprt) locus (CHO/HGPRT assay) (3, 4). Our early studies showed that X irradiation is weakly mutagenic (5).

Recently, by transforming an HGPRT-deficient CHO cell with the plasmid vector pSV2*gpt*, we isolated a transformant, AS52. AS52 cells carry a single copy of the *gpt* gene, the bacterial equivalent of the mammalian gene *hgprt*, stably integrated into the host DNA (6). We found that X irradiation is equitoxic to AS52 cells and the parental K1-BH4 cells; however, X rays are up to eight times more effective in inducing mutation to 6-thioguanine (TG) resistance to AS52 cells than to K1-BH4 cells (6-8). Thus, AS52 cells are hypersensitive to X irradiation-induced gene mutation.

Like radiation, radiomimetic chemicals are known to exhibit their biological effects through the mediation of reactive oxygens (1, 2). If reactive oxygens were the common mediators for both radiation and radiomimetic chemicals, compounds such as streptonigrin and bleomycin, which produce reactive oxygens as their major mechanism of action, would be expected to be equitoxic to both K1-BH4 cells and AS52 cells, but more mutagenic to AS52 cells than to K1-BH4 cells. Likewise, potent oxidizing agents such as potassium superoxide and hydrogen peroxide would be more mutagenic to AS52 cells and equitoxic to both cell types. Experiments were performed to test these hypotheses.

RESULTS

We used subclone K1-BH4 of the CHO-K1 cell line (3) and AS52 cells (6) for all the experiments. The cytotoxicity and mutagenicity as affected by physical and chemical agents in these cells were determined using the standard procedure (3, 4). Cytotoxicity, the effect of mutagen treatment on cellular cloning efficiency, is expressed as % survival relative to the untreated controls. To quantify mutations at either hgprt or gpt locus, the treated cells were allowed to fully express their mutant phenotypes followed by selection in a hypoxanthine-free medium containing 10 μM TG. Mutation frequency is calculated as the number of TG-resistant (TG^r) colonies per 10^6 clonable cells at the end of the expression period.

We found that X rays are equitoxic to K1-BH4 cells and AS52 cells; X irradiation is up to eight times more mutagenic to AS52 cells than to K1-BH4 cells (6-8). Thus, AS52 cells are hypersensitive to X ray-induced mutations. Southern blot analyses showed that the majority of the X-irradiation-induced TG^r mutants had deletions in the structural gene of gpt. Similarly, most TG^r mutants derived from X irradiation of K1-BH4 cells showed a deletion of the hgprt gene (8). The classical missense mutagens including ethyl methanesulfonate (EMS), ICR-191 and UV light were found to induce TG^r mutants with no detectable alteration in the structural genes of hgprt and gpt in K1-BH4 cells and AS52 cells, respectively (8, 9) (Table 1).

The pattern of the hypermutability of AS52 cells is not unique to X rays since a differential mutagenic response between K1-BH4 cells and AS52 cells was also produced by irradiation with neutrons. Neutron irradiation resulted in nearly identical lethality to both cell types; however, neutrons are up to ten times more effective in inducing mutation in AS52 cells than in K1-BH4 cells (10) (Fig. 1).

If reactive oxygen species were to mediate the mutagenic effects of radiations such as X rays and neutrons, then radiomimetic chemicals such as streptonigrin and bleomycin, which are known to produce superoxide and hydroxy radicals, would be expected to be equitoxic to both cell types and more mutagenic to AS52 cells than to K1-BH4 cells. Likewise, oxidizing compounds such as potassium superoxide and hydrogen peroxide, which are reactive oxygen species, should be more mutagenic in AS52 cells but equitoxic to both cell types. Our experiments fulfilled such expectations (10) (Fig. 1). Agents such as EMS, ICR 191 and UV light, whose major modes of action are not associated with the production of reactive oxygens, should be equally toxic and mutagenic to both cell types. Our experiments also fulfilled these predictions (10) (Table 1). The combination of these experiments support the view that reactive oxygens induce gene mutations in mammalian cells.

DISCUSSION

Since X rays induce predominantly deletion mutations (8), the low mutagenic activity of X irradiation in K1-BH4 cells where hgprt gene exists in a hemizygous state, is likely due to lethal events associated with multi-locus deletions that reduce the viability of the induced TG^r mutants. The gpt gene may be stably integrated in the diploid state in AS52 cells. Thus, a multi-locus deletion induced by X irradiation could produce a hemizygous state in the vital genes that

Table 1. Interrelationships of Reactive Oxygen Species, Hypermutability and Oxidative DNA Damage/Deletion Mutation

Agents	Production of Reactive Oxygen Species	Hypermutability to AS52 Cells	Induction of Oxidative DNA Damage and/or Deletion Mutation
	Major Mode of Actions		
Chemical			
Radiomimetic			
Streptonigrin	+	+	+
Bleomycin	+	+	+
Oxidizing			
Potassium Superoxide	+	+	+
Hydrogen Peroxide	+	+	+
Missense Mutagenic			
Ethyl Methanesulfonate	-	-	-
ICR 191	-	-	-
Physical			
Ultraviolet Light	-	-	-
X Rays	+	+	+
Neutrons	+	+	+

Adapted from reference 1

may flank the *gpt* gene resulting in a higher probability of recovering induced TG^r mutants. In contrast, a deletion larger than the *hgprt* gene, which exists in a functionally haploid state in K1-BH4 cells, could result in a loss of vital genes flanking the *hgprt* gene causing the production of non-viable TG^r mutants. By inference from experiments with X rays, the low mutagenic response in K1-BH4 cells to other agents which produce reactive oxygens may also be due to low mutant recovery rather than to reduced mutation induction. This leads to the prediction that agents, which are known to produce reactive oxygens should produce predominantly deletion mutations and/or chromosomal aberrations; such agents could be detected using AS52 cells as biological dosimeter (10).

It is known that many naturally occurring chemicals can be reduced by intracellular reductases, and, in turn, reduce oxygen to reactive oxygens via a redox cycling mechanism (11). These natural compounds together with background radiations and man-made reactive oxygen producers may ultimately cause spontaneous (deletion) mutations and possibly also cancer and aging.

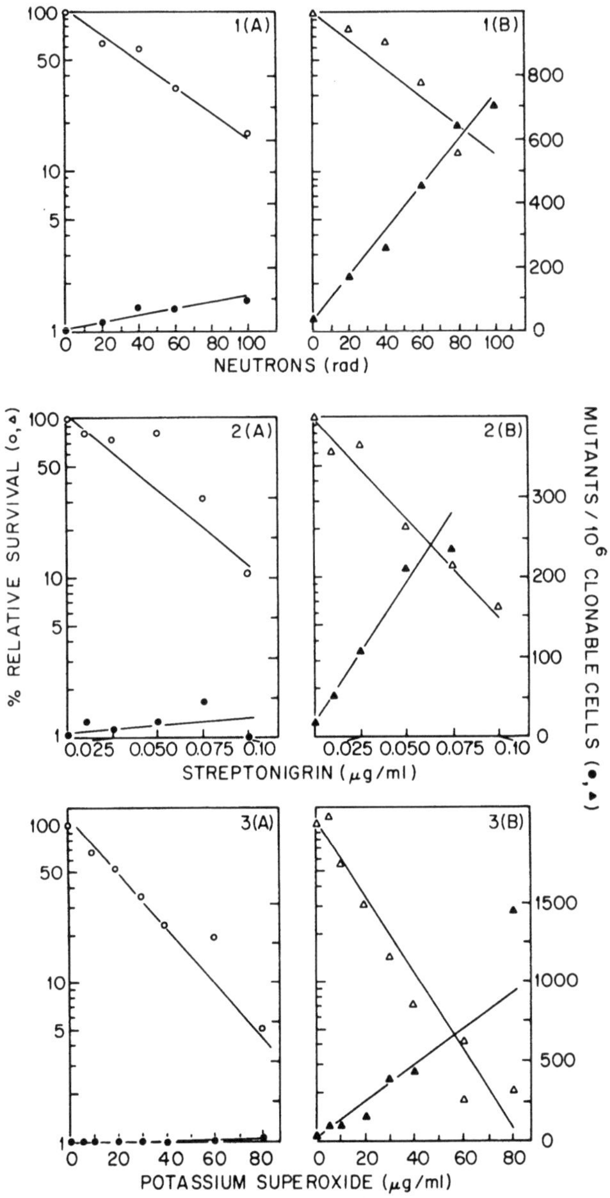

Figure 1. The cytotoxic (o--o, Δ--Δ) and mutagenic (o--o, ▲--▲) effects of neutron irradiation (1), of streptonigrin (2) and of potassium superoxide (3) to CHO-K1-BH4 cells (A) and to AS52 cells (B). Adapted from reference 1.

ACKNOWLEDGMENTS

I thank R.L. Schenley and L.C. Waters of Oak Ridge National Laboratory (ORNL) for reviewing the draft, J.F. Young (ORNL) for preparing the manuscript as well as F.T. Kenney (ORNL) and M.A. Waters (EPA) for hosting me at ORNL and EPA, respectively. The mutagenesis experiments described in this article were conducted in collaboration with L. Recio, D.S. Katz, C.Q. Lee, M. Wagner and R.L. Schenley while I was affiliated with ORNL.

I am supported in part as a Distinguished Visiting Scientist of EPA under interagency agreement DOE 40-1601-85 with the Department of Energy under contract (DE-AC05-840R2140) with Martin Marietta Energy Systems, Inc. Although the research described in this article has been funded by EPA, it has not been subjected to EPA review and therefore, does not necessarily reflect the views of EPA, and no official endorsement should be inferred.

REFERENCES

1. J.E. Coggle, Biological Effects of Radiation, 247. International Publications Service, p. 247. Taylor and Francis, 1983.
2. T.L. Dormandy, An approach to free radicals. Lancet ii, 1010-1014 (1983).
3. A.W. Hsie, P.A. Brimer, T.J. Mitchell and D.G. Gosslee. The dose-response relationship for ethyl methanesulfonate-induced mutations at the hypoxanthine-guanine phosphoribosyltransferase locus in Chinese hamster ovary cells. Somatic Cell Genet. 1, 247-261 (1975).
4. A.W. Hsie, D.A. Casciano, D.B. Couch, D.R. Krahn, J.P. O'Neill and B.L. Whitfield, The use of Chinese hamster ovary cells to quantify specific locus mutation and to determine mutagenicity of chemicals. Mutat. Res. 86, 193-214 (1981).
5. J.P. O'Neill, D.B. Couch, R. Machanoff, J.R. San Sebastian, P.A. Brimer and A.W. Hsie, A quantitative assay of mutation induction at the hypoxanthine-guanine phosphoribosyltransferase locus in Chinese hamster ovary cells (CHO/HGPRT system): utilization with a variety of mutagenic agents. Mutat. Res. 45, 103-109 (1977).
6. K.R. Tindall, L.F. Stankowski, Jr., R. Machanoff and A.W. Hsie, Detection of deletion mutations in pSVgpt-transformed cells. Mol. Cell. Biol. 4, 1411-1415 (1984).
7. K.R. Tindall, L.F. Stankowski, Jr., R. Machanoff and A.W. Hsie, Analyses of mutation in pSVgpt-transformed CHO cells. Mutat. Res. 160, 121-131 (1986).
8. L.F. Stankowski, Jr. and A.W. Hsie, Quantitative and molecular analyses of radiation-induced mutation in AS52 cells. Radiat. Res. 105, 37-48 (1986).
9. L.F. Stankowski, Jr., K.R. Tindall and A.W. Hsie, Quantitative and molecular analyses of ethyl methanesulfonate and ICR-induced mutation in pSVgpt-transformed CHO cells. Mutat. Res. 160, 133-147 (1986).
10. A.W. Hsie, L. Recio, D.S. Katz, C.Q. Lee, M. Wagner and R.L. Schenley, Evidence for reactive oxygen species inducing mutations in mammalian cells. Proc. Natl. Acad. Sci., USA 83, 9616-9620 (1986).
11. H. Kappus and H. Sies, Toxic drug effects associated with oxygen metabolism: redox cycling and lipid peroxidation. Experientia 37, 1233-1241 (1981).

THE EXPRESSION OF BACTERIAL DNA ALKYLATION REPAIR ENZYMES IN MER$^-$ HUMAN CELLS

Leona Samson, Patrick Carroll, Bruce Derfler and William Rebeck

Charles A. Dana Laboratory of Toxicology
Harvard School of Public Health
Boston, MA 02115

INTRODUCTION

Simple monofunctional alkylating agents such as methylmethane sulphonate (MMS) and *N*-methyl-*N*'-nitro-*N*-nitrosoguanidine (MNNG) cause a variety of alkylated lesions in DNA, and these lesions can cause the induction of mutation and cell death in both human and bacterial cells. The repair of these DNA alkylation damages reduces cell killing and the induction of mutations and chromosome damage. In *E. coli*, the repair of O6-methylguanine (O6MeG) and O4-methylthymine (O4MeT) specifically prevents these lesions from causing G:C to A:T and A:T to G:C transition mutations, because if left unrepaired these lesions mispair during replication (1-6). On the other hand, the repair of N3-methylpurines (N3MeA and N3MeG) and O2-methylpyrimidines (O2MeC and O2MeT) in *E. coli* specifically prevents cell killing (7-9). The repair of DNA alkylation damage in mammalian cells is somewhat less well understood, and it is not yet clear which alkylated lesions cause mutation and which cause cell death. Like *E. coli*, mammalian cells can repair O6alkylG, N3alkylA, N3alkylG, O4alkylT, O2alkylT and O2alkylC lesions (10-21). However, because observations have been made with a variety of cell types the results have not always been consistent. For instance, some studies indicated that O4MeT is repaired by rat liver cells (13,20,21), while others did not (22-24). Not all cell lines are able to repair all the alkylated lesions; for instance, CHO and V79 cells (25,26) and some human tumor cell lines (27-29) are unable to repair O6MeG. The ability of these particular cell lines to repair alkylated pyrimidines has not yet been measured. Other human cells, however, have been shown to repair O6MeG (19,27-29), O4alkylT, O2alkylT, O2alkylC (11), N3alkylA, N7alkylG (11,14,15), N3alkylG (15,30) and N7alkylA (15).

The determination of the biological effects of particular alkylated lesions in *E. coli* was made possible by the isolation of isogenic strains that were mutated in different DNA alkylation repair genes and therefore differed in their ability to carry out specific types of DNA alkylation repair. Efforts to determine the biological effects of particular alkylated lesions in human cells, and thus to understand how alkylation damage induces human cell death, mutation and chromosome damage, have thus far been severely limited by the lack of isogenic cell strains that have lost or gained alkylation sensitivity. In view of this, we recently adopted a novel approach to the generation of such isogenic cell lines.

We have now shown that one can begin to apply the information gained through studying DNA alkylation repair in E. coli to the elucidation of the mechanisms of alkylation toxicity in human cells. We have demonstrated that E. coli DNA alkylation repair enzymes can be stably expressed in human cells and that their activity can completely suppress the alkylation sensitivity of Mer$^-$ human cells to the induction of the ultimate effects of DNA alkylation -- killing, mutation and chromosome damage (31). We therefore, in principle, have the means to endow human cells with the ability to repair one or two specific alkylated DNA lesions more efficiently, by engineering the over-expression of particular bacterial enzyme species. Thus, we can generate isogenic human cell lines of a predetermined alkylation repair genotype.

EXPRESSION OF THE E. COLI ADA-ALKB OPERON IN HELA S3 CELLS

The ada-alkB operon helps to protect E. coli against the effects of many alkylating agents. The ada gene product, a DNA methyltransferase (Ada protein), removes abnormal methyl groups from three different lesions; namely O6MeG and O4MeT whose methyl groups protrude into the major groove of the DNA double helix, and methylphosphotriesters (32). Since unrepaired O6MeG and O4MeT lesions cause transition mutations the ada gene product specifically protects bacteria against alkylation-induced mutation. DNA methyltransferase directly transfers the offending methyl group from O6MeG or O4MeT to a cysteine residue in the carboxy-terminal half of the enzyme; methyl groups from phosphtriesters are transferred to another cysteine residue in the amino-terminal half of the enzyme (32). The reaction is rather unusual because, once both active sites are methylated the DNA methyltransferase enzyme is inactivated, i.e., it is a suicide enzyme (33-35). In this alkylated form the Ada protein assumes a second function; it becomes a positive regulator for the transcription of the genes involved in the adaptive response of E. coli to alkylating agents (32). The alkB gene encodes a 27kDa protein, the specific function of which is not yet known, although it is known to be involved in the protection of bacteria from both the lethal and the mutagenic effects of alkylation damage (36,37). We subcloned the E. coli ada-alkB operon (generously provided by Dr. Graham Walker) into a mammalian expression vector such that the operon comes under the control of the SV40 early promotor; this construction generated the plasmid pSV2ada-alkB. We cotransfected this plasmid into alkylation-sensitive HeLa S3 Mer$^-$ human cells (30,38) with the plasmid pSV2neo that carries the E. coli neo gene which makes mammalian cells resistant to the aminoglycoside G418. Ten G418 resistant transformant cell lines were isolated (S3-1 through S3-10) and screened for the presence of E. coli DNA methyltransferase activity. We detected very high levels of methyltransferase activity in cell extracts of one of the ten G418 resistant transformants, namely S3-9, and intermediate levels in four other transformants. That the DNA methyltransferase activity in S3-9 cells is from the E. coli ada gene is supported by the following observations: (1) Southern blot analysis showed that DNA from the S3-9 transformant contains several copies of pSV2ada-alkB integrated in a tandem array; (2) Western blot analysis S3-9 cell-free extracts detected a 39kDa protein that was recognized by polyclonal antibodies directed against the E. coli DNA methyltransferase; (3) In bacterial extracts the 39kDa ada gene product quickly breaks down to form a 19kDa fragment that can only repair O6MeG and O4MeT. The methyltransferase in S3-9 extracts is also 39kDa and it too breaks down to an active 19kDa fragment, though not as quickly (4). We, and others, have shown that, unlike the bacterial enzyme, the human DNA methyltransferase cannot repair methylphosphotriester lesions (39). We found that the methyltransferase species in the S3-9 transformant cell line can repair methylphosphotriester lesions *in vitro*, confirming that the activity in S3-9 cells can not be the human enzyme.

It should be stressed that, since the function of the *alkB* gene is not known, and since we do not have antisera directed against the *alkB* protein, we were unable to determine whether this gene, which lies downstream from the ada gene, is also being expressed in S3-9 cells. However, Northern blot analysis of mRNA in S3-9 cells indicated that *alkB* is indeed being transcribed; i.e., we detected the presence of mRNA species that hybridized to both ada and alkB sequences.

ANALYSIS OF THE PHENOTYPE OF S3-9 CELLS

Having demonstrated that the ada-alkB operon was integrated into the genome of S3-9 cells, that this operon is efficiently transcribed, and that at least the *ada* gene is translated to produce active, stable enzyme in these human cells, the next step was to determine the effect of this expression upon the sensitivity of S3-9 cells to alkylating agents. We compared the alkylation sensitivity of the S3-9 cell line and one of the methyltransferase-negative cell lines (S3-7). We found that expression of the *ada* gene (and possibly also the *alkB* gene) conferred tremendous resistance to cell killing and sister chromatid exchange (SCE) induction by MNNG and *N*,*N*'-bis(2-chloroethyl)-*N*-nitrosourea (BCNU) (31) and MMS, and also conferred tremendous resistance to mutation induction by MNNG (BCNU and MMS were not tested). BCNU is a commonly used cancer chemotheraputic agent and several lines of evidence indicate that both the mammalian and bacterial DNA methyltransferases prevent primary BCNU lesions from forming potentially lethal DNA interstrand crosslinks; we were therefore not surprised that S3-9 cells were resistant to BCNU-killing. In E. coli, Ada does not provide MNNG-killing resistance by the repair of alkylated lesions *per se*, but rather provides killing resistance through the induction of at least one other DNA repair gene whose product repairs other methylated lesions, namely *alkA* (8,9). It was therefore a surprise to find that S3-9 cells were tremendously resistant to MNNG killing and this has led us to consider (among other things) that *alkB* is being expressed in S3-9 cells. The characterization of several pSV2ada-alkB transformants (S3-2,3, and 6) showed that the level of resistance to alkylation-induced killing and SCE induction (mutation not yet tested) is proportional to the level of ada expression; however, note that this resistance may also be proportional to the level of alkB expression.

DETERMINATION OF THE RELATIVE CONTRIBUTIONS OF ADA AND ALKB TO THE S3-9 PHENOTYPE

Growth of cells in the presence of the O6MeG free base depletes DNA methyltransferase activity from Mer^+ human cells, presumably because O6MeG acts as a weak substrate for the enzyme (40,43). We have shown that the O6MeG free base also depletes Ada activity from S3-9 cells; to our surprise we found that both the O6MeG and MePT repair activities were both reduced by over 90%. Thus, by removing Ada activity we can assess the contribution of alkB to the alkylation resistant S3-9 phenotype. DNA methyltransferase depletion grossly sensitized S3-9 cells to SCE induction by both MNNG and BCNU. However, depletion of methyltransferase activity did *not* sensitize cells to killing by MNNG. Moreover, Ada-depleted cells that were not senstized to MNNG-killing were sensitized to BCNU-killing. These results are consistent with the notion that the E. coli DNA methyltransferase itself does not protect HeLa S3 cells against MNNG-induced cell death, and suggests that Mer^- sensitivity to MNNG killing might be suppressed by *alkB* expression.

The next steps in this study are to engineer the expression of the ada gene alone, and the alkB gene alone, in the same HeLa Mer$^-$ S3 cell line. Some recent reports have addressed this point. Ishizaki et al. (44) transfected a plasmid carrying both the ada gene and the neo gene into the HeLa Mer$^-$ cell line. They found that no G418 resistant transformants expressed methyltransferase activity and so challenged their G418 resistant population of cells with a dose of ACNU (a compound similar to BCNU) that gave less than 0.01% survival; among the survivors they found two transformants that expressed very high levels of ada-derived methyltransferase activity. The authors conclude that Mer$^-$ human cells expressing the E. coli ada gene gain resistance to killing by MNNG. While this may indeed turn out to be the case, it is hard to justify this conclusion since the ada transformants were initially selected on the very basis of being alkylation resistant. Kataoka et. al. report a similar study selecting for transfectants expressing the ada gene in CHO cells (45). Brennand and Margison (46) have reported the cloning of the ada gene into a retrovirus derived vector that also carries the selectable neo gene. In this case, however, they screened G418 resistant transfectants of Chinese hamster V79 cells to find that one out of eight clones had a very high level of Ada activity. This transfectant was resistant to killing by MNNG. Thus the expression of the E. coli Ada protein in V79 cells does confer resistance to alkylation-induced cell killing. Moreover, we recently showed that the expression of the ada gene alone in HeLa Mer$^-$ cells also confers resistance to alkylation-induced killing. These findings contradict our results in HeLa cells using the O6MeG free base to deplete Ada DNA methytransferase activity (see above). Is it possible that the Ada protein has some other function in mammalian cells that does not depend upon the activities depleted by the O6MeG free base? At this point we can not rule out the possibility that the suppression of killing sensitivity in S3-9 cells might be conferred by some action of the Ada protein that does not depend upon its ability to scavenge methyl groups from O6MeG and methylphosphotriester lesions. This question is currently being addressed.

ACKNOWLEDGMENTS

This work was supported by American Cancer Society Research Grant NP448 and National Institute of Environmental Health Sciences Fund Grant 1-P01-ES03926. L.S. was supported by an American Cancer Society Scholar Award, P.C. by the Charles A. Dana Foundation, and W.R. by an NSF Graduate Fellowship.

REFERENCES

1. A. Loveless, Possible relevance of 0-6 alkylation of deoxyguanosine to the mutagenicity and carcinogenicity of nitrosamines and nitrosamides. Nature (London) 223, 206-207 (1969).
2. P. F. Schendel and P. E. Robins, Repair of O6-methylguanine in adapted Escherichia coli. Proc. Natl. Acad. Sci. USA 75, 6017-6020 (1978).
3. P. Karran, T. Lindahl, and B. Griffin, Adaptive response to alkylating agents involves alteration in situ of O6-methylguanine residues in DNA. Nature (London) 280, 76-77 (1979).
4. L. A. Dodson, R. S. Foote, S. Mitra, and W. E. Masker, Mutagenesis of bacteriophage T7 in vitro by incorporation of O6-methylguanine during DNA synthesis. Proc. Natl. Acad. Sci. USA 79, 7440-7444 (1982).

5. E. L. Loechler, C. L. Green, and J. M. Essigman, In vivo mutagenesis by O6-methylguanine built into a unique site in a viral genome. Proc. Natl. Acad. Sci. USA 81, 6271-6275 (1984).
6. B. D. Preston, B. Singer, and L. A. Loeb, Mutagenic potential of O4-methylthymine in vivo determined by an enzymatic approach to site-specific mutagenesis. Proc. Natl. Acad. Sci. USA 83, 8501-8505 (1986).
7. P. Karran, T. Lindahl, I. Ofsteng, G. B. Evensen, and E. Seeberg, Escherichia coli mutants deficient in 3-methyladenine-DNA glycosylase. J. Mol. Biol. 140, 101-127 (1980).
8. P. Karran, T. Hjelmgren, and T. Lindahl, Induction of a DNA glycosylase for N-methylated purines is part of the adaptive response to alkylating agents. Nature (London) 296, 770-773 (1982).
9. G. Evensen and E. Seeberg, Adaptation to alkylation resistance involves the induction of a DNA glycosylase. Nature (London) 296, 773-775 (1982).
10. J. Shackleton, W. Warren, and J. J. Roberts, The excision of N-methyl-N-nitrosourea-induced lesions from the DNA of Chinese hamster cells as measured by the loss of sites sensitive to an enzyme extract that excises 3-methylpurines but not O6-methylguanine. Eur. J. Biochem 97, 425-433 (1979).
11. W. J. Bodell, B. Singer, G. H. Thomas, and J. E. Cleaver, Evidence for removal at different rates of O-ethyl pyrimidines and ethylphosphotriesters in two human fibroblast cell lines. Nucleic Acids Res. 6, 2819-2829 (1979).
12. T. P. Brent, Partial purification and characterization of a human 3-methyladenine-DNA glycosylase. Biochemistry 8, 911-916 (1979).
13. B. Singer, S. Spengler, and W. J. Bodell, Tissue-dependent enzyme-mediated repair or removal of O-ethyl pyrimidines and ethyl purines in carcinogen-treated rats. Biochemistry 18, 911-916 (1981).
14. R. Cathcart and D. A. Goldthwait, Enzymatic excision of 3-methyladenosine and 7-methylguanine by a rat liver nuclear fraction. Biochemistry 19, 273-280 (1981).
15. B. Singer and T. P. Brent, Human lymphoblasts contain DNA glycosylase activity excising N-3 and N-7 methyl and ethyl purines but not O6-alkylguanines or 1-alkyladenines. Proc. Natl. Acad. Sci. USA 78, 856-860 (1981).
16. G. P. Margison, and A. E. Pegg, Enzymatic release of 7-methylguanine from methylated DNA by rodent liver extracts. Proc. Natl. Acad. Sci. USA 78, 861-865 (1981).
17. J. M. Bogden, A. Eastman, and E. Bresnik, A system in mouse liver for the repair of O6-methylguanine lesions in methylated DNA. Nucleic Acids Res. 9, 3089-3092 (1981).
18. J. R. Mehta, D. B. Ludlum, A. Renard, and W. G. Verly, Repair of O6-ethylguanine in DNA by a chromatin fraction from rat liver: transfer of the ethyl group to an acceptor protein. Proc. Natl. Acad. Sci. USA 78, 6766-6770 (1981).
19. A. E. Pegg, M. Roberfroid, C. vonBahr, R. S. Foote, S. Mitra, H. Bresil, A. Likkacher, and R. Montesano, Removal of O6-methylguanine from DNA by human liver fractions. Proc. Natl. Acad. Sci. USA 79, 5162-5165 (1982).
20. R. A. Becker and R. Montesano, Repair of O^4-methyldeoxythymidine residues in DNA by mammalian liver extracts. Carcinogenesis 6, 313-317 (1985).
21. L. D. Engelse, G. J. Menkveld, R. J. D. Brij, and A. D. Tates, Formation and stability of alkylated pyrimidines and purines (including amidazole ring-opened 7-alkylguanine) and alkyl-phosphotriesters in liver DNA of adult rats treated with ethylnitrosourea or dimethylnitrosamine. Carcinogenesis 7, 393-403 (1986).

22. M. E. Dolan, D. Sciccitano, B. Singer, and A. E. Pegg, Comparison of repair of methylated pyrimidines in poly(dT) by extracts from rat liver and Escherichia coli. Biochem. Biophys. Res. Commun. 123, 324-330 (1984).
23. M. E. Dolan and A. E. Pegg, Extent of formation of O4-methylthymidine in calf thymus DNA methylated by N-methyl-N-nitrosourea and lack of repair of this product by rat liver O6-alkylguanine-DNA-alkyltransferase. Carcinogenesis 6, 1611-1614 (1985).
24. N. Huh and M. F. Rajewsky, Enzymatic elimination of O6-ethylguanine and stability of O^4-ethylthymine in the DNA of malignant neural cell lines exposed to N-ethyl-N-nitrosourea in culture. Carcinogenesis 7, 435-439 (1986).
25. W. Warren, A. R. Crathorn, and K. V. Shooter, The stability of methylated purines and of methylphosphotriesters in the DNA of V79 cells after treatment with N-methyl-N-nitrosourea. Biochem. Biophys. Res. Commun. 563, 82-88 (1979).
26. R. Goth-Goldstein, Inability of Chinese hamster ovary cells to excise O6-alkylguanine. Cancer Res. 40, 2623-2624 (1980).
27. R. S. Day, III, C. H. J. Ziolkowski, D. A. Scuidero, S. A. Meyer, A. S. Lubiniecki, A. J. Girardi, S. M. Galloway, and G. D. Bynum, Defective repair of alkylated DNA by human tumour and SV40-transformed human cell strains. Nature (London) 288, 724-727 (1980).
28. R. S. Day, III, C. H. J. Ziolkowski, D. A. Scuidero, S. A. Meyer, and M. R. Mattern, Human tumor cell strains defective in the repair of alkylation damage. Carcinogenesis 1, 21-31 (1980).
29. R. Sklar and B. Strauss, Removal of O6-methylguanine from DNA of normal and xeroderma pigmentosum-derived lymphoblastoid lines. Nature (London) 289, 417-420 (1981).
30. L. Samson and S. Linn, DNA alkylation repair and the induction of cell death and sister chromatid exchange in human cells. Carcinogenesis 8, 227-230 (1987).
31. L. Samson and J. L. Schwartz, Evidence for an adaptive DNA repair pathway in CHO and human skin fibroblast cell lines. Nature (London) 287, 861-863 (1980).
32. I. Teo, B. Sedgewick, M. W. Kilpatrick, T. V. McCarthy, and T. Lindahl, The intracellular signal for induction of resistance to alkylating agents in E. coli. Cell 45, 315-324 (1986).
33. P. Robins and J. Cairns, Quantitation of the adaptive response to alkylating agents. Nature (London) 280, 74-76 (1979).
34. T. Lindahl, B. Demple, and P. Robins, Suicide inactivation of the E. coli O6-methylguanine-DNA methyltransferase. EMBO J. 1, 1359-1363 (1982).
35. B. Demple, A. Jacobsson, M. Olsson, P. Robins, and T. Lindahl, Repair of alkylated DNA in Escherichia coli. J. Biol. Chem. 257, 13776-13780 (1982).
36. H. Kataoka, Y. Yamamoto, and M. Sekiguchi, A new gene (alkB) of scherichia coli that controls sensitivity to methyl methane sulfonate. J. Bacteriol. 153, 1301-1307 (1983).
37. Y. Nakabeppu, H. Kondo, and M. Sekiguchi, Cloning and characterization of the alkA gene of Escherichia coli that encodes 3-methyladenine DNA glycosylase II. J. Biol. Chem. 259, 13723-13729 (1984).
38. R. M. Baker, W. C. VanVoorhis, and L. A. Spencer, HeLa cell variants that differ in sensitivity to monofunctional alkylating agents, with independence of cytotoxic and mutagenic responses. Proc. Natl. Acad. Sci. USA 76, 5249-5253 (1979).
39. D. B. Yarosh, A. J. Fornace, and R. S. Day, O6alkylguanine-DNA alkyltransferase fails to repair O4methylthymine and methylphosphotriesters in DNA as efficiently as does the alkyltransferase from Escherichia coli. Carcinogenesis 6, 949-953 (1985).

40. P. Karran, Possible depletion of a DNA repair enzyme in human lymphoma cells by subversive repair. Proc. Natl. Acad. Sci. USA 82, 5285-5289 (1985).
41. P. Karran and S. A. Williams, The cytotoxic and mutagenic effects of alkylating agents on human lymphoid cells are caused by different DNA lesions. Carcinogenesis 6, 789-792 (1985).
42. J. Domoradzki, A. E. Pegg, M. E. Dolan, V. M. Maher, and J. J. McCormick, Depletion of 06-methylguanine-DNA-methyltransferase in human fibroblasts increases the mutagenic response to N-methyl-N'-nitro-N-nitrosoguanidine. Carcinogenesis 6, 1823-1826 (1985).
43. M. E. Dolan, C. D. Corsico, and A. E. Pegg, Exposure of HeLa cells to 06-alkylguanines increases sensitivity to the cytotoxic effects of alkylating agents. Biochem. Biophys. Res. Commun. 132, 178-185 (1985).
44. K. Ishizaki, T. Tsujimura, H. Yawata, C. Fujio, Y. Nakabeppu, M. Sekiguchi, and M. Ikenaga, Transfer of the E. coli 06-methylguanine methyltransferase genes into repair deficient human cells and restoration of cellular resistance to N-methyl-N'-nitro-N-nitrosoguanidine. Mutat. Res. 166, 135-141 (1986).
45. H. Kataoka, J. Hall, and P. Karran, Complementation of sensitivity to alkylating agents in Escherichia coli and Chinese hamster ovary cells by expression a cloned bacterial DNA repair gene. EMBO J. 5, 3195-3200 (1986).
46. J. Brennand and G.P. Margison, Reduction of the toxicity and mutagenicity of alkylating agents in mammalian cells harboring the Escherichia coli alkyltransferase gene. Proc. Natl. Acad. Sci. USA 83, 6292-6296 (1986).

DRPASE, A NEW ACTIVITY IN THE DNA BASE EXCISION REPAIR PATHWAY

William A. Franklin

Imperial Cancer Research Fund, Clare Hall Laboratories
South Mimms, Hertfordshire, EN6 3LD, England

Agents that damage DNA are known to cause both mutagenesis and carcinogenic effects to cells. One specific type of DNA damage results in the loss of a purine or pyrimidine base, resulting in the formation of a base-free sugar or so-called apurinic/apyrimidinic (AP) site. Agents responsible for the induction of AP sites include gamma radiation (1), alkylating agents (2), and several other environmental and chemical carcinogens (3). AP sites can also form spontaneously in DNA (4), and they often result from the action of specific DNA glycosylases, such as uracil-DNA-glycosylase, which act to remove unusual or damaged DNA bases from DNA (5). The repair of AP sites can result in the formation of a mutation at the AP site, and such mutations may ultimately be responsible for the induction of the carcinogenic process.

The repair of DNA containing AP sites is thought to occur via a four step process: (1) the nicking of DNA at the AP site by an AP endonuclease; (2) the subsequent removal of the base-free sugar phosphate; (3) the filling-in of the gapped DNA strand by a DNA polymerase; and (4) the closure of the nicked DNA strand by a DNA ligase. Most of the steps of this pathway have been well characterized, particularly in E.coli. AP endonucleases have been isolated and purified from both E.coli and human cells (5,6). Two classes of AP endonuclease exist: those that cleave DNA at the 3' side of an AP site (class I enzyme), and those that cleave DNA at the 5' side (class II) (7). Following strand cleavage by an AP endonuclease, it is presumed that there is then a release of the remaining base-free sugar-phosphate residue, either as 2-deoxyribose-5'-phosphate or 2-deoxyribose-3'-phosphate. Once the sugar moiety is removed, then DNA polymerase can fill in the resulting gap, and DNA ligase can seal the nicked DNA resulting in the repair of the molecule.

The removal of the base-free sugar-phosphate residue following cleavage by an AP endonuclease has not been well characterized. In E.coli, more than 85% of the measurable AP endonuclease activity is due to a function of exonuclease III (8,9), a class II enzyme. Other AP endonucleases found in E.coli include endonuclease IV, a class II enzyme (7,10), and endonuclease III, a class I enzyme (7,11). Following cleavage of AP site-containing DNA by a class I enzyme, it has been demonstrated that the 3'->5' exonuclease activity of DNA polymerase I is able to remove the deoxyribose-3'-phosphate moiety (12,13). However, the enzyme was unable to remove the deoxyribose-phosphate residue following cleavage of DNA with the AP endonuclease (14). It has been proposed that DNA polymerase I can remove the sugar-phosphate residue following cleavage with a class II enzyme by virtue of its associated 5'->3' exonuclease activity. It has been demonstrated in vitro that this is not the case (13); instead, strand displacement occurs resulting in the formation of a branch structure originating at the AP site. It has also been suggested that removal of a sugar-phosphate can occur by the action of both class I and class II enzymes (13,14). This has been demonstrated in vitro using purified preparations of both types of AP endonucleases, yet no in vivo evidence suggests that this mechanism for deoxyribose-phosphate removal occurs in the cell. Indeed, the M.luteus and T4 UV-specific endonucleases and endonuclease III of E.coli are class I AP endonucleases that are associated with DNA glycosylase activities (11,14,15). These glycosylase activities (removal of pyrimidine cyclobutane dimers and oxidatively damaged bases) are probably the principal activities of these enzymes in vivo.

Is there an activity in E.coli that removes deoxyribose-5'-phosphate following cleavage by a class II (5') AP endonuclease? Such an activity may be predicted as the removal of an AP site from DNA seems to involve a short-patch repair process that involves the excision of one or a very few nucleotide residues (16-18). Below are described recent experiments from our laboratory suggesting that such an activity exists in E.coli and in human cells as well.

THE DRPASE ACTIVITY

In order to determine if an activity exists that removes deoxyribose-5'-phosphate residues from AP sites-containing DNA, a specific assay system was developed, as is shown in Figure 1. Either DNA or an alternating polynucleotide, poly (dA:dT), was nick-translated with [α ^{32}P]-dUTP and DNA polymerase I to allow incorporation of uracil residues in place of thymine in a small (<0.1%) percentage of the total bases. This uracil-containing substrate was then treated with a purified preparation of uracil-DNA-glycosylase, resulting in the removal of the uracil base from the DNA and the formation of AP site-containing DNA. Following the uracil-DNA-glycosylase treatment, the substrate was then treated with endonuclease IV, a class II (5') AP endonuclease that will cleave the DNA at AP sites. The resulting compound is then used as a substrate to measure the removal of the deoxyribose-5'-phosphate (dRp) residue. The amount of dRp released can be measured by determination of the amount of [^{32}P]-labelled dRp release following separation by either reverse-phase or anion exchange HPLC. A simpler assay can be employed where the reaction is stopped by the addition of trifluoroacetic acid and Norit activated-charcoal, resulting in the precipitation of the substrate, and the binding of any small oligonucleotides released to the Norit (19,20).

It was important to establish if a deoxyribose-phosphatase or DRPase activity existed in E.coli and whether the activity belonged to an enzyme that had been characterized previously. Cell extracts were prepared from several strains of E.coli that were deficient in one or more known exonuclease or endonuclease activities. The strains tested are shown in Table 1. All of the strains tested had DRPase activity, as determined by the release of radioactive material in the Norit assay and by direct determination by HPLC analysis of the amount of dRp released.

The analysis of the E.coli strains reveals some features of the DRPase activity. As the activity was present in strains deficient in endonuclease III (BW372 and BW415), this suggests that the presence of both class I and class II AP endonucleases is not required for the removal of dRp. The removal of dRp in a strain deficient in both DNA polymerase I and exonuclease III (AB3027) suggests that these two activities are not required for DRPase activity. Figure 2 shows the removal of dRp in this strain by HPLC analysis. Furthermore, purified preparations of both exonuclease III and DNA

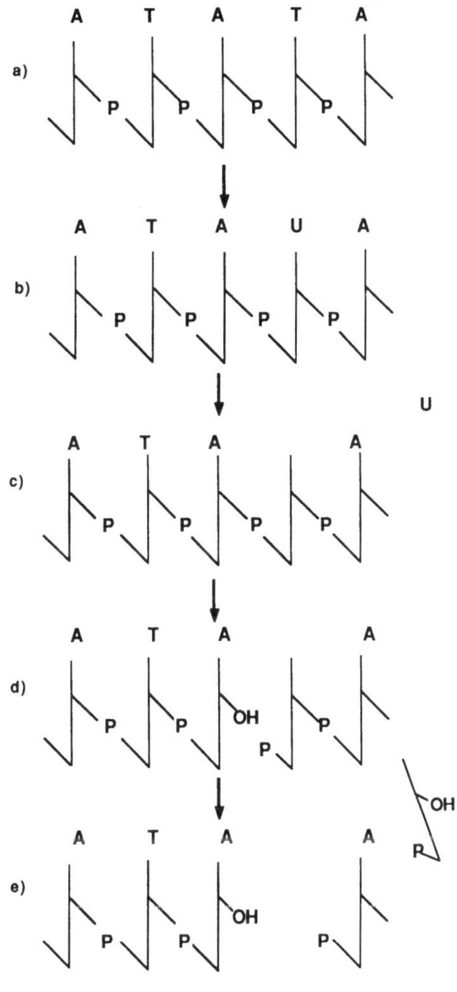

Fig. 1. Assay for the measurement of DRPase activity. a) DNA or poly(dA:dT) is used as the initial substrate. b) Nick-translation with DNA polymerase I and [α ^{32}P]-dUTP results in the incorporation of uracil residues in place of some of the thymine residues. c) Most of the uracil residues are removed by treatment with uracil-DNA-glycosylase. d) Treatment of the AP site-containing substrate with endonuclease IV results in the exposure of dRp residues. e) Release of dRp following treatment of the substrate with cell extracts or partially purified DRPase activity

Table 1. E.coli Strains Tested for DRPase Activity

Strain	Gene	Activity
AB1157	-	wild type
AB3027	xth polA	DNA polymerase I, exonuclease III
BW327	nth	endonuclease III
BW415	nth	endonuclease III
BW531	nth+	wild type
NH5033	xon	endonuclease I
KLC381	xse	exonuclease VII

polymerase I, as well as endonuclease IV, were found not to have an associated DRPase activity. This data suggests that DRPase is a new activity that is not associated with the DNA repair activities described in Table 1.

Purification of DRPase activity from E.coli is currently in progress. Strain AB3072 (xth polA) was chosen as the starting material as this strain is deficient in both DNA polymerase I and exonuclease III, two possible contaminants. Preliminary evidence suggests that the activity corresponds to a protein with a molecular weight of 30k daltons, based on Aca-54 and Superose 12 gel filtration chromatography. The activity has an absolute requirement for Mg^{++} cation.

DRPase activity has also been found to exist in a series of human cell lines, including those associated with a defect in normal DNA repair functions (21). Cell lines derived from patients with the diseases ataxia telangiectasia, Bloom's syndrome, and Fanconi's anaemia are all positive for

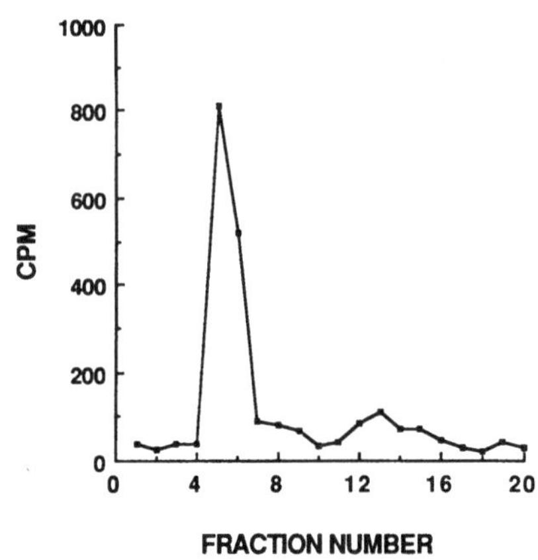

Fig. 2. Measurement of DRPase activity in E.coli strain AB3072. A sample of poly (dA:dT) nick-translated with [α ^{32}P]-dUTP and treated with uracil glycosylase and endonuclease IV containing 10,000 dpm was treated with 1μg of a AB3072 cell extract in 50 mM HEPES pH7.8, 10 mM $MgCl_2$, 5 mM dithiothreitol for 30 minutes at 37°. The entire reaction mixture (100μl) was then injected onto a Varian MCH-10 (reverse phase) HPLC column. The column was eluted with 50 mM KH_2PO_4 pH 4.5 at a flow rate of 1ml/min. Fractions were collected every 0.5 min. Deoxyribose-5-phosphate elutes just after the column void volume, in fractions 5 and 6.

DRPase activity, as are cell lines considered "normal" for DNA repair including HeLa cells and Raji, a human T-cell line.

Does the DRPase activity responsible for removal of dRp act on midified sugar groups? Agents that cause oxidative damage to DNA, such as gamma radiation and hydrogen peroxide, are known to produce sugar damage in the phosphotriester backbone (1, 22-24). Damage that occurs at 3' termini can be removed by activities associated with endonuclease IV and exonuclease III (22,25). It is not known which if any of the previously characterized DNA repair enzymes act at damaged sugar residues that could occur at 5' termini of DNA subject to oxidative damage. DRPase is a likely candidate for such an activity.

ACKNOWLEDGEMENTS

I wish to thank Tomas Lindahl for continued support and enthusiasm for this project. This material is based upon work supported by the North Atlantic Treaty Organization under a Grant awarded in 1985.

REFERENCES

1. J.F. Ward, Adv. Radiat. Biol. 5, 181-239 (1975).
2. P.D. Lawley and P. Brooks. Biochem. J. 89, 127-138 (1963).
3. P.C. Hanawalt, P.K. Cooper, A.K. Ganesan and C.A. Smith. Annu. Rev. Biochem. 48, 783-836 (1979).
4. T. Lindahl and B. Nyberg. Biochemistry 11, 3610-3618 (1972).
5. T. Lindahl. Progr. Nucleic Acid Res. Mol. Biol. 22, 135-192 (1979).
6. B. Demple and S. Linn. Nature (London) 257, 203-208 (1980).
7. S. Linn. In Nucleases (S.M. Linn and R.J. Roberts, Eds.), pp 291-309. Cold Spring Harbor Laboratory, Cold Spring Harbor, New York (1982).
8. D.M. Yajko and B. Weiss. Proc. Natl. Acad. Sci. U.S.A. 72, 688-692 (1975).
9. B. Weiss. J. Biol. Chem. 251, 1896-1901 (1976).
10. S. Ljungquist. J. Biol. Chem. 252, 2808-2814 (1977).
11. L.H. Breimer and T. Lindahl. J. Biol. Chem. 259, 5543-5548 (1984).
12. V. Bailly and W.G. Verly. FEBS Letts. 178, 223-224 (1984).
13. D.W. Mosbaugh and S. Linn. J. Biol. Chem. 257, 575-583 (1982).
14. L.K. Gordon and W.A. Haseltine. J. Biol.Chem. 256, 6608-6616 (1981).
15. L.K. Gordon and W.A. Haseltine. J. Biol.Chem. 255, 12047-12050 (1980).
16. J.D. Regan and R.B. Setlow. Cancer Res. 34, 3318-3325 (1974).
17. P. Karran, N.P. Higgins and B. Strauss. Biochemistry 16, 4483-4490 (1977).
18. H. Kataoka and M. Sekiguchi. J. Biochem. 92, 971-973 (1982).
19. R.H. Grafstrom, N.L. Shaper and L. Grossman. J. Biol. Chem. 257, 13459-13465 (1982).
20. P. Whitcome, K. Fry and W. Salser. Methods Enzymol. 29, 295-321 (1974).
21. E.C. Friedberg. DNA Repair, pp 505-574. W.H. Freeman, San Francisco, (1974).
22. W.D. Henner, S.M. Grunberg and W.A. Haseltine. J. Biol. Chem. 258, 15198-15205 (1983).
23. W.D. Henner, L.O. Rodriguez, S.M. Hecht and W.A. Haseltine. J. Biol. Chem. 258, 711-713 (1983).
24. M.F. Janicek, W.A. Haseltine and W.D. Henner. Nucleic Acids Res. 13, 9011-9029 (1985).
25. B. Demple, A. Johnson and D. Fung. Proc. Natl. Acad. Sci. U.S.A. 83, 7731-7735 (1986).

MECHANISM OF ACTION OF EXONUCLEASE III FROM ESCHERICHIA COLI

Yoke W. Kow, Hiroshi Ide and Susan S. Wallace

Department of Microbiology and Immunology
New York Medical College
Valhalla, New York 10595

INTRODUCTION

Reactive oxygen species such as hydrogen peroxide, superoxide radical, and hydroxyl radical are formed during aerobic respiration (1). These reactive oxygen species can react with vital cell components such as the cell membrane, protein, as well as DNA, potentially leading to mutagenesis and cell death (1). In Escherichia coli, repair endonucleases, such as exonuclease III (2,4) and endonuclease IV (3) appear to be involved in the repair of oxidative damages in DNA. Further, endonuclease III was shown to recognize oxidative base damages (4).

Apart from its function as a repair nuclease, exonuclease III has been shown to be involved in other biological responses such as heat shock (5) and mutagenesis (5-7). Xth mutants are unable to synthesize heat shock proteins upon abrupt temperature shift from 37°C to 50°C (5). Plumbagin-induced mutagenesis in aerobically grown E. coli requires the presence of the wild type gene product of xth (6). Further, sensitivity to killing by plumbagin is greatly reduced in xth mutants as compared to wild type (6). Similarly, mutagenesis in a sod A sod B mutant of E. coli grown under elevated oxygen tension, requires the active participation of wild type exonuclease III (7). Xth mutants also show hypersensitivity to hydrogen peroxide (8) and long wave UV irradiation (9).

Exonuclease III is a small monomeric protein having a molecular weight of about 28,000 daltons (10). Despite its small size, exonuclease III exhibits a number of enzymatic activities including a 3' → 5' exonuclease activity that releases deoxyribonucleoside 5'-monophosphates from the 3' end of a duplex DNA molecule. The exonuclease activity is processive at low temperatures, but becomes distributive at elevated temperatures (11). Furthermore, the enzyme works well only at blunt or recessive 3' ends, and has no activity towards a protruding 3' end with more than four bases (12). Further, damaged sugar residues such as glycolate residues are also substrates for exonuclease activity (13). Exonuclease III is a class II apurinic/apyrimidinic endonuclease that incises on the 5' side of an abasic site, leaving a 3' OH and a 5'-phosphate on the baseless sugar (10). It has a 3'-phosphatase activity that releases orthophosphate from the 3' end of a duplex DNA molecule. It is also a 3'→ 5' RNase H that releases ribonucleoside monophosphates from a DNA-RNA hybrid (10). Apparently, the

Table I. Substrate Specificity of Exonuclease III

Modified DNA bases recognized by exonuclease III*	Modified DNA bases not recognized by exonuclease III
Apurinic/apyrimidinic site	Thymine glycol
Urea	Thymine dimer
Reduced apurinic site	Formamidopyrimidine
O-methyl hydroxylamine	Dihydrothymine
O-ethyl hydroxylamine	β-ureidoisobutylic acid
O-allyl hydroxylamine	
O-carboxymethyl hydroxylamine	
O-benzyl hydroxylamine	
O-4-nitrobenzyl hydroxylamine	

*Substrate DNAs are arranged in descending order according to their rate of nicking by exonuclease III using the same substrate concentration.

enzyme binds to the DNA strand, and has the ability to discern the presence or absence of a 2' OH group (10).

Recently we have shown that urea residues, a degradation product of thymine glycol or thymine hydroperoxides, are recognized by exonuclease III, and that the enzyme makes an incision at the 5' side of the damage (2). According to the Weiss model for the mode of action of exonuclease III (10), the recognition of the damage is due to the existence of a 'space' left in the duplex DNA, created either by a missing base or by the presence of a small fragmented base such as a urea residue (2). In order to examine this proposed mechanism further, PM2 DNA containing different sizes of residues attached to C-1' of deoxyribose was prepared by chemical modification of the C-1' of an apurinic sugar residue using O-alkyl hydroxylamines (15). This modification converts the alkali-labile apurinic PM2 DNA to an alkali-stable form containing O-alkyl hydroxylamine residues. The reaction is quantitative, selective and rapid. O-Alkyl hydroxylamine will also react with cytosine and thymine, but the reaction requires high concentrations of reactant and the rate of reaction is slow. Under the conditions used for the modification of apurinic sites, no modification of cytosine or thymine was observed. Table I summarizes the substrate specificity of exonuclease III on these modified AP derivatives. It can be seen that the rate of incision of substrates containing O-alkyl hydroxylamine residues was affected by the size of the residue. As the size of the residue was increased, a slower rate of incision was observed. Further, the endonucleolytic activity of exonuclease III required M++. Other divalent cations such as Ca++ could only replace Mg++ when the substrate was apurinic DNA.

If 'space' is what is recognized by exonuclease III, then how can the enzyme recognize a DNA molecule containg an O-4-nitrobenzyl hydroxylamine residue, where apparently very little interstrand 'space' would be available between the duplex DNA molecule. Conversely, the enzyme does not recognize DNA containing β-ureidoisobutylic acid residue where 'more interstrand space' should be present due to the smaller fragment size of these residues. When the structures of all the damages introduced into PM2 were examined, it appeared that there were two factors that governed whether or not a particular damage was a substrate for exonuclease III. The first was the absence of a secondary amine at the N-glycosylic linkage. Second, when the N-glycosylic linkage was a secondary amine, the rate of incision was proportional to the size of the modified residue. Since the presence of an hydrogen at the N1 position of the damage site appeared to play a major role

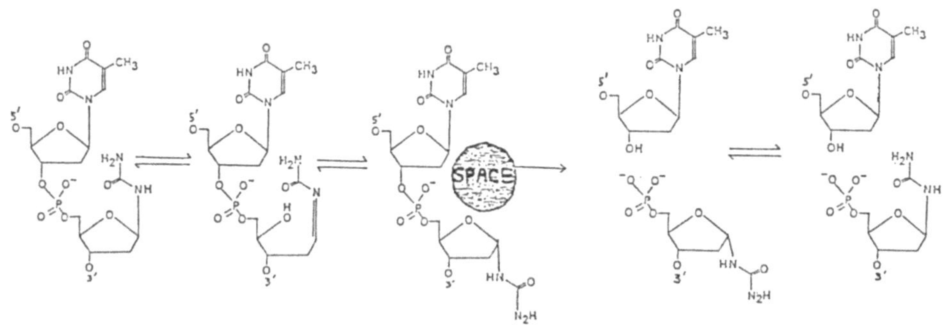

Fig. 1. Proposed mechanism of action of exonuclease III.

in determining whether the damage was going to be a substrate for exonuclease III, we proposed that the function of this secondary amine at the glycosylic linkage might be to facilitate the ring opening of the deoxyribose at the site of attachment of the damage. If an hydrogen is absent from the glycosylic linkage (such as for thymine glycol), ring opening of the deoxyribose would not be favored since this would involve the formation of an unstable imminium ion. Anomerization at C-1' would occur upon ring closure, so that a mixture of alpha and beta anomers of the damage would be formed. We further proposed that alpha anomers might provide the necessary 'space' for exonuclease III recognition. This model is presented in Figure 1.

With this extended 'space' model, one can attempt to explain some of the properties of exonuclease III. For example, exonuclease III only releases nucleoside 5'-monophosphates (10). Exonuclease III should not be able to release dinucleotides or trinucleotides since anomerization at the penultimate or tertiary base would be sterically hindered by the 3' base, which is always in the beta position. This steric hinderance would also explain why exonuclease III does not release a dinucleotide or oligonucleotide from the 3' end even if it is less than 4 protruding bases; why exonuclease III does not incise at multiple mismatches (14), even though there is appearence of a 'space'; and why exonuclease III does not nick at hairpins (10).

ACKNOWLEDGEMENT

This research was supported by grant GM 37216, DHHS.

REFERENCES

1. G. Cohen and R.A. Greenwald, Oxy Radicals and Their Scavenger Systems. Vol II. Elsevier Biomedical, New York, 1983.
2. Y.W. Kow and S.S. Wallace, Exonuclease III recognizes urea residues in oxidized DNA. Proc. Natl. Acad. Sci. (USA) 82:8354-8358 (1985).
3. R.P. Cunningham, S.M. Saporito, S.G. Spitzer and B. Weiss, Endonuclease IV (nfo) mutant of Escherichia coli. J. Bacteriol. 168:1120-1127 (1986).
4. H.L. Katcher and S.S. Wallace, Characterization of the Escherichia coli X-ray endonuclease, endonuclease III. Biochemistry 22:4071-4080 (1983).
5. K-H. Paek and G.C. Walker, Defect in expression of heat-shock proteins at high temperature in xthA mutants. J. Bacteriol. 164:1309-1316 (1986).

6. S.B. Farr, D.O. Natvig and R. Kagoma, Toxicity and mutagenesis of plumbagin and the induction of a possible new DNA repair pathway in Escherichia coli. J. Bacteriol. 165:763-770 (1985).
7. S.B. Farr, R. D'Ari and D. Touati, Oxygen-dependent mutagenesis in Escherichia coli lacking superoxide dismutase. Proc. Natl. Acad. Sci. (USA) 83:8268-8272 (1987).
8. B. Demple, H.L. Halbrook and S. Linn, Escherichia coli xth mutants are hypersensitive to hydrogen peroxide. J. Bacteriol. 153:1079-1082 (1983).
9. L.J. Sammartano and R. W. Tuveson, Escherichia coli xthA mutants are sensitive to inactivation by broad spectrum near-UV (300- to 400-nm) radiation. J. Bacteriol. 156:904-906 (1983).
10. B. Weiss. In: The Enzymes, (P.D. Boyer Ed.) Vol. XIV pp. 203-231. Academic Press, New York, 1981.
11. R. Wu, G. Ruben, B. Siegel, E. Jay, P. Spielman and C.D. Tu, Synchronous digestion of SV40 DNA by exonuclease III. Biochemistry 15:734-739, (1976).
12. S. Henikoff, Unidirectional digestion with exonuclease III creates targeted breakpoints for DNA sequencing. Gene 28:351-359 (1986).
13. B. Demple, A. Johnson and D. Fong, Exonuclease III and endonuclease IV remove 3' blocks from DNA synthesis primers in H_2O_2-damaged Escherichia coli. Proc. Natl. Acad. Sci. (USA) 83:7731-7735 (1986).
14. R. Roychoudbury and R. Wu, Novel properties of Escherichia coli exonuclease III. J. Biol. Chem. 252:4786-4789 (1977).
15. M. Liuzzi and M. Talpaert-Borle, A new approach to the study of the base-excision repair pathway using methoxyamine. J. Biol. Chem. 260:5252-5258 (1985).

THE ISOLATION AND PRELIMINARY CHARACTERIZATION OF ENDONOCLEASE VIII FROM ESCHERICHIA COLI

Robert J. Melamede, Yoke Wah Kow, and Susan S. Wallace

Department of Microbiology & Immunology, New York Medical College Valhalla, New York 10595

INTRODUCTION

Free radicals are highly reactive chemical species that are generated by a number of different mechanisms, including x-rays, certain chemical carcinogens and even aerobic metabolism. In order to survive the damages caused by free radical species, cells have evolved a number of protective mechanisms (5). Since DNA is an important target for free radicals, the ability to repair radical-induced DNA damages should constitute a significant cellular survival response. For a number of years our laboratory has been involved in identifying, quantitating, and assessing the biological impact of X-ray-induced DNA damages (26,27), many of which are produced by free radicals (2,10). Since the 5-6 double bond of thymine is particularly susceptible to free radical attack (5), enzymatic, chemical and immunological methods have been used to focus our efforts on radical-induced oxidation products of thymine (26,27). Endonuclease III from E. coli has been used as a tool to probe x-irradiated DNA for thymine ring saturation (15) and fragmentation (16) products. Furthermore, both thymine glycols and urea residues (a fragmentation product of thymine hydroperoxides) act as replicative blocks in vitro (4,9,12,25) and are lethal lesions in vivo (12,23).

The biological function of endonuclease III on damaged thymine residues was investigated by using E. coli mutants that lack this enzyme (6). Thymine glycols can be repaired by endonuclease III in vivo since ϕX-174 RF DNA containing thymine glycols was inactivated at a greater rate in E. coli mutants defective in endonuclease III than in wildtype hosts (18,19). Yet, endonuclease III deficient mutants of E. coli do not exhibit any increased sensitivity to ionizing radiation over wildtype (6). A possible explanation for this observation is the existence of alternative repair pathways. For example, E. coli may have another enzyme activity that can initiate the repair of the damages normally repaired by endonuclease III. Although we have not yet demonstrated its biological role, we now present preliminary data on the purification and characterization of such an enzyme activity that we have tentatively called endonuclease VIII.

MATERIALS and METHODS

Bacteria and Bacteriophage: The following strains were kindly provided by Dr. B. Weiss: BW9109 (AB1157 d(xth)), BW435 (BW9109 d(man-nth)), and BW434 (BW9109 nth::kan). PM2 bacteriophage and host Altermonas espejiana were grown as described (27).

Buffers: TE buffer is 10 mM Tris-HCl, pH 7.5 containing 1 mM EDTA. HE buffer is 10 mM HEPES pH 7.5 containing 1 mM EDTA.

Enzymes: Exonuclease III was purchased from Pharmacia. Endonuclease III and endonuclease VIII were purified by a new procedure using FPLC (Pharmacia, manuscript in preparation). Briefly, cells were mechanically disrupted using a Braun Homogenizer. The crude lysate was precipitated with 15% (final concentration) polyethylene glycol (PEG) 6000 to remove DNA. The supernatant was then loaded directly onto an iced FPLC Mono S column equilibrated with HE buffer by pumping through a Milton/Roy Constametric pump. The enzyme was eluted with a 0-0.5 M NaCl gradient over 50 minutes. At this stage endonuclease VIII was typically purified greater than 600 fold. For further purification, samples were applied to the Mono Q and Superose 12 columns by injection.

Substrates: PM2 DNA was prepared as previously described (15). The production of thymine glycols (1,3,7), urea residues (24), apurinic (20), methoxyamine-modified apurinic sites (21), and dihydrothymines (7,11) in DNA has been described. β-ureidoisobutylic acid-containing was prepared by treating dihydrothymine-containing DNA with alkali (manuscript in preparation).

Endonuclease VIII Assays: Most reactions were performed using the standard reaction conditions for endonuclease III. A 100 μl reaction contained 500 ng of DNA in TE buffer with 100 mM NaCl. Incubation, otherwise indicated, was for 5 minutes at 37°C. Enzyme activity was determined using a fluorometric assay as previously described (17).

RESULTS

When a PEG precipitated crude extract was applied to a Mono S a broad asymetric peak of endonuclease specific for thymine glycol-containing DNA was observed (Figure 1). While this work was in progress, a number of

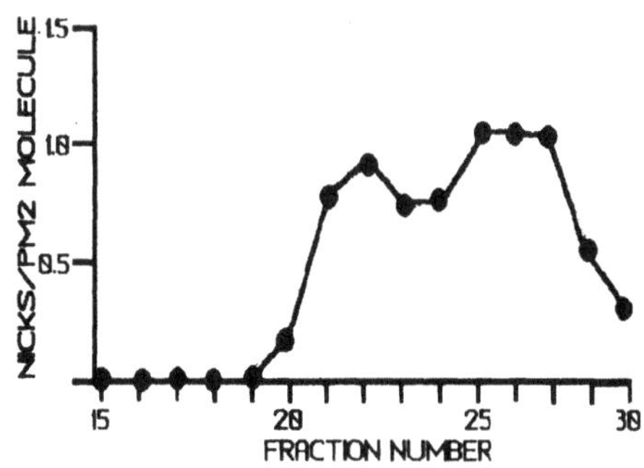

Figure 1. Mono S FPLC chromatography of a PEG precipitated crude extract of BW 9109 [AB 1157d(xth)]. Endonuclease activity was assayed under conditions previously described (16) using osmium tetroxide-treated supercoiled PM2 DNA (5). Nicking of the substrate was measured using the fluorometric assay of Kowalski (19).

mutants that lacked endonuclease III were kindly provided by Dr. B. Weiss. The use of these mutant strains facilitated the identification and purification of endonuclease VIII. The PEG-precipitated crude extracts of the mutant strains had about 5-10% of the nicking activity against thymine glycol-containing DNA as did a similar preparation from a strain that contained endonuclease III. The existence of two unique enzymes was clearly demonstrated when endonuclease III and endonuclease VIII were, individually and in combination, chromatographed on a Mono Q FPLC column (Figure 2). Endonuclease VIII eluted between 0.15 and 0.19 M NaCl whereas endonuclease III eluted between 0.19 and 0.23 M NaCl. However, the elution times of endonuclease III and endonuclease VIII from an FPLC Superose sizing column were the same, yielding a molecular weight of approximately 26,000 (data not shown).

In addition to having similar physical properties, endonuclease III and endonuclease VIII share many common substrates. Figure 3 shows the known structures recognized by endonuclease VIII. Endonuclease VIII incised DNA containing thymine glycols, urea, apurinic sites, methoxyamine-modified apurinic sites, dihydrothymine, and the alkali-cleaved dihydro product, β-ureidoisobutylic acid. Endonuclease VIII activity was not dependent on the presence of Mg^{++} with any of the substrates tested.

Endonuclease VIII was distinguishable from endonuclease III by its activity on DNA containing dihydrothymine or β-ureidoisobutylic acid residues. Both of these modified thymine residues were poor substrates for endonuclease III, in contrast, endonuclease VIII exhibited full activity on both substrates (data not shown).

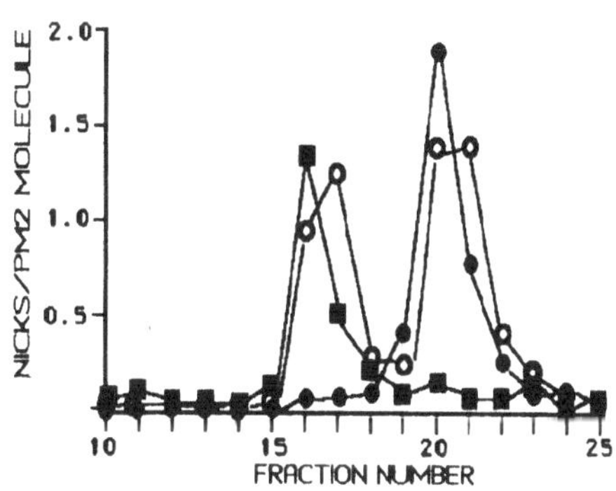

Figure 2. Mono Q FPLC chromatography of endonuclease III and endonuclease VIII. Endonuclease III was purified from BW 9109 [AB 1157 d(xth)] by Mono S FPLC chromatography and applied to a Mono Q column (●). Endonuclease VIII was purified from BW 435 [BW 9109 d(man-xth)] by Mono S FPLC chromatography and applied to a Mono Q column (■). A mixture of equivalent amounts of Mono S purified endonucleases III and VIII were also chromatographed by Mono Q FPLC (○). Endonuclease activity was assayed under conditions previously described (16) using osmium tetroxide-treated supercoiled PM2 DNA as the substrate (5). Nicking of substrate DNA was measured using the fluorometric assay of Kowalski (19).

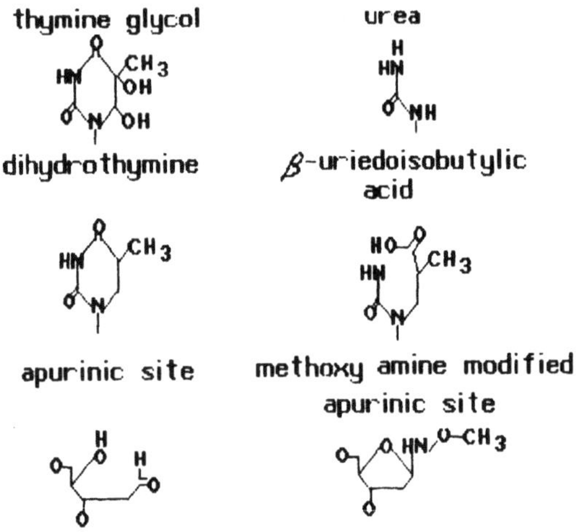

Figure 3. Modified DNA nucleosides recognized by endonuclease VIII.

CONCLUSIONS

The identification of a novel activity in E. coli, endonuclease VIII, that may partially substitute for endonuclease III in the initiation of excision repair of thymine ring saturation and fragmentation products, may explain the lack of an X-ray-sensitive phenotype for endonuclease III deficient mutants. However, this observation raises new questions. Why does E. coli code for two repair enzymes having such similar properties? This question assumes added relevance when one considers that there appears to be only about 1/20 the activity of endonuclease VIII as compared to endonuclease III. If endonuclease VIII has the same turnover rate as endonuclease III, there would only be about 10 molecules of endonuclease VIII per cell. Perhaps such low levels of endonuclease VIII are present because it is an inducible enzyme. Our preliminary data suggest however, that endonuclease VIII is not induced by activated Rec A or paraquat. The understanding of the biological role of endonuclease VIII may have to await the isolation of mutants lacking this enzyme, a project that we are currently pursuing.

ACKNOWLEDGEMENTS

This research was supported by grants CA 33657 and GM 37216, DHHS.

REFERENCES

1. O. Baudisch and D. Davidson, The mechanism of oxidation of thymine. 5,6-dihydroxydihydrothymine (thymine glycol). J. Biol. Chem. 64, 233-239 (1955).
2. J. Cadet and R. Teoule, Irradiation gamma de la thymine en milieu desaere: Isomers cis et trans de 1 hydroxy-6-dihydro-5,6 thymine. Int. J. Appl. Radiat. Isotopes 22, 273-280 (1971).
3. J. Cadet, J. Ulrich and R. Teoule, Isomerization and new specific synthesis of thymine glycol. Tetrahedron 31, 2057-2061 (1975).
4. J.M. Clark and G.P. Beardsley, Thymine glycol lesions terminate chain elongation by DNA polymerase I in vitro. Nucleic Acids Res. 14, 737-749 (1986).
5. G. Cohen and R.A. Greenwald, Oxy Radicals and Their Scavenger Systems. Vol II Elsevier Biomedical, New York, 1983.

6. R.P. Cunningham and B. Weiss, Endonuclease III (nth) mutants of Escherichia coli. Proc. Natl. Acad. Sci. (U.S.A.) 82 474-478 (1985).
7. P.V. Hariharan, Determination of thymine ring saturation products of the 5,6-dihydroxydihyrothymine type by alkaline degradation assay. Radiat. Res. 81, 496-498 (1980).
8. P.V. Hariharan, P.M. Achey and P.A. Cerutti, Biological effect of thymine ring saturation in coliphage ϕX-174. Radiat. Res. 69, 375-378 (1977).
9. R.C. Hayes and J.E. LeClerc, Sequence dependence for bypass of thymine glycols in DNA by DNA polymerase I. Nucleic Acids Res. 14, 1045-1061 (1986).
10. F. Hutchinson, Chemical changes induced in DNA by ionizing radiation. Prog. Nucleic Acid Res. 32, 115-154 (1985).
11. H. Ide, R.J. Melamede and S.S Wallace, Synthesis of dihydrothymidine and thymidine glycol triphosphates and their ability to serve as substrates for Escherichia coli DNA polymerases I. Biochemistry 26, 964-969 (1987).
12. H. Ide, Y.W. Kow and S.S. Wallace, Thymine glycols and urea residues in M13 DNA constitute replicative blocks in vitro. Nucleic Acids Res. 13, 8035-8052 (1985).
13. H. Ide, R.J. Melamede, Y.W. Kow. and S.S. Wallace, Incorporation of dihydrothymidine and its triphosphate during DNA replication: An implication for the biological consequence of thymine C5-C6 bond saturation. (These proceedings).
14. S. Iida, and H. Hayatsu, The permanganate oxidation of thymine. Biochim. Biophys. Acta 213, 1-13 (1970).
15. H.L. Kateher, and S.S. Wallace, Characterization of the Escherichia coli X-ray endonuclease, Endonuclease III. Biochemistry 22, 4071-4080 (1983).
16. Y.W. Kow, and S.S. Wallace, Exonuclease III recognizes urea residues in oxidized DNA. Proc. Natl. Acad. Sci. (U.S.A.) 82, 8354-8358 (1985).
17. D. Kowalski, A procedure for the quantitation of relaxed closed circular DNA in the presence of supercoiled DNA: An improved fluorometric assay for nicking-closing enzyme. Anal. Biochem. 93, 346-354 (1979).
18. M.F. Laspia and S.S. Wallace, Survival of X-174 transfecting DNA containing thymine glycols or urea residues in hosts deficient in endonuclease III or exonuclease III. Abstract. 34th Annual Meeting of the Radiat. Res. Society, Las Vegas, Nevada, (1986), p. 38.
19. M.F. Laspia and S.S. Wallace, Cellular repair of unique DNA base damages produced by ionizing radiation. Abstract. 35th Annual Meeting of the Radiat. Res. Soc., Atlanta, Georgia, (1987), p. 106.
20. T. Lindahl and A. Andersson, Rate of chain breakage at apurinic sites in double-stranded deoxyribonucleic acid. Biochemistry, 11, 3618-3623 (1972).
21. M. Liuzzi and M. Talparet-Borle, A new approach to the study of the base excision repair pathway using methoxyamine. J. Biol. Chem. 260, 5252-5258 (1985).
22. T. Maniatis, E.F. Fritsch and J. Sambrook, Molecular Cloning, A Laboratory Manual, pp. 109-112. Cold Spring Harbor Laboratory, New York, 1982.
23. E. Moran and S.S. Wallace, The role of specific DNA base damage in the X-Ray-induced inactivation of bacteriophage PM2. Mutat. Res. DNA Repair Reports 146, 229-240 (1985).
24. S. Murahashi, H. Yuki, K. Kosai and F. Doura, Methylenehydantoin and related compounds. I Reaction of pyruvic acid and urea. Synthesis of 5-methylenehydantoin. Bull. Chem. Soc. Japan 39, 1559-1562 (1966).
25. P. Rouet and J.M. Essigman, Possible role for thymine glycol in the selective inhibition of DNA synthesis on oxidized DNA templates. Cancer Res. 45, 6113-6118 (1985).

26. S.S. Wallace, Detection and repair of DNA base damage produced by ionizing radiation. Environ. Mutagen. 5, 769-788 (1983).
27. S.S. Wallace, The biological consequences of oxidized DNA bases. Br. J. Cancer, 55, Suppl. VIII, 118-128 (1987).
28. S.S. Wallace, H.L. Katcher and P.R. Armel, Measurement of X-ray damage in supercoiled DNA by means of enzyme probes. In: DNA Repair, A Laboratory Manual of Research Procedures (E. Friedberg, and P. Hanawalt Eds.) Vol 1A 113-125. Marcel Dekker, New York, 1981.

INCORPORATION OF DIHYDROTHYMIDINE AND ITS TRIPHOSPHATE DURING DNA REPLICATION: AN IMPLICATION FOR THE BIOLOGICAL CONSEQUENCE OF THYMINE C5-C6 BOND SATURATION

Hiroshi Ide, Robert J. Melamede, Yoke W. Kow and Susan S. Wallace

Department of Microbiology and Immunology
New York Medical College
Valhalla, New York 10595

INTRODUCTION

It has been shown by extensive product analysis that the saturation of the C5-C6 bond of thymine and the concomitant loss of planarity of the thymine ring are notable characteristics of the DNA lesions produced by ionizing radiation (1). Recently, we (2) and other laboratories (3-5) have found that thymine glycols present in a DNA template constitute replicative blocks to DNA synthesis <u>in vitro</u>. Interestingly a base, presumably adenine, is incorporated opposite thymine glycol; however, DNA synthesis is arrested at the site. In order to obtain a more complete picture of the biological consequence of thymine C5-C6 bond saturation, we have focused on dihydrothymine. Although dihydrothymine is produced by ionizing radiation preferentially under anoxic conditions (6,7), it shares common structural characteristics with thymine glycol: the thymine rings of both compounds are no longer planar due to the saturation of C5-C6 bond and assume a half chair conformation with C5 and C6 significantly out of the plane of the other four ring atoms (8). Since it is difficult to selectively introduce dihydrothymine into DNA by ionizing radiation or chemical reagents, we have used the following alternative approaches. In the first approach, DNA containing dihydrothymine was prepared by <u>in vitro</u> DNA synthesis catalyzed by DNA polymerase in the presence of dihydrothymidine 5'-triphosphate (DHdTTP). In the second approach, we developed an <u>in vivo</u> system in which exogeneously added dihydrothymidine was incorporated into DNA. Thus we could obtain information not only about the coding property of dihydrothymine and the effect of dihydrothymine on the local structure of DNA, but also about the initial interactions of the molecule with DNA polymerases.

PREPARATION OF MODIFIED NUCLEOSIDE TRIPHOSPHATES

DHdTTP was synthesized by catalytic hydrogenation of dTTP. For comparison, thymidine glycol 5'-triphosphate (dTTP-GLY) was also prepared by bromination of dTTP followed by treatment with silver oxide. The modified nucleotides were extensively purified by anion exchange HPLC. Details of the synthesis and HPLC purification have been described elsewhere (9). HPLC analysis of the alkaline phosphatase digested DHdTTP revealed that DHdTTP was a mixture of C5 diastereoisomers [(5S)- and (5R)-DHdTTP].

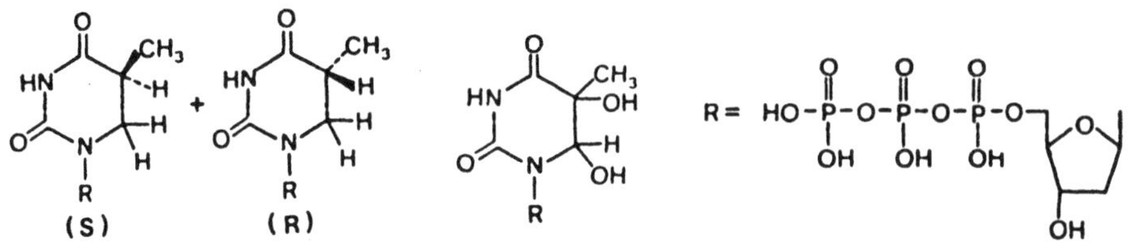

Figure 1. Structures of dihydrothymidine (DHdTTP) and thymidine glycol (dTTP-GLY) 5'-triphosphates.

ABILITY OF DHdTTP AND dTTP-GLY TO SERVE AS SUBSTRATES FOR DNA POLYMERASES

Despite the significant distortion of the thymine ring, DHdTTP was incorporated in place of dTTP during primer elongation on an M13mp11 DNA template in a reaction catalyzed by E. coli DNA polymerase I (Pol I) (9). The rate of incorporation of DHdTTP was about 13-fold lower than that of dTTP. DHdTTP replaced only dTTP among the four normal nucleotides. The ratio of the isomers incorporated into DNA (S:R = 73:27) was virtually the same as that of the DHdTTP substrate (S:R = 79:21). This result suggests that Pol I uses both isomers of DHdTTP as substrates and that the overall efficiency of incorporation is primarily determined by the concentration of the isomers in the nucleotide pool. In contrast to DHdTTP, dTTP-GLY did not replace dTTP and no elongation of the primer was observed. We also examined the ability of DHdTTP and dTTP-GLY to serve as substrates for T4 DNA polymerase, that contains an active proofreading (3'→ 5' exonuclease) activity, and for AMV reverse transcriptase, that lacks proofreading activity (10). Neither of these polymerases utilized DHdTTP or dTTP-GLY as substrates. Further, DHdTTP was converted to the corresponding monophosphate during Pol I-catalyzed idling-turnover but dTTP-GLY was not. With T4 DNA polymerase, no idling-turnover of DHdTTP or dTTP-GLY was observed. These results strongly suggest that the distortion of the thymine ring by saturation of the C5-C6 bond and the substituent group at this position differentially exert effects on binding of the triphosphates to DNA polymerases complexed with a primer terminus on DNA.

SEQUENCE DEPENDENCE OF DHdTTP INCORPORATION

The sequence specificity of DHdTTP incorporation catalyzed by Pol I was analyzed using M13mp11 DNA as a template (10). Most single adenine sites in the template were capable of templating DHdTTP, however, DNA synthesis was partially arrested at multiple (doublet and triplet) adenine sites. These data on the sequence specificity of DHdTTP incorporation indicate that, irrespective of the distortion in the thymine ring, the presence of dihydrothymine in the primer terminus site does not produce enough disorder to inhibit subsequent polymerization. However, multiple incorporation of DHdTTP residues in the newly synthesized strand might produce enough strain so as to partially inhibit polymerization at these sites. This is probably due to hindered stacking interactions between distorted thymine rings. When manganese (II), that is known to reduce the fidelity of DNA synthesis (11), was replaced for magnesium (II) as a cofactor, the doublet adenine site was bypassed almost completely, but the triplet site was not.

INCORPORATION OF DIHYDROTHYMIDINE INTO PHAGE PM2 DNA *IN VIVO*

In addition to the *in vitro* incorporation of DHdTTP into DNA, we also found that exogeneously added dihydrothymidine was incorporated into phage PM2 DNA *in vivo*. For the preparation of PM2 DNA containing dihydrothymine, host *Altermonas espejiana* was grown in Bal broth at 25°C until the optical density at 600 nm reached 0.8. Then PM2 phage was added (MOI~10). After 10 min of infection, dihydrothymidine was added to the culture medium at a concentration of 0.25-3 μg/ml every 3 min for 1 hour. Incubation was continued for 3 hours and then the lysate was refrigerated at 4°C overnight. PM2 phage was purified by reported procedures (12,13) with slight modification. Bacterial debris was removed by centrifuging the lysate for 15 min at 10,000 g. To the clear supernatant, polyethylene glycol (molecular weight 8000) and NaCl were added to the final concentration of 8% and 1 M, respectively. The solution was kept at 4°C with stirring for 1 hour. The precipitated phage was pelleted by centrifuging at 10,000 g for 15 min, and the phage pellet was resuspended in a solution containing 2 M NaCl, 10 mM Tris (pH 8.0), and 1 mM EDTA. Then PM2 phage was purified by 5 → 20% sucrose gradient sedimentation [the sucrose solution contained 2 M NaCl, 10 mM Tris (pH 8.0), and 1 mM EDTA]. Centrifugation was carried out with an SW 25.1 rotor (25,000 rpm) for 70 min. The light blue phage band was collected and PM2 DNA was extracted with a mixture of chloroform and 1-butanol (5:1) several times until the interphase became clear. Finally, the aqueous phase containing PM2 DNA was dialyzed three times against a solution containing 10 mM Tris•HCl (pH 7.5) and 1 mM EDTA.

The number of dihydrothymines in PM2 DNA was determined based on the conversion of form I to form II DNA using saturating amounts of *E. coli* endonuclease III (endo III) that has been reported to have dihydrothymine glycosylase activity as well as thymine glycol glycosylase activity (14). In a typical assay, PM2 DNA containing dihydrothymine (0.2 μg) was incubated with endo III in a solution (200 μl) containing 10 mM Tris•HCl (pH 7.5) and 1 mM EDTA at 37°C for 10 min. It should be noted that the relative rate of recognition of dihydrothymine in PM2 by endo III was significantly lower than that of thymine glycol, so that about 10-fold more endo III than was used for thymine glycol was required to achieve complete recognition of all dihydrothymine. The percentages of form I and II DNAs were determined using the fluorometric assay of Kowalski (15). The number of dihydrothymines in PM2 DNA increased with added amounts of dihydrothymidine and then leveled off (Table 1). The maximum amount of dihydrothymidine incorporated under the present conditions was 1.38 per PM2 DNA molecule, which corresponds to 0.018% of thymine in PM2. Interestingly, the percentage of form I DNA in the DNA preparation and the total yield of DNA decreased with increasing amounts of dihydrothymidine added. It is likely that a presumptive DNA repair enzyme from host (*Altermonas espejiana*) produced nicks at dihydrothymine sites in PM2 DNA so that percentage of form I and the yield of PM2 DNA were reduced simultaneously.

We also checked the incorporation of dihydrothymidine analogues, thymidine glycol and deoxydihydrouridine. However, these compounds were not incorporated into PM2 DNA. Recently, a novel activity that recognizes thymine glycol in DNA has been isolated from *E. coli* (tentatively called endo VIII, R.J. Melamede *et al.*, see these following proceedings.) The γ-endonuclease of *M. luteus* has also been shown to recognize thymine glycol (16,17). These enzymes were tested against PM2 DNA containing dihydrothymine, and both enzymes produced nicks converting form I PM2 DNA to form II. Further study with γ-endonuclease revealed that this enzyme contained an associated dihydrothymine glycosylase activity in addition to a thymine glycol glycosylase activity. The release of tritium-labeled

Table 1. Incorporation of dihydrothymidine into PM2 DNA

DHTd (μg/ml)[a]	DHT/PM2[b]	Form I (%)	DNA yield (%)
0	0	95	100
0.25	0.86	97	91
0.5	1.17	87	107
1.0	1.16	78	96
1.5	1.28	76	69
3.0	1.38	70	65

[a]Amount of dihydrothymidine added every 3 min.

[b]The number of dihydrothymines per PM2 DNA molecule obtained by endo III assay assuming random incorporation of dihydrothymidine (Poisson distribution). The average number (N) of dihydrothymine per PM2 DNA molecule is given by $N = -\ln[1.3/(0.3 + B/A)]$, where B and A are the fluorescence intensities before (B) and after (A) heat treatment (for detail see ref 15).

dihydrothymine from PM2 DNA, that was prepared by nick translation of PM2 DNA in the presence of tritium-labeled DHdTTP and three normal nucleotides, was confirmed by HPLC analysis.

CONCLUSIONS

Based on the present study it seems reasonable to predict that, in contrast to thymine glycol which has been shown to be a replicative block *in vitro* (2-5), dihydrothymine in the template strand should not constitute a replicative block unless it is a multiple dihydrothymine site. It appears that a dihydrothymine does not produce significant disorder in local DNA structure. In the case of thymine glycol, a nucleotide inserted following the thymine glycol site may not be stabilized through stacking and hydrogen bonding due to possible disorder in the template and primer terminus. As a result, the inserted nucleotide would be efficiently excised by 3' → 5' exonuclease activity associated with Pol I and T4 DNA polymerases. An alternative explanation is that the primer terminus containing thymine glycol in the template strand is not a good primer terminus so that DNA polymerases may dissociate from the primer terminus after inserting a nucleotide opposite thymine glycol. This could be clarified by studying idling-turnover of nucleotides catalyzed by DNA polymerases on a template containing thymine glycol. Such experiments have been done with DNA containing apurinic/apyrimidinc (AP) sites (18).

In addition, neither dihydrothymine nor thymine glycol should be mutagenic since they appear to retain the ability to pair with the proper base, adenine. Although it is unlikely that dihydrothymine is mutagenic or produces significant disorder in DNA, it is removed from DNA by at least three DNA repair enzymes (*E. coli* endonucleases III and VIII, *M. luteus* γ-endonuclease), suggesting that it might have a biological effect. These results could be rationalized by the fact that the N-glycosylic bond of dihydrothymidine is relatively easily hydrolyzed compared with thymidine, thereby producing an apyrimidinic site in DNA.

ACKNOWLEDGEMENTS

We are grateful to T.J. Jorgensen and W.D. Henner for the generous gift of M. luteus γ-endonuclease. This research was supported by grant CA 33675 awarded by National Cancer Institute and a grant from the U.S. Department of Energy.

REFERENCES

1. R. Teoule and J. Cadet, Radiation-induced degradation of the base component in DNA and related substances-final products. In: Effects of Ionizing Radiation on DNA (J. Hutterman, W. Kohnlein and R. Teoule Eds.) pp. 171-203. Springer-Verlag, New York, 1978.
2. H. Ide, Y.W. Kow and S.S. Wallace, Thymine glycols and urea residues in M13 DNA constitute replicative blocks in vitro. Nucleic Acids Res. 13:8035-8052 (1985).
3. R. Rouet and J.M. Essigmann, Possible role of thymine glycol in the selective inhibition of DNA synthesis on oxidized DNA templates. Cancer Res. 45:6113-6118 (1985).
4. R.C. Hayes and J.E. LeClerc, Sequence dependence for bypass of thymine glycols in DNA by DNA polymerase I. Nucleic Acids Res. 14:1045-1061 (1986).
5. J.M. Clark and G.P. Beardsley, Thymine glycol lesions terminate chain elongation by DNA polymerase I in vitro. Nucleic Acids Res. 14:737-749 (1986).
6. E.A. Furlong, T.J. Jorgensen and W.D. Henner, Production of dihydrothymidine stereoisomers in DNA by γ-irradiation. Biochemistry 25:4344-4349 (1986).
7. S. Nishimoto, H. Ide, K., Nakamichi and T. Kagiya, Radiation-induced reduction of thymidine in aqueous solution: Isolation and characterization of a novel dimeric product. J. Am. Chem. Soc. 105:6740-6741 (1983).
8. I.L. Karle, Crystal and molecular structure of photoproducts from nucleic acids. In: Photochemistry and Photobiology of Nucleic Acids (S.Y. Wang Ed.) Vol I pp. 483-519. Academic Press, New York, 1976.
9. H. Ide, R.J. Melamede and S.S. Wallace, Synthesis of dihydrothymidine and thymidine glycol 5'-triphosphates and their ability to serve as substrates for Escherichia coli DNA polymerase I. Biochemistry 26:964-969 (1987).
10. H. Ide, and S.S. Wallace, Dihydrothymine and thymidine glycol 5'-triphosphates as substrates for DNA polymerases: Differential recognition of thymine C5-C6 bond saturation and sequence specificity of incorporation. Submitted to Nucleic Acids Res.
11. R.A. Beckman, A.S. Mildvan, and L.A. Loeb, On the fidelity of DNA replication: Manganese mutagenesis in vitro. Biochemistry 24:5810-5817 (1985).
12. J.P. Richardson, Mechanism of ethidium bromide inhibition of RNA polymerase. J. Mol. Biol. 78:703-714 (1973).
13. S.S. Wallace, H.L. Katcher, and P.R. Armel, Measurement of x-ray damage in supercoiled DNA by means of enzyme probes. In: DNA Repair: A Laboratory Manual of Research Procedures (E. Friedberg and P. Hanawalt Eds.) Vol 1A pp. 113-125. Marcel Dekker, New York, 1981.
14. B. Demple and S. Linn, DNA-glycosylases and UV repair. Nature (London) 287:203-208 (1980).
15. D. Kowalski, A procedure for the quanitation of relaxed closed circular DNA in the presence of superhelical DNA: An improved fluorometric assay for nicking-closing enzyme. Anal. Biochem. 93:346-354 (1979).

16. P. Hentosh, W.D. Henner and R.J. Reynolds, Sequence specificity of DNA cleavage by Micrococcus luteus gamma-endonuclease. Radiat. Res. 102:119-129 (1985).
17. T.J. Jorgensen, Y.W. Kow, S.S. Wallace and W.D. Henner, Mechanism of action of Micrococcus luteus gamma-endonuclease. Biochemistry in press, 1987.
18. D. Sagher and D. Strauss, Insertion of nucleotides opposite apurinic/apyrimidinc site in deoxyribonucleic acid during in vitro synthesis: Uniqueness of adenine nucleotides. Biochemistry, 22 4518-4526 (1983).

OXIDATIVE DAMAGE:

DNA REPAIR AND INDUCIBLE RESISTANCE

Bruce Demple, Yasmin Daikh,
Jean Greenberg and Arlen Johnson

Department of Biochemistry and Molecular Biology
Harvard University
Cambridge, Massachusetts 02138

INTRODUCTION

Free radicals and activated oxygen species are generated in a variety of ways. For example, most of the biological effects of ionizing radiations can be ascribed to the oxygen radicals they produce, such as superoxide anion ($O_2^{\cdot-}$) and hydroxyl radical ($\cdot OH$) (1). Such radicals are also produced deliberately by macrophages during inflammatory responses, evidently as a means of destroying invading cells (2). It seems likely that the normal reduction of O_2 to water by cytochrome C oxidase might inadvertently release intracellular activated oxygen. Indeed, bacteria devoid of superoxide dismutase, which destroys $O_2^{\cdot-}$, display a strongly increased spontaneous mutation frequency in aerated but not in hypoxic culture (3). This apparent production of oxygen radicals under normal aerobic conditions underscores the need to understand the nature of the cellular defenses against oxidative damage. We are investigating the mechanisms that cells employ to avert oxidative damage and to correct such damage when it does occur.

DNA REPAIR OF RUPTURED DEOXYRIBOSE

A Physiological Role for Escherichia coli Exonuclease III

Free-radical attack on DNA produces an array of damages, including modified bases and the prominent strand breaks (1). Radiation-induced DNA single-strand breaks are rapidly repaired in most organisms (1), although they cannot be used directly by DNA polymerase or ligase (4). Such breaks instead probably require processing by specific nucleases.

An important clue to the enzymatic functions required for repair of free radical-induced DNA strand breaks came with the observation that E. coli mutants deficient in exonuclease III (xth mutants) are extraordinarily sensitive to the toxicity of H_2O_2 (5). We have now shown that upon exposure to H_2O_2, such mutants accumulate single-strand breaks (nicks) in their chromosomal DNA at a much greater rate than wild-type bacteria (6). DNA isolated from such peroxide-challenged xth mutants contains nicks whose 3' termini do not support E. coli DNA polymerase I. The priming activity of this damaged DNA can be restored by treatment with even low levels of purified exonuclease III, while such activation is not required for DNA isolated from peroxide-challenged wild-type bacteria (6). The number of DNA strand breaks does not increase even in samples where DNA synthesis is stimulated strongly. Purified DNA that is damaged in vitro by H_2O_2 (in the presence of trace amounts of $CuCl_2$), also has strand breaks that do not support DNA polymerase I activity, but which can be activated substantially by purified exonuclease III.

Exonuclease III allows DNA repair synthesis by removing blocking groups from 3' primer termini (Fig. 1). The 3' termini formed in DNA by ionizing radiation bear phosphomonoesters or esters of phosphoglycolic acid, a fragment of deoxyribose, and the same or similar damages may block 3' ends after attack by peroxide-generated radicals. Such blocked termini would then require an activity like exonuclease III for DNA repair synthesis (Fig. 1), and are probably lethal when they accumulate, as in xth mutants. We are currently developing chemical methods to identify the broken termini caused by oxidative attack both in vitro and in vivo.

Multiple Enzymes in E. coli That Remove 3' Blocks

We have developed synthetic substrates that contain radioactively labeled 3'-terminal phosphoglycolaldehyde (PGA) (6), and these can be used to monitor diesterase activities that deblock 3' termini. Exonuclease III accounts for about 97% of the diesterase that hydrolyzes PGA esters in E. coli. Of the xth-independent activity, about half persists in the absence of Mg^{++} and the presence of EDTA. An EDTA-resistant nuclease that removes 3'-terminal deoxyribose-5-phosphate was known in the form of the enzyme endonuclease IV (7). Purified preparations of endonuclease IV activated DNA synthesis on DNA from H_2O_2-challenged xth mutants, and released phosphate or PGA from the artificial polymers (6), but these activities could be due to other polypeptides present in the endonuclease IV preparation.

Genetic experiments revealed that the 3'-phosphatase and 3'-PGA diesterase activities are indeed functions of endonuclease IV. Extracts of strains with deletions of the endonuclease IV (nfo) gene (8) had no detectable EDTA-resistant 3'-phosphatase or 3'-PGA diesterase activity; a strain carrying a multicopy plasmid with the nfo gene overproduced the 3'-phosphatase and diesterase activities. Thus, as with exonuclease III, 3'-monoesterase and 3'-diesterase activities reside in the same polypeptide.

For radiation- and H_2O_2-mediated damages, endonuclease IV probably functions in much the same way as exonuclease III. Consistent with this hypothesis, mutation of the nfo gene increases the H_2O_2-sensitivity of xth mutants but not of wild-type cells; for ionizing radiation, neither of the single mutants are sensitized, but xth nfo double mutants are several-fold more sensitive (8).

About 2% of the 3'-PGA diesterase in E. coli still persists in xth nfo double mutants. This diesterase activity absolutely requires Mg^{++}. Gel filtration chromatography reveals two peaks of activity, corresponding to putative molecular weights of about 55,000 and 30,000, respectively. It may well be that the higher molecular-weight species is a dimeric form of the ca. 30,000-dalton protein. This third E. coli PGA diesterase can be distinguished from the known bacterial nucleases (9), however, by either genetic or biochemical criteria. The true substrate specificity of this third enzyme remains to be determined.

Eukaryotic 3'-PGA Diesterases

We have identified activities in both the yeast Saccharomyces cerevisiae and human cells that hydrolyze 3'-PGA esters. A yeast enzyme

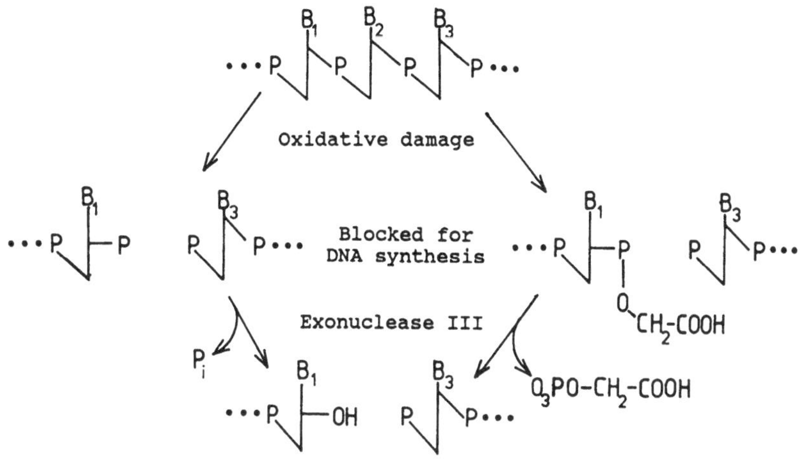

Fig. 1. Proposed role for exonuclease III in the repair of oxidative strand breaks in DNA.

has been partly purified that has a molecular weight of about 40,000 daltons (by gel filtration chromatography). The partly purified enzyme displays a requirement for divalent metals, although such a requirement is not apparent in crude cell extracts. Several lines of evidence tentatively indicate a second species of PGA diesterase present at lower levels. There is a substantial amount of diesterase activity in crude extracts of yeast, that does not vary appreciably with the stage of growth. The main activity also seems to be present at normal levels in radiation-senstive yeast mutants of the rad50 through rad56 series. Current efforts are directed at purification of the enzyme to allow its detailed enzymological characterization and molecular cloning of the structural gene.

A major DNA diesterase of HeLa cells attacks 3'-PGA termini. This protein has a native molecular weight of 40,000 to 45,000 daltons (gel filtration) and displays an absolute requirement for divalent metals that is satisfied by Mg^{++}. This activity does not appear to be a function of the major HeLa endonuclease for base-free (AP) sites (10): the purified AP endonuclease exhibited only a trace of activity on the 3'-PGA substrate. The diesterase activity can be readily detected in cultured human fibroblasts. This has allowed us to determine that cells from patients with the radiation-sensitive syndrome ataxia telangiectasia are not obviously deficient in this activity. Whether other known human genetic diseases are caused by a deficiency in this type of activity awaits additional experiments.

THE ADAPTIVE RESPONSE TO OXIDATIVE DAMAGE IN ESCHERICHIA COLI

Nontoxic levels of H_2O_2 induce in E. coli an impressive resistance to the toxicity of much higher peroxide levels (11). This increased resistance to H_2O_2 is accompanied by induced resistance to a variety of agents including gamma rays (11), organic peroxides, heat and (independent of glutathione) the sulfhydryl reagent N-ethyl maleimide (12). The array of inducible resistances is accompanied by the induction of at least 30 proteins in E. coli (13; J. Greenberg and B. Demple, unpublished data) and Salmonella (14). The induction of nine of the proteins, primarily those with scavenging activities such as catalase, is under the control of a gene called oxyR (13, 14).

We have been interested in the mechanisms of induced resistance to oxidative damage other than increased scavenging of active oxygen species. Several strains with deletions of the oxyR gene (ΔoxyR mutants) induce considerable H_2O_2 resistance (J. Greenberg and B. Demple, unpublished data). As a means of examining the oxyR-independent pathways of inducible resistance, we have mutagenized ΔoxyR mutants with nitrosoguanidine and selected isolates that grow in otherwise toxic levels of H_2O_2. The mutants fall into four basic categories as summarized below.

Class I Mutants: Regulation that Supercedes OxyR

The first class of peroxide-resistant mutants (suppressing the sensitivity of ΔoxyR strains) contain gene alterations that affect the regulation of several proteins. These mutants regain the resistance of wild-type (oxyR+) bacteria to H_2O_2, cumene hydroperoxide, menadione, paraquat, and $CdCl_2$. The mutants also regain the ability to induce catalase and alkyl hydroperoxide reductase in response to H_2O_2. Indeed, these bacteria have the general phenotype of $oxyR^+$ cells. The oxyR deletion is still present in these mutants, as shown by its transducibility from the mutants back into wild-type cells. Two of the mutations responsible for this unexpected phenotype map between 65 and 73 minutes on the E. coli genome, far away from oxyR at 90 minutes (14). The mechanism of this control that overcomes the lack of a functional oxyR gene is not yet known. It seems possible, for example, that another regulatory protein that responds to H_2O_2 might have altered its specificity so as to activate genes (e.g., katG) that are normally activated by the OxyR protein.

Class II Mutants: Intermediate Resistance

Class II mutants exhibit resistance to H_2O_2 between wild-type bacteria and ΔoxyR mutants, but some are exceptionally resistant to cumene hydroperoxide. The mutations responsible for these phenotypes have not yet been mapped. From the phenotype, some of these mutants may contain promoter-constitutive mutations in the ahp gene that encodes alkyl hydroperoxide reductase. Others are clearly more complicated, since they can induce catalase in response to H_2O_2, and might be regulatory mutants.

Class III Mutants: Catalase Overexpressers

Class III mutants exhibit resistance to H_2O_2 that is greater than $oxyR^+$ cells, while retaining the ΔoxyR sensitivity to $CdCl_2$ and cumene hydroperoxide. These mutants are more resistant to paraquat and menadione than are ΔoxyR mutants, but less resistant to these agents than are wild-type bacteria. Crude extracts of selected isolates of this family contain 35 times the normal level of catalase activity. These strains therefore most likely contain promoter-constitutive mutations in one of the kat genes, perhaps the normally oxyR-controlled katG.

Class IV: Other Regulatory Mutants?

Class IV mutants exhibit unusual growth patterns around a disc containing H_2O_2, consisting of multiple rings and haloes. These rings are found at both intermediate and ΔoxyR-like distances from the disc, with the actual zone of killing varying between different isolates. Class IV mutants exhibit moderately increased resistance to cumene hydroperoxide and menadione, but not to paraquat or $CdCl_2$. Some of these isolates can induce catalase in response to hydrogen peroxide and therefore are candidates for strains with mutated regulatory genes that have taken over the control of some genes from oxyR.

We are using two-dimensional gel electrophoresis to analyze in detail the proteins that are inducible in the mutants, and the most recent experiments support the conclusion that the Class I mutants are regulatory mutants. These mutants will give us new insight into the complex pathways of regulation that control the adaptive response to oxidative damage in E. coli, and ultimately will help reveal the multiple mechanisms that cells employ to counteract the toxicity of oxidative damage.

ACKNOWLEDGEMENT

We are grateful to other members of the laboratory for help and discussions. This work was supported by grants to Bruce Demple from the National Cancer Institute and from the National Institute of Environmental Health Sciences.

REFERENCES

1. F. Hutchinson, Chemical changes induced in DNA by ionizing radiation. Prog. Nucl. Acids Res. Mol. Biol. 32, 115-154 (1985).
2. P. I. Fields, R. V. Swanson, C. G. Haidaris and F. Heffron, Mutants of Salmonella typhimurium that cannot survive within the macrophage are avirulent. Proc. Natl. Acad. Sci. (USA) 83, 5189-5193 (1986).
3. S. B. Farr, R. D'Ari and D. Touati, Oxygen-dependent mutagenesis in Escherichia coli lacking superoxide dismutase. Proc. Natl. Acad. Sci. (USA) 83, 8268-8272 (1986).
4. T. Noguti and T. Kada, Studies on DNA repair in Bacillus subtilus: II. Partial purification and mode of action of an enzyme enhancing the priming activity of gamma-irradiated DNA. Biochem. Biophys. Acta 395, 294-305 (1975).
5. B. Demple, J. Halbrook and S. Linn, Escherichia coli xth mutants are hypersensitive to hydrogen peroxide. J. Bacteriol. 153, 1079-1083 (1983).
6. B. Demple, A. Johnson and D. Fung, Exonuclease III and endonuclease IV remove 3' blocks from DNA synthesis primers in H_2O_2-damaged Escherichia coli. Proc. Natl. Acad. Sci. (USA) 83, 7731-7735 (1986).
7. H. R. Warner, B. Demple, W. A. Deutsch, C. M. Kane, and S. Linn, Apurinic/apyrimidinic endonucleases in repair of pyrimidine dimers and other lesions in DNA. Proc. Natl. Acad. Sci. (USA) 77, pp. 4602-4606 (1980).
8. R. P. Cunningham, S. M. Saporito, S. G. Spitzer and B. Weiss, An Endonuclease IV (nfo) mutant of Escherichia coli. J. Bacteriol. 168, 1120-1127 (1986).
9. S. Linn. The deoxyribonucleases of Escherichia coli. In Nucleases (S. Linn and R. J. Roberts, Eds.) pp. 291-310. Cold Spring Harbor Laboratory, New York (1985).
10. C. M. Kane and S. Linn, Purification and characterization of a apurinic/apyrimidinic endonuclease from HeLa cells. J. Biol. Chem. 256, 3405-3414 (1981).
11. B. Demple and J. Halbrook, Inducible repair of oxidative DNA damage in Escherichia coli. Nature (London) 304, 466-468 (1983).
12. J. T. Greenberg and B. Demple, Glutathione in Escherichia coli is dispensable for resistance to H_2O_2 and ionizing radiation. J. Bacteriol. 168, 1026-1029 (1986).
13. R. A. VanBogelen, P. M. Kelley and, F. C. Neidhardt, Differential induction of heat shock, SOS, and oxidation stress regulons and accumulation of nucleotides in Escherichia coli. J. Bacteriol. 169, 26-32 (1987).
14. M. F. Christman, R. W. Morgan, F. S. Jacobson and B. N. Ames, Positive control of a regulon for defenses against oxidative stress and some heat shock proteins in Salmonella typhimurium. Cell 41, 753-762 (1985).

5-HYDROXYMETHYLURACIL IN CELLULAR DNA IS REPAIRED AND SENSITIZES CELLS TO INHIBITORS OF POLY(ADP-RIBOSE) SYNTHESIS

Robert J. Boorstein, Dan D. Levy, and George W. Teebor

Departments of Pathology and Environmental Medicine
New York University School of Medicine
New York, NY 10016.

INTRODUCTION

5-Hydroxymethyluracil (HmUra) can be formed in DNA from thymidine by the action of ionizing radiation or by activated leukocytes (1-7). HmUra can be removed from DNA *in vitro* through the action of HmUra-DNA glycosylase (8,9) The repairability of HmUra suggests that it is deleterious to cells but the nature of its effects on cell function are uncertain. HmUra can also be introduced into cellular DNA as a result of the incorporation of the nucleoside, 5-hydroxymethyl-2'-deoxyuridine (HmdUrd) (10-12). HmdUrd has been reported to be toxic to cells in culture and to animals (11,13,14) but the mechanism of this toxicity is unknown.

To study the effects of HmUra in DNA on cell growth, cells were exposed to increasing concentrations of HmdUrd to determine whether toxicity increased as a function of the number of HmUra residues in DNA. Since the HmUra residue could be radiolabelled, its disappearance from DNA could be followed as a measure of *in vivo* HmUra-DNA glycosylase mediated repair.

The toxicity of ionizing radiation also can be potentiated by 3-aminobenzamide (3AB), (15-19) an inhibitor of the synthesis of poly(ADP-ribose) (20). This potentiation of toxicity has been attributed to interference with DNA repair processes which require synthesis of poly(ADP-ribose) (21-24). The repairable DNA modification produced by ionizing radiation which sensitizes cells to 3AB has not yet been identified. Therefore, effects of 3AB on cells treated with HmdUrd were studied to determine whether the potentiation of radiation toxicity by 3AB might, at least in part, result from interference with the repair of radiogenically formed HmUra.

METHODS AND RESULTS

Incorporation and Toxicity of HmdUrd

V79 cells grown for 24 hours in medium containing [^{3}H]HmdUrd incorporated ^{3}H into TCA precipitable material over a wide range of concentrations (Table 1). This incorporation was nearly linear, falling off at high concentrations which produced up to 10-20% substitution of Thy in DNA by HmUra. Cell survival (Table 1) was not inhibited at levels of substitution of 1 HmUra/3600 Thy. Confirmatory experiments using HPLC demonstrated that the

Table 1. Incorporation and toxicity of HmdUrd

HmdUrd concentration (uM)	HmdUrd incorporated pmoles/10^6 cells	Substitution HmdUrd/dThy	Survival %
0.1	1.6	1/3600	101
1	39	1/150	4.2
10	470	1/12	0.1
100	1070	1/5.3	0.0

HmdUrd incorporation into V79 cells during a 24 hour incubation as a function of initial HmdUrd concentration was determined as described elsewhere (Boorstein et al, in press), and the degree of substitution for thymidine was determined. Survival in a colony formation assay was determined in dishes treated in parallel with non-radioactive HmdUrd.

^{3}H containing TCA precipitable incorporated into DNA was exclusively [^{3}H]HmdUrd (data not shown).

Removal of HmdUrd from DNA

To measure the overall removal rate of HmUra from DNA, V79 cells were treated with non-toxic concentrations of [^{3}H]HmdUrd. To distinguish repair from alternate causes for the loss of ^{3}H-containing material from the DNA of HmdUrd treated cells, the cellular DNA was prelabelled with [^{14}C]dThd. During the two days following the removal of HmdUrd from the growth medium, 30-40% of the [^{3}H]HmUra was removed per day (Figure 1) relative to [^{14}C]dThd.

Effects of 3AB on the Toxicity and Repair of HmdUrd

3-AB substantially enhanced the lethality of HmdUrd in V79 cells as assayed by colony formation (Figure 2). Doses of HmdUrd which were minimally toxic (0.1-0.4 uM) were made up to 95% toxic by subsequent treatment of the cells with 4 mM 3AB. By itself 4 mM 3AB was non-toxic and is known to have relatively little effect on other cellular functions (25, 26). A

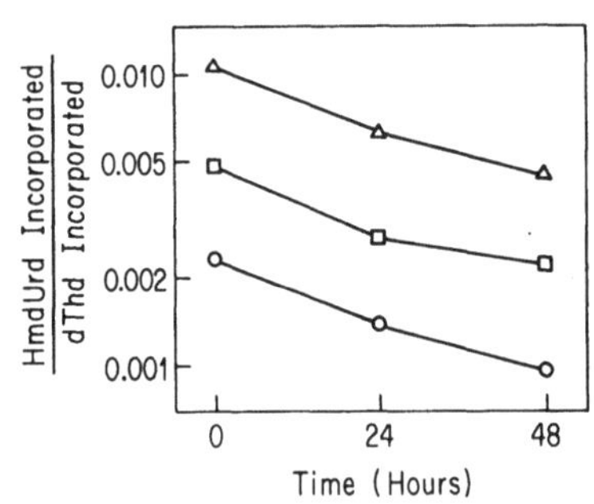

Fig. 1. Loss of ^{3}H radioactivity from [^{3}H]HmdUrd treated V79 cells. V79 cells prelabelled with [^{14}C]dThd were treated with 0.05 uM HmdUrd (O), 0.1 uM HmdUrd (□), or 0.2 uM HmdUrd (Δ), for 24 hours. The cells were washed and reicubated in complete medium. At 0, 24, and 48 hours following the removal of HmdUrd, the ratios of [^{3}H]HmdUrd incorporated (pmoles) to [^{14}C]dThd incorporated (pmoles) were determined. Data presented is an average of duplicate determinations whose range varied less than 5%.

lower concentration of 3AB (1 mM) also increased cell killing of HmdUrd treated cells.

To determine whether 3AB affected the initial step of removal of HmUra from DNA, cells were treated with [^{3}H]HmdUrd following prelabelling with [^{14}C]dThd. Repair was allowed to proceed in the presence or absence of 4mM 3AB. At the three concentrations of HmdUrd employed, no inhibition of removal of [^{3}H]HmUra from DNA was seen (data not shown). This finding is similar to that of Cleaver (27) who reported that 3AB did not inhibit the initial step of repair of alkylated bases.

DISCUSSION

HmdUrd is toxic to cells in culture and to mice (13,14). We have shown here that substitution in DNA at the level of 1 HmUra/150 Thy was cytotoxic. However, at levels of substitution of 1 HmUra/3600 Thy, HmUra was not a substantial block to replication . The growth rate (data not shown), colony forming ability and cell cycle distribution (data not shown) of V79 cells treated with HmdUrd to produce this level of substitution were unchanged when compared to untreated control cells.

The removal of 30-40% per day of [^{3}H]HmUra from DNA was measured in cells grown in [^{3}H]HmdUrd after prelabelling with [^{14}C]dThd. The removal of HmUra from cellular DNA is presumably mediated through the action of HmUra-DNA glycosylase (8,9). The result of our experiments provide the first evidence that HmUra is actively removed from cellular DNA. The similarity of the rates of *in vivo* repair in cells whose DNA was substituted with from 1-4 HmUra residues/10^{4} Thy residues (data not shown) indicates that repair was not saturated at these levels of substitution.

The results of our experiments are consistent with the hypothesis that it is the enzymatic removal of HmUra from DNA, not the presence of HmUra in DNA, which is primarily responsible for the toxicity of HmdUrd. Cells

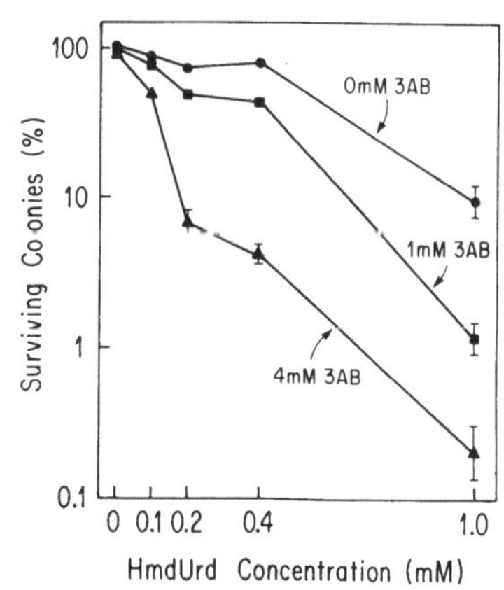

Fig. 2. Cytotoxicity of 3AB following HmdUrd pretreatment. V79 cells were treated for 24 hours with indicated concentrations of HmdUrd and replated in the presence of 0 mM 3AB (●), 1 mM 3AB (■), or 4 mM 3AB (▲) to determine colony forming ability. After 24 hours the cells were washed with complete medium and grown for an additional nine days to determine survival. Each point represents the average plating efficiency (+ or - the standard error of the mean) of between 6 and 9 dishes.

tolerated high levels of substitution, up to 1 HmUra/1000 Thy, and still replicated normally, removing HmUra from their DNA at a slow rate. However, toxic effects were evident at higher levels of substitution of HmUra in DNA. This might be due to the excision of HmUra residues at close proximity on opposite DNA strands, followed by apurinic/apyrimidinic (AP) endonucleolytic incision and exonucleolytic digestion producing lesions that are essentially lethal double strand breaks. A similar mechanism of toxicity has been proposed for radiogenically formed single strand breaks that are in close proximity on complementary DNA strands (28). An additional contribution to HmdUrd toxicity may have resulted from the cyclic incorporation of HmdUrd into DNA repair patches followed by removal of HmUra by the DNA glycosylase and further excision repair. Such a mechanism is similar to that proposed for the action of methotrexate in which uracil residues in DNA are continuously being excised through the action of uracil-DNA glycosylase (29).

When HmUra was produced in the DNA of Hela cells by ionizing radiation, the level of substitution was 1.2 HmUra/10^9 Thy/0.01 Gy (7). V79 cells tolerated amounts of HmUra in their DNA equal to that formed by 5000 gy. Therefore, HmUra in DNA cannot, by itself, contribute significantly to the toxic effects of ionizing radiation, since such radiation kills cell sat doses 1000 times lower.

Having established that cells tolerated high levels of HmUra substitution in their DNA without significant toxicity and that HmUra was repaired *in vivo*, we demonstrated that 3AB was toxic to cells containing HmUra in their DNA. These experiments tested the hypothesis that the lethality of 3AB results from interference with the completion of excision repair of DNA base adducts which are themselves non-lethal (26). 3AB inhibits the synthesis of poly(ADP-ribose) which may be necessary for normal ligation of DNA to occur during excision repair (30, 31). HmdUrd treated cells are a good model system in which to study the function of poly(ADP-ribose) in base excision repair because high levels of one type of repairable modified base may be introduced into cellular DNA without significant toxicity. In addition, in contrast to alkylated purines which readily undergo ring opening or depurination, HmdUrd is relatively stable (32).

Cells treated with HmdUrd were sensitized to 3AB to a greater extent than has been reported for cells exposed to alkylating agents or ionizing radiation (15, 18, 26, 31, 35-37). This is probably because much higher levels of repairable DNA lesions (HmUra residues) were achieved without the significant toxicity which accompanies treatment of cells by those other agents. Current models explaining 3AB action propose that poly(ADP-ribose) is synthesized in response to strand breakage mediated by repair endonucleases (26, 33, 34). The fact that HmUra was removed at a normal rate in the presence of 3AB supports this hypothesis. If HmUra-DNA glycosylase activity were inhibited by 3AB and the base remained on the DNA backbone, AP endonuclease action could not take place and the stimulus to poly(ADP-ribose) synthesis would not occur.

Ionizing radiation has also been shown to lower cellular NAD concentration, stimulate poly(ADP-ribose) synthesis and sensitize cells to 3AB (18, 35-37). However, in general, the cytotoxic effects of 3AB on irradiated cells have been found to be modest (15, 31). Most investigators have been forced to use highly toxic doses of radiation producing 90-99.9% lethality and very high concentrations of 3AB to detect significant effects. (18, 35). We suggest that interference with repair of radiogenically formed HmUra may contribute to the potentiation of radiation toxicity caused by 3AB. The reason one needs to employ high doses of radiation to observe 3AB sensitization is that the yield from ionizing irradiation of repairable DNA base modifications is relatively small (38). Only when there are significant amounts of such damage present in DNA may the effects of 3AB be seen.

ACKNOWLEDGEMENTS

This work was supported by grants ES 03847 (g.w.t. and r.j.b.) and CA 16669 (g.w.t.) from the National Institutes of Health and by postdoctoral fellowship PF 2893 (r.j.b.) from the American Cancer Society.

REFERENCES

1. G. Scholes, The radiation chemistry of pyrimidines, purines and related substances. In Photochemistry and Photobiology of Nucleic Acids (S.Y. Wang, Ed.), Vol. I, pp. 521-577. Academic Press, New York, 1976.
2. J. G. Lewis and D. O. Adams, Induction of 5,6-ring-saturated thymine bases in NIH-3T3 cells by phorbol ester stimulated macrophages: Role of reactive oxygen intermediates. Cancer Res., 45, 1270-1275 (1985).
3. K. Frenkel, K. Chrzan, W. Troll, G. W. Teebor, and J. J. Steinberg, Radiation-like modification of bases in DNA exposed to tumor promoter-activated polymorphonuclear leukocytes. Cancer Res., 46, 5533-5540 (1986).
4. P. A. Cerutti, Base damage induced by ionizing radiation. In Photochemistry and Photobiology of Nucleic Acids (S.Y. Wang, Ed.), Vol. II, pp. 375-403. Academic Press, New York (1976).
5. H. L. Lewis, D. R. Muhleman, and J. F. Ward, Serologic assay of DNA base damage. I. 5-Hydroxymethyldeoxyuridine, a radiation product of thymidine. Radiat. Res., 75, 305-316 (1978).
6. G. W. Teebor, K. Frenkel, and M. S. Goldstein, Ionizing radiation and tritium transmutation both cause formation of 5-hydroxymethyl-2'-deoxyuridine in cellular DNA. Proc. Natl. Acad. Sci. USA, 81, 318-321 (1984).
7. K. Frenkel, A. Cummings, J. Solomon, J. Cadet, J. J. Steinberg, and G. W. Teebor, Quantitative determination of the 5-(Hydroxymethyl) uracil moiety in the DNA of gamma-irradiated cells. Biochemistry, 24, 4527-4533 (1985).
8. M. C. Hollstein, P. Brooks, S. Linn, and B. N. Ames, Hydroxymethyluracil DNA glycosylase in mammalian cells. Proc. Natl. Acad. Sci. USA, 81, 4003-4007 (1984).
9. R. J. Boorstein, D. D. Levy, and G. W. Teebor, 5-Hydroxymethyluracil-DNA glycosylase activity may be a differentiated function of mammalian cells. Mutat. Res., 183, 257-263 (1987).
10. E. Matthes, D. Barwolff, and P. Langen, Altered DNA-protein interactions induced by 5-hydroxymethyldeoxyuridine in Ehrlich ascites carcinoma cells. Stud. Biophys., 67, 115-116 (1978).
11. L. I. Kahilainen, D. E. Bergstrom, and J. A. Vilpo, 5-Hydroxymethyl-2'-deoxyuridine. Cytotoxicity and DNA incorporation studied by using a novel [2-^{14}C]derivative with normal and leukemic human hematopoietic cells. Acta Chem. Scand. B, 39, 477-483 (1985).
12. E. R. Kaufman, Biochemical analysis of toxic effects of 5 hydroxymethyl-2'-deoxyuridine in mammalian cells. Somat. Cell Molec. Genet., 12, 501-512 (1986).
13. S. Waschke, J. Reefschlager, D. Barwolff, and P. Langen, 5 Hydroxymethyl-2'-deoxyuridine, a normal DNA constituent in certain *Bacillus subtilis* phages, is cytostatic for mammalian cells. Nature, 225, 629-630 (1975).
14. J. B. Meldrum, V. S. Gupta, N. R. Lowes, and A. R. P. Paterson, Toxicologic and antitumor studies on 5-hydroxymethyldeoxyuridine. Toxicol. Appl. Pharmacol., 79, 423-435 (1985).
15. N. Nduka, C. J. Skidmore, and S. Shall, The enhancement of cytotoxicity of N-methyl-N-nitrosourea and of gamma-radiation by inhibitors of poly(ADP- ribose) polymerase. Eur. J. Biochem., 105, 525-530 (1980).
16. J. Lunec, A. M. George, M. Heges, W. A. Cramp, W. J. D. Whish, and B.

Hunt, Post-irradiation sensitization with the ADP-ribosyltransferase in hibitor 3-acetamidobenzamide. Br. J. Cancer, 49, Suppl. VI, 19-25 (1984).
17. I. Szumiel, D. Wlodek, K. J. Johnson, and S. Sundell-Bergman, ADP-ribosylation and post-irradiation recovery in two strains of L5178Y cells. Br. J. Cancer, 49, Suppl. VI, 33-38 (1984).
18. P. Thraves, K. L. Mossman, T. Brennan, and A. Dritschilo, Radiosensitization of human fibroblasts by 3-aminobenzamide: an inhibitor of poly(ADP-ribosylation). Radiat. Res., 104, 119-127 (1985).
19. A. M. Ueno, O. Tanaka, and H. Matsudaira, Inhibition of gamma-ray dose-rate effects by D_2O and inhibitors of poly(ADP-ribose) synthetase in cultured mammalian L5178Y cells. Radiat. Res., 98, 574-582 (1984).
20. M. Purnell and W. J. D. Whish, Novel inhibitors of poly(ADP-ribose) synthetase. Biochem. J., 185, 775-777 (1980).
21. T. Lindahl, Worthy of a detour. Nature, 298, 424-425 (1982).
22. T. Sugimura, and M. Miwa, Poly(ADP-ribose) and cancer research. Carcinogenesis, 4, 1503-1506 (1983).
23. J. Lunec, Introductory review: involvement of ADP-ribosylation in cellular recovery from some forms of DNA damage. Br. J. Cancer, 49, Suppl. VI, 13-18, (1984).
24. N. E. Berger, Poly(ADP-ribose) in the cellular response to DNA damage. Radiat. Res., 101, 4-15, (1985).
25. K. M. Milam and J. E. Cleaver, Inhibitors of poly(adenosine diphosphate-ribose) synthesis: Effect on other metabolic processes. Science, 233, 589-591, (1984).
26. R. J. Boorstein and A. B. Pardee, Factors modifying 3-aminobenzamide cytotoxicity in normal and repair-deficient human fibroblasts. J. Cell Physiol., 120, 335-344, (1984).
27. J. E. Cleaver, K. M. Milam, and W. F. Morgan, Do inhibitor studies demonstrate a role for poly(ADP-ribose) in DNA repair. Radiat. Res., 101, 16-28, (1985).
28. D. I. Dugle, C. J. Gillespie, and J. D. Chapman, DNA strand breaks, repair and survival in x-irradiated mammalian cells. Proc. Natl. Acad. Sci USA, 73, 809-812, (1976).
29. M. Goulian, B. Bleile, and B. Y. Tseng, Methotrexate-induced misincorporation of uracil into DNA. Proc. Natl. Acad. Sci. USA, *77*, 1956-1960, (1980).
30. D. Criessen and S. Shall, Regulation of DNA ligase activity by poly(ADP-ribose). Nature, 296, 271-272 (1982).
31. M. R. James and A. R. Lehmann, Role of Poly(adenosine diphosphate ribose) in deoxyribonucleic acid repair in human fibroblasts. Biochemistry, 21, 4007-4113, (1982).
32. N. K. Kochetkov and E. I. Budovskii, Organic Chemistry of Nucleic Acids. Plenum Press, London, 1972.
33. R. C. Benjamin and D. M. Gill, Poly(ADP-ribose) synthesis *in vitro* programmed by damaged DNA. J. Biol. Chem., 255, 10502-10508, (1980).
34. A. R. Lehmann and B. C. Broughton, Poly(ADP-ribosylation) reduces the steady-state level of breaks in DNA following treatment of human cells with alkylating agents. Carcinogenesis, 5, 117-119, (1984).
35. N. L. Oleinick and H. N. Evans, Poly(ADP-ribose) and the response of cells to ionizing radiation. Radiat. Res., 101, 29-46, (1985).
36. E. Ben-Hur, H. Utsumi, and M. M. Elkind, Inhibitors of poly (ADP-ribose) synthesis enhance radiation response by differentially affecting repair of potentially lethal damage. Br. J. Cancer, 49, Suppl. VI, 39-42, (1984).
37. D. M. Brown, J. W. Evans, and J. M. Brown, The influence of inhibitors of poly(ADP-ribose) polymerase on X-ray-induced potentially lethal damage repair. Br. J. Cancer, 49, Suppl. VI, 27-31, (1984).
38. R. B. Painter, Nonconservative Replication of Damaged DNA in Mammalian Cells. In Genetic Concepts and Neoplasia, pp. 593-599. Williams and Wilkins, Baltimore, 1970.

METABOLIC PROTECTION AGAINST RAPID CELL DEATH

INDUCED BY HIGH LEVEL DNA DAMAGE

Nathan A. Berger, Sosamma J. Berger and Donna C. Sudar

Hematology/Oncology Division
Departments of Medicine and Biochemistry
Ireland Cancer Center, University Hospitals of Cleveland
Case Western Reserve University School of Medicine
Cleveland, Ohio 44106

In response to DNA damage, the cell initiates a series of enzymatic reactions designed to repair its genetic structure and restore its normal function. In some cases, the metabolic consequences of these reactions can so drastically perturb cellular homeostasis that they lead to cell death. It is important to understand these biochemical processes in order to design rational approaches to alter their course.

Figure 1 provides an example of a biochemical pathway by which DNA damage leads to cell death. The pathway is initiated by any type of DNA damage that causes DNA strand breaks. These breaks can be produced as a direct effect of ionizing radiation, by the action of endonucleases and/or by the chemical or physical alteration of bases that activate glycosylases

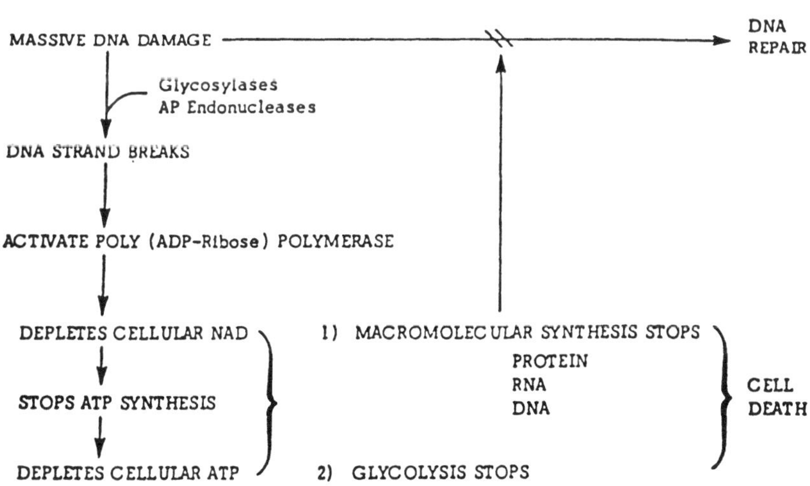

Fig. 1. Suicide Response of Poly(ADP-Ribose) Polymerase

Table 1. Effect of Treatment with 10μg/ml MNNG on NAD and ATP Levels (pmol/10^6 Cells) in Different Cell Lines

Time	L1210		E63 Myocytes	
(HR)	NAD	ATP	NAD	ATP
0	424	3,920	2,120	13,400
6	0.6	79	750	17,200
24	1.2	83	1,300	8,490

and endonucleases (1). DNA strand breaks activate the chromosomal enzyme, poly(ADP-ribose) polymerase which cleaves its substrate NAD with liberation of nicotinamide moieties and attachment of ADP-ribose moieties into long chains of poly(ADP-ribose) (2). The polymer is usually covalently attached to the enzyme itself but sometimes it is found attached to other chromosomal proteins (2). Poly(ADP-ribose) polymerase is activated in proportion to the level of DNA strand breaks (3) and high levels of DNA damage lead to such increases in enzyme activity that cellular pools of NAD are consumed (3). As a result of the depletion of NAD the cell loses its ability to utilize NAD dependent dehydrogenases. Since these enzymes are important at several steps in glycolysis and the tricarboxylic acid cycle, the utilization of glucose for energy production stops and cellular ATP pools are depleted (3,4).

Some consequences associated with the depletion of ATP include drastic alterations in nucleotide pools, macromolecular synthesis and cell size. For example, we have shown that cells treated with increasing doses of MNNG undergo a dose-dependent decrease in deoxynucleoside triphosphates. At 20 μM MNNG, dGTP pools are depleted within hours. Since dGTP is the limiting deoxynucleotide for DNA synthesis (5), its rapid depletion as an indirect metabolic consequence of DNA damage provides a metabolic mechanism that halts DNA synthesis. Treatment with MNNG also results in drastic reductions in the use of radioactive thymidine, uridine and leucine for the synthesis of DNA, RNA and protein (1). Cells undergoing the NAD-ATP depletion pathway also undergo a process of cell shrinkage and death which appears to be correlated with a lack of ability to maintain ion pumps and membrane integrity. Thus, DNA damage which causes DNA strand breaks can lead to a dose-dependent depletion of NAD and ATP, cessation of all energy dependent functions and consequent cell death.

Examination of the pathway described above provides insight as to why some cells or tissues escape the toxic effects of DNA damaging agents. Table 1 shows the NAD and ATP alterations that occur when L1210 cells and E63 myocytes are treated in culture with MNNG. The L1210 cells clearly undergo NAD and ATP depletion leading to cell death, whereas the myocytes recover their NAD and ATP levels and survive.

One possibility that may help protect the myocytes from the metabolic consequences of DNA damage is that they have a five-fold lower concentration of poly(ADP-ribose) polymerase. Thus they are not as active as L1210 cells in lowering their NAD. In addition, they start out with higher intrinsic levels of NAD and ATP and are therefore less likely to suffer the complete depletion reaction.

Table 2. Metabolite Preservation of Cell Viability at 24 Hrs. Following 5 µg/ml MNNG

Treatment	% Viable
MNNG, 5 µg/ml	16
+2mM Pyruvate	11
+2mM Succinate	42
+2mM Aspartate	24
+2mM Glutamate	10
+2mM Glutamine	14
+2mM Adenosine	12
+2mM Fructose	39

The mechanisms responsible for the higher NAD and lower poly(ADP--ribose) polymerase activity in myocytes compared to lymphoblasts are not clearly defined. The possibility certainly exists that the higher NAD levels are directly related to the lower activity of poly(ADP-ribose) polymerase. Nonetheless the differences between these cell types provide some suggestions for interfering with the DNA strand break-induced depletion of NAD. In some cases, treatment of cells with nicotinamide will result in an increased intracellular concentration of NAD (6). Nicotinamide also serves as a competitive inhibitor of poly(ADP-ribose) polymerase (6). Thus nicotinamide serves the dual function of raising NAD levels and interfering with the function of poly(ADP-ribose) polymerase (6). Our studies have shown that treatment of L1210 cells with 2mM nicotinamide blocks the rapid depletion of NAD and ATP that occurs consequent to DNA damage (3,6). By interfering with the poly(ADP-ribose) polymerase reaction, nicotinamide and other enzyme inhibitors also prevent or retard the DNA damage induced alterations in nucleotide metabolism, carbohydrate metabolism and macromolecular synthesis (3-6). By preserving ATP pools, nicotinamide treatment also stimulates an increase in DNA repair synthesis compared to that which occurs in cells subjected to damage alone (7).

Another approach to maintaining cell viability following high levels of DNA damage is to provide the cells with metabolites that may lead to a higher capacity for generation of ATP. Table 2 shows the results of a survey in which cells were pretreated with the indicated metabolites for 18 hrs, then exposed to 5µg/ml MNNG and then incubated in the continued presence of the metabolite. 24 hrs after treatment with MNNG alone only 16% of cells were viable as determined by exclusion of Trypan Blue. Higher levels of viability were obtained in cells treated with succinate or fructose. We subsequently evaluated the effect of these two agents along with nicotinamide for their ability to preserve ATP pools and viability in MNNG treated L1210 cells. Table 3 shows the effects of treating L1210 cells with 2 µg/ml MNNG in the presence or absence of the metabolites noted above. MNNG treatment produced a marked decrease in ATP levels. As anticipated, this decrease was partially blocked by the presence of 2mM nicotinamide. Neither succinate nor fructose were able to protect against this ATP depletion. However, the combination of 2mM each nicotinamide, succinate and fructose was able to maintain the ATP level for 24 hrs and retard the fall for as long as 48 hrs.

Table 4 shows that fructose, succinate and nicotinamide also help maintain the viability of L1210 cells treated with 2µg/ml MNNG. Within

Table 3. Metabolite Preservation of ATP Pools in L1210 Cells Treated with 2μg/ml MNNG

Treatment	Time After MNNG (Hrs)			
	0	6	24	48
	ATP pmol/10^6 Cells			
MNNG, (2μg/ml)	4,350	3,110	2,720	660
+2mM Nicotinamide		4,890	2,680	1,240
+2mM Succinate		2,270	2,340	875
+2mM Fructose		1,430	2,120	510
+2mM N + S + F		4,460	4,050	2,160

24 hrs of treatment, the cells exposed to MNNG alone were 58% viable whereas those treated with nicotinamide or with a combination of nicotinamide plus succinate plus fructose were 90% viable. After 96 hrs the cells treated with MNNG were 22% viable, whereas those treated with nicotinamide alone or in combination with succinate and fructose were 40 to 45% viable. Thus the combination of nicotinamide plus fructose plus succinate helps preserve the ATP pools and viability of cells treated with MNNG. The combination is more effective at preserving ATP than is nicotinamide alone, however the combination appears to be equally effective at preserving viability.

These studies illustrate that selective metabolites can protect against the rapid depletion of ATP, and loss of cell viability that occurs in response to some types of DNA damaging agents. Additional studies are required to determine whether this preservation of metabolite pools will have any effect on long term survival of the cells in culture. Further studies are required to determine whether partial preservation of cell viability will have any biologic importance at the level of the whole organism.

Table 4. Metabolite Preservation of Cell Viability Following Treatment with 2μg/ml MNNG

Treatment	Time After MNNG (Hrs)				
	0	6	24	48	96
	% Viable				
MNNG, (2μg/mL)	97	87	58	46	22
+2mM Nicotinamide	98	98	86	86	45
+2mM Succinate	97	93	74	72	31
+2mM Fructose	97	83	69	62	28
+2mM N + S + F	97	98	91	82	40

ACKNOWLEDGEMENT

These studies were supported by NIH Grants GM32647, CA35983 and CA43703.

REFERENCES

1. N. A. Berger, Poly(ADP-ribose) in the cellular response to DNA damage. Radiat. Res. 101, 4-15 (1985).
2. J. J. Cohen, D. M. Catino, S. J. Petzold, and N. A. Berger, Activation of poly (adenosine diphosphate ribose) polymerase by SV40 minichromosomes: Effects of DNA damage and histone H-1. Biochemistry 21, 4931-4940 (1982).
3. J. L. Sims, S. J. Berger, and N. A. Berger, Poly(ADP-ribose) polymerase inhibitors preserve nicotinamide adenine dinucleotide and adenosine 5'-triphosphate pools in DNA-damaged cells: Mechanism of stimulation of unscheduled DNA synthesis. Biochemistry 22, 5188-5194 (1983).
4. S. J. Berger, D. C. Sudar, and N. A. Berger, Metabolic consequences of DNA damage: DNA damage induces alterations in glucose metabolism by activation of poly(ADP-ribose) polymerase. Biochem. Biophys. Res. Commun. 134, 227-232 (1986).
5. S. K. Das and N. A. Berger, Alterations in Deoxynucleoside triphosphate metabolism in DNA damaged cells: Identification and consequences of poly(ADP-ribose) polymerase dependent and independent processes. Biochem. Biophys. Res. Commun. 137, 1153-1158 (1986).
6. J. L. Sims, G. W. Sikorski, D. M. Catino, S. J. Berger, and N. A. Berger, Poly(adenosinediphosphoribose) polymerase inhibitors stimulate unscheduled deoxyribonucleic acid synthesis in normal human lymphocytes. Biochemistry 21, 1813-1821 (1982).
7. N. A. Berger and G. W. Sikorski, Nicotinamide stimulates repair of DNA damage in human lymphocytes. Biochem. Biophys. Res. Commun. 94, 67-72 (1980).

TUMOR PROMOTION: A PROBLEM OF DIFFERENTIAL RESPONSES OF NORMAL AND NEOPLASTIC CELLS TO TROPHIC STIMULI

Stuart H. Yuspa[1], Henry Hennings[1], Takashi Sako[1], George R. Pettit[2], John Hartley[3] and Peter M. Blumberg[1]

1Laboratory of Cellular Carcinogenesis and Tumor Promotion National Cancer Institute, Bethesda, Maryland 20892
2Cancer Research Institute and Department of Chemistry Arizona State University, Tempe, Arizona 85287
3Department of Chemistry, University of Alberta Edmonton, Alberta, Canada T6G2G

Increased understanding of the tumor promotion phase of chemical carcinogenesis has resulted from the discovery of pure tumor promoting-agents and the development of animal models in which tumor promoters modify experimental carcinogenesis (1). Simultaneously, interest in this area of cancer research has been stimulated by epidemiological studies indicating that a promotion phase is important in the development of human cancer (2). Tumor promoters cause or allow the expression of the latent tumor phenotype induced in some cells by limited doses of carcinogens. Commonly, the model systems where tumor promotion plays an important role in neoplastic development are those involving epithelial tissues, in particular those composed of more than one cell type or cells in different states of maturation such as skin, breast and bronchus. As a classical model for carcinogenesis in a lining epithelium, mouse skin has provided most of the conceptual framework regarding the biology of tumor promotion (1,3).

Certain generalizations regarding tumor promotion have evolved from studies in a variety of model systems. Promoting agents must be repeatedly administered to produce their biological effects. In most experimental models, the pathological consequence of the promoting stimulus is the formation of preneoplastic or premalignant lesions. Often preneoplastic lesions occur in large numbers following exposure to the promoting stimulus, although few actually progress to malignancies. Many promoter-induced pathological changes actually regress upon withdrawal of the promoting stimulus, and prolonged administration of most tumor promoters does not enhance neoplastic progression to malignancy (4-6).

The reversibility of individual promoter exposures, the requirement for repeated frequent exposures to tumor promoters to accomplish promotion, and the regression of promoter-induced lesions when promotion is terminated all suggests that the major mechanisms involved in tumor promotion are epigenetic. In mouse skin, an important class of tumor promoters which have been useful for establishing the mechanisms of tumor promotion have been the phorbol esters. Pharmacologically, phorbol esters demonstrate a strict specificity of structure, pleiotrophic effects at nanomolar concentrations, and effectiveness after brief exposures (7). The discovery that these agents bind to and activate a cellular enzyme, protein kinase C, provided further evidence for the epigenetic basis of tumor promotion (8,9). Protein kinase C is implicated as a second messenger in the action of a large class of hormones and other exogenous stimuli which function through membrane receptors. The endogenous ligand for protein kinase C, diacylglycerol, competitively inhibits phorbol ester binding to the enzyme and reproduces many of the pharmacological effects of phorbol esters (9,10). Thus a common, epigenetic, primary action may exist for several classes of tumor promoters including certain hormones.

By analyzing the effects of phorbol esters on mouse skin and in cultures of mouse epidermal basal cells, we concluded that the underlying basis for tumor promotion by these agents involved cell selection (11). Support for this concept included the facts that tumors promoted by phorbol esters are monoclonal, and the effectiveness of the promoting stimulus is dependent on both the frequency and duration of exposure to these tumor promoters. Furthermore, the response pattern of epidermal cells to multiple exposures to phorbol esters is different from that seen with a single exposure, suggesting that a cell type with a particular response pattern was being selected. In cell cultures, apparently homogeneous basal keratinocytes respond heterogeneously to tumor-promoting phorbol esters (12). One subpopulation is induced to terminally differentiate while another is stimulated to proliferate. Distinct subclasses of phorbol ester receptors were found in both mouse skin and cultured epidermal cells (13,14), suggesting a pharmacological basis for the heterogeneous response pattern. In vivo, phorbol esters accelerate differentiation of a subclass of keratinocytes (15) and such cells are rapidly lost to the differentiating compartment, leading to a regenerative epidermal hyperplasia (16). The consequences of this disruption of normal epidermal homeostasis is the clonal expansion of cell populations which resist the induction of terminal differentiation and are stimulated to proliferate.

Recently it has been determined that initiated epidermal cells are uniformly resistant to the induction of differentiation by phorbol esters (17,18). These results are consistent with the initiated cell compartment becoming selectively enriched with each promoter exposure to ultimately yield a benign tumor, which represents a clonal expansion of initiated cells (11). Several predictions evolve from this scheme of tumor promotion. Agents that interfere with either the induction of epidermal differentiation or the stimulation of proliferation by phorbol esters could inhibit tumor promotion in skin. Retinoids, moderately potent inhibitors of tumor promotion by phorbol esters, are a class of compounds which inhibit the induction of epidermal terminal differentiation by phorbol esters (19). Corticosteroids are potent inhibitors of epidermal mitogenesis and block the stimulation of proliferation induced by phorbol esters, thus also acting as promotion inhibitors (20).

A novel natural product that provides a powerful test of the hypothesis that a divergent response pattern in mouse keratinocytes is essential for the promoting action of phorbol esters is bryostatin 1. This compound, a macrocyclic lactone isolated from the marine bryozoan Bugula neritina (21), was shown to activate protein kinase C and to induce some phorbol ester responses such as mitogenesis of Swiss 3T3 cells (22). In contrast, however, bryostatin 1 blocks other effects of phorbol esters, particularly the induction of differentiation of HL-60 cells (23) or the inhibition of differentiation of Friend erythroleukemia cells (24). The effects of bryostatin 1 were particularly interesting in cultured mouse keratinocytes (25). In this system, the compound induced ornithine decarboxylase activity, a marker of the proliferative response. Correspondingly, in mouse skin in vivo, bryostatin 1 induced a moderate hyperplasia and a moderate to marked inflammatory response (26). In cultured keratinocytes, however, bryostatin 1 did not induce epidermal transglutaminase or increase the number of cornified cells, suggesting that this activator of protein kinase C was incapable of inducing the differentiative responses characteristic of the phorbol esters in these cells (25). Furthermore, combined treatment with bryostatin 1 and phorbol esters blocked the induction of differentiation seen with phorbol esters alone. In vivo multiple combined exposures resulted in a hyperplastic response which was slightly greater than bryostatin alone.

The characteristics of the biological and biochemical responses of mouse epidermis in vivo and in vitro to bryostatin 1 are particularly useful for testing the cell selection hypothesis. Since the hypothesis requires that both cell loss and cell proliferation are essential for the promoting effects of phorbol esters, it was predicted that combined exposure to bryostatin 1 and a phorbol ester would inhibit tumor promotion. Furthermore, it was predicted that bryostatin 1 alone would be ineffective as a tumor promoter although it is a moderately potent stimulent of skin hyperplasia. After initiation with 7,12-dimethylbenz[a]-anthracene, mice were repeatedly exposed to bryostatin 1 or bryostatin 1 followed 5 minutes later by the potent tumor promoter 12-O-tetradecanoylphorbol-13-acetate (TPA)(26). Bryostatin 1 was ineffective as a complete tumor promoter. Furthermore, combined exposure of mice to both bryostatin 1 and TPA resulted in a substantial inhibition of promotion by TPA. These results are consistent with the requirement for a heterogeneous response pattern in tumor promotion and an epigenetic mechanism of action for the phorbol esters.

While phorbol esters have been extremely useful for formulating theories regarding the mechanism of tumor promotion, the generality of the cell selection hypothesis has not been tested widely with other promoting agents. Benzoyl peroxide is a particularly relevant tumor promoter since there is substantial exposure to this compound in the human environment. Benzoyl peroxide and several related peroxides were shown to promote papillomas in SENCAR mice (27,28). Benzoyl peroxide, although not a complete carcinogen, enhances malignant conversion of papillomas, a characteristic not exhibited by phorbol esters (29). The high concentrations required for benzoyl peroxide promoting activity belie a mechanism mediated by a specific cellular receptor. Furthermore, the enhancement of malignant conversion, previously only shown to be accomplished by mutagens, suggests that benzoyl peroxide could have direct genetic effects. In epidermal cell culture, benzoyl peroxide was shown to produce single-strand breaks in DNA and to effect substantial cell killing of normal cultured keratinocytes (30). Surprisingly, when initiated cells or cells derived from chemically induced papillomas or carcinomas were exposed to benzoyl peroxide in culture, they demonstrated a partial resistance to both DNA strand breakage and cytotoxicity (30, 31). This differential resistance to DNA strand breaks and cytotoxicity among normal and carcinogen-altered keratinocytes could form the biological basis for the promoting action of benzoyl peroxide. Enhanced survival of initiated cells would provide such cells with a growth advantage in a promoting environment where substantial cell death of normal keratinocytes was caused by exposure to benzoyl peroxide. Furthermore, sublethal DNA damage in preneoplastic or neoplastic keratinocytes could also account for the potency of benzoyl peroxide in causing malignant conversion.

Amplification of the initiated cell population, based on the differential responses to promoting agents among normal and preneoplastic cells, could provide a cellular basis for tumor promotion common to a number of tissues. Other laboratories have provided data comparing the effects of tumor promoters on carcinogen-altered (initiated), preneoplastic or benign, and malignant tumor cells compared to effects on normal cells of the same tissue. Studies of four tissues (skin, liver, colon, and bronchus) of several species and a variety of promoting agents indicate that during the process of carcinogenesis, preneoplastic or neoplastic cells become more resistant to both specific and nonspecific cytocidal or cytostatic effects of certain promoters (32). Additional examples have been provided to indicate that some altered cells become more sensitive to the growth-stimulatory effects of specific promoting agents than their normal cellular counterparts. Either response would produce the same biological effect, a selective growth of cells with a carcinogen-altered phenotype. The promoting environment would produce selective growth conditions favorable for the neoplastic cell phenotype. The clonal expansion of the initiated population would then result in a clinically evident premalignant lesion. Clonal expansion of premalignant cells would increase the probability that further changes relevant to neoplastic progression would occur in the population already at risk because of a preexisting genetic defect. This experimental model is consistent with the occurrence of cancer in human populations (2).

REFERENCES

1. T.J. Slaga, Mechanisms of Tumor Promotion, Vols. I, II, III, IV. CRC Press, Boca Raton, 1983..
2. S.H. Moolgavkar and A.G. Knudson, Mutation and cancer: A model for human carcinogenesis. J. Natl. Cancer Inst. 66, 1037-1052 (1981).
3. S.H. Yuspa, Cutaneous chemical carcinogenesis. J. Am. Acad. Dermatol. 15, 1031-1044 (1986).
4. H. Hennings, R. Shores, M.L. Wenk, E.F. Spangler, R. Tarone, and S.H. Yuspa, Malignant conversion of mouse skin tumours is increased by tumour initiators and unaffected by tumour promoters. Nature (London) 304, 67-69 (1983).
5. T.L. Goldsworthy and H.C. Pitot, The quantitative analysis and stability of histochemical markers of altered hepatic foci in rat liver following initiation by diethylnitrosamine administration and promotion with phenobarbital. Carcinogenesis 6, 1261-1269 (1985).
6. W.K. Kaufman, S.A. Mackenzie, and D.G. Kaufman, Quantitative relationship between hepatocytic neoplasms and islands of cellular alteration during hepatocarcinogenesis in the male F344 rat. Am. J. Pathol. 119, 171-174 (1985).
7. P.M. Blumberg, In vitro studies on the mode of action of the phorbol esters, potent tumor promoters: part I. CRC Crit. Rev. Toxicol. 8, 153-197 (1980).
8. K.L. Leach, M.L. James, and P.M. Blumberg, Characterization of a specific phorbol ester aporeceptor in mouse brain cytosol. Proc. Natl. Acad. Sci. USA 80, 4208-4212 (1983).
9. C.L. Ashendel, The phorbol ester receptor: a phospholipid-regulated protein kinase. Biochim. Biophys. Acta. 822, 219-242 (1985).
10. Y. Nishizuka, Studies and perspectives of protein kinase C. Science 233, 305-312 (1986).
11. S.H. Yuspa, H. Hennings, and U. Lichti, Initiator and promoter induced specific changes in epidermal function and biological potential. J. Cell Biochem. 17, 245-257 (1981).

12. S.H. Yuspa, T. Ben, H. Hennings, and U. Lichti, Divergent responses in epidermal basal cells exposed to the tumor promoter 12-O-tetradecanoyl phorbol-13-acetate. Cancer Res. 42, 2344-2349 (1982).
13. J.A. Dunn and P.B. Blumberg, Specific binding of [20-3H]12-deoxyphorbol 13-isobutyrate to phorbol ester receptor subclasses in mouse skin particulate preparations. Cancer Res. 43, 4632-4637 (1983).
14. J.A. Dunn, A.Y. Jeng, S.H.. Yuspa, and P.B. Blumberg, Heterogeneity of [3H]phorbol 12,13 dibutyrate binding in primary mouse keratinocytes at different stages of maturation. Cancer Res. 45, 5540-5546 (1985).
15. J.J. Reiners and T.J. Slaga, Effects of tumor promoters on the rate and commitment to terminal differentiation of subpopulations of murine keratinocytes. Cell 32, 247-255 (1983).
16. T.S. Argyris, Regeneration and the mechanism of epidermal tumor promotion. CRC Crit. Rev. Toxicol. 14, 211-258 (1985).
17. S.H. Yuspa, D. Morgan, U. Lichti, E.F. Spangler, D. Michael, A. Kilkenny, and H. Hennings, Cultivation and characterization of cells derived from mouse skin papillomas induced by an initiation promotion protocol. Carcinogenesis 7, 949-958 (1986).
18. H. Hennings, D. Michael, U. Lichti, and S.H. Yuspa, Response of carcinogen-altered mouse epidermal cells to phorbol ester tumor promoters and calcium. J. Invest. Dermatol. 88, 60-65 (1987).
19. S.H. Yuspa and U. Lichti, Retinoids and skin carcinogenesis: a mechanism of anticarcinogenesis by the modulation of epidermal differentiation. In, Retinoids: New Trends in Research and Therapy (J.H. Saurat, Ed.), pp. 56-64. S. Karger AG, Basel, 1985.
20. T.J. Slaga, U. Lichti, H. Hennings, K. Elgjo, and S.H. Yuspa, Effects of tumor promoters and steroidal anti-inflammatory agents on skin of newborn mice in vivo and in vitro. J. Natl. Cancer Inst. 60, 425-431 (1978).
21. G.R. Pettit, C.L. Herald, D.L. Doubek, E. Arnold, and J. Clardy, Isolation and structure of bryostatin 1. J. Am. Chem. Soc. 104, 6846-6848 (1982).
22. J.A. Smith, L. Smith, and G.R. Pettit, Bryostatins: potent, new mitogens that mimic phorbol ester tumor promoters. Biochem. Biophys, Res. Commun. 132, 939-945 (1985).
23. A.S. Kraft, J.B. Smith, and R.L. Berkow, Bryostatin, an activator of the calcium phospholipid-dependent protein kinase, blocks phorbol ester-induced differentiation of human promyelocytic leukemia cells HL-60. Proc. Natl. Acad. Sci. USA 83, 1334-1338 (1986).
24. M.L. Dell'Aquila, H.T. Nguyen, C.L. Herald, G.R. Pettit and P.M. Blumberg, Bryostatin 1 inhibits phorbol ester-induced blockage of differentiation in hexamethylene bisacetamide-treated Friend erythroleukemia cells. Submitted.
25. T. Sako, S.H. Yuspa, C.L. Herald, G.R. Pettit, and P.M. Blumberg, Bryostatin 1 both mimics and blocks effects of phorbol ester tumor promoters on primary mouse epidermal cells. Cancer Res., in press.
26. H. Hennings, P.M. Blumberg, G.R. Pettit, C.L. Herald, R. Shores, and S.H. Yuspa, Bryostatin 1, an activator of protein kinase C, inhibits tumor promotion by phorbol esters in Sencar mouse skin. Carcinogenesis, in press.
27. T.J. Slaga, A.J.P. Klein-Szanto, L.L. Triplett, and L.P. Yotti, Skin tumor-promoting activity of benzoyl peroxide, a widely used free radical-generating compound. Science 213, 1023-1025 (1981).
28. A.J.P. Klein-Szanto and T.J. Slaga, Effects of peroxides on rodent skin: epidermal hyperplasia and tumor promotion. J. Invest. Dermatol. 79, 30-34 (1982).

29. J.F. O'Connell, A.J.P. Klein-Szanto, D.M. DiGiovanni, J.W. Fries, and T.J. Slaga, Enhanced malignant progression of mouse skin tumors by the free-radical generator benzoyl peroxide. Cancer Res. 46, 2863-2865 (1986).
30. J.A. Hartley, N.W. Gibson, L.A. Zwelling, and S.H. Yuspa, The association of DNA strand breaks with accelerated terminal differentiation in mouse epidermal cells exposed to tumor promoters. Cancer Res. 45, 4864-4870 (1985).
31. J.A. Hartley, N.W. Gibson, A. Kilkenny, and S.H. Yuspa, Mouse keratinocytes derived from initiated skin or papillomas are resistant to DNA strand breakage by benzoyl peroxide: a possible mechanism for tumor promotion mediated by benzoyl peroxide. Submitted.
32. S.H. Yuspa, Tumor promotion. In, Accomplishments in Cancer Research 1986 (J.G. Fortner and J.E. Rhoads, Eds.) pp. 169-182. Lippincott Co., Philadelphia, 1987.

IMPLICATION OF SUPEROXIDE RADICAL ANION IN PROMOTION OF NEOPLASTIC TRANSFORMATION IN MOUSE JB6 CELLS BY TPA

John L. Seed, Yoshiuki Nakamura, and Nancy H. Colbourne

Cell Biology Section
Laboratory of Viral Carcinogenesis
National Cancer Institute, FCRF
Frederick, MD 21701-1013

INTRODUCTION

Specific biochemical and genetic events that appear to be essential to promotion of transformation in JB6 cells have been identified.[1] The two earliest events appear to be the activation of protein kinase C (PKC) (and the subsequent phosphorylation of its substrates) and an increase in oxidant stress. There is a significant body of evidence that implicates reactive oxygen, and the superoxide radical anion in particular, as an important mediator of tumor promotion.[2] This evidence includes studies with nonphorbol tumor promoters as well as TPA, several tissues _in vivo_, and a number of different model systems _in vitro_. For example, copper(II)(3,5-diisopropylsalicylate)$_2$ (CuDIPS) is a potent inhibitor of phorbol ester-mediated tumor promotion in the epidermis of CD-1 mice.[3] CuDIPS is a lipophilic copper complex that catalyzes the dismutation of the superoxide anion radical at a rate similar to that of SOD, its biological counterpart _in vivo_.[4] In _in vitro_ studies with JB6 P^+ cells, SOD was observed to inhibit promotion of transformation by TPA. Other _in vitro_ studies have demonstrated that SOD will inhibit TPA-dependent radiogenic transformation of C3H10T1/2 cells.[5,6]

These studies provide substantive and direct _in vitro_ and _in vivo_ evidence for a critical role of SOD in tumor promotion. Alternatively, in studies with biochemical markers of tumor promotion, SOD and catalase were observed to inhibit the induction of ornithine decarboxylase (ODC) by TPA in mouse mammary tumor cells.[7] CuDIPS inhibits the induction of ODC by TPA in mouse epidermis as well and also inhibits the induction of ODC by the nonphorbol tumor promoters mezerein and anthralin.[3] Thus, the inhibitory effects of SOD on tumor promotion may be applicable to tumor promoters other than TPA. More direct evidence for a critical role of reactive oxygen, and superoxide radical anion in particular, in promotion has been obtained using oxygen radical-generating systems. SOD-inhibitable reactive oxygen generated by the xanthine/xanthine oxidase system can substitute for TPA in promotion of transformation in C3H10T1/2 cells[8] as well as in JB6 P^+ cells.

The generation of superoxide radical anion appears to be primarily important as an early event in tumor promotion. Inhibition of the induction of ODC by CuDIPS is ineffective if administered more than 2 hours after application of TPA.[3] The short duration of a protective effect for SOD and early appearance of reactive oxygen species in the cascade of events following exposure to a tumor promotor suggest that the superoxide radical anion may function primarily as a signal transducing element.

In the JB6 cell system, the availability of promotion-sensitive and promotion-insensitive variants in addition to unique promotion sensitivity genes[9] that control cellular susceptibility to promoters provides a valuable opportunity to identify those oxidant stress-related events which may be promotion-relevant.

RESULTS AND DISCUSSION

TPA-dependent promotion of transformation in JB6 cells is inhibited by SOD

Studies from this laboratory have established a role for reactive oxygen, and the superoxide radical anion in particular, in the transformation of JB6 cells by tumor promoters. SOD was found to be a very efficient inhibitor of JB6 cell transformation by TPA, inhibiting soft agar colony formation by as much as 90%.[10] The JB6 P^+ cell line Cl 22, which had a twofold lower level of baseline SOD activity than the P^+ cell line Cl 41 (Table 1), was twice as sensitive to the inhibitory effects of externally added bovine erythrocyte CuZn-SOD. Thus inhibition of TPA transformation by SOD in P^+ cells was closely correlated with both endogenous and exogenously added activity.

An early SOD-inhibitable event is a critical determinant of promotion of transformation by TPA

The time course for inhibition of promotion of transformation by SOD was found to be critically dependent on the time of SOD addition. Maximal inhibition of TPA-induced transformation of JB6 P^+ cells was observed over a time period extending 18 hours before TPA exposure and up to 1 hour after (Figure 1). Addition of SOD more than 2 hours after addition of TPA results in a greater than 50% loss of antipromoting activity. Thus an early SOD-sensitive event is an important mediator of promotion of neoplastic transformation in JB6 cells. This event is likely to be the generation of superoxide radical anion, but the reactive nature of superoxide and its short lifetime preclude the possibility of obtaining with existing techniques, direct evidence for its generation in cells other than inflammatory cells. However, the use of specific enzymatic inhibitors such as SOD and an unambiguous approach to the generation of superoxide radical anion has made it possible to obtain substantive evidence in support of an essential role for this reactive oxygen species in the promotion of transformation in JB6 cells.

Reactive oxygen generated by xanthine/xanthine oxidase can drive promotion of transformation in JB6 cells

SOD was observed to inhibit transformation of JB6 cells by reactive oxygen generated by the xanthine/xanthine oxidase system.[11] The latter system generates superoxide radical anion and hydrogen peroxide during the reduction of xanthine to urate. Because reactive oxygen generated by the xanthine/xanthine oxidase system alone is capable of driving promotion of transformation in JB6 cells, its inhibition by SOD provides particularly strong evidence for an essential role of superoxide in promotion of transformation.

Table 1. Endogenous superoxide dismutase activities in mouse JB6 epidermal cells.

Cell line	Cell phenotype	SOD activity[a] (Units/10^6 cells)	Average activity per phenotype
Cl 22	P+	0.43 ± 0.10	0.8 ± 0.4
Cl 41	P+	1.16 ± 0.44	
Cl 25	P−	0.93 ± 0.36	1.4 ± 0.4
Cl 30	P−	1.78 ± 0.22	
T36271	Tx	0.76 ± 0.27	N/A[b]

[a]Logarithmically growing cells in monolayer culture were trypsinized with 0.06% trypsin, washed with PBS and homogenized in PBS on ice at a concentration of 10^6 cells/ml. The homogenates were precipitated with ethanol/chloroform, the supernatents dialyzed against PBS for 12 hours and assayed for superoxide dismutase activity as described by Imanari et al.[12] Data are expressed as the mean ± S.D.
[b]N/A, Not applicable.

Singlet oxygen and hydrogen peroxide are not candidates for reactive oxygen-dependent promotion of transformation

Evidence for the involvement of reactive oxygen species other than superoxide radical anion in the promotion of JB6 cells was mixed. Hydrogen peroxide did not appear to be essential to promotion of transformation because neither catalase nor glutathione peroxidase could be demonstrated to inhibit it (Table 2). Although interpretation of the catalase results was complicated by its stimulatory effect on the expression of transformation, glutathione peroxidase did not affect expression of the transformed phenotype and did not inhibit the promoting effects of TPA. In the case of Cl 41 cells, glutathione peroxidase actually stimulated promotion of transformation. A singlet oxygen scavenger also did not appear to affect promotion of transformation by TPA (Table 2).

Hydroxyl radical scavengers and lipoxygenase inhibitors inhibit promotion of transformation

Some hydroxyl radical scavengers and lipoxygenase inhibitors were noted to have inhibitory activity (Table 2). However, the effects of these inhibitors were less pronounced than those observed with SOD. The marked efficiency of the inhibitory effect of SOD and its association with early events in promotion of transformation suggest a model in which these other oxygen radical events may be secondary to an event initiated by the superoxide radical anion.

TPA treatment is associated with inhibition of antioxidant enzyme activities

The period between initial exposure to TPA and the end of the effective period for inhibition of transformation by SOD was associated with a marked decline in the activity of endogenous SOD, as well as other antioxidant enzymes such as catalase and glutathione peroxidase in JB6 P^+ cells.[11] A similar decrease in SOD activity is observed in P^- cells. This early effect of TPA is also seen in mouse epidermis following treatment with TPA.[13]

Recovery of antioxidant enzyme activity following TPA treatment: Phenotypic specificity in P^+ and P^- cells

The time course for recovery of antioxidant enzyme activities following TPA treatment appears to differ between P^+ and P^- cell lines. In Cl 41 cells (P^+), exposure to TPA results in a sustained decrease in SOD activity which gradually returns to baseline after 48 hours. In Cl 30 (P^-) cells, enzyme activity recovers to baseline levels within a few hours. Similar changes were also noted with catalase.

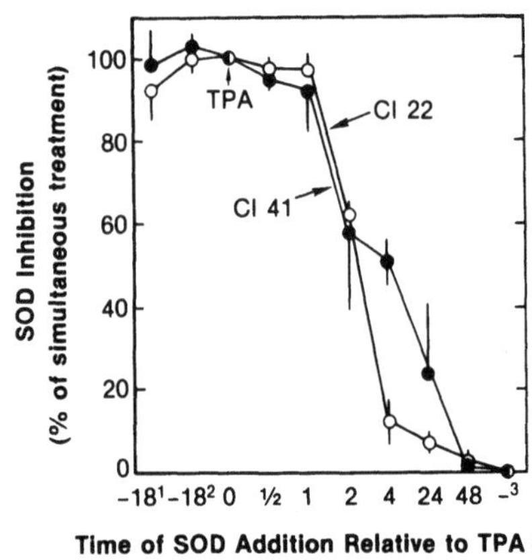

Figure 1. Time course for specific inhibition by SOD of the TPA-induced promotion of neoplastic transformation in mouse epidermal JB6 cells. The cells were plated in soft agar with TPA and/or SOD and scored for anchorage-independent (AI) colony formation. When SOD pretreatment was used, monolayer cells in complete medium were treated with 600 U/ml of SOD. Eighteen hours after treatment, cells were plated in agar as above, with or without SOD. When SOD post-treatment was used, cells were plated with TPA as above, and then at the indicated times, 0.3 ml of 0.33% agar containing 200,000 U/ml of SOD were added as a second layer of top agar. This amount of SOD gives a final concentration of 600 U/ml assuming homogeneous distribution throughout top and bottom agar layers. Results are expressed as percent of inhibitory effect of simultaneous treatment with TPA and SOD. Each value represents the mean $\pm$ one half the range from two separate experiments run in duplicate. The TPA control gave AI colonies of 1062 $\pm$ $50/10^4$ cells in Cl 22 and 1450 $\pm$ $124/10^4$ cells in Cl 41. These were decreased by the simultaneous addition of 600 U/ml CuZn-SOD to 23 $\pm$ 4.2 in Cl 22 and 51.4 $\pm$ 2.5 in Cl 41. -18^2 and -18^1 indicate pretreatment experiments followed by simultaneous exposure with or without SOD, respectively.

Table 2. Summary of effect of reactive oxygen modifiers on promotion of transformation by TPA.

Reactive oxygen modifier[b]	Dose	Known activity	Inhibition of AI colony induction (% of TPA-control) C1 22	C1 41	Average inhibition of $T^{3}6274$ & JB8 AI colony formation	Promotion-specific inhibitory activity[a]
CuZn SOD	600 units/ml	$O_2\cdot$ H_2O_2	90	49	-4	+++
CuDIPS	5 μM	$O_2\cdot$ H_2O_2	61	43	20	++
Catalase	200 units/ml	H_2O_2 H_2O	Increased	Increased	Enhanced	0
GSH-Px	5 units/ml	H_2O_2 H_2O	-40	11	3	0
D-Mannitol	10 mM	·OH scavenger	32	37	8	++
Na-Benzoate	1 mM	·OH scavenger	27	48	19	+
TMU	1 mM	·OH scavenger	35	14	4	+
DABCO	1 mM	$^{1}O_2$ quencher	-26	31	-1	0/+
n-PG	0.5 μM	Antioxidant	43	49	4	++
Tannic Acid	1.5 μg/ml	Antioxidant	75	60	13	++
BHA	10 μM	Antioxidant	42	39	9	++
BHT	20 μM	Antioxidant	16	17	12	0
NDGA	10 μM	Lipoxygenase inhibitor	53	53	14	++
	5 μM	Lipoxygenase inhibitor	31	37	0	+
Quercetin	5 μM	Lipoxygenase inhibitor	14	34	2	+
Indomethacin	5 μM	Cycloxygenase inhibitor		8	ND	$\pm$

[a]Promotion inhibitory activity is expressed as strong (+++), moderate (++), weak (+), marginal ($\pm$), or none (0), according to % inhibition of AI colony induction by TPA not attributable to inhibition of AI growth of tumor cells or to inhibition of monolayer growth of C1 22 or C1 41.

[b]Reactive oxygen modifiers were tested for inhibition of promotion of transformation. Dose response relationships were established for each modifier. The doses presented in this table represent the maximal inhibitory concentrations tested.

Taken together, the above observations provide substantive evidence for a role of superoxide radical anion in the promotion of transformation in JB6 cells. However, it is also clear that *pro* genes and superoxide anion generation interact in an as yet undefined manner during promotion of transformation. The inhibitory effects of TPA on antioxidant enzyme activity are observed in P^-, as well as in P^+ cells, but only the latter progress to the transformed phenotype. However, the recovery of SOD activity is considerably delayed in P^+ cells. It would appear that both an activated *pro* gene and superoxide radical anion generation are necessary components of promotion of transformation and that they are not independent. Further studies will be required to elucidate the molecular basis of their interaction.

Cotransfer of SOD sensitivity with activated pro genes

Events regulated by SOD appear to be closely related to events controlled by promotion sensitivity genes. P^- cells transfected with activated *pro*-1 or *pro*-2 are not only sensitive to promotion by TPA, but also retain their sensitivity to inhibition of promotion by SOD (Colburn et al., submitted). In addition, P^- cells, which appear to have higher levels of basal SOD activity, have lower basal levels of a *pro*-1 homologue (Garrity et al., submitted).

CONCLUSIONS

The observation that an SOD-sensitive event is an early and essential step in promotion of transformation in JB6 cells establishes its importance as a signal transducing event and provides the basis for developing an understanding of the apparently ubiquitous role of reactive oxygen in tumor promotion. Further studies with SOD in JB6 cell variants will establish the relationship between the induction of SOD-sensitive events and other signal transducing events, such as activation of PKC and will provide insights into the relationship between oxidant stress and the action of promotion sensitivity genes.

REFERENCES

1. N. H. Colburn, Genes and membrane signals involved in neoplastic transformation. In Huberman, E. (Ed.): The Role of Chemicals and Radiation in the Etiology of Cancer, 1985 (E. Huberman, Ed.) pp. 235-248. Raven Press, New York.

2. T. W. Kensler, Oxygen free radicals in chemical carcinogenesis. In Superoxide Dismutase, Vol.3, 1986 (L. W. Oberley, Ed.) pp. 191-236. CRC Press, Boca Raton.

3. T. Kensler, D. M. Bush, and W. J. Kozumbo, Inhibition of tumor promotion by a biomimetic superoxide dismutase. Science 221, 75-77 (1983).

4. E. Lengfelder and U. Weser, Superoxide dismutation by low molecular weight copper complexes. Bull. Eur. Physiopathol. Resp. 17S, 73-90 (1981).

5. C. Borek and W. Troll, Modifiers of free radicals inhibit in vitro the oncogenic actions of x-rays, bleomycin, and the tumor promoter 12-O-tetradecanoylphorbol-13-acetate. Proc. Natl. Acad. Sci. USA 80, 1304-1307 (1983).

6. J. B. Little, A. R. Kennedy, and H. Nagasawa, Involvement of free-radical intermediates in oncogenic transformation and tumor promotion in vitro. In Radioprotectors and Anticarcinogens, 1983 (O. F. Nygaard and M. G. Simic, Eds.) pp. 487-502, Academic Press, Orlando.

7. J. Freidman and P. Cerutti, The induction of ornithine decarboxylase by phorbol 12-myristate-13-acetate or by serum is inhibited by antioxidants. Carcinogenesis 4, 1425-1427 (1983).

8. R. Zimmerman and P. Cerutti, Active oxygen acts as a promoter of transformation in mouse embryo C3H/10T1/2/C18 fibroblasts. Proc. Natl. Acad. Sci. USA 81, 2085-2087 (1984).

9. M. I. Lerman, G. A. Hegamyer, and N. H. Colburn, Cloning and characterization of putative genes that specify sensitivity to neoplastic transformation by tumor promoters. Int. J. Cancer 37, 293-302 (1986).

10. Y. Nakamura, N. H. Colburn, and T. D. Gindhart, Role of reactive oxygen in tumor promotion: Implication of superoxide anion in promotion of neoplastic transformation in JB6 cells by TPA. Carcinogenesis 6, 229-235 (1985).

11. Y. Nakamura, T. D. Gindhart, D. Winterstein, J. L. Seed, I. Tomita, and N. H. Colburn, Early superoxide dismutase sensitive event promotes neoplastic transformation in mouse epidermal JB6 cells. Carcinogenesis, in press (1987).

12. T. Imanari, M. Hirota, M. Miyazaki, K. Hayakawa, and Z. Tamura, Improved assay procedure for the determination of superoxide dismutase activity. Igaki no Ayumi 101, 497-497 (1977).

13. V. Solanki, R. S. Rana, and T. J. Slaga, Diminution of mouse epidermal superoxide dismutase and catalase activities by tumor promoters. Carcinogenesis 2, 1141-1146 (1981).

EXPRESSION OF OXIDANT STRESS-RELATED GENES IN TUMOR PROMOTION OF MOUSE EPIDERMAL CELLS JB6

Dana Crawford and Peter Cerutti

Department of Carcinogenesis
Institute for Experimental Cancer Research
1066 Epalinges, Switzerland

INTRODUCTION AND SUMMARY

The evidence is convincing that oxidants and agents which induce a cellular prooxidant state can act as carcinogens, in particular as promoters and progressors (1-7). Bona fide oxidants with promotional activity include H_2O_2, superoxide, ozone, hyperbaric oxygen, peroxyacetic acid, chlorobenzoic acid, benzoyl-peroxide, decanoylperoxide, cumene-hydroperoxide, p-nitro-perbenzoic acid and periodate (8,9). Infiltrated phagocytes represent a major source of oxidants in inflamed tissues (10,11) and in several instances inflammation appears to be a prerequisite for promotion (12-14). Tumor promotion results in the clonal expansion of initiated cells at the cost of the surrounding tissue. This can involve the modulation of the expression of growth- and differentiation-related genes and selective cytostatic effects on non-initiated cells. It is of obvious interest to elucidate the mechanism(s) by which oxidants exert their promotional activity. In addition, the question arises whether other classes of promoters which lack oxidizing properties might owe at least part of their promotional activity to the induction of a cellular prooxidant state (1). Support for this notion has been obtained for the phorbol-ester promoter phorbol-12-myristate-13-acetate (PMA). PMA induced an increase in the ratio of oxidized over reduced glutathione in mouse epidermal cells (5). Furthermore, several antioxidants, in particular CuZn-superoxide dismutase (SOD), inhibit cellular reactions induced by PMA as well as promotion in vivo and in vitro implicating the superoxide radical $O_2^{\bar{}}$ in its mechanism of action (1,4,9,15).

In this article we summarize our recent work on the modulation of the expression of the protooncogenes c-fos and c-myc and the antioxidant genes Cu,Zn-superoxide dismutase (SOD), catalase (CAT) and glutathione peroxidase (GP_x) by PMA and active oxygen (AO) in mouse epidermal cells JB6. PMA and AO (a mixture of $O_2^{\bar{}}$ and H_2O_2 generated extracellularly by xanthine/xanthine-oxidase, X/XO), selectively stimulated the anchorage independent growth of JB6 clone 41 but not of clone 30 in soft agar (4,9; I. Zbinden & P. Cerutti, unpublished). A selective promotional effect of AO for clone 41 cells could also be demonstrated in monolayer cultures. AO treatment resulted first in the retardation of the growth of both clones but only clone 41 exhibited compensatory growth stimulation after prolonged incubation. On the other hand, PMA transiently inhibited the growth of monolayer cultures of both clones to a similar degree (R. Larsson and P. Cerutti, unpublished).

An interesting difference between the JB6 clones was discovered in the constitutive, steady-state concentrations of c-myc mRNA. c-Myc was persistently expressed in the promotable clone 41 but little or no expression was observed in the non-promotable clone 30. This difference is compatible with the notion that clone 41 has progressed to a more advanced stage of transformation than clone 30 (16-20). While AO induced c-myc and c-fos in both clones c-fos-induction was more pronounced in clone 30. The fact that AO has the capacity to induce these genes is intriguing. c-Myc and c-fos may play a role in the acquisition of competence by quiescent cells to enter the cell cycle (21). Bona fide oxidants or AO released by infiltrated inflammatory leukocytes may serve as a "competence signal" in wound healing and tumor promotion. Alternatively, it is conceivable that AO induces differentiation-related functions more efficiently in clone 30 than in clone 41.

Insight into the basis for the differential growth response to AO of clone 41 and clone 30 was obtained from measurements of enzyme activities, protein levels and mRNA concentrations of the major antioxidant enzymes SOD, CAT and GPx. We found that the constitutive levels of SOD and CAT were considerably higher in the promotable clone 41. The higher antioxidant defence of clone 41 may protect it from excessive AO-toxicity and allows the induced proto-oncogenes to exert their functions.

As mentioned, above there is considerable evidence that the phorbolester promoter PMA, which does not possess oxidizing properties, may induce a cellular prooxidant state (1,4,5,9,15). Support for this notion derives from the observation that PMA temporarily caused a decrease in the activities of SOD and CAT but not GPx in mouse skin (22). Our results with JB6 cells indicate that the down-regulation of these genes occurs on the level of the steady-state mRNA concentrations and is part of the complex reprogramming of the expression of multiple families of genes in response to PMA. In contrast to PMA, AO treatment had little or no effect on the levels of SOD, CAT or GPx.

Expression of the proto-oncogenes fos and myc in mouse epidermal cells JB6

The steady state levels of c-fos, c-myc and β-actin mRNA were compared for the clones 30 and 41 by Northern blot analysis of total RNA from cells which had been cultured in monolayers for 3 days in medium containing only 0.25% fetal calf serum. The probes were SP6 recombinant transcripts from plasmids containing c-fos, human c-myc and mouse β-actin inserts, respectively. From the untreated controls in Figure 1 it is evident that the constitutive myc mRNA concentration is higher in clone 41 than in clone 30 while little or no c-fos message is detectable in either clone. Treatment with an extracellular burst of AO (generated by 20 ug/ml X and 2 ug/ml XO) induced myc more strongly in clone 30 than in clone 41. In the latter an increase in mRNA is detectable at 4 hrs against the background of the high persistent messenger concentration. Similarly, the lower half of Figure 1 shows that the fos mRNA concentration is increased more dramatically in clone 30 than in clone 41. As is the case for serum, phorbolesters and certain growth factors in various types of cells (23, 24) c-fos induction by AO is more rapid than c-myc induction (21) and lasts for a shorter time. AO had no effect on the steady state concentrations of the β-actin message in JB6 cells nor rasH (data not shown). We are presently carrying out nuclear runon experiments in order to further define the mechanism which results in the accumulation of fos- and myc-message in response to AO. As expected c-fos and c-myc were also induced

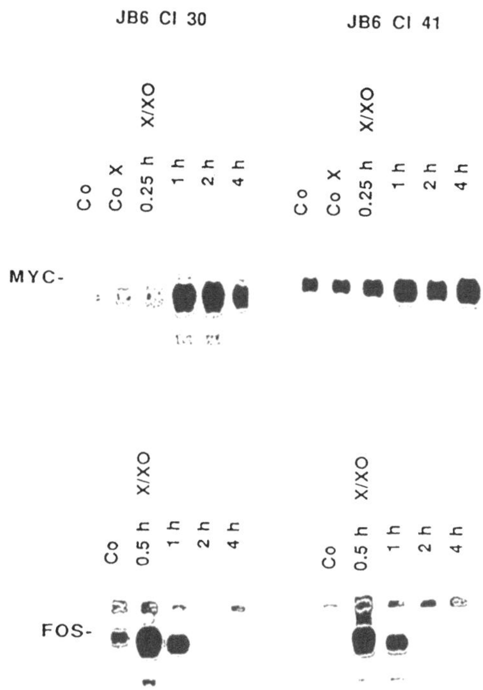

INDUCTION OF C-MYC AND C-FOS BY ACTIVE OXYGEN IN MOUSE EPIDERMAL CELLS JB6.

Fig. 1 Effect of active oxygen on the expression of c-fos, c-myc and β-actin in JB6 cells

JB6 cells were passaged on day 0 in monolayer cultures and grown 3 days in MEM supplemented with 8% fetal calf serum followed by 3 days in medium containing only 0.25% fetal calf serum. At this time, cells were exposed to xanthine-xanthine oxidase (20 ug/ml X - 2ug/ml XO) for the time indicated. RNA was extracted and run out an formadelhyde gels. After transfer to gene screen, filters were prehybridized and hybridized with appropriate RNA probes from SP6 recombinants containing c-fos or human c-myc, respectively.

by the phorbolester promoter PMA in promotable and non-promotable JB6 cells (data not shown).

To our knowledge this is the first demonstration that an extracellular burst of AO can modulate the expression of c-fos and c-myc. The significance of our findings could reach beyond tumor promotion. Many tissues are exposed to AO under physiological conditions, as a consequence of inflammatory processes in acute and chronic diseases and as a consequence to the exposure to certain xenobiotics (1). However, the biological functions of c-fos and c-myc have not been clearly identified. In the most general terms fos induction may be an "immediate early" response to multiple forms of cellular stress (25,26), including "oxidative stress" according to the present results. Depending on the cell type c-fos induction may represent an early step in differentiation or in the acquisition of competence for cell proliferation. Expression of c-myc appears to be more specifically related to cell growth (21). Evidently AO

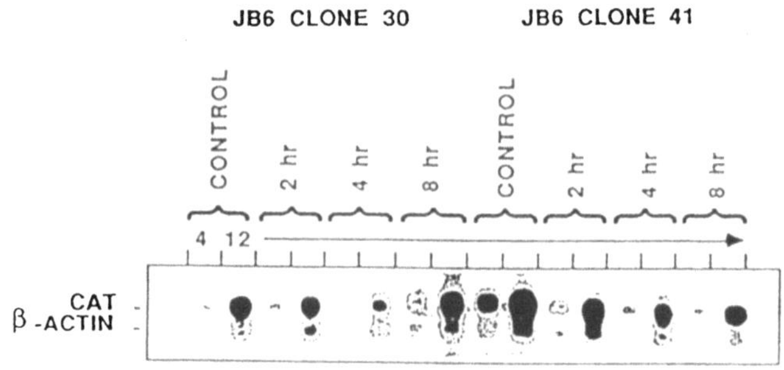

Fig. 2 Effect of PMA on catalase mRNA levels in JB6 cells
JB6 cells were passaged on day 0, fed on day 3 and then exposed to PMA (10 ng/ml) on day 5 for the times indicated. A concentration of 8% fetal calf serum in MEM was used throughout. RNA was then extracted as above and gene screen filters probed with RNA transcripts of SP6 recombinants containing inserts of human CAT and mouse β-actin.

does not interact with a specific cell surface receptor. Its capacity to induce c-fos and c-myc may be related to the observation that an increase in Ca^{2+} entry via voltage dependent channels induced these genes in PC12 phaeochromocytoma cells (27,28). Since certain oxidants are known to disturb the cellular Ca^{2+} metabolism (29,30) we are presently investigating AO-induced changes in cytosolic Ca^{2+}-concentrations in JB6 clone 41 and clone 30. The mechanisms of induction of fos by AO described here and by ultraviolet light reported earlier (31) could be related.

Antioxidant defence capacity of mouse epidermal cells JB6

The cellular antioxidant defence is expected to attenuate the effects of AO. The major defence systems consist of the enzymes SOD, CAT and GPx and the low molecular weight sulfhydryl compound glutathione (GSH). Differences in the antioxidant capacities between the promotable clone 41 and the non-promotable clone 30 might in part form the basis for their different biological responses to AO exposure. Therefore, we compared GSH concentrations and antioxidant enzyme activities of both clones. Total and reduced GSH concentrations were comparable in both clones. In contrast, clone 41 contained approximately twice the activity of Cu,Zn-SOD and CAT than clone 30 (expressed as units per mg proteins) while GPx activities were comparable. In agreement with the higher activity immunoblots indicated that the CAT protein concentration was approximately twice as high in clone 41 than in clone 30.

In order to address the question whether these differences between the two JB6 clones are at the level of gene expression we measured steady-state messenger concentrations for SOD, CAT and GPx by Northern blot analysis. As probes we used transcripts from SP6 constructions containing fragments of cDNA of human Cu,Zn-SOD (32), human CAT (33) and bovine GPx (34). Figure 2 indicates that clone 41 contains considerably higher messenger concentrations for SOD and CAT than clone 30 while their content in GPx message was comparable. Experiments are presently being carried out to distinguish between the alternative possibilities (1) that the two JB6 clones contain different numbers of SOD and CAT genes (2) that these genes are more efficiently transcribed in clone 41 (3) that the messages for SOD and CAT are more stable in clone 41 than in clone 30.

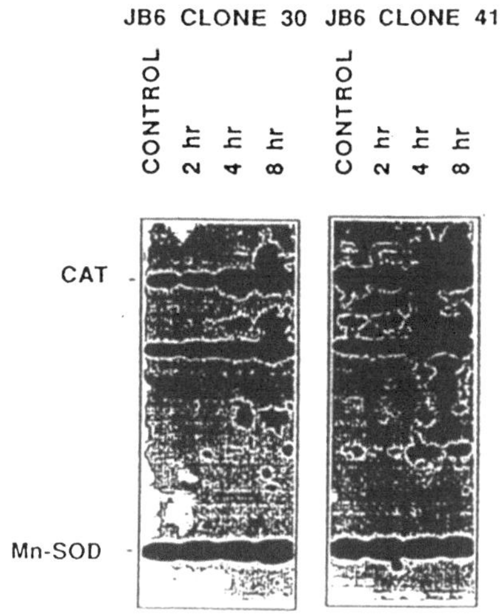

Fig. 3 Effect of active oxygen on the concentrations of catalase- and Mn-superoxide dismutase protein in JB6 cells
Cells were grown as outlined in the legend to Figure 2 and exposed to xanthine-xanthine oxidase (20 ug/ml X - 2ug/ml XO) on day 5 for the times indicated. Cell sonicates were then separated on 12.5% SDS polyacrylamide gels and transferred to immobilon PVDF membranes (Millipore) by electrotransfer. Membranes were first hybridized to rabbit anti-human CAT and Mn-SOD, washed, and then hybridized to I^{125}-labelled goat anti-rabbit IgG.

Down-regulation of the expression of Cu,Zn-superoxide dismutase and catalase by PMA in mouse epidermal cells JB6

PMA activates protein kinase C and many of its pleiotropic effects are probably unrelated to the metabolism of AO. However, numerous observations suggest that in addition, or as a consequence of protein kinase C activation, PMA also induces a cellular prooxidant state which may play a role in some aspects of its promotional activities (1,4,5,9,15). A cellular prooxidant state can result either from the overproduction or the deficient destruction of AO. The observation that PMA caused a decrease in the activities of SOD and CAT in mouse skin (22) and JB6 cells in culture (19) supports the second alternative. We were interested in the question as to whether the decrease in the activities of these enzymes was part of the complex program of modulation of gene expression induced by PMA or due to post-translational inactivation. Therefore, we measured the enzyme activities and steady-state messenger concentrations of SOD, CAT and GPx in JB6 cells following treatment with PMA (20 ng/ml). We found that the activities of SOD and CAT decreased by approximately 30% within 8 hrs in clone 30 and by approximately 40% in clone 41 while the GPx activities remained unchanged. These changes in enzyme activities were preceded by decreases in mRNA concentrations. As shown in Figure 2 in clone 30 CAT-mRNA concentrations reached a minimum 4 hrs after PMA treatment. The CAT message also decreased dramatically in clone 41 while the decrease in SOD message was less pronounced. In agreement with our measurements of GPx enzyme activity no change was observed in its messenger concentration. Our results indicate that PMA down-regulates the expression of SOD and CAT in JB6 cells on the transcriptional level but the exact mechanism of this regulation remains to be elucidated. AO generated extracellularly as described above, in contrast to PMA, had little effect on the steady state messenger levels of SOD, CAT and GPx in JB6 cells. In contrast to PMA, active oxygen generated extracellularly had little effect on the steady state messenger levels of CuZn-SOD, CAT and GPx in JB6 cells. The same held for catalase and MnSOD at the protein level as is shown in the Western blot of Fig. 3. It is also of interest to note that both SOD and CAT mRNAs were refractory to active oxygen or PMA mediated modulation in cultured human keratinocytes.

PMA has been shown to induce a family of "immediate-early genes" which include c-fos, c-myc and β-actin by a mechanism which appears to involve the binding of a nuclear protein to a defined serum responsive element (35,36). It also activates a second family of genes. For the activation of this second family of genes which contains collagenase, stromelysin, metallothionein IIA and interleukin II the binding of the transcription factor API to a short "TPA responsive element" appears to be required (37). The experiments which allowed the identification of these regulatory systems in response to PMA, serum factors, certain growth factors and carcinogens were designed to detect genetic induction rather than down-regulation. It will be of particular interest to elucidate whether the down-regulation of SOD and CAT described here uses similar or unrelated genetic mechanisms.

CONCLUDING REMARKS

JB6 clone 41 but not clone 30 is stimulated to grow in response to AO. Major differences in the constitutive expression of the antioxidant genes SOD and CAT and the protooncogene and c-myc were discovered between the two clones. The AO promotable clone 41 possessed higher persistent expression of SOD, CAT and c-myc. Probably as a consequence of the higher antioxidant defence capacity which attenuates the AO signal more effectively in clone 41 c-fos and c-myc induction was weaker than in clone 30. On the basis of the results presented here and additional data from our laboratory which indicates that AO induces less DNA strand breakage and poly ADP-ribosylation in clone 41 than in clone 30 we are considering two working models (1) AO is more cytostatic to clone 30 because of the more severe depletion of the DNA synthesis precursor dGTP as a consequence of excessive poly ADP-ribosylation. Under these conditions even the strong induction of c-fos and c-myc is ineffective. In contrast, the higher antioxidant defence of clone 41 protects it from prolonged cytostatic effects and allows c-fos and c-myc to exert their functions. (2) AO acts as a signal for cell differentiation. This signal is less attenuated in clone 30 than in clone 41.

It will be important to explore whether there exist relationships between the regulation of the protooncogenes c-fos and c-myc and the antioxidant genes SOD and CAT in JB6 cells and the pro-genes identified by Colburn and collaborators which upon transfection restore promotability to clone 30 and other non-promotable JB6 clones (9).

REFERENCES

1. P. Cerutti, Prooxidant States and Tumor Promotion. Science 227, 375-381, (1985).
2. W. Kozumbo, M. Trush, and T. Kensler, Are free radicals involved in tumor promotion. Chem. Biol. Interaction 54, 199-207 (1985).
3. P. Cerutti, Oxidant Tumor Promoters. In "Growth Factors, Tumor Promoters and Cancer Genes" (eds. N. Colburn, H. Moses and E. Stanbridge), Alan R. Liss, N.Y. in press (1987).

4. Y. Nakamura, N.H. Colburn, and T. D. Gindhart, Role of reactive oxygen in tumor promotion: implication of superoxide anion in promotion of neoplastic transformation in JB6 cells by TPA. Carcinogenesis 6, 229-235 (1987).
5. J.P. Perchellet, E.M. Perchellet, D.K. Orten and B.A. Schneider, Decreased ratio or reduced/oxidized glutathione in mouse epidermal cells treated with tumor promoters. Carcinogenesis, 7, 503-506 (1986).
6. S.M. Fisher, J.K. Baldwin, and L.M. Adams, Effects of anti-promoters and strain of mouse on tumor promoter-induced oxydants in murine epidermal cells. Carcinogenesis 7, 915-918 (1986).
7. J.F. O'Connell, A.J.P. Klein-Szanto, D.M. Di Giovanni, J.W. Fries, T.J. and Slaga, Enhanced progression of mouse skin tumors by the free radical generator benzoyl peroxide. Cancer Res. 46, 2863-2865 (1986).
8. T., Slaga, V. Solanki, and M. Logani. In "Radioprotectors and Anticarcinogens" (eds. O. Nygaard and M. Simic) pp 471-485, Academic Press, N.Y. (1983).
9. T. Gindhart, Y. Nakamura, L. Stevens, G. Hegameyer, M. West, B. Smith, and N. Colburn. In "Cancer of the Respiratory Tract", Carcinogenesis - A comprehensive Survey Vol. 8 (eds. M. Mass, D. Kaufman, J. Siegfried, V. Steele S. Nesnow) pp 341-367, Raven Press, N.Y. (1985).
10. S.A. Weitzman, A.B. Weitberg, E.P. Clark, and T. P. Stossel, Phagocytes as carcinogen: malignant transformation produced by human neutrophils. Science 227, 1231-1233 (1985).
11. J.G. Lewis, and D.O. Adams, Enhanced release of hydrogen peroxide and metabolites of arachidonic acid by macrophages from SENCAR mice following stimulation with phorbol-ester. Cancer Res. 46, 5696-5700 (1986).
12. T. Slaga, S. Fisher, A. Viaje, D. Berry, W. Bracken, S. LeClerc, and D. Miller. In Carcinogenesis 2, "Mechanisms of Tumor Promotion and Cocarcinogenesis" (eds. T. Slaga, A. Sivak and R. Boutwell) pp. 173-, Raven Press, N.Y. (1978).
13. J. Holland, E. Perkinds, L. Gipson, Proc. Am. Ass. Cancer Res. 18, 10 (1977).
14. J. DiGiovanni, Banbury Report 25-Genotoxic Oxidant Tumor Promoters in "Nongenotoxic mechanisms in carcinogenesis", (eds. T. Slaga, B. Butterworth), Cold Spring Harbor Press, in press (1987).
15. W. Kozumbo, and P. Cerutti. In "Antimutagenesis and anticarcinogenesis mechanisms", Antioxidants as Antitumor Promoters (eds. D. Shankel, P. Hartman, T. Kada and A. Hollaender) pp. 491-506, Plenum Press, N.Y. (1986).
16. S. Shall, M. Tavassoli, and A. Cockrell, Changes in c-myc expression with cellular immortalization. EMBL Conference on "Oncogenes and Growth Factors", Heidelberg; April 1987, Abstr. p. 161.
17. T. Shuin, T. Billings, P. Lillehaug, J., Patierno, S., Roy-Burman, P. and J. Landolph, Enhanced expression of c-myc and decreased expression of c-fos protooncogenes in chemically and radiation transformed C3H/10T1/2 C18 mouse embryo fibroblasts. Cancer Res. 46, 5302-5311 (1986).
18. G. Dotto, M. Gilman, M. Maruyama, and R. Weinberg, c-Myc and c-fos expression in differentiating mouse primary keratinocytes. EMBO J. 5, 2853-2857 (1986).
19. E. Mougneau, C. Cerni, F. Tillier, and F. Cuzin, Transfer of a myc oncogene into FR3T3 fibroblasts induces a high risk of further transformation. EMBL Conference on "Oncogenes and Growth factors", Heidelberg, April 1987, Abstr. 127.
20. S. Cory, A. Harris, W. Langdon, M. Graham, L. Cocoran, W. Alexander and J. Adams, Lessons from translocations and transgenic mice: constitutive c-myc expression predisposes to neoplasia. Progress in Immunology VI. pp 675-682, Academic Press, Inc. N.Y. (1986).

21. M. Cole, The myc oncogene: its role in transformation and differentiation. Ann. Rev. Genet. 20, 361-384 (1986).
22. V. Solanki, R. Rana, and T. Slaga, Diminution of mouse epidermal superoxide dismutase and catalase activities by tumor promoters. Carcinogenesis 2, 1141-1446 (1981).
23. R. Muller, Cellular and viral fos genes: structure, regulation of expression and biological properties of their encoded products. Biochim. Biophys. Acta 823, 207-225 (1986).
24. I. Verma, Proto-oncogene fos: a multifaceted gene. TIG, April 1986, p. 93-96 (1986).
25. L. Lan, and D. Nathans, Expression of a set of growth-related immediate early genes in Balb 3T3 cells: coordinate regulation with c-fos and c-myc. Proc. Natl. Acad. Sci. 84, 1182-1186 (1987).
26. H. Rahmsdorf, A. Schönthal, P. Angel, M. Litfin, U. Rüther, and P. Herrlich, Post-transcriptional regulation of c-fos mRNA expression. Nucl. Acid Res., in press (1987).
27. M. Greenberg, E. Ziff, and L. Greene, Stimulation of neuronal acetylcholine receptors induces rapid gene transcription. Science 234, 80-82 (1986).
28. J. Morgan, and T. Curran, Role of ion flux in the control of c-fos expression. Nature 322, 552-555 (1986).
29. G. Bellomo, S. Jewell, and S. Orrenius, The metabolism of menadione impairs the ability of rat liver mitochondria to take up and retain calcium. J. Biol. Chem. 257, 11558-11562 (1982).
30. C. Richter, B. Frei, and P. Cerutti, Mobilization of mitochondrial Ca^{2+} by hydroperoxy eicosatetraenaoic acid. Biochem. Biophys. Res. Comm. 143, 609-616 (1987).
31. P. Angel, H. Rahmsdorf, A. Pöting and P. Herrlich, c-fos mRNA levels in primary human fibroblasts after arrest in various stages of cell cycle. Cancer Cells 3, 315-319 (1985).
32. L. Sherman, D. Levanon, J. Lieman-Hurwitz, N. Dafui, and Y. Groner, Nucleotide sequence and expression of human chromosome 21-encoded superoxide dismutase mRNA. Proc. Natl. Acad. Sci. USA 80, 5465-5469 (1984).
33. R. Korneluk, F. Quan, W. Lewis, K. Guise, H. Willard, M. Holmes, and R. Gravel, Isolation of human fibroblast catalase cDNA clones. J. Biol. Chem. 259, 13829-13823 (1984).
34. Received from Dr. G. Mullenbach, Chiron Corp., CA.
35. M. Greenberg, Z. Siegfried, and E. Ziff, Mutation of the c-fos gene dyad symmetry element inhibits serum inducibility of transcription in vivo and the nuclear regulatory factor binding in vitro. Mol. Cell. Biol. 7, 1217-1225 (1987)
36. R. Treisman, Identification of a protein binding site that mediates transcriptional response of the c-fos gene to serum factors. Cell 46, 567-574 (1986).
37. P. Angel, M. Imagawa, R. Chin, B. Stein, J. Imra, H. Rahmsdorf, C. Jonat, P. Herrlich, and M. Karin, Phorbol ester inducible genes contain a common cis element recognized by a TAP-modulated trans-acting factor. Cell, in press.

ACKNOWLEDGEMENTS

We would like to thank Dr. A. Baret for supplying antibodies against catalase and Mn-superoxide dismutase and Ms I. Zbinden for expert technical assistance. This work was supported by the Swiss National Science Foundation, the Swiss Association of Cigarette Manufacturers and the Association for International Cancer Research.

ACTIVATION OF ORGANIC HYDROPEROXIDE TUMOR PROMOTERS TO FREE RADICALS IN TARGET CELLS

Bonita G. Taffe§, Thomas W. Kensler§, Naohiro Takahashi¶, and Ronald P. Mason¶

§Department of Environmental Health Sciences, Johns Hopkins School of Hygiene and Public Health, Baltimore, Maryland 21205 and ¶Laboratory of Molecular Biophysics, National Institute of Environmental Health Sciences, Research Triangle Park, North Carolina 12233

INTRODUCTION

Although the molecular mechanisms involved in carcinogenesis are poorly understood at this time, intermediary events in the development of cancer have been described through the selective actions of discrete chemical agents (1). These processes have been studied in various model systems and have been termed initiation, promotion and progression (2). Initiation appears to involve the modification of cellular DNA, resulting in genotypically altered cells, while promotion encompasses a continuum of events allowing for the selection and clonal expansion of the initiated cells (3). Finally, progression completes the conversion of pre-malignant cells to malignant cells. Substantial evidence supports the involvement of free radical species, especially those derived from molecular oxygen, in multiple aspects of these processes (4).

Five major lines of experimental evidence indicate a role for free radicals in tumor promotion. Briefly, [1] potential free radical-generating compounds such as organic hydroperoxides are tumor promoters, [2] some tumor promoters stimulate the production of reactive oxygen species from endogenous sources in a variety of cell types, [3] reactive oxygen species can mimic the biochemical actions of tumor promoters, [4] tumor promoters modulate antioxidant defense systems, and [5] free radical scavengers and detoxifiers inhibit tumor promotion (4). One approach to defining the role of free radicals in the process of tumor promotion involves both the demonstration and identification of the radical species generated by the promoting agent and the identification of the molecular targets of these reactive species.

A variety of organic peroxides and hydroperoxides function as tumor promoters and progressors in mouse skin previously treated with an initiating agent such as 7,12-dimethylbenzanthracene (5). A discussion of the actions of these agents in the process of tumor progression is found elsewhere in these proceedings. Although these peroxides and hydroperoxides are active as promoters and progressors, they appear to be inactive as initiators or complete carcinogens (6,7). The promoter activity of the organic hydroperoxides is generally ascribed to the generation of free radicals from these compounds, although they are remarkably stable to uncatalyzed unimolecular homolysis at body temperature (8). The study described here involves the detection and identification via electron spin resonance of the free radical spin-trapped adducts which are generated following the incubation of tert-butyl hydroperoxide or cumene hydroperoxide with isolated murine keratinocytes. The species trapped include alkyl and alkoxyl radicals, whose formation is dependent on a heat-labile factor predominantly associated with the cytosol of the epidermal cell. Possible molecular targets for these radical species are also discussed.

METHODS

Epidermal keratinocytes were isolated from SENCAR neonatal mice, 1-3 days of age via the trypsin flotation method (9). Briefly, the animals were anesthetized on ice, a dorsal longitudinal incision made through the skin, the spinal column severed, and the limbs and then the skin removed. The skins were washed first in distilled water and then in modified phosphate-buffered saline (PBS, without calcium or magnesium salts, with 0.02% EDTA) and subsequently floated overnight at 4°C in 0.25% trypsin/modified PBS. The epidermis was separated from the dermis, minced and stirred in Dulbeccos Modified Eagles Medium with 10% fetal calf serum for 1 hour at room temperature. This suspension was filtered through nylon gauze to obtain a single-cell suspension which was then pelleted and resuspended in PBS twice. The cells were resuspended to a final concentration of 20 x 10^7/ml, and 0.2 ml aliquots were used for electron spin resonance (ESR) analysis. Cell preparations were fractionated by Dounce homogenization followed by centrifugation to produce a 3,000 x g crude nuclear pellet, and a 105,000 x g cytosol and pellet. Where indicated, the viable cells were heat-inactivated in a closed vessel by heating in a boiling water bath for 30 minutes. Electron spin resonance conditions are described in the figure legends.

RESULTS

Incubation of isolated murine keratinocytes with the spin trap, DMPO (5,5-dimethyl 1-pyrroline N-oxide), and tert-butyl hydroperoxide results in the ESR spectrum found in Fig. 1A while omission of tert-butyl hydroperoxide from the mixture results in no detectable radical spectrum (Fig. 1B). The principal radical adducts of DMPO have been identified as the tert-butoxyl radical-DMPO adduct (Fig. 1A, X), which rapidly decays, and the more stable alkyl radical-DMPO adduct (Fig. 1A, arrows). When cumene hydroperoxide was incubated with cells and DMPO, the alkyl radical-DMPO adduct but not the cumoxyl radical-DMPO adduct was detected (data not shown). Prior heat-inactivation of the cells results in no generation of radicals on incubation with either tert-butyl hydroperoxide (Fig. 1C) or cumene hydroperoxide (not shown).

To further identify the alkyl radicals which were formed in the epidermal cells, the spin trap tert-butyl nitrone (tNB) was used. This spin trap provides greater spectral information as to the chemical structure of the radical adduct due to the addition of the radical directly to the nitrogen atom of the spin trap molecule. The spectrum generated by the incubation of cells with tNB and tert-butyl hydroperoxide is shown in Fig. 2A. The principal radical adducts identified in this spectrum include: the methyl radical-tNB adduct (Fig. 2A, closed circles), the ethyl-tNB adduct (Fig. 2A, closed diamonds), and a hydrogen atom-tNB adduct (Fig. 2A, apostrophes). These same radical species were identified on incubation of cells, tNB and cumene hydroperoxide (data not shown). Once again heat-inactivation resulted in no radical generation on incubation of cells with

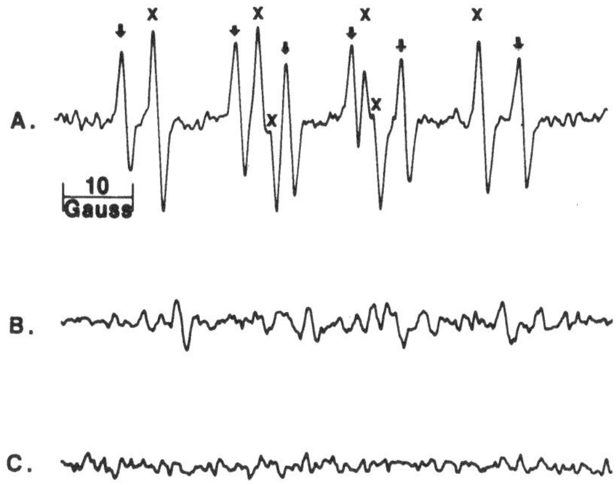

Fig. 1. The ESR spectra of DMPO radical adducts formed on incubation of <u>tert</u>-butyl hydroperoxide with isolated murine keratinocytes. Incubations consisted of: 7 x 10^7 cells in 0.2 ml PBS, DMPO (90 mM), and where indicated, <u>tert</u>-butyl hydroperoxide (30 mM). A. ESR spectrum of cells 4 min following incubation with DMPO and <u>tert</u>-butyl hydroperoxide. B. Incubation of cells with DMPO but without <u>tert</u>-butyl hydroperoxide results in no radical spectrum generated. C. Heat-inactivation of cells resulted in loss of radical generating ability of the cells when incubated with DMPO and <u>tert</u>-butyl hydroperoxide. Instrumental conditions (using a Varian Century Series E-109 spectrometer and TM_{110} cavity) were: power = 20 mW, modulation amplitude = 0.33 G, time constant = 0.5 sec., scan time = 4 minutes, scan range = 100 G, gain = 8 x 10^4.

tNB and either hydroperoxide (Fig. 2B represents <u>tert</u>-butyl hydroperoxide incubation). To attempt to localize the radical-generating activity of the epidermal cells, these cells were fractionated as described in Methods. ESR spectroscopy of incubations of the various subcellular fractions with DMPO and either hydroperoxide demonstrated the only active fraction to be the cytosol (105,000 x g supernatant); the nuclear (3,000 x g) and microsomal/mitochondrial (105,000 x g) pellet fractions were not active.

DISCUSSION

The observation of alkyl and alkoxyl radical formation in epidermal cells demonstrates that these cells are capable of activating hydroperoxides to form free radicals. Pryor and colleagues (10) have suggested that radical production from hydroperoxides may occur through a metal-mediated reaction between an endogenous reductant such as ascorbate or glutathione and the hydroperoxide. However, in epidermal cells the activation of hydroperoxides may be enzymatic in that it is heat labile. In such a case, activation of the organic hydroperoxide to a free radical is postulated to involve the enzymatic activity of an iron-centered protein, such as a peroxidase, which through redox activity can generate both the peroxyl (ROO·) and alkoxyl (RO·) radicals from the hydroperoxide (11):

$$ROOH + PROTEIN(Fe^{3+}) \dashrightarrow ROO\cdot + PROTEIN(Fe^{2+})$$

$$ROO\cdot + PROTEIN(Fe^{2+}) \dashrightarrow RO\cdot + PROTEIN(Fe^{3+})$$

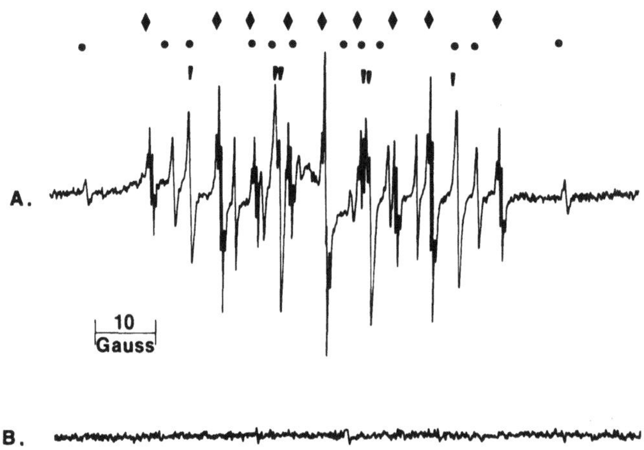

Fig. 2. The ESR spectra of tNB radical adducts formed on incubation of tert-butyl hydroperoxide with isolated murine keratinocytes. Incubations consisted of: 7 x 10^7 cells in 0.2 ml PBS containing 1 mg tNB/ml buffer, and where indicated tert-butyl hydroperoxide (30 mM). A. ESR spectrum of cells 26 minutes following incubation with tNB and tert-butyl hydroperoxide. B. Heat-inactivation of cells resulted in loss of radical generating ability of the cells when subsequently incubated with tNB and tert-butyl hydroperoxide. Instrumental conditions were: power = 20 mW, modulation amplitude = 0.33 G, time constant = 0.5 sec., scan time = 16 min, scan range = 100 G, gain = 5 x 10^4.

The formation of the methyl radical ($CH_3\cdot$) on incubation with either tert-butyl or cumene hydroperoxide can be explained through the instability of the alkoxyl radicals formed in this system. These tertiary alkoxyl radicals undergo β-scission to form the corresponding ketone and the methyl radical (8):

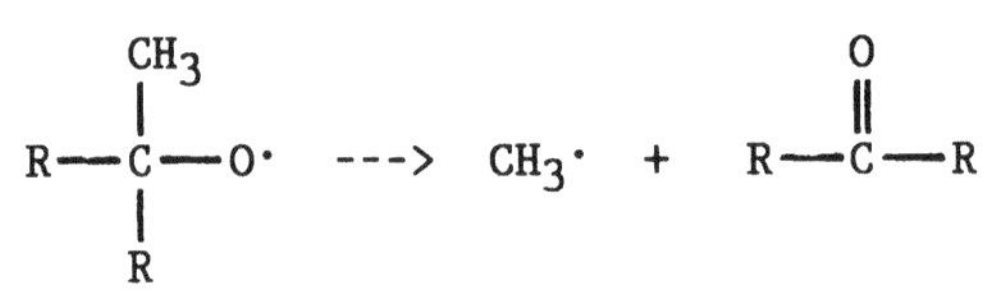

Formation of the ethyl radical is unexpected and could be derived from lipid peroxidation; however, this radical adduct of tNB has also been identified in cell-free peroxide-metabolizing systems in which no lipids are present (data not shown). Most likely, the ethyl radical adduct is formed through the molecular addition of two methyl radicals to the spin trap and is therefore of no direct biological significance.

The demonstration that incubation of target cells, the epidermal keratinocytes, with organic hydroperoxides (known tumor promoters) can result in the generation of alkyl radicals raises some interesting questions concerning the action of these radicals in the process of tumor promotion. Alkyl radicals can undergo a variety of biologically important interactions including hydrogen abstraction, addition, and substitution reactions (12). Various cellular targets in these reactions could include lipid, protein, RNA or DNA. Given the high reactivity of alkyl radicals, the localization of the radical-generating activity in these epidermal cells to the cytosol would indicate that adjacent, nongenetic targets may be the most likely sites for radical modification of biomolecules. This hypothesis is supported by the fact that although *tert*-butyl hydroperoxide and cumene hydroperoxide are weakly mutagenic in bacterial test systems (13), they do not function as complete carcinogens in mouse skin (6). Benzoyl peroxide, another putative free radical-generating tumor promoter in the mouse skin, causes DNA damage in the form of strand breaks (14). However, incubation of radiolabeled benzoyl peroxide with epidermal cells results in covalent binding of the phenyl ring to protein, but not to DNA (data not shown). Thus DNA strand breakage induced by peroxides may result from an indirect action that is not dependent on localized free radical generation adjacent to DNA. Ochi and Cerutti (15) have recently observed that DNA strand breakage by hydroperoxy fatty acids is calcium-dependent. Hydroperoxides such as *tert*-butyl hydroperoxide and cumene hydroperoxide are well known to effectively mobilize intracellular calcium (16) that could participate in the activation of nucleases (15).

The major radical species trapped following keratinocyte activation of these organic hydroperoxides is the methyl radical. Methylation reactions are well described in chemical carcinogenesis. Methylating carcinogens, particularly the mono- and di-substituted hydrazines, have been shown to generate alkyl radicals on metabolic activation (17). Studies which assess the binding to cellular macromolecules on exposure to dimethylhydrazine have demonstrated methylation of DNA, RNA, and protein, with the highest specific activity reportedly found in the covalently modified cellular RNA (18). Nuclear protein methylation by dimethylhydrazine includes both histone and the high mobility group (HMG) proteins with the predominant covalent modification occurring in the HMG proteins. Amino acid residues that have been reported to be methylated are the basic amino acids, lysine, arginine, and histidine (17-20). One functional effect of methylation of these basic amino acid residues in histones may be to alter their binding to DNA (21), which may in turn contribute to alteration of gene expression. The effects of methylation on alteration of the functions of other proteins, such as the cytosolic proteins, has not been determined. Alkylation of cytosolic constituents could result in effects ranging from cell death by direct interference with protein synthesis to modulation of phenotypic expression through covalent modification of RNA or proteins involved in DNA transcriptional control.

There is little evidence in these spin trapping studies for the presence of peroxyl or alkoxyl radical generation, but these negative results may be largely artifactual. Formation of the methyl radical from the tertiary hydroperoxides (*tert*-butyl hydroperoxide and cumene hydroperoxide) implies the intermediacy of alkoxyl radicals. Carbon centered radicals readily react with molecular oxygen to form peroxyl radicals which subsequently react quite slowly with the spin traps used in this study. Peroxyl and alkoxyl radicals may play a greater role in the molecular mechanisms involved in tumor promotion through an indiscriminant reaction with cellular macromolecules (22), resulting in cell death. One aspect of the role which cell differentiation and cell death may play in the process of promotion is through cell loss in the basal layer of the epidermis, thus allowing for the facilitated clonal expansion of the initiated cells. Peroxyl radicals may also stimulate pathways involved in arachidonate metabolism (23), thus modulating signal transduction in the cell and possibly affecting cellular differentiation.

The results obtained in this study demonstrate the generation of free radical species by the organic hydroperoxide tumor promoters in a target tissue, mouse epidermis. These observations raise many questions including: whether the radicals generated play a significant role in tumor promotion, which radicals are involved in this process, and what are the molecular mechanisms through which these radicals participate in the process of tumor promotion.

ACKNOWLEDGEMENTS

This work was supported by grants from the National Institutes of Health (CA 44530, CA 01230, and ES 00454) and the American Cancer Society (SIG-3).

REFERENCES

1. R.M. Hicks, Pathological and biochemical aspects of tumour promotion. Carcinogenesis 4, 1209-1214 (1983).
2. S.H. Yuspa, T. Ben, H. Hennings, and U. Lichti, Divergent responses in epidermal basal cells exposed to the tumor promoter 12-O-tetradecanoylphorbol-13-acetate (TPA). Cancer Res. 42, 2344-2349 (1982).
3. T.J. Slaga, (Ed.), Mechanisms of Tumor Promotion, Vol. 1-4, CRC Press, Boca Raton, 1984.
4. T.W. Kensler and B.G. Taffe, Free radicals in tumor promotion. Adv. Free Radical Biol. Med. 2, 347-387 (1986).
5. J.F. O'Connell, A.J.P. Klein-Szanto, D.M. DiGiovanni, J.W. Fries, and T.J. Slaga, Enhanced malignant progression of mouse skin tumors by the free-radical generator benzoyl peroxide. Cancer Res. 46, 2863-2865 (1986).
6. P. Watts, Peroxides, genes and cancer. Food Chem. Toxicol. 23, 957-960 (1985).
7. T.J. Slaga, A.J.P. Klein-Szanto, L.L. Triplett, L.P. Yotti, and J.E. Trosko, Skin tumor-promoting activity of benzoyl peroxide, a widely used free radical-generating compound. Science 213, 1023-1025 (1981).
8. W.A. Pryor, Free Radicals. McGraw-Hill Book Company, New York, 1966.
9. S.H. Yuspa, P. Hawley-Nelson, J.R. Stanley, and H. Hennings, Epidermal cell culture. Transplant. Proc. 12S, 114-122 (1976).
10. C.H. Kennedy, W.A. Pryor, G.W. Winston, and D.F. Church, Hydroperoxide-induced radical formation in liver mitochondria. Biochem. Biophys Res. Commun. 141, 1123-1129 (1986).
11. B. Kalyanaraman, C. Mottley, and R.P. Mason, A direct electron spin resonance and spin-trapping investigation of peroxyl free radical formation by hematin/hydroperoxide systems. J. Biol. Chem. 258, 3855-3858 (1983).
12. W.A. Pryor, D.L. Fuller, and J.P. Stanley, Reactivity patterns of the methyl radical. J. Am. Chem. Soc. 94, 1632-1638 (1972).
13. D.E. Levin, M. Hollstein, M.F. Christman, E.A. Schwiers, and B.N. Ames, A new Salmonella tester strain (TA 102) with A-T base pairs at the site of mutation detects oxidative mutagens. Proc. Natl. Acad. Sci. U.S.A. 79, 7445-7449 (1982).
14. H.L. Gensler, and G.T. Bowden, Evidence suggesting a dissociation of DNA strand scissions and late-stage promotion of tumor cell phenotype. Carcinogenesis 4, 1507-1511 (1983).
15. T. Ochi, and P.A. Cerutti, Clastogenic action of hydroperoxy-5,8,11,13-eicosatetraenoic acids on the mouse embryo fibroblasts C3H/10T½. Proc. Natl. Acad. Sci. U.S.A. 84, 990-994 (1987).
16. H. Sies, P. Graf, and J.M. Estrela, Hepatic calcium efflux during cytochrome P-450 dependent drug oxidations at the endoplasmic reticulum in intact liver. Proc. Natl. Acad. Sci. U.S.A. 78, 3358-3362 (1981).
17. O. Augusto, L.R. Du Plessis, and C.L.V. Weingrill, Spin trapping of methyl radical in the oxidative metabolism of 1,2-dimethylhydrazine. Biochem. Biophys. Res. Commun. 126, 853-858 (1985).
18. L.C. Boffa, R.J. Gruss, and V.G. Allfrey, Aberrent and nonrandom methylation of chromosomal DNA-binding-proteins of colonic epithelial cells by 1,2-dimethylhydrazine. Cancer Res. 42, 382-388 (1982).
19. S. Ichimura, K. Mita, and M. Zama, Essential role of arginine residues in the folding of deoxyribonucleic acid into nucleosome cores. Biochemistry 21, 5329-5334 (1982).

20. W.K. Paik, P. Dimaria, S. Kim, P.N. Magee, and P.D. Lotlikar, Alkylation of protein by methyl methane sulfonate and 1-methyl-1-nitrosourea in vitro. Cancer Lett. 23, 9-17 (1984).
21. C. Turberville and V.M. Craddock, Methylation of nuclear proteins by dimethylnitrosamine and by methionine in the rat in vivo. Biochem. J. 124, 725-739 (1971).
22. R.L. Willson, Organic peroxy free radicals as ultimate agents in oxygen toxicity. In: Oxidative Stress (H. Sies, Ed.) pp. 41-72, Academic Press, London, 1985.
23. M.E. Hemler, H.W. Cook, and W.E.M. Lands, Prostaglandin biosynthesis can be triggered by lipid peroxides. Arch. Biochem. Biophys. 193, 340-345 (1979).

IN VIVO STIMULATION OF OXY RADICALS BY MOUSE SKIN TUMOR PROMOTERS

G. Witz, B. Czerniecki, S.C. Gad* and B.D. Goldstein

UMDNJ-Robert Wood Johnson Medical School
675 Hoes Lane, Piscataway, NJ
*G. D. Searle, Skokie, IL 60077

Our original suggestion in 1980 (1,2), that free radicals may be involved in tumor promotion was based on linkage of a number of puzzling observations about tumor promoters with newer insights into the biological and pathological role of free radicals. Of note was that the then known skin tumor promoters in the classic two-stage mouse skin tumor promotion system were all inflammatory agents. However, the lack of correlation among phorbol esters of tumor promoting ability with inflammation, as measured by standard techniques such as number of infiltrating inflammatory cells, seemed to preclude a role for inflammation in tumor promotion. Yet it was hard to ignore the many clinical observations, dating back to the nineteenth century, that inferentially seemed to point to chronic inflammation having some role in tumor development, observations ranging from the frequency of basal cell tumors on the bridge of the nose of eye glass wearers, to colorectal carcinoma in patients with ulcerative colitis. Our first approach to a possible role for free radicals in tumor promotion was to consider that the source of such free radicals in the mouse skin tumor initiation-promotion model might be derived from the activation of phagocytic cells, a known source of free radicals and active states of oxygen. Further, hepatic models of tumor promotion tend to feature compounds that are powerful inducers of cytochrome P-450 or of peroxisomes, both conceivably sources of free radicals or other active states of oxygen.

In collaboration with Dr. W. Troll, we evaluated the possibility that proteolytic enzymes might be involved in activating the burst of oxygen consumption which occurs in stimulated phagocytic cells. There is an obligatory lag period of about 30 seconds between initial stimulation and this burst of oxygen consumption. The NADPH oxidase responsible for

Address all correspondence to: Bernard D. Goldstein, M.D., Professor and Chairman, Depart. of Environmental & Community Medicine, UMDNJ-Robert Wood Johnson Medical School, 675 Hoes Lane, Piscataway, NJ 08854-5635 this process is located on the external surface of the cell membrane and leads to the formation of highly destructive active states of oxygen capable of effects ranging from perturbing the membrane, to DNA alterations and gross cytotoxicity (3). We observed that the burst of oxygen consumption, as well as the resultant production of $O_2^{\overline{\cdot}}$ and other active species, occurring in activated human polymorphonuclear leukocytes is inhibited by various compounds which have in common the ability to antagonize the effects of proteolytic enzymes. This effect of protease inhibitors was observed with a variety of stimuli, both phagocytic and non-phagocytic, used to activate $O_2^{\overline{\cdot}}$ production in human polymorphonuclear leukocytes (4). Among the stimuli was phorbol myristate acetate (PMA) whose tumor-promoting activity in mouse skin is diminished by anti-protease compounds (5). Inhibition of $O_2^{\overline{\cdot}}$ production was also noted in rat polymorphonuclear leukocytes and alveolar macrophages. The results indicated that proteolysis may be involved in activating the burst of oxygen consumption following stimulation of phagocytic cells.

The fact that protease inhibitors are also capable of inhibiting the tumor promoting abilities of PMA supported the hypothesis, shown in Figure 1, that the production of $O_2^{\overline{\cdot}}$ and related species by stimulated phagocytic cells might play a role in the mode of action of phorbol esters in tumor promotion. In our studies on $O_2^{\overline{\cdot}}$ production by human polymorphonuclear leukocytes (PMN), the rate of $O_2^{\overline{\cdot}}$ production was found to correlate with the tumor-promoting activity of the phorbol esters as opposed to their inflammatory activity. Mezerein and teleocidin B were slightly better stimulators of $O_2^{\overline{\cdot}}$ production than was PMA. Acetic acid was inactive (6).

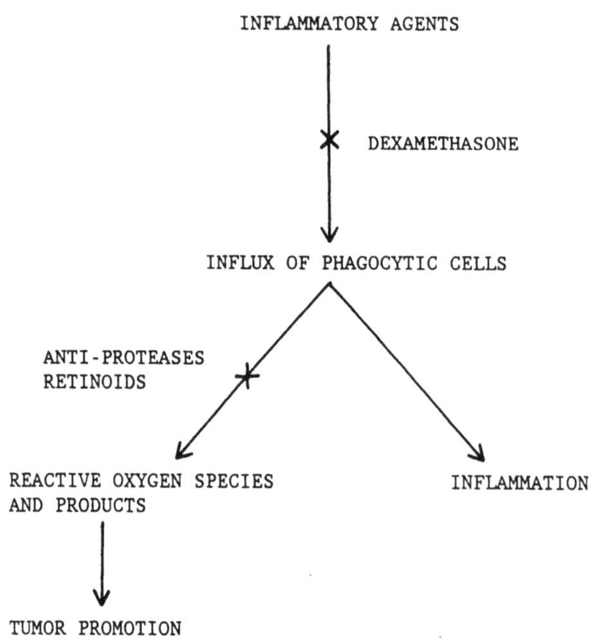

Figure 1. Hypothetical Scheme for the Involvement of Reactive Oxygen Species in Tumor Promotion

Retinoids also have been reported to interfere with tumor promotion (7). We tested various retinoid derivatives for their ability to inhibit PMA-stimulated $O_2^{\overline{\cdot}}$ production by human granulocytes. A dose-responsive inhibition was observed with all-trans-retinol, retinyl acetate and retinoic acid (8). Preliminary evidence that PMA produced hydrogen peroxide in mouse skin was also obtained (2). However this required the use of a cumbersome indirect assay not suitable for the type of manipulations required to assess whether free radicals were involved in tumor promotion in vivo, rather than in the limited *in vitro* systems described above.

In a series of ongoing studies we have further addressed the question of whether free radicals generated from inflammatory cells might be responsible for tumor promotion in vivo. Peritoneal macrophages constitute a ready population of phagocytic cells with the potential of producing oxy radicals upon stimulation. Coupled with the nitroblue tetrazolium (NBT) assay as a means of detecting cells actively producing superoxide anion radicals, murine peritoneal macrophages treated *in vivo* with tumor promoters and subsequently assessed *in vitro* by the NBT assay presented itself as a system for exploring the *in vivo* production of active states of oxygen. The latter is extremely difficult to demonstrate in tissues due to the transitory nature of reactive oxygen species. Indirect methods, however, such as the formation of dihydroxybenzoic acids from salicylate which acts as a trap for reactive hydroxyl radicals (9) potentially promise to be useful for the detection of oxygen radicals during the promotion process.

Resident unmanipulated female CD-1 mice (6 weeks old) or mice injected six days before the experiment with 1 ml 3% thioglycollate were treated intraperitoneally (i.p.) with phorbol myristate acetate (PMA), phorbol diacetate (PdA), 4-O-Me-PMA, phorbol dibutyrate (PdB), mezerein, or the calcium ionophore A23187, in 1 ml sterile phosphate buffered saline (PBS), two hours before sacrifice. The peritoneal exudate cells (PEC) were harvested and the NBT assay was performed as described previously (10). In this assay, NBT is reduced to an insoluble, blue-black formazan precipitate which accumulates in the cytoplasm. To determine that the reduction of NBT was specifically caused by superoxide anion radicals, the assay was also performed on samples to which superoxide dismutase (SOD, 0.1mg/ml final concentration) had been added to the PEC during the *in vitro* incubation with NBT. Using light microscopy, cells producing $O_2^{\cdot}$, or formazan (+) cells, were scored as those cells containing blue-black formazan granules in the cytoplasm. At least 250-500 cells were counted per sample.

The peritoneal exudate cell population from unmanipulated mice was found to consist of macrophages (70-80%, which will be referred to as resident macrophages), lymphocytes (25-30%), and less than 1% neutrophils, eosinophils and basophils as determined by light microscopy following acridine orange staining of the cells. At a dose of 0.1ug, only PMA stimulated superoxide anion radical production in PEC of resident mice, with PdB, 4-O-Me-PMA, mezerein, calcium ionophore A23187, PdA, and phorbol being inactive at this dose (Table 1). At doses 50-100 times greater than 0.1ug, 4-O-Me-PMA and mezerein stimulated oxy radical production in resident PEC, while PdA, PdB, the parent compound phorbol, and calcium ionophore A23187 were inactive (Witz and Czerniecki, unpublished data).

In contrast to the results obtained with PEC from resident mice, all promoters including the incomplete but first stage promoters 4-O-Me-PMA and the calcium ionophore A23187 and the incomplete but second stage promoter mezerein stimulated superoxide anion radical production at a 0.1ug dose in thioglycollate-elicited PEC (Table 1). These results indicate that oxy radical production depends not only on the nature of the initiating agent, but also on the physiological state of the

Table 1. Effect of Tumor Promoters and Non-Promoting Analogs Administered In Vivo on Peritoneal Exudate Cells O_2^- Production

Treatment (0.1 ug)	Resident mice, number formazan (+) PEC/250 PEC	Thioglycollate pretreated mice[b], number formazan (+) PEC/250 PEC
Phorbol-12-myristate-13-acetate	49	63
Phorbol-12,13-dibutyrate	0	71
4-O-methyl-PMA	2	30
Mezerein	2	80
Calcium ionophore A23187	1	55
Phorbol-12,13-diacetate	1	10
Phorbol	0	ND
PBS control	0	4

[a]Measured 2 hr after in vivo treatment.
[b]Mice injected with 1 ml 3% thioglycollate 6 days prior to treatment.

phagocytic cell involved, in this case, the peritoneal macrophage. The lack of an effect by mezerein at doses comparable to PMA in PEC from resident mice is notable, since, with very few exceptions such as its weak promoting activity and its minimal ability to stimulate the production of dark cells in mouse epidermis (11,12), PMA and mezerein have been reported to elicit responses of equal magnitude *in vivo* in mouse skin as well as *in vitro* (13,14,6).

In mouse skin, short chain phorbol esters and mezerein have been reported to inhibit promotion by PMA (11,15). It was therefore of interest to determine whether there is a correspondence between the mouse skin model tumor promotion system and the peritoneal macrophage oxy radical production system. Coadministration of PMA (0.1ug) with 1-1000ng PdB or 4-O-Me-PMA to resident CD-1 mice resulted in a dose-dependent decrease in the number of formazan-positive PEC relative to the number observed after treatment with PMA alone. At equimolar doses of PMA and 4-O-Me-PMA or PdB, the number of formazan positive PEC was reduced by about 50%. Coadministration with PdA or mezerein did not reduce the number of formazan-positive PEC resulting from PMA treatment in resident mice (16). In thioglycollate-pretreated mice, the number of formazan-positive PEC obtained after i.p. injection of 0.1ug mezerein was found to be decreased by coadministration of PdA, resulting in about a 50% decrease at equimolar doses of mezerein and PdA. A similar effect was not observed when PdA was coadministered with PMA to thioglycollate-pretreated CD-1 mice (10). Thus, while there is a certain correspondence between the inhibition of PMA promotion and the inhibition of the stimulation of oxy radical production in murine peritoneal exudate cells, this correspondence is not absolute.

Since PdA was found to inhibit oxy radical production stimulated by mezerein, it was of interest to determine whether PdA could affect second stage promotion by this compound, and whether the mouse peritoneal system could predict the findings in a bioassay. Phorbol diacetate had previously not been investigated for its effect on first or second stage promotion. Our results (10) indicate that PdA is an effective inhibitor of second stage promotion by mezerein in SENCAR mice (Table 2). Of interest also is the finding that coadministration of PdA with PMA (group #6) during second stage promotion did not appear to significantly affect the number of tumors/mouse, based on that obtained for the positive control (group #2). It was previously reported that PdA strongly inhibits promotion by PMA in NMRI mice in a 2-stage carcinogenesis experiment (15) while no effect was observed in a similar experiment in SENCAR mice initiated with DMBA and promoted with PMA/PdA (11). Our data suggest that the biochemical sequence of events during second-stage promotion by mezerein differs from that caused by PMA in the SENCAR mouse during the second stage promotional time period.

Table 2. Two-Stage Promotion in SENCAR Mice[a]

Group[b]	Treatment	Avg. no. of papillomas/ mouse (% inhibition)	% of mice w/papillomas	No. of mice w/carcinomas
1	DMBA, PMA, acetone	3.0	33	0
2	DMBA, PMA, mezerein	22.3	100	0
3	DMBA, PMA, mezerein/ PDA 2ug	13.8 (38)[c]	93	0
4	DMBA, PMA, mezerein/ PDA 20ug	12.4 (44)[c]	97	1
5	DMBA, PMA, PDA 20ug	1.4	33	2
6	DMBA, PMA, PMA/PDA 20ug	26.3	100	0

[a]Adapted from Czerniecki et al., Ref. 11.
[b]Groups 1-5 consisted of 30 mice per group and group 6 consisted of 20 mice. Promotion was terminated at 14 weeks.
[c]Significantly different from group 2 as determined by chi square analysis, $p < 0.05$.

The idea of a role for free radicals and active states of oxygen in tumor promotion has met the test of being a useful speculation in that it has led to a series of testable hypotheses and some interesting research findings from a number of laboratories, as has recently been extensively reviewed by Cerutti (17) and by Kensler and Taffe (18). While the overall hypothesis remains far from proven, it is not at all unlikely that free radicals are involved in at least certain aspects of tumor promotion, including progression to frank carcinoma. It must be emphasized that cancer can be considered to be a final common biological pathway with many different routes, each with a multiplicity of steps, leading to a clinically recognized cancer. The question is whether free radicals or related active species, and their products, play a role in any of these routes leading to human cancer. It is distinctly unlikely that all, or even most, of these routes, include a free radical step.

In conclusion, our studies implicate a role for active states of oxygen in the sequence of events leading to the induction of tumors by promoters. They also indicate that the peritoneal macrophage system has the potential for being useful in the identification of promoters and inhibitors of promotion, as well as in mechanistic studies on the biochemical effects caused by promoters in vivo.

Acknowledgements: The authors thank Mrs. N. Esterline and Miss F. Fountain for help in the preparation of this manuscript. These investigations were supported in part by PHS grant number CA33270 awarded by the National Cancer Institute, DHHS, and by a gift from the Harrington Foundation.

REFERENCES

1. G. Witz, B. D. Goldstein, M. A. Amoruso, D. S. Stone, and W. Troll, Stimulation of human polymorphonuclear leukocyte (PMN) superoxide anion radical (O_2^{-}·) production by tumor promoters. Proc. Am. Assoc. Cancer Res. 21, 112 (1980).
2. B. D. Goldstein, G. Witz, J. Zimmerman and C. Gee, Free radicals and reactive oxygen species in tumor promotion. In: Oxy Radicals and Their Scavenger Systems, Vol. 2: Cellular and Medical Aspects. (R. A. Greenwald and G. Cohen, Eds.) pp. 321-325. Elsevier, New York (1983).
3. B. M. Babior, Oxidants from phagocytes: Agents of defense and destruction. Blood 64, 959-966 (1984).
4. B. D. Goldstein, G. Witz, M. A. Amoruso and W. Troll, Protease inhibitors antagonize the activation of polymorphonuclear leukocyte oxygen consumption. Biochem. Biophys. Res. Commun. 88, 854-860 (1979).
5. W. Troll, A. Klassen and A. Janoff, Tumorigenesis in mouse skin: Inhibition by synthetic inhibitors of proteases. Science 169, 1211-1213 (1970).
6. B. D. Goldstein, G. Witz, M. A. Amoruso, D. S. Stone and W. Troll, Stimulation of human polymorphonuclear leukocyte superoxide anion radical production by tumor promoters. Cancer Lett. 11, 257-262 (1981).
7. G. Witz, B. D. Goldstein, M. A. Amoruso, D. S. Stone and W. Troll, Retinoid inhibition of superoxide anion radical production by human polymorphonuclear leukocytes stimulated with tumor promoters. Biochem. Biophys. Res. Commun. 97, 883-888 (1980).
8. A. K. Verma, B. G. Shapas, H. M. Rice and R. K. Boutwell, Correlation of the inhibition by retinoids of tumor promoter-induced mouse epidermal ornithine decarboxylase activity and of skin tumor promotion. Cancer Res. 39, 419-425 (1979).
9. R. A. Floyd and J. J. Watson, Sensitive assay of hydroxyl free radical formation utilizing high pressure liquid chromatography with electrochemical detection of phenol and salicylate hydroxylation products. J. Biochem. Biophys. Methods 10, 221-235 (1985).
10. B. Czerniecki, S. C. Gad, C. Reilly, A. C. Smith and G. Witz, Phorbol diacetate inhibits superoxide anion radical production and tumor promotion by mezerein. Carcinogenesis 7, 1637-1641 (1986).
11. T. J. Slaga, S. M. Fischer, K. Nelson and G. L. Gleason, Studies on the mechanism of skin tumor promotion: Evidence for several stages in promotion. Proc. Natl. Acad. Sci. U.S.A. 77, 3659-3663 (1980).

12. A. J. Klein-Szanto, S. M. Major and T. J. Slaga, Induction of dark keratinocytes by 12-O-tetradecanoylphorbol-13-acetate and mezerein as an indicator of tumor promoting efficiency. Carcinogenesis 1, 399-403 (1980).
13. R. K. Boutwell, The function and mechanism of promoters of carcinogenesis. CRC Crit. Rev. Toxicol. 2, 419-443 (1974).
14. P. M. Blumberg, In vitro studies on the mode of action of the phorbol esters, potent tumor promoters. CRC Crit. Rev. Toxicol. 8, 153-193 (1980).
15. R. Schmidt and E. Hecker, Simple phorbol esters as inhibitors of tumor promotion by TPA in mouse skin. In: Carcinogenesis. Vol. 7. (E. Hecker et al., Eds.) pp. 57-63. Raven Press, New York (1982).
16. G. Witz and B. Czerniecki, In vivo inhibition by phorbol esters of oxy radical production by mouse preritoneal exudate cells stimulated with phorbol myristate acetate. Ann. N.Y. Acad. Sci. 736, in press.
17. P. A. Cerutti, Prooxidant states and tumor promotion. Science 277, 375-381 (1985).
18. T. W. Kensler and B. G. Taffe. Free radicals in tumor promotion. Adv.Free Radical Biol. Med. 2, 347-387, 1986.

UVB PREVENTION OF MOUSE SKIN TUMOR INDUCTION BY THE INITIATION-PROMOTION PROTOCOL

H.L. Gensler

Department of Radiation Oncology and Cancer Center
University of Arizona Health Sciences Center
Tucson, Arizona 85724

INTRODUCTION

Repeated exposure of mice to UVB (280 to 320 nm) radiation results in the development of a state of tumor-specific suppression that renders the host incapable of mounting an effective immunological rejection response to antigenic determinants expressed by UV-induced skin tumor cells (1,2). In this system, normal mice can reject transplanted UV-induced tumor cells, whereas UV-treated mice cannot. The objective of this research was to determine if UVB irradiation influences mouse skin tumorigenesis induced by an initiation-promotion protocol.

METHODS

Six week old female $B6D2F_1/J$ mice were used for the initiation-promotion regimen. BALB/c mice were used for the complete carcinogenesis protocol. The mice were housed on a shelf 20 cm below a bank of 6 Westinghouse FS40 sunlamps. The dorsal fur of the mice was shaved weekly during the UV treatment period. The Westinghouse FS40 lamps delivered an average dose of 4.8 Jm^{-2} per sec over the wavelength range of 280-340 nm. This range includes approximately 80% of the total energy output of the lamps. A group of 56 mice were UVB-irradiated with five 30 min dorsal treatments per week for 11.5 weeks. The total UV dose received by the mice during the 11.5 week period, measured under the wire cage top at the level of the backs of the mice was approximately 4.9×10^5 Jm^{-2}. Four days after cessation of the ultraviolet irradiation treatments, 100 μg of 7,12-dimethylbenz[a]anthracene (DMBA) was applied to the shaved ventral surface of 56 irradiated mice and 41 unirradiated, age-matched control mice. Beginning 4 days after initiation with DMBA, the mice were treated with 5 μg of 12-O-tetradecanoylphorbol-13-acetate three times weekly throughout the experiment. In the complete carcinogenesis protocol, benzo[a]pyrene (BP) was applied to the shaved ventral surface of BALB/c mice twice weekly. Mice received 1.0 mg BP for 10 weeks, 0.1 mg BP twice weekly for 20 weeks, or solvent for 20 weeks.

RESULTS

The mean number of ventral tumors in mice treated ventrally with DMBA and TPA is shown for selected times during promotion in Figure 1. After 20 weeks of tumor promotion, the mean number of tumors per mouse was 4.5. Pretreatment of such animals on the dorsal surface with UV radiation dramatically decreased the development of ventral tumors to a mean value of 1.2 Tumor occurrence did not follow a normal distribution, so that a nonparametric analysis of the data was required (3). Wilcoxon rank-sum tests were performed on the mean number of tumors per mouse between the UV and non-UV groups to test the hypothesis that there was a significant different between the medians of the two groups. While the difference between medians was not yet significant at 9 weeks, by week 12, there was a significant different ($p < 0.001$) and that difference continued throughout the duration of the experiment.

The percentage of mice which were ventral tumor-bearers at different time points is depicted in Figure 2. This initiation-promotion protocol resulted in a tumor incidence of 80% after 12 weeks of promotion. Pre-treatment with dorsal UV irradiation reduced the tumor incidence to 23% at the 12 week time point. After 20 weeks of tumor promotion, there were still 20% fewer tumor bearers in the irradiated than in the unirradiated mice. These results were analyzed with a censored rank order test devised by Breslow (4). A significant difference in the duration of tumor-free survival was found between the UV-irradiated and unirradiated

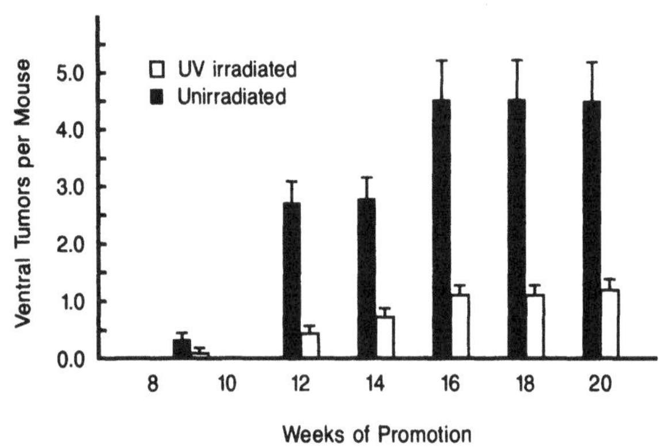

Figure 1. The prevention of two-stage skin tumorigenesis by previous exposure of distant skin to UVB-radiation. Mice were initiated on the ventral surface with 100 μg DMBA and promoted with 5 μg TPA three times weekly.

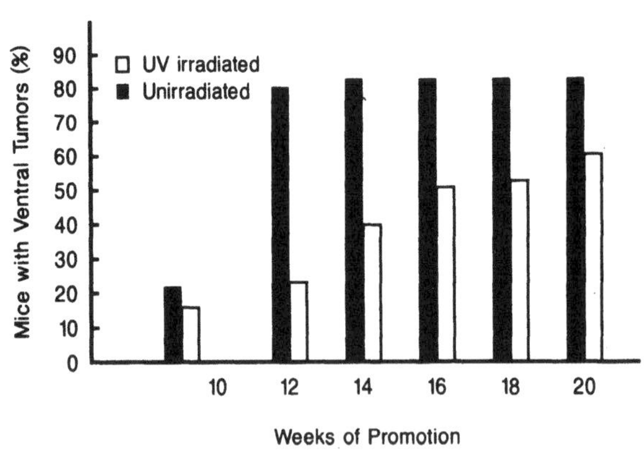

Figure 2. The influence of prior UVB-irradiation upon the time required for mice to become tumor-bearers after initiation and promotion.

mice ($p < 0.001$). The results presented in Figure 3 demonstrate that pretreatment of mice with dorsal UVB-irradiation enhanced rather than reduced the incidence of ventral tumors induced by benzo[a]pyrene.

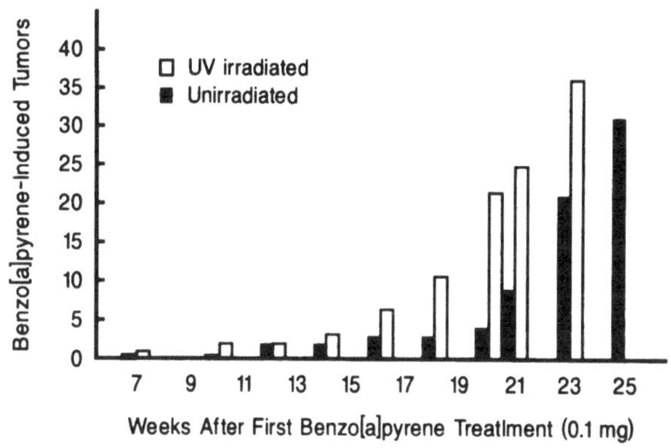

Figure 3. The effect on benzo[a]pyrene-induced mouse skin tumorigenesis of previous exposure of distant skin to UVB-radiation. A group of 18 BALB/c mice received UVB-irradiation consisting of five 30-minute dorsal treatments per week for 13 weeks. Eighteen mice received no UV irradiation before carcinogenesis. Benzo[a]pyrene (0.1 mg) was applied ventrally twice weekly for 20 weeks.

DISCUSSION

UVB irradiation of dorsal skin significantly reduced tumorigenesis by subsequent chemical initiation and promotion on ventral skin. The preventive action must have occurred by a systemic mechanism in view of the dorsal exposure to UV radiation and the ventral exposure to initiation and promotion. Conversely, dorsal UVB-irradiation enhanced subsequent ventral tumor induction by benzo[a]pyrene in a complete carcinogenesis protocol. These last results are similar to those reported by Roberts and Daynes (5). They irradiated C3Hf/HeN mice for 29 weeks, starting 3 weeks before the benzo[a]pyrene treatments (0.1 mg BP twice weekly). They found that development of BP-induced tumors was accelerated in UV-irradiated mice, relative to unirradiated controls. It is noteworthy that the results were similar in experiments involving 5 (present study) or 29 (Roberts and Daynes) weeks of UVB-irradiation, as if the total acceleration is due to the 3-5 weeks of exposure before carcinogenesis. The enhancing effect of UVB radiation may be immunologic, as has been shown in UV-tumorigenesis (1,2). Whether the UVB irradiation inhibited initiation, promotion or both in this study, was not determined. UVB irradiation has been found to enhance 2-stage carcinogenesis resulting from initiation with UVB radiation and promotion with TPA (6). However, the apparent UV enhancement of promotion by TPA may have resulted from a reduction in immunological surveillance of antigenic tumors, as has been demonstrated in UV carcinogenesis (7). Tumors induced by UVB irradiation are highly antigenic (8), so that tumors initiated with multiple UVB treatments may have some antigenicity. Tumors induced by low levels of chemical carcinogens are not antigenic (9). There is precedence for agents which enhance complete carcinogenesis by DMBA but inhibit initiation and promotion by DMBA and TPA. These agents are 1α,25-dihydroxyvitamin D_3 (10-12) and retinoic acid (13). They are both antipromoters.

In conclusion, UVB irradiation inhibited tumorigenesis by an initiation-promotion protocol. A systemic mechanism was involved in the prevention.

REFERENCES

1. M.L. Kripke, The immunology of skin cancer. In Immunology and Cancer, The University of Texas M.D. Anderson Symposium on Fundamental Cancer Research, Vol. 38, 1986 (M.L. Kripke and P. Frost, Eds.) pp. 113-120, University of Texas Press, Austin, Texas.
2. C.W. Spellman and R.A. Daynes, Ultraviolet light induced murine suppressor lymphocytes dictate specificity of anti-ultraviolet tumor immune responses. Cell. Immunol. 38, 25-34 (1978).
3. W.J. Conover, Practical Nonparametric Statistics, 1980, John Wiley and Sons, New York.
4. N. Breslow, A generalized Kruskal-Wallis test for comparing K samples subject to unequal patterns of censorship. Biometrika 57, 579-594 (1970).
5. L.K. Roberts and R.A. Daynes, Modification of the immunogenic properties of chemically induced tumors arising in hosts treated concomitantly with ultraviolet light. J. Immunol. 125, 438-447 (1980).
6. P.T. Strickland, D. Creasia and M.L. Kripke, Enhancement of two-stage skin carcinogenesis by exposure of distant skin to UV radiation. J. Natl. Cancer Inst. 74, 1129-1134 (1985).
7. M.S. Fisher and M.L. Kripke, Suppressor T lymphocytes control the development of primary skin cancers in ultraviolet-irradiated mice. Science 216, 1133-1134 (1982).
8. M.L. Kripke, Latency, histology, and antigenicity of tumors induced by ultraviolet light in three inbred mouse strains. Cancer Res. 37, 1395-1400 (1977).
9. L.K. Roberts and R.A. Daynes, Modifications of the immunogenic properties of chemically induced tumors arising in hosts treated concomitantly with ultraviolet light. J. Immunol. 124, 438-444 (1980).
10. A.W. Wood, R.L. Chang, M.T. Huang, M. Uskokovic and A.H. Conney, 1α,25-dihydroxyvitamin D_3 inhibits phorbol ester-dependent chemical carcinogenesis in mouse skin. Biochem. Biophys. Res. Comm. 116, 605-611 (1983).
11 A.W. Wood, R.L. Chang, M.T. Huang, E. Baggiolini, J.J. Partridge, M. Uskokovic and A.H. Conney, Stimulatory effect of 1α,25-dihydroxyvitamin D_3 on the formation of skin tumors in mice treated chronically with 7,12-dimethylbenz[a]anthracene. Biochem. Biophys. Res. Comm. 130, 924-931 (1985).
12. K. Chida, H. Hashiba, M. Fukushima, T. Suda and T. Kuroki, Inhibition of tumor promotion in mouse skin by 1α,25-dihydroxyvitamin D_3. Cancer Res. 45, 5426-5430 (1985).
13. A.K. Verma, E.A. Conrad and R.K. Boutwell, Differential effects of retinoic acid and 7,8-benzoflavone on the induction of mouse skin tumors by the complete carcinogenesis process and by the initiation-promotion regimen. Cancer Res. 42, 3519-3525 (1982).

A POSSIBLE ROLE FOR FREE RADICALS IN TUMOR PROGRESSION

J.B. Rotstein, J.F. O'Connell and T.J. Slaga

The University of Texas System Cancer Center
Science Park-Research Division, Smithville, TX 78957

INTRODUCTION

The possible role of free radicals during experimental chemical carcinogenesis has been recently reviewed and speculated on by several authors (1-4). Much of the direct evidence for free-radical participation in carcinogenesis has come from the mouse skin model system.

The mouse skin system has been one of the most studied models for determining the sequence of events that occur during the carcinogenic process. The model consists of two operational steps. In the first, the "initiation" stage, the backs of mice are treated with a single, subcarcinogenic dose of a topically applied carcinogen, such as 7,12-dimethylbenz[a]anthracene (DMBA). This process alters a segment of the cell population such that it can respond to the second operation. In the second step, the "promotion" stage, mice receive multiple treatments of an agent such as 12-O-tetradecanoylphorbol-13-acetate (TPA) that amplifies the "initiated" cell population. This 2-step protocol can generate a relatively large number of benign lesions (papillomas) which may progress to squamous cell carcinoma, a malignant lesion. This progression is investigator-independent.

In this paper we will focus on the evidence suggesting the role of free radicals during tumor promotion and tumor progression. Possible mechanisms of tumor progression involving free radicals are also discussed.

FREE RADICALS AND TUMOR PROMOTION

Tumor promotion is a process by which an initiated cell population is expanded to form visible lesions. The lesions formed are not malignant, but they can participate in the subsequent steps leading to malignancy. Generally, the promoting agent is nonmutagenic and does not generate tumors, either benign or malignant, in the absence of other agents. Although the vast majority of carcinomas (90%) form from papillomas, typically, very few (5-10%) papillomas go on to form carcinomas.

The possibility that free radicals are important in tumor promotion was suggested by the fact that, in the mouse skin system, promoting agents are also inflammatory agents. Therefore, it was assumed that the infiltrating neutrophils were a source for free radicals. It was later shown that TPA, the most commonly used skin tumor promoter, could generate free radicals from neutrophils. However, more direct evidence of free radical generation by neutrophils came from studies in the mouse skin system. Particularly convincing was the demonstration that benzoyl peroxide (BzPo), a free radical generator, could be used in a promotion regimen to generate papillomas (4). Subsequent studies by Slaga and coworkers (1) showed that several free radical generators, such as lauroyl peroxide, decanoyl peroxide, cumene peroxide, chloroperbenzoic acid, P-nitroperoxybenzoic acid, tertbutyl-hydroperoxide, methylethylketone peroxide and hydrogen peroxide, could promote tumor formation in mouse skin. These studies were followed up by experiments in which antioxidants, such as butylated hydroxyanisole (BHA), butylated hydroxytoluene (BHT) or disulfiram, were used concomitantly with the tumor promoter, either TPA or benzoyl peroxide, to determine if inhibition of the possible free radical effect could prevent tumor formation. The data showed that tumor formation was significantly reduced (1).

The possibility of free radical involvement during promotion was supported by Solanki et al. (5), who demonstrated a decrease in superoxide dismutase and catalase activity after mouse skin was exposed to TPA. Complementary to this finding, Solanki showed that the application of a compound containing copper diisopropylsalicylate (CuDIPS), a superoxide dismutase-mimetic agent, along with TPA could inhibit TPA's promoting ability (6). A similar, though more extensive, study was carried out by Egner and Kensler (7). Their results imply a role for reactive oxygen species during promotion.

Further investigations by Kensler and coworkers (8,9) into the relationship between free radicals and tumor promotion involved the use of antioxidants. It was found that BHA and BHT were most effective in inhibiting TPA promotion; less effective were α-tocopherol, propyl gallate and disulfiram; L-ascorbate was inactive. These findings were followed up by an extensive structural study of the potency of BHA isomers at inhibiting TPA tumor promotion (9). The findings showed that the hydroxyl and t-butyl substituents were critical to BHA inhibitory action.

More recently, endogenous free radical scavengers have been used to inhibit papilloma formation. Perchellet has shown that vitamin E, glutathione and selenium have both individual and additive effects in preventing the generation of TPA-induced tumors (10,11). They have also provided evidence suggesting that the way these compounds work is by their ability to enhance glutathione-dependent free radical scavenging systems.

Studies using tissue culture model systems have also given credence to the theory of free racidal participation during promotion. Kennedy and coworkers (12) demonstrated that the addition of catalase to a culture of C3H 10T1/2 cells prevented the enhancement of TPA-induced transformation. This work indicated that hydrogen peroxide might be important in malignant transformation. Using a similar system, Borek et al (13) showed that the addition of selenium and/or vitamin E to a culture prior to TPA exposure significantly inhibited malignant transformation. Interestingly, the addition of selenium boosted the levels of cellular glutathione, glutathione peroxidase and catalase, suggesting that selenium may work by activating or augmenting cellular free-radical-scavenging systems.

Colburn and colleagues (14) have also investigated the role of free radicals during promotion. In these studies JB6 mouse epidermal cells were exposed to several hydroxyl radical scavengers, antioxidants, eliminators of H_2O_2, or superoxide dismutases, in order to prevent TPA-induced transformation. The results implicated a requirement for superoxide anion. Other studies by that group (15) demonstrated that BzPo could induce transformation in the JB6 cell line. BzPo treatment and the subsequent transformation resulted in a reduction in the level of a major epidermal ganglioside, trisialoganglioside, in the culture. The addition of the ganglioside to the culture prior to BzPo exposure prevented transformation. Consistent with this finding was that BzPo was able to inhibit metabolic cooperation between cultured human epidermal keratinocytes (16).

In recent studies, using freshly isolated mouse keratinocytes, it was demonstrated that several tumor promoters decrease the ratio of reduced to oxidized glutathione within the cell (17), suggesting that tumor promoters initiate a free radical attack. The effect can be inhibited by supplementary amounts of reduced glutathione. This finding is consistent with earlier work showing a decrease in superoxide dismutase and catalase levels after TPA exposure (5).

Recently, it has also been shown that keratinocytes can act as a source for free radicals (18,19). This is a fundamental observation, since it had been assumed that neutrophils were necessary intermediaries in the induction of free radical damage to keratinocytes. This study demonstrated that TPA could induce chemiluminescence (CL), an indication of membrane perturbation in mouse epidermal cells. Using inhibitors of arachidonic acid metabolism, the research showed that most of the CL induced by TPA was due to the metabolism of arachidonic acid. More recent studies have shown that several antioxidants are effective in preventing TPA-induced CL.

More basic has been the examination of free radical generators. Previous work was based on the assumption that free radical generators do form free radicals under experimental conditions. However, many free radical generators are stable at body temperature (20) and, therefore, hypothetically unlikely to generate radicals. Recently, it has been possible to show, using electron spin traps, that cumene hydroperoxide does generate free radicals in mouse keratinocyte cytosolic fractions (21).

In summary, free radicals appear to have a role in promotion because free radical generators can act as tumor promoters and because free radical scavengers and antioxidants can inhibit the effects of promoters. These effects have been demonstrated in vivo and in vitro.

FREE RADICALS AND TUMOR PROGRESSION

Tumor progression can be described as a phase of carcinogenesis in which a population of premalignant lesions acts as a source for cancer formation. The process occurs without the need of experimental intervention. The study of tumor progression is not only of academic interest, but is also potentially relevant to practical medicine, since this part of experimental carcinogenesis parallels the clinical situation.

There are several reasons why free radicals may be involved in tumor progression. The premalignant lesion contains numerous inflammatory cells. Macrophages and polymorphonucleocytes are known sources of free radicals. In addition, the premalignant lesion is characterized by necrosis. Dying or damaged cells not only attract inflammatory cells, but can also generate free radicals. In addition, premalignant populations have a relatively large amount of cell proliferation. Since DNA undergoing replication is particularly sensitive to damage, free radicals, which are normal cell constitutents, may be endogenous agents for DNA damage during cell division. This possibility is consistent with the appearance of chromosomal abnormalities, which become more bizzare during the progression process. Because inflammatory and necrotic cells can act as sources of free radicals and the large number of proliferating cells present in a premalignant lesion are susceptible to them, free radicals may be involved in tumor progression.

Until recently, very few studies had been carried out on experimental tumor progression. However, much of the work that has been done has used the 2-stage mouse skin model, which has at least two major advantages for the study of tumor progression. First, the model permits direct visualization of the transition from papilloma to carcinoma. Thus, the investigator knows whether the carcinoma that appears came from preexisting papillomas or arose de novo. Second, it allows modulating agents to be applied directly to the premalignant population. In this way, the target tissue is predominantly exposed and systemic effects are lessened.

Using the 2-stage mouse skin model, Hennings et al. (22) demonstrated that malignant conversion could be increased if tumor-bearing mice were repeatedly exposed to carcinogenic doses of N-methyl-N'-nitro-N-nitrosoguanidine (MNNG), 4-nitroquinoline-N-oxide or urethane. All of these agents can be used as initiators in the 2-stage model. The authors concluded that, like initiation, malignant conversion is probably also a genetic event, since of all the agents tested, only the initiating agents were capable of increasing malignant conversion. The promoting agent TPA was unable to enhance the process. This conclusion was supported by the results of O'Connell et al. (23). In this study, tumor-bearing mice were exposed to four noncarcinogenic doses of either ethylnitrosourea (ENU), MNNG or TPA. Thereafter, all groups were repeatedly treated with TPA. The results were consistent with the earlier study: ENU and MNNG, agents that can initiate the carcinogenic process, also increased malignant conversion; TPA, a noninitiator, did not increase the number of cancers. Further studies (J.F. O'Connell, unpublished data) have examined the progressing potential of mezerein, a second-stage promoter, and of calcium ionophore A23187, a first-stage promoter; both of these compounds were shown to be weak progression agents. These findings are consistent with the hypothesis that progression involves a mutation-like event, since these compounds are nonmutagenic and inactive as both initiating agents and complete carcinogens.

Recently, O'Connell and coworkers (24) examined the possibility that free radicals may be involved in tumor progression. In that experiment, papillomatous mice were treated with BzPo twice-a-week for 20 weeks. BzPo, a free-radical-generating agent, is not a complete carinogen, and cannot initiate the carcinogenic process in the 2-stage mouse skin model (4). Though it is only moderately active compared with TPA, it can be used to promote papilloma formation. The results showed that BzPo-treated mice had an increased carcinoma incidence (70% vs 37%) and a higher cumulative number of cancers than the TPA-treated groups of mice (40 vs 11 carcinomas, respectively). These results were the first direct evidence that free radicals could enhance tumor progression. The findings have been repeated and extended by Rotstein and coworkers (25). BzPo was demonstrated to be a more potent enhancer of progression after 4, 12 and 20 weeks of twice-a-week application (carcinoma incidences of 61, 74 and 81%, respectively) than acetone (carcinoma incidence 41%). Thus, even a relatively brief exposure to BzPo can significantly enhance tumor progression. In the same study, the potency of hydrogen peroxide (H_2O_2) as a tumor progressing agent was tested. H_2O_2 is neither a carcinogen nor an initiating agent; it is an extremely weak promoting agent in this model (26). When H_2O_2 was applied twice-a-week at a dose of 15 or 5% to the backs of tumor-bearing mice for 20 weeks, the carcinoma incidence increased to 68 and 64%, respectively, from 41% in the acetone controls. The conversion rate of papilloma to carcinoma in the higher H_2O_2 dose group was 12%, compared with 11% in the lower H_2O_2 dose group and 8% in the acetone-treated group. This was a second example of a free-radical-generating agent enhancing tumor progression. The data from a follow-up study suggest that limiting BzPo treatments to 1 or 4 exposures results in an increased tendency to form carcinomas, whereas limited application of 15% H_2O_2 (1, 4 or 8 exposures) is not effective in enhancing progression (J.B. Rotstein, unpublished data). It is interesting that the potency of a progressing agent does not necessarily parallel its promoting ability (Table 1). This seems to be true for both BzPo and H_2O_2. These experiments illustrate one

Table 1. Activity of Promoting Agents Throughout Carcinogenesis

Chemical	Complete Carcinogenesis	Initiation	Promotion	Progression
Acetone	—	—	—	Weak
TPA	—	—	Strong	Weak
Mezerein	—	—	Moderate (2nd stage)	Weak
Benzoyl peroxide	—	—	Moderate	Strong
Hydrogen peroxide	—	—	Weak	Strong
Acetic acid	—	—	Weak	Strong
Calcium Ionophore A23187	—	—	Moderate (1st Stage)	Weak
Diethylhexyl-phthalate	—	—	Moderate	Weak

Dash indicates no activity

approach for exploring the role of free radicals during progression.

A second approach used in our laboratory to demonstrate the possible role of free radicals during progression utilized free radical scavengers and antioxidants to inhibit cancer formation. In this experiment, copper diisopropylsalicylate, vitamin E, BHA, glutathione (GSH) or acetone was applied topically to papillomatous mice. The doses used were similar to those that had been demonstrated to be effective in decreasing papilloma formation during TPA-induced promotion (1). The preliminary results showed the acetone-treated control group to have a 70% cancer incidence. The other groups had similar incidences, except the GSH-treated group, which had only a 50% cancer incidence. Consistent with these results was the slightly lower conversion rate in the GSH group. These findings are further evidence that free radicals may play a role in tumor progression, since topical application of a free radical scavenger like GSH inhibited cancer formation. These results are summarized in Table 2.

Table 2. Inhibitors of Promotion Tested for Activity During Progression

Chemical	Promotion	Progression
Butylated hydroxyanisole	Moderate	—
Vitamin E	Moderate	—
Copper Diisopropylsalicylate	Strong	—
Sodium Benzoate	Moderate	—
Glutathione	Moderate	Moderate

Dash indicates no activity.

A third approach to test the role of free radicals during progression uses chemicals that deplete or overwhelm free radical defense mechanisms. In this study, papilloma-bearing mice were treated with one of the following agents: acetic acid (AA), which would overwhelm free radical defenses by cytotoxicity; aminotriazole (AT), which would decrease catalase activity; diethylmaleate (DEM), which would decrease cellular GSH levels; diethylhexylphthalate (DEHP), a peroxisome proliferator that would generate excess H_2O_2 within the cell and overcome free radical defenses; or acetone as a control. The results are summarized in Table 3. The data show that DEM and AA are effective enhancers of progression (77% with DEM; 80% with AA; 47% with acetone), whereas DEHP and AT are not effective at the doses given. This experiment suggests that at least one inhibitor of the free radical defense system, DEM, can enhance tumor progression and that by overwhelming these defenses, as in the case with AA, cancer formation can be increased.

Table 3. Effect of Modifiers of Free Radical Defenses During Progression

Chemical	Effect	Progression
Diethylameate	↓ Glutathione	Strong
Aminotriazole	↓ Catalase	Weak
Diethylhexylphthalate	Peroxisome proliferator	Weak
Acetic Acid	Toxicity	Strong

In this section we have detailed three experimental approaches to examine the hypothesis that free radicals have a role during tumor progression. The data presented here are consistent with the hypothesis for several reasons: (1) free radical generators, when applied to premalignant populations, enhance cancer formation; (2) at least one free radical scavenger can inhibit the formation of malignant lesions; and (3) when free radical defenses are lowered or overwhelmed, the development of carcinoma is enhanced. The possible mechanisms by which free radicals increase cancer formation are discussed in the next section.

FREE RADICALS AND POSSIBLE MECHANISMS FOR ENHANCING TUMOR PROGRESSION

One possible mechanism for the enhancement of tumor progression by free radicals is genetic. By directly acting on the genome, free radicals may induce chromosomal abnormalities, which have been observed to increase in degrees of aberration throughout tumor progression (27). Consistent with this possibility is the fact that both benzoyl peroxide and H_2O_2 can induce DNA strand breakage in human epithelial cells (28).

Interestingly, it has been shown that DNA strand breaks created by H_2O_2 are more slowly repaired than breaks induced by ionizing radiation (29). Thus, an opportunity may exist for these longer-lasting errors to have a more permanent effect on the genome owing to a subsequent base change or chromosome rearrangement. The consequences of free radical interaction with DNA could be a mutation; a subtler possibility is a change in the state of methylation. Preliminary data suggest that there is a gradation from methylated to undermethylated to highly undermethylated DNA as normal mouse epidermis changes from papillomatous to cancerous (J.N. La Peyre, J.F. O'Connell, F.F. Becker, and T.J. Slaga, unpublished results). The effect of a mutation or altered methylation may explain, in part, the number of changes in protein expression that occur during tumor progression, such as the lack of high-molecular-weight keratins (30) and the presence of gamma-glutamyltransferase (30,31).

Strand breakage has been associated with accelerated terminal differentiation in mouse epidermal cells (32). It is possible, therefore, that DNA strand breaks and DNA cross-linking are part of a programmed cell death, which can be induced by free radical damage. Thus, free radicals may enhance tumor progression as a consequence of cytotoxic effects. Free radicals are known to be capable of damaging not only DNA but also proteins and lipids. Cells within a papilloma may be killed by free radical damage; as a result, more resistant cells could form a regenerative response leading to a more aggressive tumor. It is also possible that free radicals are killing the cells surrounding the papilloma, allowing the more resistant cells of the papilloma to expand into the surrounding area.

A third possible mechanism of tumor growth enhancement consists of a combination of the first two. Free radicals could kill some cells within a tumor, but others suffering nonlethal mutagenic damage could give rise to a regenerative response. This response could result in a chromosomally aberrant cell population, which would become more aberrant with tumor progression.

SUMMARY

The evidence for free radical involvement during chemical carcinogenesis is outlined in Figure 1. Free radicals can aid in carcinogen metabolism; therefore, they may play a role during initiation. Free radicals can also interact with DNA, which suggests that they may be able to create a critical mutation. This action could initiate a population of cells and begin the carcinogenic process. Most of the work on free radicals and cancer development has examined the role of free radicals during tumor promotion. In this stage of the process, it has been demonstrated that inflammatory cells and epidermal cells can act as a source of free radicals; these radicals can be stimulated by the most commonly used skin tumor promoter, TPA. Furthermore, free radical generators, such as benzoyl peroxide, can promote tumor formation. Other evidence supporting this hypothesis is that antioxidants and free radical scavengers can inhibit the tumor formation process. Similarly, during tumor progression, it has been shown that free radical generators can increase cancer for-

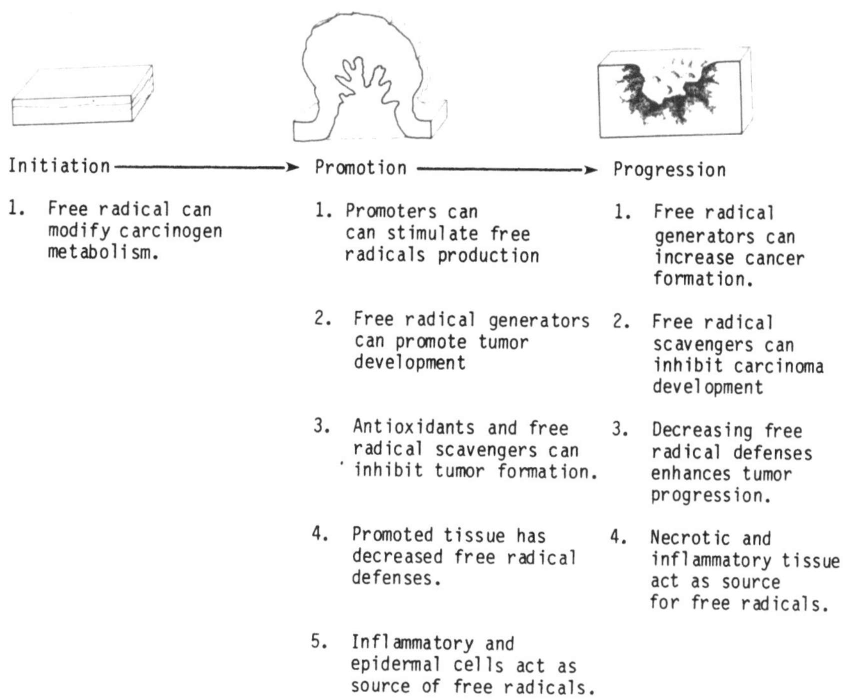

Figure 1. Free Radicals and Carcinogenesis

mation and that at least one free radical scavanger can inhibit the development of carcinomas. In addition, if the level of free radical defense is decreased or overwhelmed, then tumor progression is enhanced.

It can be argued that free radicals are more likely to influence tumor progression than any other stage of carcinogenesis. This hypothesis can be supported by the fact that progression is the longest phase of the carcinogenic process and that free radicals are continuously present, even in normal cells. In addition, the observation that premalignant lesions have a large number of proliferating cells suggests that the target for free radicals is greatest during this part of the process. The consequences of such an interaction are consistent with the ever-increasing chromosomal abnormalities seen during tumor progression (27) and with the large extent of cell death. The involvement of free radicals during tumor progression is an interesting possibility and worthy of further exploration as a component of cancer development.

ACKNOWLEDGEMENTS

This work was supported in part by NIH grants CA-34962 and CA-43278. The authors wish to thank Christie Hoy and Karen Engel for preparing this manuscript.

REFERENCES

1. T.J. Slaga, V. Solanki and M. Logani, Studies on the mechanism of action of antitumor promoting agents: suggestive evidence for the involvement of free radicals in promotion. In: Radioprotectors and Anticarcinogenesis. (O.F. Nygaard and M.G. Simic, Eds.), pp. 471-485 Academic Press, New York, 1983.
2. P.A. Cerutti, Prooxidant states and tumor promotion. Science 227:375-381 (1985).
3. W. Troll and R. Wiesner, The role of oxygen radicals as a possible mechanism of tumor promotion. Annu. Rev. Pharmacol. Toxicol. 25: 509-528 (1985).

4. T.J. Slaga, A.J.P. Klein-Szanto, L.L. Triplett, L.P. Yotti and J.E. Trosko, Skin tumor promoting activity of benzoyl peroxide, a widely used free radical generating compound. Science 213: 1023-1025 (1981).
5. V. Solanki, R.S. Rana and T.J. Slaga, Diminution of mouse epidermal superoxide dismutase and catalase activities by tumor promoters. Carcinogenesis 2: 1141-1146 (1981).
6. V. Solanki, L. Yotti, M.K. Logani and T.J. Slaga, The reduction of tumor initiating activity and cell mediated mutagenicity of dimethylbenz[a]-anthracene by a copper coordination compound. Carcinogenesis 5: 129-131 (1984).
7. P.A. Egner and T.W. Kensler, Effects of a biomimetic superoxide dismutase on complete and multistage carcinogenesis in mouse skin. Carcinogenesis 6: 1167-1172 (1985).
8. W.J. Kozumbo, J.L. Seed and T.W. Kensler, Inhibition of 2(3)-tert-Butyl-4-hydroxyanisole and other antioxidants of epidermal ornithine decarboxylase activity induced by 12-O-tetradecanoylphorbol-13-acetate. Cancer Res. 43: 2555-2559 (1983).
9. W.J. Kozumbo, M.A. Trush and T.W. Kensler, Are free radicals involved in tumor promotion? Chem-Biol Interact. 54: 199-207 (1985).
10. J.P. Perchellet, M.D. Owen, T.D. Posey, D.K. Orten and B.A. Schneider, Inhibitory effects of glutathione level-raising agents and D-α-tocopherol on ornithine decarboxylase induction and mouse skin tumor promotion by 12-O-tetradecanoylphorbol-13-acetate. Carcinogenesis 6: 567-573 (1985).
11. J.P. Perchellet, N.L. Abney, R.M. Thomas, Y.L. Guislan and E.M. Perchellet, Effects of combined treatments with selenium, glutathione, and vitamin E on glutathione peroxidase activity, ornithine decarboxylase induction and complete and multistage carcinogenesis in mouse skin. Cancer Res. 47: 477-485 (1987).
12. A.R. Kennedy, W. Troll and J.B. Little, Role of free radicals in the initiation and promotion of radiation transformation in vitro. Carcinogenesis 5: 1213-1218 (1984).
13. C. Borek, A. Ong. H. Mason, L. Donahue and J.E. Bigalow, Selenium and vitamin E inhibit radiogenic and chemically induced transformation in vitro via different mechanisms. Proc. Natl. Acad. Sci. U.S.A. 83: 1490-1949 (1986).
14. Y. Nakamura, N.H. Colburn and T.D. Ginhart, Role of reactive oxygen in tumor promotion: implication of superoxide anion in promotion of neoplastic transformation in JB-6 cells by TPA. Carcinogenesis 6: 229-235 (1985).
15. T.D. Ginhart, L. Srinivas and N.H. Colburn, Benzoyl peroxide promotion of transformation of JB-6 mouse epidermal cells: inhibition by ganglioside G_T but not retinoic acid. Carcinogenesis 6: 309-311 (1985).
16. N.J. Lawrence, E.K. Parkinson and A. Emmerson, Benzoyl peroxide interferes with metabolic co-operation between cultured human epidermal keratinocytes. Carcinogenesis 5: 419-421 (1984).
17. J.P. Perchellet, E.M., Perchellet, D.K. Orten and B.A. Schneider, Decreased ratio of reduced/oxidized glutathione in mouse epidermal cells treated with tumor promoters. Carcinogenesis 7: 503-506 (1986).
18. S.M. Fischer and L.M. Adams, Suppression of tumor promoter-induced chemiluminescene in mouse epidermal cells by several inhibitors of arachidonic acid metabolism. Cancer Res. 45: 3130-3136 (1985).
19. S.M. Fischer, J.K. Baldwin and L.M. Adams, Effects of anti-promoters and strain of mouse on tumor promoter-induced oxidants in murine epidermal cells. Carcinogenesis 7: 915-918 (1986).
20. W.A. Pryor, The role of free radical reactions in biological systems. In: Free Radicals in Biology, Vol. 1 (W.A. Pryor, Ed.), pp. 1-49. Academic Press, New York, 1975.
21. B.G. Taffe, N. Takahashi, T.W. Kensler and R.P. Mason, Generation of free radicals from organic hydroperoxide tumor promoters by mouse epidermal cells. Pharmacologist 28:175 (1986).
22. H. Hennings, R. Shores, M.L. Wenk, E.F. Spangler, R. Tarone and S.H. Yuspa, Malignant conversion of mouse skin tumours is increased by tumor initiators and unaffected by tumour promoters. Nature (London) 304: 67-69 (1983).
23. J.F. O'Connell, A.J.P. Klein-Szanto, D.M. DiGiovanni, J.W. Fries and T.J. Slaga, Malignant progression of mouse skin papillomas treated with ethylnitrosourea, N-methyl-N^1-nitrosoguanidine, or 12-O-tetradecanoylphorbol-13-acetate. Cancer Lett. 30: 269-271 (1986).

24. J.F. O'Connell, A.J.P. Klein-Szanto, D.M. DiGiovanni, J.W. Fries and T.J. Slaga, Enhanced malignant progression of mouse skin tumors by the free-radical generator benzoyl peroxide. Cancer Res. 46: 2863-2865 (1985).
25. J.B. Rotstein, J.F. O'Connell and T.J. Slaga, The enhanced progression of papillomas to carcinomas by peroxides in the 2-stage mouse skin model. Proc. Am. Assoc. Cancer Res. 27: 143 (1986) [Abstract].
26. A.J.P. Klein-Szanto and T.J. Slaga, Effects of peroxides on rodent skin: epidermal hyperplasia and tumor promotion. J. Invest. Dermatol. 79: 30-34 (1982).
27. C.J. Conti, C.M. Aldaz, J.F. O'Connell, A.J.P. Klein-Szanto and T.J. Slaga, Aneuploidy, an early event in mouse skin tumor development. Carcinogenesis 7: 1845-1848 (1986).
28. A.J. Saladino, J.C. Willey, J.F. Lechner, R.C. Grafstrom, M. LaVeck and C.C. Harris, Effects of formaldehyde, acetaldehyde, benzoyl peroxide and hydrogen peroxide on cultured normal human bronchial epithelial cells. Cancer Res. 45: 2522-2526 (1986).
29. H.C. Birnboim, DNA strand breaks in human leukocytes induced by superoxide anion, hydrogen peroxide and tumor promoters are repaired slowly compared to breaks induced by ionizing radiation. Carcinogenesis 7: 1511-1517 (1986).
30. A.J.P. Klein-Szanto, K.G. Nelson, Y. Shah and T.J. Slaga, Simultaneous appearance of keratin modifications and γ-glutamyltransferase activity as indictors of tumor progression in mouse skin papillomas. J. Natl. Cancer Inst. 70: 161-168 (1986).
31. M. Chiba, M.A. Maley and A.J.P. Klein-Szanto, Sequential study of γ-glutamyltransferase during complete and two-stage mouse skin carcinogenesis. Cancer Res. 46: 259-263 (1986).
32. J.A. Hartley, N.W. Gibson, L.A. Zwelling and S.H. Yuspa, Association of DNA strand breaks with accelerated terminal differentiation in mouse epidermal cells exposed to tumor promoters. Cancer Res. 45: 4864-4870 (1985).

DIFFERENTIAL EFFECTS OF 12-O-TETRADECANOYLPHORBOL-13-ACETATE (TPA) ON NORMAL INITIATED AND TRANSFORMED CELLS OF A SINGLE MOUSE KERATINOCYTE LINEAGE

Molly Kulesz-Martin

Grace Cancer Drug Center
Roswell Park Memorial Institute
666 Elm St., Buffalo, NY 14263

INTRODUCTION AND METHODS

Previous studies of the differential response of normal, initiated and tumor cells to TPA have been performed on cell lines derived from normal epidermis, epidermis which had been treated in vivo with chemical carcinogen, or disaggregated tumor tissue. Thus, the cell lines were heterogeneous in cellular origin. In the present study, derivatives of a single cell lineage were examined for responsiveness to TPA at various stages of tumorigenic potential after chemical treatments in vitro. The advantage of this approach is that differences in target cell response can be related to defined stages of transformation, rather than individual variation among cells of different parental origins.

The derivation and characterization of the parental cell line, 291, and its derivatives have been reported elsewhere (1,2). The normal characteristics of the parental line include keratin production, terminal differentiation of the line in response to 1.4 mM Ca^{2+}, and failure to produce papillomas even after TPA treatment of cultured cell grafts in athymic nude mice (3, and unpublished observations). The treatment, passage history and tumorigenic potential of each of the designated sub-lines of 291 are presented in Figure 1. Initiated cells were defined as those which continued to proliferate in culture medium containing standard Ca^{2+} concentration (1.4 mM), as reported previously (1). Other characteristics suggestive of initiated cells were the dose-response of this phenotype to certain chemical carcinogens, the heritability of the phenotype, and the production of tumor derivatives after further passage in vitro. All experiments to test tumor promoter responsiveness of each cell line were performed according to the same protocol, as follows: cells were plated in the growth medium optimized for the parental cell line, which required low Ca^{2+} concentration (0.02 mM), conditioning on primary syngeneic dermal cell cultures, and 10 ng/ml epidermal growth factor (EGF) (1). Before the first TPA treatment, fresh medium without conditioning or EGF was applied at the indicated Ca^{2+} concentration. Tumor promoter was added in 10 µl acetone per 3 ml culture medium on the 3rd, 5th, 7th and 10th day after plating. On the 12th day dishes were fixed in methanol, stained with Giemsa and colonies containing a minimum of 10 cells were counted. Results were expressed as the number of colonies per dish in test groups, relative to the number of colonies per dish in the solvent-control dishes, normalized to 100%.

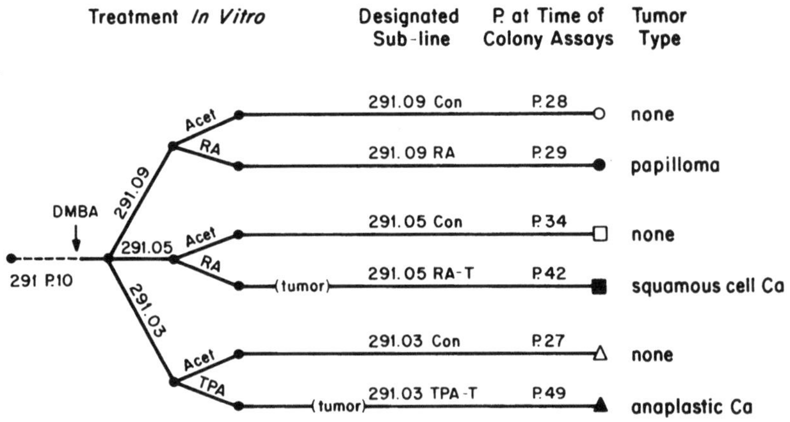

Fig. 1. Treatment history of parental line 291 and its designated derivatives. P, cumulative passage in vitro; RA, 1×10^{-8} to 5×10^{-9}M retinoic acid; Acet, 0.3% acetone; TPA, 10 ng/ml 12-O-tetradecanoylphorbol-13-acetate; DMBA, 50-100 ng/ml 7,12-dimethylbenz[α]-anthracene x 24 h.

RESULTS AND DISCUSSION

Under low Ca^{2+} conditions, TPA reduced the number of parental cell colonies by about 50%, similarly to primary epidermal cultures studied by Yuspa et al. (4). In contrast to the report of Yuspa et al. (5) however, initiated sub-lines also exhibited reduced colony numbers in response to TPA (Figure 2a,b,c). This effect was more pronounced under standard Ca^{2+} conditions in two of the sub-lines (Figure 2d,e). Furthermore, the papilloma-producing derivative, 291.09RA, but not the squamous cell and anaplastic carcinoma tumor derivatives, displayed similar colony reduction by TPA under standard Ca^{2+} conditions. The colony reduction effect was inversely proportional to TPA dose. Thus the number of initiated cell colonies in dishes treated with 100 ng/ml TPA was approaching that of solvent control, whereas the number of colonies in dishes treated with 1 ng/ml TPA was reduced by as much as 90% (Figure 2d). The appearance of the cells was observed because epidermal cell morphology changes during differentiation. Proliferating cells are generally smaller than differentiating cells and more basophilic in staining due to a higher nuclear/cytoplasmic ratio. Microscopic examination of the 291.09RA colonies indicated that colony size, and the proportion of small, basophilic cells with high nuclear/cytoplasmic ratio increased in direct relation to TPA dose (data not shown). Conversely, the number of large cells with low nuclear/cytoplasmic ratio decreased with increasing TPA dose. While approximately half of colonies treated with acetone contained a sub-population of these large cells, colonies treated with 1 ng/ml TPA were nearly all composed of such cells. The proportion of the large cells decreased with increasing TPA dose until at 100 ng/ml most colonies contained only the smaller basophilic cells.

The present results indicated that initiated and papilloma-producing cells remained sensitive to a colony-reducing effect of TPA. The results appear to differ from those reported by Yuspa et al. (5) using cell lines derived from pooled papillomas which had been chemically induced in Sencar mice. Six of their cell lines exhibited between 1.5 and 25-fold enhancement of colony-forming efficiency by treatment with 100 ng/ml TPA when tested under standard Ca^{2+} conditions. These cells also were demonstrated to lack certain differentiation markers induced in primary cell cultures by

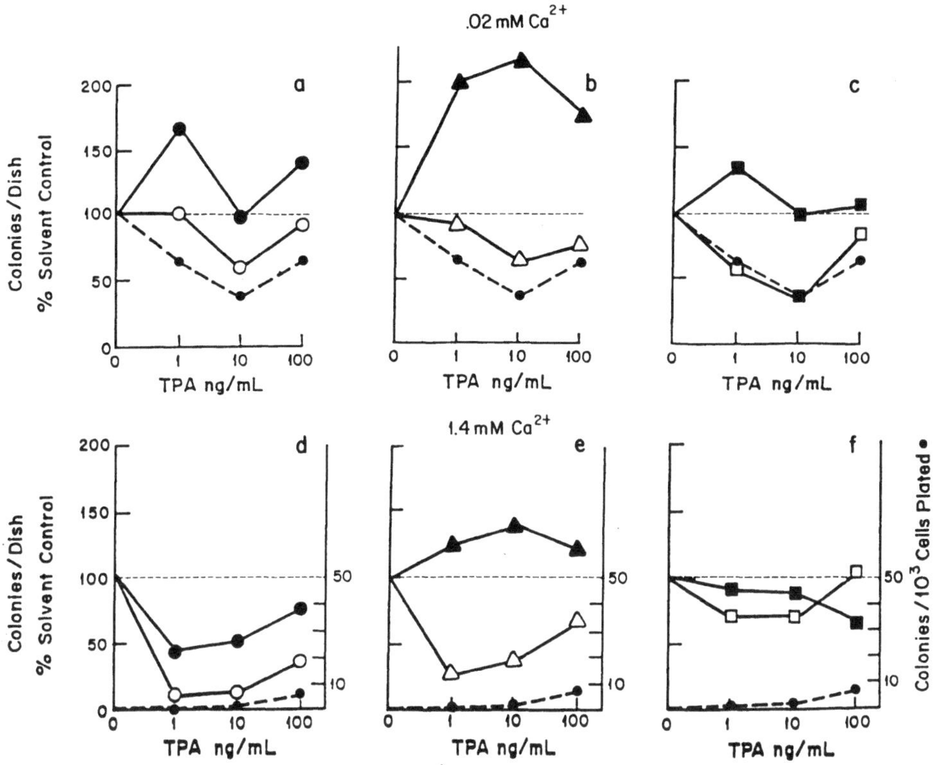

Fig. 2. Response to TPA of normal (•-- •--), three initiated (○ ,□ , △) and three tumor derivative sublines (●, ■ , ▲) under low (a,b,c) and standard (d,e,f) Ca^{2+} conditions. Symbols corresponding to each sub-line are presented in Fig. 1. See text for methods. Because parental line 291 did not grow under standard Ca^{2+} conditions without TPA, absolute colony number per 1000 plated cells was reported rather than relative colony number. Colonies per dish (% Solvent control) = $\overline{X}$ number colonies per dish in test group divided by $\overline{X}$ number colonies per dish in acetone control group multiplied by 100. The colony-forming efficiencies of the low Ca^{2+} solvent controls in these experiments were in the range of 6% for the parental line, 25 to 32% for the initiated lines, and 5, 15 and 16% for the tumor derivatives 291.03 TPA-T, 291.09 RA, and 291.05 RA-T, respectively. The corresponding values for standard Ca^{2+} solvent controls were equivalent in the initiated lines, were increased 2-fold in the tumor derivatives and were decreased to negligible levels in the parental line.

TPA. However, the lower doses of 1 and 10 ng/ml TPA were not tested on the papilloma lines. The high stimulation of colony number found in these lines may have been associated with the comparatively low colony forming efficiency of these cells under standard Ca^{2+} conditions (less than 20% of that under low Ca^{2+} conditions for each line).

These data were suggested by the authors to support the hypothesis that tumor promotion occurs by selection of initiated cells which are refractory to the TPA differentiation signal, and against normal cells which are sensitive to this signal (6,7). In the present study, a selective effect of TPA on normal, initiated and tumor cell derivatives was observed, in that the colony-reducing effect of TPA generally decreased in association with increasing malignant potential of the target cells. This result is consistent with that of Parkinson et al. (8) who found that SV40-transformed keratinocyte cultures exposed to TPA showed smaller increases in differentiation markers than did normal human keratinocyte cultures. However, selection for proliferative initiated cells, and against normal cells, did not appear to be the major effect of tumor promoters in the present study. The most potent colony reduction of the initiated and papilloma derivatives by TPA occurred under standard Ca^{2+} conditions, analogous to the spinous cell compartment in vivo. Under these conditions, normal cells are not proliferative, and a few parental cell colonies were actually stimulated to grow by TPA. Furthermore, teleocidin, reportedly a more potent promoter of mouse skin carcinogenesis than TPA, exhibited a greater colony-reducing effect than TPA on both the initiated and tumor derivatives tested. The data for the tumor derivatives are shown in Figure 3. These data are consistent with the view that tumor promoters exert selective pressure on initiated cells directly by selecting against a sub-population sensitive to growth inhibition by TPA, and for a sub-population less sensitive to this effect. This is further supported by the finding that initiated cell lines treated with TPA or retinoic acid in vitro became tumorigenic sooner than their acetone-treated replicate cultures (2,3).

The dose-response curve, and the morphological observations described above, suggest that TPA colony reduction was not due to toxicity. This is in contrast to the report of Nettesheim et al. (9) that 10 to 100 ng/ml TPA was toxic to immortalized, pre-neoplastic rat tracheal epithelial cell lines. The present findings (Figures 2d and 3b), however, were similar to the results of Nettesheim et al. in that normal cell colonies were stimulated under standard Ca^{2+} conditions. Induction of differentiation in sub-populations of the initiated 291 keratinocyte sub-lines by TPA is consistent with the data, but remains to be established directly in future studies. The finding that initiated and tumor cell derivatives of squamous epithelium can remain sensitive to colony reduction by certain agents favors the investigation of new therapeutic approaches using differentiation-inducing factors.

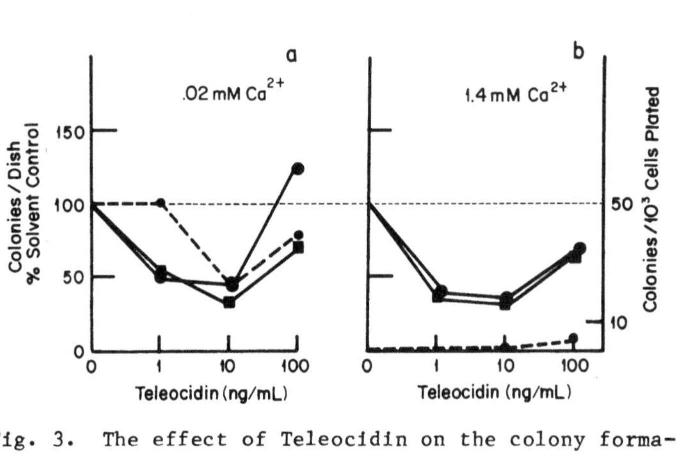

Fig. 3. The effect of Teleocidin on the colony formation of normal 291 and tumor cell derivatives. The symbols and methods are the same as in Fig. 2.

ACKNOWLEDGEMENTS

The technical expertise of Ms. Barbara Lisafeld and the secretarial assistance of Ms. Mae Brown are gratefully acknowledged. My thanks to Dr. H. Fujiki for providing the teleocidin, and to Dr. Remedios Penetrante for evaluation of tumor histopathology. Supported by NIH grants CA-31101 and CA-24538.

REFERENCES

1. M. Kulesz-Martin, M.A. Yoshida, L. Prestine, S.H. Yuspa and J.S. Bertram, Mouse cell clones for improved quantitation of carcinogen-induced altered differentiation. Carcinogenesis 6, 1245-1254 (1985).
2. M. Kulesz-Martin, L. Blumenson and B. Lisafeld, Retinoic acid enhancement of an early step in the transformation of mouse epidermal cells in vitro. Carcinogenesis 7, 1425-1429 (1986).
3. C.J. East and M. Kulesz-Martin, Retinoic acid enhanced malignant potential of an initiated mouse keratinocyte cell line. Proc. Am. Assoc. Cancer Res. 28, 172 (1987).
4. S.H. Yuspa, T. Ben, H. Hennings and U. Lichti, Divergent responses in epidermal basal cells exposed to the tumor promoter 12-0-tetradecanoyl-phorbol-13-acetate. Cancer Res. 42, 2344-2349 (1982).
5. S.H. Yuspa, D. Morgan, U. Lichti, E.F. Spangler, D. Michael, A. Kilkenny and H. Hennings, Cultivation and characterization of cells derived from mouse skin papillomas induced by an initiation-promotion protocol. Carcinogenesis 7, 949-958 (1986).
6. S.H. Yuspa, H. Hennings, M. Kulesz-Martin and U. Lichti, The study of tumor promotion in a cell culture model for mouse skin-a tissue that exhibits multistage carcinogenesis in vivo. In: Carcinogenesis, Vol. 7 (E. Hecker, N.E. Fusenig, W. Kunz, F. Marks and H.W. Thielmann, Eds.), pp. 217-230. Raven Press, New York, 1982.
7. J.J. Reiners and T.J. Slaga, Effects of tumor promoters on the rate and commitment to terminal differentiation of subpopulations of murine keratinocytes. Cell 32, 247-255 (1983).
8. F.K. Parkinson, P. Grabham and A. Emmerson, A subpopulation of cultured human keratinocytes which is resistant to the induction of terminal differentiation-related changes by phorbol, 12-myristate, 13-acetate: evidence for an increase in the resistant population following transformation. Carcinogenesis 4, 857-861 (1983).
9. P. Nettesheim, T. Gray and J.C. Barrett, The toxic response of preneoplastic rat tracheal epithelial cells to 12-0-tetradecanoyl-phorbol-13-acetate. Carcinogenesis 6, 1427-1434 (1985).

ONCOGENES AND MALIGNANT TRANSFORMATION OF HUMAN KERATINOCYTES

Norbert E. Fusenig, Petra Boukamp, Dirk Breitkreutz, Siegfried Karjetta, and Rule T. Petrusevska

Division of Differentiation and Carcinogenesis in vitro
Institute of Biochemistry, German Cancer Research Center
Im Neuenheimer Feld 280, 6900 Heidelberg, FRG

Activated oncogenes, particularly of the ras gene family, have been detected in a variety of solid human tumors and interpreted as causal agents in the development of neoplasia (1, 2). Immortalized and aneuploid mesenchymal cells from the rodent NIH 3T3 cell line could be rendered tumorigenic by transfection oncogenes (3-5) while normal fibroblasts required two cooperating oncogenes or additional enhancers (6-9). So far, it has not been possible to malignantly transform human fibroblasts with oncogenes in vitro (10, 11).

Epithelial animal cells have also not yet been malignantly transformed by transfection with a single oncogene. Primary mouse skin keratinocyte cultures could be transformed to papilloma-forming cell lines by infection with the viral Ha-ras oncogene (12) but malignant conversion by the ras oncogene was only feasible with immortalized papilloma cell lines (13). Malignant transformation of human epithelial cells by oncogenes has only been reported twice: i) by sequential infection with two oncogenic viruses (14) and ii) by transfection with a viral Ha-ras oncogene (15). In both cases the initial event seemed to be immortalization of the cells followed by neoplastic conversion. As postulated for rodent fibroblasts (16), immortalization also seems to be a prerequisite for malignant conversion of human cells by the ras oncogene.

We have therefore established two immortalized human keratincoyte cell lines: i) after transfection with a plasmid containing origin-defective SV40 DNA (17), an immortalized cell line developed similar to those described by others (18, 19), and was designated HaSV; ii) the second cell line immortalized spontaneously after long-term growth in primary culture in low Ca^{++} (0.2 mM) medium and at 38.5°C, and was termed HaCaT. The primary cultures for the latter line were originally derived from histologically normal skin at the periphery of a melanoma and their identity with the donor was proved by DNA fingerprinting using Jeffreys method (20, 21).

Both cell lines exhibit a transformed phenotype in vitro, have remained non-tumorigenic up to high passage levels but differ in their growth and differentiation behaviour, as summarized in Table 1. Both cell lines exhibit an aneuploid karyotype with cell line specific stable structural rearrangements (marker chromosomes) present in 100 % of metaphases (22). The appearance of these markers in all cells of early passages and their persistance strongly indicates the clonal origin of the cell lines and the essential role of these chromosomal rearrangements for the altered phenotype. The spontaneously immortalized cell line HaCaT originated as a hypodiploid population from which a hypotetraploid clone having improved growth capacity under different culture conditions (including clonogenicity on plastic and in agar) (21, 22) emerged.

Cells from both cell lines were able to form well-organized epithelia after transplantation, but HaSV cells, even at early passages, tended to grow invasively, although no progressively growing tumors developed after s.c. injection of cells into nude mice. Expression of specific differentiation products such as involucrin, filaggrin and suprabasal keratins was detected in the transplants of both cell lines by immunofluorescence and/or gel electrophoresis, but was always quantitatively less and more regular in localization in HaSV cells (21, 22). This close similarity in differentiation compared with normal keratinocytes rendered these cell lines, particularly the HaCaT cell line, excellent model systems to study the role of cellular oncogenes in the malignant conversion of human epithelial cells. At low to medium passage levels (HaSV passage 14, HaCaT passage 30) both lines were transfected with a plasmid containing the neomycin resistance gene (pSV-neo) and the 6.6 kb Bam HI fragment of the activated cellular Ha-ras oncogene (EJ) (23), using the calcium phosphate precipitation technique (24).

Table 1. Characteristics of the Transformed Human Keratinocyte Cell Lines HaSV and HaCaT

Parameters	HaSV (SV40)	HaCaT (spontaneous)
Cell morphology	epithelial	epithelial
Cytoskeleton	keratin	keratin
T-AG	positive	negative
PD-times	22 h	21 h
Cloning efficiency	7 %	13 %
Agar C.E.	1.7 %	0.4 %
karyotype	hypotetraploid	hypotetraploid
marker chromos.	stable	stable
keratins (Cat.No.) in vitro	5,6,14,16/7,8,18,19	(1,10),5,6,14,16/(7,8,19)
keratins (Cat.No.) in vivo	5,6,(10),14,16	1,10,5,6,14,16
growth in vivo	organized epithelium	organized epithelium
non-tumorigenic	< 60 passages	< 80 passages
ras transfection	passage 14	passage 30

While attempts to select stably transfected clones of normal keratinocytes consistently failed, both immortalized cell lines could reproducibly be transfected with a rather high efficiency (e.g. 1.7 x 10^4 cells per 10 ug DNA for the HaCaT cell line). The selected neo-resistant cell clones (HaCaT-EJ) and derived sublines did not differ in morphology and growth behaviour from the parental lines, so that a number of randomly picked clones were investigated for altered biological behaviour. From two experiments with the HaCaT cell line, a total of 36 HaCaT-EJ clones were further analyzed by the Southern blot analysis.

Interestingly, in more than one fourth of these clones no integration of ras, but only of the neo gene, was noticed, indicating fragmentation of the plasmid before or during integration. The other HaCaT-EJ clones (as well as two randomly selected HaSV-EJ clones) showed additional bands on Southern blots when probed with a 2.9 kb SstI ras fragment (22, 25) following Bam HI digestion. This was even more evident after restriction analysis with Eco RI which distinguishes the different bands including one of approximately 12.3 kb corresponding to the size of the complete plasmid. The intensity of this 12.3 kb band in several HaCaT and HaSV clones indicated multiple integration of tandem repeats. The lack of additional bands (other than that of the protooncogene at 23 kb) in those clones where only the neo-gene could be detected after Bam HI digestion, was further proof for the absence of integrated exogeneous ras sequences.

While the parental HaCaT cells (at early to medium passage level) showed no ras RNA expression and were essentially negative upon Northern blot analysis, all Ha-ras oncogene-containing clones gave rise to a single 1.2 kb band when probed with the 2.9 kb SstI c-Ha-ras fragment. Clones which showed the 12.3 kb Eco RI fragment usually had higher expression at the RNA level (22, 25) and expressed the p21 ras product.

Two randomly picked HaSV-EJ clones, which showed integrated ras on Southern blots, were tumorigenic in nude mice, and formed well differentiated carcinomas. When the transfected HaCaT clones were tested for tumorigenicity 5 sublines containing neo only formed small differentiating nodules which regressed within a few weeks similar to those formed after s.c. injection of (up to 10^7) cells of the parental HaCaT cell line of 18 selected HaCaT EJ clones, six clones, which had no 12.3 kb Eco RI bands and only 1 or 2 Bam HI extra bands, gave rise only to small regressing cystic nodules. Five further clones which exhibited higher ras integration levels formed, in addition to regressing cysts, slowly enlarging but histologically benign nodules. The last group of 7 clones gave rise (in some or most cases) to highly differentiated squamous cell carcinomas and 6 of these showed the intense 12.3 kb Eco RI band. Tumors could be regrown in culture and reinjected, leading to tumor formation, sometimes within shorter latency periods, and, in one case, even to spontaneous lymph node metastases (22, 25). The tumorigenic and non-tumorigenic HaCaT-EJ clones analyzed so far exhibited distinct karyotype patterns containing, in addition to the HaCaT marker chromosomes, new chromosomal rearrangements which were unique for each clone.

All HaCaT-ras clones, whether tumorigenic or not, had largely maintained the high differentiation capacity of the HaCaT cell line. This was documented by their potential, upon transplantation onto nude mice, to form regularly structured squamous epithelia, which still expressed differentiation specific keratins and other differentiation markers (21, 22).

With these studies we have demonstrated for the first time that immortalized human keratinocyte cell lines can be reproducibly converted to carcinoma-forming cells following transfection with a single human cellular oncogene. Whether malignant conversion was caused by the ras oncogene product alone or by independent events such as additional chromosomal rearrangements, as a consequence of the ras integration, or by a combination of both effects, cannot be answered at present. Since the parental HaSV and HaCaT cell lines remained non-tumorigenic for more than two years and up to 60 and 80 passages, respectively, and transfection was performed done with early passaged cells, spontaneous transformation events are less probable. Also since those clones which exhibited higher integration of ras genes and increased ras RNA expression were capable of prolonged growth in vivo (forming benign and malignant tumors), this altered growth behaviour may be due to the oncogene product.

These results indicate that immortalized human epithelial cells with a high degree of differentiation are sensitive to ras oncogene transformation. This further argues for immortalization as an obligatory prerequisite for the induction of malignant cell transformation by oncogene transfection, not only in rodent (16), but also in human cells. On the other hand, major alterations or defects in the differentiation process are obviously neither essential nor early steps in keratinocyte carcinogenesis (26, 27). Moreover, such cell lines, particularly the spontaneously immortalized HaCaT cells, may provide more relevant model systems (than rodent mesenchymal cells) in which to detect dominantly acting oncogenes in human epithelial malignancies and to study their mechanism of action in malignant cell transformation.

REFERENCES

1. D.J. Slamon, J.B. de Kernion, I.M. Verma, M.J. Cline, Expression of cellular oncogenes in human malignancies. Science 224, 256-262 (1984).
2. S.A. Aaronson, S.R. Tronick, The role of oncogenes in human neoplasia. In: Important Advances in Oncology (V.T. De Vita, S. Hellman, and S.A. Rosenberg, Eds.) pp. 3-15. Lippineott Ca, Philadelphia, Pennsylvania, (1985).
3. S. Pulciani, E. Santos, A.V. Lauver, L.K. Long, K.C. Robbins, M. Barbacid, Oncogenes in human tumor cell lines: molecular cloning of a transforming gene from human bladder carcinoma cells. Proc. Natl. Acad. Sci. USA 79, 2845-2849 (1982).
4. M.J. Murray, B.-Z. Shilo, C. Shih, D. Cowing, H.W. Hsu, R.A. Weinberg, Three different human tumor cell lines contain different oncogenes. Cell 25, 355-361 (1981).
5. C.S. Cooper, D.G. Blair, M.K. Oskarsson, M.A. Tainsky, L.A. Eader, G.F. Vande Woude, Characterization of human transforming genes from chemically transformed, teratocarcinoma, and pancreatic carcinoma cell lines. Cancer Res. 44, 1-10 (1984).
6. H. Land, L.F. Parada, R.A. Weinberg, Tumorigenic conversion of primary embryo fibroblasts requires at least two cooperating oncogenes. Nature 304, 596-602 (1983).
7. H.E. Ruley, Adenovirus early region 1A enables viral and cellular transformation genes to transform primary cells in culture. Nature 304, 602-606 (1983).
8. M. Oshimura, T.M. Gilmer, J.C. Barrett, Chemically induced aneuploidy in mammalian cells: mechanisms and biological significance in cancer. Nature 316: 636-639 (1985).
9. D.A. Spandidos, N.M. Wilkie, Malignant transformation of early passage rodent cells by a single mutant human oncogene. Nature 310, 469-475 (1984).
10. R. Sager, K. Tanaka, C.C. Lau, Y. Ebina, A. Anisowicz, Resistance of

human cells to tumorigenesis induced by cloned transforming genes. Proc. Natl. Acad. Sci. USA 80, 7601-7605 (1983).
11. J.A. DiPaolo, Relative difficulties in transforming human and animal cells in vitro. J. Natl. Cancer Inst. 70, 3-8 (1983).
12. D.R. Roop, D.R. Lowy, P.E. Tambourin, J. Strickland, J.R. Harper, M. Balaschak, E.F. Spangler, S.H. Yuspa, An activated Harvey ras oncogene produces benign tumours on mouse epidermal tissue. Nature 323, 822-824 (1986).
13. J.R. Harper, D.R. Roop, S.H. Yuspa, Transfection of the EJ ras^{Ha} gene into keratinocytes derived from carcinogen-induced mouse papillomas causes malignant progression. Mol. Cell. Biol. 6, 3144-3149 (1986).
14. J.S. Rhim, G. Jay, P. Arnstein, F.M. Price, K.K. Sanford, S.A. Aaronson, Neoplastic transformation of human epidermal keratinocytes by AD12-SV40 and Kirsten sarcoma virus. Science 227, 1250-1252 (1985).
15. G.H. Yoakum, J.F. Lechner, E.W. Gabrielson, B.E. Korba, L. Malan-Shibley, J.C. Willey, M.G. Valerio, A.M. Shamsuddin, B.F. Trump, C.C. Harris, Transformation of human bronchial epithelial cells transfected by Harvey ras oncogene. Science 227, 1174-1179 (1985).
16. R.F. Newbold, R.W. Overell, Fibroblast immortality is a prerequisite for transformation by EJ c-Ha-ras oncogene. Nature 304, 648-651 (1983).
17. Y. Gluzman, R.J. Frisque, J.F. Sambrook, Origin-defective mutants of SV40. Cold Spring Havor Symposium, Quant. Biol. 44, 293-299 (1980).
18. S.P. Banks-Schlegel, P.M. Howley, Differentiation of human epidermal cells transformed by SV40. J. Cell Biol. 96, 330-337 (1983).
19. K.W. Brown, E.K. Parkinson, Extracellular matrix components produced by SV-40 transformed human epidermal keratinocytes. Int. J. Cancer 33, 257-263 (1984).
20. A.J. Jeffreys, V. Wilson, S.L. Thein, Hypervariable "minisatellite" regions in human DNA. Nature 314, 67-73 (1985).
21. P. Boukamp, R.T. Dzarlieva-Petrusevska, D. Breitkreutz, J. Hornung, A. Markham, N.E. Fusenig, Essential defects in differentiation are not prerequisites for transformation of human skin keratinocytes. (submitted).
22. P. Boukamp, Untersuchungen zum Studium der Carcinogenese menschlicher Epidermiszellen: Etablierung und Charakterisierung von Primtrkulturen und immortalen Zellinien sowie die maligne Transformation durch das zellultre Harvey-ras Onkogen. Ph.D. Thesis, University of Kaiserslautern (1987).
23. L.F. Parada, C.J. Tabin, C. Shih, R.A. Weinberg, Human EJ bladder carcinoma oncogene is homologue of Harvey sarcoma virus ras gene. Nature 297, 474-478 (1982).
24. F.L. Graham, A.J. Van der Eb, A new technique for the assay of human adenovirus 5 DNA. Virology 52, 456-469 (1973).
25. P. Boukamp, E.J. Stanbridge, P. Cerutti, N.E. Fusenig, Malignant conversion of immortalized human skin keratinocytes by the cellular Harvey-ras oncogene. (in preparation).
26. N.E. Fusenig, R.T. Dzarlieva-Petrusevska, D. Breitkreutz, Phenotypic and cytogenetic characteristics of different stages during spontaneous transformation of mouse keratinocytes in vitro. In: Carcinogenesis Vol. 9, (J.C. Barrett, R.W. Tennant, Eds.), pp. 293-326. Raven Press, New York (1985).
27. D. Breitkreutz, J. Hornung, J. P¶hlmann, L. Brown-Bierman, A. Bohnert, P.E. Bowden, N.E. Fusenig, Environmental induction of differentiation-specific keratins in malignant mouse keratincoyte lines. Europ. J. Cell Biol. 42: 255-267 (1986).

NUCLEOPHILES AS ANTICARCINOGENS

Lee W. Wattenberg, J. Bradley Hochalter, Usha D. G. Prabhu, and Arthur R. Galbraith

Department of Laboratory Medicine and Pathology
University of Minnesota School of Medicine
Minneapolis, Minnesota 55455

INTRODUCTION

The objective of the present work was to obtain chemopreventive compounds that can trap direct-acting carcinogens within the lumen of the gastrointestinal tract and thus prevent these carcinogens from attacking tissues of the host. Many direct-acting carcinogens are electrophiles (19, 36). One possible strategy for blocking their action is by trapping them with nucleophiles (electron donors). In the studies to be presented, emphasis has been placed on trapping direct-acting carcinogens in two sites, i.e., the stomach and the large intestine. To some extent different considerations pertain for each.

The major source of carcinogenic electrophiles in the stomach is likely to be from food. In addition, carcinogens can also originate in the lung, be trapped in the mucous stream and swallowed. They also may arise in the oral cavity or enter the upper alimentary tract from the blood stream through the mucosa. The conditions for trapping electrophiles in the stomach are unusual in that the gastric contents have an acid pH. Many nucleophiles are inactive in acid solution. This fact places a particular stricture on the number of available compounds that would be active in the stomach. The acidity of the stomach varies with species and with food consumption. Even in species with a very low fasting gastric acidity, as the human, the pH will rise rapidly to 4 or above shortly after the start of a meal (18). While ideally it would be desirable to have nucleophiles effective in the low pH range, those maintaining significant nucleophilicity in the higher range could still be useful.

Work will be presented dealing with two compounds that are effective nucleophiles within the stomach. These compounds are sodium thiosulfate (STS) and 4-mercaptobenzenesulfonate (4-MBS). STS is a potent nucleophile with very low toxicity (11,29). In *in vitro* studies, STS has been shown to react with beta-propiolactone (BPL), methyl bromide, methyl-bis (β-chloroethyl)amine hydrochloride (HN2) and chlorambucil (6,12,19,38). It has been shown to inhibit mutagenesis of Salmonella *typhimurium* strain TA-100 resulting from exposure to BPL and styrene oxide (34). In in vivo studies, protection against toxicity from HN2 by simultaneous administration of STS has been reported (3,5,13). Other related alkylating reagents whose toxicity is reduced by STS include mustard gas (bis(2-chloroethyl)sulfide) and

N,N'-ethylene-bis-[5(2-bromoethylthiol)valeramide] (8,39). STS has also been shown to inhibit toxicity of cis-diaminedichloroplatinum in humans and rodents (15,16,30). It also has been used in the therapy of cyanide poisoning (9,37). An older literature exists describing the use of STS for diminishing the toxicity of arsenicals and various metals as well as for inhibiting several carcinogens requiring metabolic activation (17,23). An overview of the major attributes of STS make it a likely inhibitor of electrophiles in the stomach. It is a potent nucleophile and retains its nucleophilicity in the acid pH range. As will be seen, STS can inhibit carcinogen-induced neoplasia of the mouse forestomach.

The second compound investigated for its inhibitory capacities on forestomach neoplasia is 4-MBS. Thiols are a widely used group of nucleophiles (22,26). Aromatic thiols, in general, are the most potent of the thiol nucleophiles. A characteristic of most thiols is that they lose their nucleophilicity in acidic media. Efforts have been made to identify aromatic thiols that would be active in acidic media. One such compound is 4-MBS. This compound has a pK_a of 5.8 and maintains activity in that pH range. Experimental work will be described showing that 4-MBS can inhibit carcinogen-induced neoplasia of the mouse forestomach.

The second major region in which trapping of electrophiles is of importance is the large intestine. Electrophiles can occur in the large bowel contents from a variety of sources. They can be ingested and remain within the alimentary tract. They can be ingested, absorbed through the gastrointestinal mucosa and excreted in the bile. In addition to electrophiles per se, compounds requiring metabolic activation may ultimately occur in the intestinal contents. Such compounds are absorbed, metabolically activated and excreted in the bile or through the mucosa often as conjugates. The large bowel flora can convert these conjugates to active electrophiles. Other types of precursor compounds can also be converted to electrophiles. A particularly important example of this latter origin is the formation of fecapentaenes within the large bowel of the human (7,14). The fecapentaenes are highly mutagenic compounds. Studies are currently in progress to ascertain their carcinogenicity.

Efforts at developing chemopreventive compounds that will trap direct-acting carcinogens within the large bowel have focussed on two attributes. The first is potency as an electrophile and the second is retention within the gastrointestinal tract. As discussed previously, aromatic thiols are potent nucleophiles. In early work, one such thiol, 4-methoxybenzenethiol (4-MBT), was found to be highly effective in reacting with electrophiles and in inhibiting carcinogen-induced mutagenesis of Salmonella typhimurium strain TA-100. Based on these findings, a decision was made to synthesize a high molecular weight aromatic thiol likely to be retained within the gastrointestinal tract. Prior work had shown that a high molecular weight phenol, 2,4,6-tri(3',5'-di-tert-butyl-4'-hydroxybenzyl)mesitylene (Ionex 330) was not absorbed from the gastrointestinal tract of dogs, rats and man after oral administration. It was quantitatively eliminated in the feces unchanged (40). Using this compound as a model, $\alpha^1,\alpha^3,\alpha^5$-tris-(4'-mercaptophenoxy)mellitene (TMM), a thiophenol with a similar structure has been synthesized (25). Its formula is shown in Fig. 1. TMM is a potent nucleophile. Preliminary studies show that it is retained within the gastrointestinal tract, but the precise magnitude of this retention has not yet been ascertained.

INHIBITION OF CARCINOGEN-INDUCED TUMOR FORMATION OF THE FORESTOMACH BY SODIUM THIOSULFATE AND 4-MERCAPTOBENZENESULFONATE

Studies have been carried out on the effects of STS and 4-MBS on car-

Fig. 1. Nucleophiles investigated in the experimental work presented. I, sodium thiosulfate (STS); II, 4-methoxybenzenethiol (4-MBT); III, 4-mercaptobenzenesulfonate (4-MBS); and IV, $\alpha^1,\alpha^3,\alpha^5$-tris-(4'-mercaptophenoxy)mellitene (TMM).

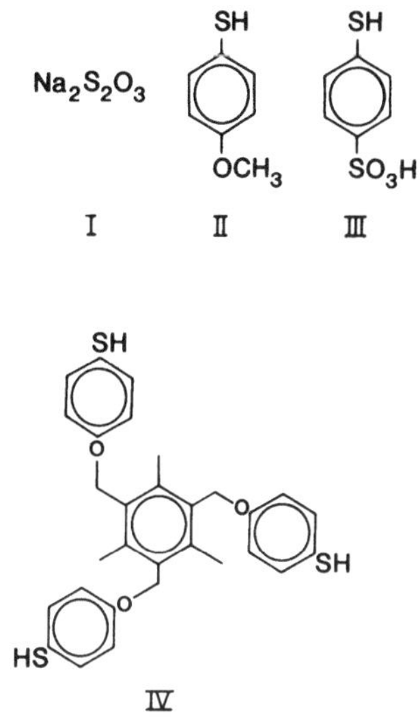

cinogenesis of the mouse forestomach induced by multiple administrations of beta-propiolactone (BPL). BPL is a low molecular weight direct-acting carcinogen with a simple chemical structure (6,19,27,31). BPL produces a neoplastic response in tissues with which it comes into direct contact. Thus it has been shown to cause cancer of the mouse epidermis as a result of topical application, of the mouse forestomach as a result of oral administration and sarcomas as a result of subcutaneous injection. Mouse tumor experiments to determine the capacity of nucleophiles to inhibit BPL-induced neoplasia of the forestomach have been performed using procedures similar to those described previously (33). Female A/J mice were randomized by weight at 8 weeks of age into groups of 15 mice and placed on a semipurified diet consisting of 27% vitamin-free casein, 59% starch, 10% corn oil, 4% salt mix (USP XIV) and a complete mixture of vitamins (Teklad, Madison, WI). One week later, administration of BPL was begun. A dose of 28 µmol of BPL in 0.1 ml H_2O was given by oral intubation twice a week x 10 weeks. Five minutes before each administration of the carcinogen, mice were given either 0.4 ml of water or the test nucleophile in water by oral intubation. One week after the last administration of carcinogen the mice were fed Purina Chow and were maintained on this diet for the duration of the protocol. The mice were weighed at intervals of 4 weeks. The experiments were terminated 26 weeks after the initial dose of carcinogen at which time the mice were autopsied. The stomachs were removed and injected with formalin so as to be fixed in the expanded state. Tumors of the forestomach were counted as described previously (33). Histopathology examination was performed to ascertain the nature of the tumors.

In an initial investigation, the capacity of STS to inhibit BPL-induced neoplasia of the forestomach was determined. The results were as follows (% of mice with forestomach tumors, average number of tumors per mouse, i.e., total number of tumors occurring in the experimental group divided by the number of mice at risk): no-STS controls - 100%, 4.3; 88 µmol STS 0%, 0; 44 µmol STS - 21%, 0.2; 22 µmol STS - 71%, 0.8. In the "no-STS controls" 20% of the mice had carcinomas of the forestomach, whereas no carcinomas were found in mice in any of the other groups. All other tumors were papillomas (34). Weight gain for all groups of mice was the same. Thus at all dose levels of STS employed, this nucleophile inhibited BPL-induced neoplasia of the mouse forestomach. In a second investigation, the effects of 4-MBS on BPL-induced neoplasia of the forestomach was determined. The results of the experiment were as follows (% of mice with forestomach tumors, average number of tumors per mouse): no-4-MBS controls - 100%, 6.8; 40 μmol 4-MBS - 57%, 0.71. The weight gain in the two groups of mice was the same. Thus this nucleophile also inhibited BPL-induced neoplasia of the forestomach (35).

STUDIES OF INHIBITION OF CARCINOGEN-INDUCED MUTAGENESIS OF *SALMONELLA TYPHIMURIUM* STRAIN TA-100 BY 4-METHOXYBENZENETHIOL AND $\alpha^1,\alpha^3,\alpha^5$-TRIS-(4'-MERCAPTOPHENOXY)MELLITENE

In these studies, the two nucleophiles being tested for inhibition of mutagenesis were preincubated with the mutagen at pH 7.4 for 10 minutes at 20° before addition to the TA-100 test strain culture. The mixture was added to top agar and poured onto minimal agar plates. The plates were incubated for 44 hours at 37°C at which time net revertants were counted (2,-20). Three mutagens were employed: BPL, N-methyl-N'-nitro-N-nitrosoguanidine and fecapentaene. The results of these studies are shown in Table 1. It will be seen that both nucleophiles inhibited the mutagenic effects of the 3 mutagens on *Salmonella typhimurium* strain TA-100.

DISCUSSION

The material presented is an initial effort to address the question of the feasibility of trapping electrophiles within the gastrointestinal tract. In order to consider such a strategy, a prerequisite is the availability of effective compounds. The work carried out thus far with STS, 4-MBS and TMM has shown efficacy in test systems. It is preliminary and much more has to be done in this regard. The toxicity of inhibitors is an important consideration for their ultimate use as chemopreventive agents. STS has very low toxicity and so is favorable in this regard. Under the experimental conditions employed in the mouse forestomach protocols described above, 4-MBS does not appear to be toxic but detailed studies of its toxicity have not been performed. 4-MBT shows considerable toxicity *in vivo*. On theoretical grounds, it would be anticipated that compounds which are not absorbed from the gastrointestinal tract might have little toxicity. TMM is likely to fall into this category. However, the experimental data remain to be obtained.

Direct-acting carcinogens that could be trapped by nucleophiles of the type under investigation include a wide variety of compounds that are electrophiles or in solution form electrophiles. These include epoxides, strained lactones such as β-propiolactone, nitrosamides, hydroxylamines, imines, fecapentaenes, alkyl sulfate esters, active halogen derivatives such as bis(chloromethyl)ether, nitrogen mustard and some of its derivatives, and additional alkylating agents that have been reviewed by others (19).

In addition to carcinogens, there are a number of mutagens that can occur within the gastrointestinal tract but whose carcinogenic effects are not known (1,4,21). It may be established after thorough investigation that some of these, such as the fecapentaenes, are full carcinogens. However, even if a full carcinogenic effect cannot be established, they could play a more limited role in the carcinogenic process.

Electrophiles have the capacity to attack DNA. Compounds of this type have been shown to initiate carcinogenesis or by themselves produce all

Table 1. Effects of 4-Methoxybenzenethiol (MBT) and $\alpha^1,\alpha^3,\alpha^5$-tris-(4'-mercaptophenoxy)mellitene on Mutagenesis of *Salmonella typhimurium* Strain TA-100[a]

Mutagen	Conc. mM	Test Compound	Conc. mM	Molar ratio test compound/ mutagen	Revertants	Revertants test/control
Beta-propio-lactone	0.83	MBT	--	--	522	1.00
			0.83	1	196	0.38
			1.66	2	56	0.11
			3.32	4	26	0.05
Beta-propio-lactone	0.83	TMM	--	--	513	1.00
			0.42	0.5	165	0.32
			0.83	1	40	0.08
			1.66	2	0	0.00
			3.32	4	0	0.00
N-Methyl-N'-nitro-N-nitroso-guanidine	0.031	MBT	--	--	1043	1.00
			0.031	1	263	0.25
			0.62	2	120	0.11
			0.124	4	43	0.04
N-Methyl-N'-nitro-N-nitroso-guanidine	0.031	TMM	--	--	497	1.00
			0.031	1	289	0.58
			0.062	2	146	0.29
			0.124	4	65	0.13
Fecapentaene	0.033	MBT	--	--	445	1.00
			0.166	5	12	0.03
Fecapentaene	0.033	TMM	--	--	1044	1.00
			0.166	5	118	0.11

[a] Mutagen and test compound or solvent were incubated together for 10 minutes at 20°C prior to addition of the bacterial tester strain. After subsequent incubation for 44 hours at 37°C. revertants were counted.

stages of carcinogenesis, i.e., initiation, promotion and progression. In addition to the implication for cancer, the presence of compounds that can alter genetic material of the host poses threats of other forms of pathology. Thus as a protective measure, the neutralization of electrophiles would appear to be desirable. As shown by the data presented, this appears achievable. However, questions as to whether such a strategy should have limited objectives in terms of specific risk factors or more general ones and the overall feasibility this type of undertaking requires considerable further exploration (32).

ACKNOWLEDGEMENTS

This work was supported by USPHS Grant CA-43285 from the National Cancer Institute and Grant SIG-5A from the American Cancer Society.

REFERENCES

1. B. Ames, The detection of environmental mutagens and potential carcinogens. In "Accomplishments in Cancer Research - 1983 Prize Year - General Motors Cancer Res. Found." (J. G. Fortner and J. E. Rhoads, Eds.), pp. 102-114. J. P. Lippincott, Philadelphia, 1983.
2. B. N. Ames, J. McCann and E. Yamasaki, Methods for detecting carcinogens and mutagens with the Salmonella/mammalian microsome mutagenicity test. Mutat. Res., 31, 347-364 (1975).
3. J. M. Anderson, The kinetics of the specific antagonism in vivo of intraarterial methyl-bis(β-chloroethyl)amine hydrochloride by intravenous solutions of hydrous sodium thiosulfate. Cancer, 16, 1281-1287 (1983).
4. L. F. Bjeldanes and H. Chow, Mutagenicity of 1,2-dicarbonyl compounds: maltol, kojic acid, diacetyl and related substances. Mutat. Res., 67, 367-371 (1979).
5. G. Bonadonna and D. Karnofsky, Protection studies with sodium thiosulfate against methyl-bis(beta-chloroethyl)amine hydrochloride and its ethylenimonium derivative. Clin. Pharmacol. Ther., 6, 50-64 (1965).
6. R. K. Boutwell, N. H. Colburn and C. C. Muckerman, In vivo reactions of β-propiolactone. Ann. N. Y. Acad. Sci., 163, 751-764 (1969).
7. W. R. Bruce, J. Batista, T. Che, R. Furrer, J. S. Ginerich, I. Grupta and J. J. Krepinsky, General structure of "fecapentaenes" - The mutagenic substances in human feces. Naturwissenschaften, 69, 557-558 (1982).
8. S. Callaway and K. W. Pearce, Protection against systemic poisoning by mustard gas, di(2-chloroethyl) sulphide, by sodium thiosulfate and thiocit in the albino rat. Br. J. Pharmacol., 13, 395-398 (1958).
9. K. K. Chen and C. L. Rose, Treatment of acute cyanide poisoning. J. Am. Med. Assoc., 162, 1154-1155 (1956).
10. T. A. Connors, A. Jeney and M. Jones, Reduction of the toxicity of "radiomimetic" alkylating agents in rats by thiol pretreatment - III. The mechanism of the protective action of thiosulfate. Biochem. Pharmacol., 13, 1545-1550 (1964).
11. D. L. Dennis and W. S. Fletcher, Toxicity of sodium thiosulfate (NSC-45624), a nitrogen mustard antagonist, in the dog. Cancer Chemother. Rep., 50, 225-257 (1966).
12. T. L. Gresham, J. E. Jensen, F. W. Shaver and J. T. Gregory, β-propiolactone. II. Reactions with salts of inorganic acids. J. Am. Chem. Soc., 70, 999-1001 (1948).
13. I. Hatiboglu, E. Mihich, G. E. Moore and C. A. Nichol, Use of sodium thiosulfate as a neutralizing agent during regional administration of nitrogen mustard: an experimental study. Ann. Surg., 156, 994-1001 (1962).
14. N. Hirai, D. G. I. Kingston, R. L. van Tassell and T. D. Wilkens, Isolation and structure elucidation of fecapentaenes-12, potent mutagens from human feces. J. Nat. Prod., 48, 622-630 (1985).
15. S. B. Howell, C. E. Pfeifle, W. E. Wung and R. A. Olshen, Intraperitoneal cis-diamminedichloroplatinum with systemic thiosulfate protection. Cancer Res., 43, 1426-1431 (1983).
16. Y. Iwamoto, T. Kawano, J. Uozumi, K. Aoki and T. Babo, "Two-route chemotherapy" using high dose i.p. cisplatin and i.v. sodium thiosulfate, its

antidote, for peritoneally disseminated cancer in mice. Cancer Treat. Rep., 68, 1367-1373 (1984).
17. W. G. Jaffe, The influence of sodium thiosulfate on experimental tumor induction. Experientia, IV, 234-236 (1948).
18. A. H. James and G. W. Pickering, The role of gastric acidity in the pathogenesis of peptic ulcer. Clin. Sci., 8, 181-210 (1949).
19. P. D. Lawley, Carcinogenesis by alkylating agents. In: Chemical Carcinogens, ACS Monograph 173, pp. 83-244. Am. Chem. Soc., Washington, DC, 1976.
20. D. M. Maron and B. N. Ames, Revised methods for the Salmonella mutagenicity test. Mutat. Res., 113, 173-215 (1983).
21. J. McCann, E. Choi, E. Yamasaki and B. Ames, Detection of carcinogens as mutagens in the Salmonella/microsome test: assay of 300 chemicals. Proc. Natl. Acad. Sci., USA, 72, 5135-5139 (1975).
22. H. Marquardt, M. D. Sapozink and M. S. Zedeck, Inhibition of cysteamine-HCl on oncogenesis induced by 7,12-dimethylbenz(a)anthracene without affecting toxicity. Cancer Res., 34, 3387-3390 (1974).
23. C. N. Myers, M. R. Groehl and C. P. Metz, Therapeutic activity of sodium thiosulfate. Proc. Soc. Exp. Biol. Med., 23, 97-101 (1925).
24. G. Owens and I. Hatibozlu, Clinical evaluation of sodium thiosulfate as a systemic neutralizer of nitrogen mustard. Ann. Surg., 154, 895-897 (1961).
25. U. D. G. Prabhu and L. W. Wattenberg. Synthesis of $\alpha^1,\alpha^3,\alpha^5$-tris-(4'-mercaptophenoxy)mellitene (unpublished).
26. D. Schmahl, M. Habs and A. M. Tacchi, Prophylaxe der Tumorentstehung in der Harnblase durch Natrium-2-mercaptoethansulfonat (Mesma). Urologie, 23, 291-296 (1984).
27. A. Segal, J. J. Solomon, J. Mignano and J. Dino. The isolation and characterization of 3-(2-carboxyethyl)cytosine following in vitro reaction of β-propiolactone with calf thymus DNA. Chem. Biol. Interact., 35, 349-361 (1981).
28. M. Shed, J. A. Koziol and S. B. Howell, Kinetics of sodium thiosulfate, a cisplatin neutralizer. Clin. Pharmacol. Ther. 35, 419-425 (1984).
29. C. G. Swain and C. B. Scott, Quantitative correlation of relative rates. Comparison of hydroxide ion with other nucleophilic reagents toward allyl halides, esters, epoxides and acyl halides. J. Am. Chem. Soc., 75, 141-147 (1953).
30. J. Uozumi, M. Ishizawa, Y. Iwamoto and T. Bab, Sodium thiosulfate inhibits cis-diamminedichloroplatinum (II) activity. Cancer Chemother. Pharmacol., 13, 82-85 (1984).
31. B. L. Van Duuren, Carcinogenic epoxides, lactones, and halo ethers and their mode of action. Ann. N. Y. Acad. Sci., 163, 633-651 (1969).
32. L. W. Wattenberg, Chemoprevention of cancer. Cancer Res., 45, 1-8 (1985).
33. L. W. Wattenberg, P. Borchert, C. M. Destafney and J. B. Coccia, Effects of p-methoxyphenol and diet on carcinogen-induced neoplasia of the mouse forestomach. Cancer Res., 43, 4747-4751 (1983).
34. L. W. Wattenberg, J. B. Hochalter and A. R. Galbraith, Inhibition of beta-propiolactone-induced neoplasia by sodium thiosulfate. Cancer Res. (submitted 1987).
35. L. W. Wattenberg, U. D. G. Prabhu, J. B. Hochalter and A. R. Galbraith, Inhibitory effects of 4-mercaptobenzene sulfonate and $\alpha^1,\alpha^3,\alpha^5$-tris-(4'mercaptophenoxy)mellitene on direct-acting carcinogens. Proc. Am. Assoc. Cancer Res., 27, 1987 (in press).
36. J. H. Weisburger and G. M. Williams, Metabolism of chemical carcinogens. In: Cancer: A Comprehensive Treatise (F. F. Becker, Ed.), Vol. 1 (2nd ed.), pp. 241-331. Plenum, New York, 1982.
37. J. Westley, Rhodanese and the sulfane pool. In: Enzymatic Basis of Detoxification, (W. B. Jakoby, Ed.), Vol. II, pp. 245-262. Academic Press, New York, 1980.

38. E. Wickstrom, Chlorambucil inhibition by dimethyl sulfoxide and thiosulfate: implications for chlorambucil chemotherapy. Med. Hypotheses, 6, 1035-1041 (1980).
39. C. E. Williamson, A. M. Seligman and B. Witten, Intracellular toxic reactions of some sulfur and nitrogen mustards. J. Pharmacol. Exp. Ther., 182, 77-82 (1942).
40. A. W. Wright, R. S. Crowne and D. E. Hathway, The fate of 2,4,6-tri-(3',5'-di-tert-butyl-4'-hydroxybenzyl)mesitylene (Ionex 330) in the dog and rat. Biochem. J., 95, 98-103 (1965).

PROTECTION FROM N-NITROSODIMETHYLAMINE MEDIATED LIVER DAMAGE BY INDOLE-3-CARBINOL, AND CORRELATION WITH NUCLEOPHILIC INDEX VALUE

H.G. Shertzer[1], M.L. Berger[2], and M.W. Tabor[1]

[1]Department of Environmental Health and [2]Department of International Medicine
(Liver Study Unit, Divivision of Digestive Diseases)
University Cincinnati Medical Center, Cincinnati, Ohio
45267-0056

Indole-3-carbinol (I-3-C), a normal constituent of the human diet via cruciferous vegetables, was examined for its ability to protect mice against 24-hour N-nitrosodimethylamine (NDMA)-mediated hepatotoxicity. NDMA (20 mg/kg body weight) alone produced extensive hemorrhagic and centrolobular necrotic lesions, with a necrotic severity index of 3.0 $\pm$ 0.4 (scale of 0-5). Treatment with 50 mg/kg body weight of I-3-C by gavage, 1 hour prior to NDMA, substantially protected against hemorrhagic lesions. Furthermore, I-3-C lowered the NDMA-mediated tissue necrotic index to 1.5 $\pm$ 0.3, by reducing the extent of tissue necrosis rather than the severity in the necrotic region. Release of liver enzymes into the blood correlated with the histopathology; I-3-C reduced NDMA-mediated elevated activities of plasma alanine transaminase and ornithine transcarbamylase by 84% and 51.3%, repectively (Table 1). Plasma activities of these enzymes were used as indicators of hepatotoxicity. Although no changes in liver non-protein sulfhydryls were evident at 24 hours after NDMA, ascorbate levels were reduced to 40% of controls values. Treatment with I-3-C prior to NDMA prevented this decline in tissue ascorbate concentrations.

In vitro, I-3-C was found to be a type II ligand for cytochrome P-450, with a K_s value of 237 uM. If such binding occurs *in vivo*, it does not protect against the approximately 60% decrease in hepatic cytochrome P-450 or 80% decrease in NDMA demethylase I activity produced by NDMA (Table 1). This implies that the histopathologic protection by I-3-C must be due to factors other than inhibiting metabolic activation of NDMA.

The mechanistic hypothesis examined in this report is: protection against hepatotoxicity is due to an enhanced ability of liver tissue to scavenge electrophilic metabolites of NDMA. A novel assay to quantify nucleophilicity as a Nucleophilic Index Value (NIV) was devleoped, based on the ability of putative nucleophiles to inhibit the methylation of cysteine by limiting concentrations of the electrophilic source, N-

Table I. Effects of I-3-C on Biochemical Parameters of NDMA hepatotoxicity

	-I-3-C		+ I-3-C	
Parameter	-NDMA (control)	+NDMA	-NDMA	+NDMA
Plasma ALT	0.02 ± 0.01	2.35 ± 0.76[a]	0.03 ± 0.01	0.41 ± 0.05a,c
Plasma OTC	10.6 ± 3.1	247 ± 55[a]	6.9 ± 0.7	122 ± 44[a,c]
Ascorbate	1.25 ± 0.08	0.50 ± 0.17[a]	1.28 ± 0.06	1.27 ± 0.07[c]
Non-protein sulfhydryls	9.7 ± 0.2	8.9 ± 0.4[b]	9.8 ± 0.3	9.6 ± 0.4
Cytochrome P-450	43.1 ± 0.7	15.8 ± 3.8[a]	47.1 ± 1.0	20.2 ± 2.5[a]
NDMA demethylase	4.6 ± 0.4	1.0 ± 0.2[a]	5.9 ± 0.4[b]	2.1 ± 0.3[a,c]

Mice were treated with 50 mg I-3-C/kg and/or 20 mg NDMA/kg. Animals were housed and treated and assays were performed, as described preiously (Shertzer *et al.* 1987). Values are expressed as means ± standard errors, with 6-8 animals per group. The units are: ALT: umol/min ml plasma; OTC: Units/l plasma; ascorbate and non-protein sulfhydryls: nmol/g tissue; cytochrome P-450: nmol/g tissue; NDMA N-demethylase: nmol/min mg protein.

[a,b]Different from control (first column) values at the p<0.01 level and p<0.05 level, respectively.

[c] Different from NDMA treated (second column) values at the p<0.01 level. Statistical analyses were performed by an Analysis of Variance (ANOVA), followed by Ducan's Multiple Range test.

methyl-N-nitrosourea (MNU). Efficacy of model and cellular nucleophiles was quantitated as nmol cysteine protected by the nucleophile from methylation by MNU/hr/mM compound. The NIVs of the pure compounds ascorbate, glutathione, 4-(4-nitrobenzyl)pyridine (NBP) and I-3-C were 2400, 1600, 3 and 0, respectively. (Calculated from the data shown in Figure 1).

When mice were treated with I-3-C by gavage at dosages of 0, 25, 50, 75 or 100 mg/kg body weight, the NIVs for ethyl acetate extracts of the livers 1 hr after treatment were 0, 33, 47, 52, and 92 nmol cysteine preserved/hr/g tissue, respectively (Figure 2). These data demonstrate that, although I-3-C itself is not sufficiently nucleophilic to compete with cysteine for methylation by MNU, it is capable of generating intracellular nucleophiles of sufficient strength to react in this NIV assay. When mice were treated with 10 mg NDMA/kg body weight 1 hr after the varying dosages of I-3-C, the 24 hr post-NDMA plasma alanine transaminase (ALT) values were decreased by I-3-C pretreatment in a dose-dependent fashion (Figure 2).

The coefficients of determination, r^2, computed from the linear least squares correlation coefficient between NIV and ALT values, were 0.80 (0-100 mg I-3-C/kg) and 0.98 (0-75 mg/kg).

It is important to recognize that the nucleophiles produced by I-3-C are in the ethyl acetate extract from liver, thereby representing more hydrophobic moieties which would tend to be localized in the membranous phase of the cell. It is attractive to speculate that such moieties represent microsomal metabolites of I-3-C, and as such would remain in the membranous fraction of the cell, a strategic location to scavenge electrophiles generated from NDMA by cytochrome P-450. Such a mechanism of intervention would account for the decrease in covalent binding of

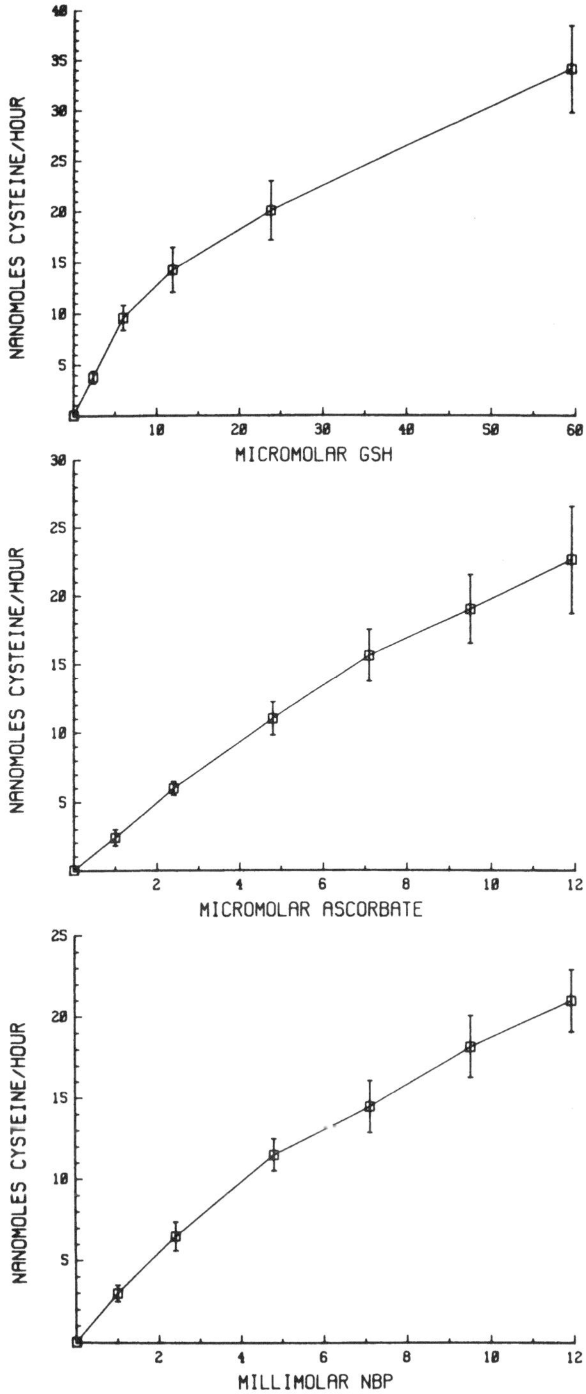

Fig. 1. Reactivity of standard chemicals in the Nucleophilic Index Value assay. GSH (top figure), ascorbate (middle figure) and NBP (bottom figure) were assayed for nucleophilicity as described in the Methods section. The results are expressed as mean nmoles of cysteine preserved/hr by the test compound versus concentration of test compound. Standard error bars are shown for three experiments.

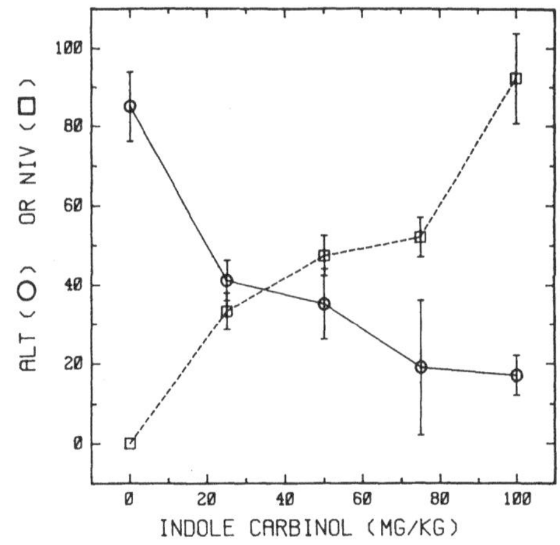

Fig. 2. Effects of I-3-C on Hepatic Nucleophilic Index values (NIV) and NDMA-mediated plasma alanine transaminase (ALT) activities. NIV and ALT are expressed as nmol cysteine preserved/hr/g tissue, and umol x 10^2/min/ml serum, respectively. Results are expressed as mean values $\pm$ SEM, with n=3.

NDMA metabolites observed when animals are pretreated with I-3-C (1). This scheme would also explain the I-3-C-mediated protection against the decreases in ascorbate and GSH observed after treatment with NDMA in mice. Thus, I-3-C metabolites would act as a first-stage non-critical targets for electrophiles generated in the endoplasmic reticulum, while the soluble nucleophiles, ascorbate and GSH, would be analogous to that of tocopherol in trapping membranous radicals, thus protecting against radical-mediated stress (2). In fact, I-3-C itself (or metabolites of I-3-C) may also function in concert with tocopherol in the membrane and act as a trap for radicals as well as nucleophiles (3). By acting in the proposed manner, I-3-C would be expected to inhibit the toxic or carcinogenic activity of many compounds which are activated by biotransformation enzymes of the endoplasmic reticulum. For example, I-3-C has been shown to reduce the incidence of chemically induced neoplasia in a number of systems (4-6). Although several mechanisms for chemoprotection by I-3-C, such as altered carcinogen metabolism, may be operating simultaneously, one mechanistic component may be the I-3-C-mediated trapping of reactive chemical species. In the situation where tissue cytotoxicity or genotoxicity is mediated by a reactive electrophile, the NIV assay on target tissue may be a useful predictor of tissue susceptibility to the chemical toxin or carcinogen.

ACKNOWLEDGEMENTS

This investigation was supported by USPHS grants ES-03373 and CA-38277. The authors thank Mrs. Diane Gorman for typing the manuscript.

REFERENCES

1. H.G. Shertzer, Indole-3-carbinol protects against covalent binding of benzo(a)pyrene and N-nitrosodimethylamine Metabolites to mouse liver macromolecules. Chem.-Biol. Interact. 48, 81-90 (1984).

2. J. Green, Vitamin E and the biological antioxidant theory. Ann. N.Y. Acad. Sci. 203, 29-44 (1972).
3. H.G. Shertzer, M.P. Niemi, F.A. Reitman, M.L. Berger, B.L. Myers, and M.W. Tabor, Protection against carbon tetrachloride hepatotoxicity by pretreatment with indole-3-carbinol. Exp. Mol. Pathol. 46, 180-189 (1987).
4. G.S. Bailey, D. Goeger, J.D. Hendricks, J.E. Nixon and N.E. Pawlowski, Indole-3-carbinol promotion and inhibition of aflatoxin B_1 carcinogenesis in rainbow trout. Proc. Am. Assoc. Cancer Res. 26, 115 (1985).
5. J.E. Nixon, J.D. Hendricks, N.E., Pawlowski, C.B. Pereira, R.O. Sinnhuber and G.S. Bailey, Inhibition of aflatoxin B_1 carcinogenesis in rainbow trout by flavone and indole compounds. Carcinogenesis 5, 615-619 (1984).
6. L.W. Wattenberg and W.D. Loub, Inhibition of polycyclic aromatic hydrocarbon-induced neoplasic by naturally occurring indoles. Cancer Res. 38, 1410-1413 (1978).

OZONE CARCINOGENESIS AND CO-CARCINOGENESIS AND ITS PREVENTION

Carmia Borek*, Augustinus Ong*, Herbert Mason*, Marco Zaider*, Gisela Witz+, and Myron A. Mehlman++

*Radiological Research Laboratory, College of Physicians & Surgeons, Columbia University, New York, N.Y., +Dept. of Environmental and Community Health, Rutgers Medical School Piscataway, N.J., and ++Mobil Oil Corporation, Princeton, N.J.

INTRODUCTION

Ozone (O_3), a reactive species of oxygen, is an important natural constituent of the atmosphere (1). Background levels of ozone in the lower atmosphere may range up to 0.1 ppm and are modified by geographic elevation, solar radiation and climatic conditions (2). Since some ozone effects are radiomimetic its actions may be enhanced in the presence of ionizing radiation from background and/or man-made sources (3,4).

While stratospheric ozone spares the earth from excess solar ultraviolet (UV) radiation (2), high levels of ozone in the environment are toxic and present a health hazard to man.

Excess environmental exposure to ozone can result from a variety of sources. Ozone is a key component in oxidant smog (2) and in the vicinity of high electric voltage equipment during operation (3). Ozone is widely used as a disinfectant for air and water, in bleaches, waxes, textiles, oils, and inorganic syntheses (3). Enhanced levels of ozone are found in planes flying at high altitudes (5).

Because of the toxic nature of ozone and its potential hazard to man, its levels in the environment are subject to government regulation (5). The current standard is set at an hourly average of 0.12 ppm, not to be exceeded more than once per year.

Urban areas with high levels of photochemical smog (e.g. Southern California) may experience high ambient ozone levels which can reach 0.5 ppm (3).

MECHANISMS

The toxicity of ozone depends upon its ozonation and oxidative properties. While the precise mechanisms of ozone damage are unclear, a variety of mechanisms have been proposed (3,6,7). Of these, the most compelling are the following, which implicate cell membranes as primary sites of ozone toxicity and suggest that ozone damage is in part induced via free radical processes (Fig. 1).

a. Ozone exerts its toxicity by oxidation of low molecular weight

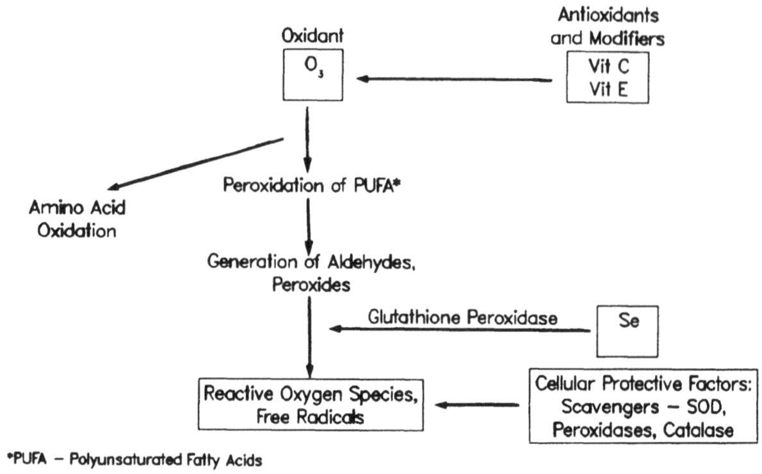

FIG. 1. Scheme of Ozone Toxicity and its Control by Antioxidants

compounds containing thiol, amine, aldehyde and alcohol functional groups and by oxidation of proteins.

b. Ozone acts by initiating peroxidation of polyunsaturated fatty acids present mainly in the cell membrane. The peroxides and secondary reactive oxygen species which ensue produce their toxicity by damaging the integrity of the cell membrane and other cellular molecules.

Both soluble peptides, such as glutathione, and proteins in lipid bilayers provide potential targets for ozone action. Protein modification takes place via oxidation of amino acids chain.

One of the major actions of ozone resides in its ability to peroxidize polyunsaturated fatty acids and produce malondialdehyde, which reacts with thiols, cross-links DNA and histones and acts as an initiator in mouse skin carcinogenesis (4).

Direct oxidation of amino acids and proteins by high ozone levels or oxidation by secondary reaction products of polyunsaturated fatty acids (PUFA) peroxidation can inhibit a variety of cellular protective systems. These include glutathione, a scavening thiol, glutathione peroxidase, superoxide dismutase and catalase, which detoxify peroxides and enzymes which supply reducing co-factors such as glucose-6-phosphate dehydrogenase (8). Both thiols and enzymes may be restored metabolically to control levels or rebound to higher protective levels following intermittent or continuous ozone exposure (9,10,11).

The degree to which ozone reacts with proteins is determined by the presence of ozone-susceptible amino acids at their active sites and the location of the amino acids in the tertiary structure of the protein.

The relationship between PUFA peroxidation and protein oxidation is further complicated by the fact that the maintenance of the tertiary structures of cellular membrane proteins is dependent on associated lipids. Any alterations in lipids surrounding the embedded proteins would result in structural alterations and changes in membrane function (12).

Peroxidation of PUFA by ozone results in the generation of fatty acid hydroperoxides. These are destroyed by glutathione peroxidase consuming GSH. Oxidized GSH is reduced by glutathione reductase consuming NADPH. Thus, the loss of GSH following ozone exposure promotes lipid peroxidation indirectly through the inhibition of glutathione peroxidase or, as mentioned earlier, by direct oxidation of GSH and its depletion.

One of the convenient tests used to assay lipid peroxidation is the thiobarbituric acid test (3,12,13). It is a spectrophotometric assay in which the red pigment formed in the reaction between 2-thiobarbituric acid (TBA) and oxidized lipids is measured by absorbance at 532 nm. The identity of the 532 nm chromogen is characterized as the adduct formed between two molecules of TBA and one molecule of malondialdehyde (MDA). It has generally been assumed that MDA and MDA precursors are the TBA-reactive substances in peroxidized lipid mixtures. Bicyclic prostaglandin-like endoperoxides were suggested as the MDA source in a mixture of oxidized lipids (13) and a variety of cyclic and acyclic peroxides have been reported as possible MDA precursors. These types of compounds also give the characteristic 532 nm absorption after reaction with TBA (12).

Following exposure to high levels of ozone, the relative importance of PUFA peroxidation and the oxidation of proteins and small molecular weight compounds depends on many factors. These include membrane composition of PUFA and proteins, which determine ozone accessibility and degree of interaction and damage, enzymatic pathways to decompose peroxides, pathways to generate thiols, the presence of antioxidants to prevent peroxide formation and to partake in scavenging free radicals arising from secondary reactions (6).

OZONE AS A CARCINOGEN AND COCARCINOGEN

In recent years, a growing volume of work has provided evidence that free radicals play a role in the initiation and promotion of neoplastic transformation (17, 18-22). The ability of ozone to generate free radical intermediates suggested that it may act as a direct environmental carcinogen or co-carcinogen (3,7).

We investigated the ability of ozone to induce cell transformation alone or in combination with ionizing radiation (4).

The cell systems used were primary cultures of diploid hamster embryo cells and the mouse C3H/10T-1/2 cell line. The two cell systems have been used extensively in radiation and chemical carcinogenesis studies and the culture conditions and transformation assays are well established (reviewed in 22).

For ozone transformation and co-carcinogenesis experiments, primary hamster cells and C3H/10T-1/2 cells were exposed to 5 ppm ozone for 5 minutes (4) after which the cells were then cloned out (4). Control cultures of both cell types not treated with ozone but exposed to air and subjected to the same conditions were cloned as descibed above.

Combined exposures to ozone and radiation were carried out by irradiating the hamster cultures 2 hours prior to ozone with 3 Gy (300 rad) gamma-rays and those of the C3H/10T-1/2 with 4 Gy (400 rad) (4).

Cell survival was assessed by colony-forming ability (22). Transformation was scored in the hamster embryo and the mouse C3H/10T-1/2 cells using morphological criteria (22).

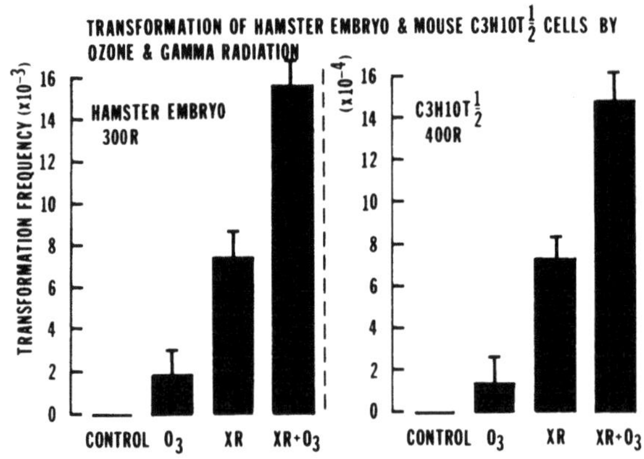

FIG. 2. Histogram Describing Ozone Carcinogenicity and Cocarcinogenicity In Vitro

The data indicated that treatment of hamster embryo and mouse C3H/10T-1/2 cells with 5 ppm of ozone for 5 minutes resulted in cell transformation (4) (Fig. 2). No transformation was observed in control untreated cells. When cells were pretreated with gamma-rays and exposed to ozone 2 hours later, transformation rates markedly enhanced indicating a synergistic interaction between the two agents (Fig. 2).

We tested the short term effects of ppm O_3 in producing lipid peroxidation products in the hamster embryo and mouse C3H/10T-1/2 cells compared to air-treated controls. The assays were done in conjunction with the transformation experiments and used cultures parallel to the ones exposed to O_3 for transformation assays (4). We found, as illustrated in Table 1, that MDA and MDA-like products were found at higher levels in O_3-exposed cells as compared to controls.

Compelling data from experiments in vivo and in vitro support the notion that free radicals play important roles in the complex course of multistep carcinogenesis (18-23). Much of the evidence emanates from the observation that antioxidants act as anticarcinogens (17-18). Agents which act at the cellular level include the enzymes superoxide dismutase and catalase (18,19). They also include the micronutrient selenium, which is a component of glutathione peroxidase, and Vitamin E, an inhibitor of lipid peroxidation (17). Our finding that lipid peroxidation is elevated in response to ozone (4) imply that free radical-mediated reactions play a role in ozone-induced cell transformation. Further evidence comes from our new findings that Vitamin E which inhibits radiation-induced transformation (17) also inhibits ozone and radiation-induced oncogenesis (Table 1) (4,24,26).

Table 1. The Effect of Vitamin E on Ozone-induced Transformation in Hamster Embryo Cells

Treatment	Transformation Frequency
Control	0
Ozone (5 ppm/5 min)	1.4×10^{-3}
Vitamin E (7 M)	0
Vitamin E + ozone	0.4×10^{-3}

Our results showing that ozone is a co-carcinogen acting synergistically with radiation enhance the implications for ozone as an environmental health hazard. Our data indicating that a nutritional factor such as Vitamin E suppresses the oncogenic effects of both agents indicates that life-style factors and suitable nutrition may act as preventive measure and inhibit the neoplastic process.

ACKNOWLEDGEMENTS

This investigation was supported by Grant No. CA-12536 awarded by the National Cancer Institute, and by a Contract from the National Foundation for Cancer Research.

REFERENCES

1. R.D. Caddle and E.R. Allen, Atmospheric photochemistry. Science 167:243-249 (1970).
2. National Research Council, Committee on Medical and Biologic Effects of Environmental Pollutants, subcommittee on ozone and other photochemical oxidants. Toxicology, Ozone and other photochemical oxidants. N.A.S.-N.R.C. Publ. 19:323-387 (1977).
3. C. Borek and M.A. Mehlman, Evaluation of health effects toxicity and biochemical mechanisms of ozone. In: The biomedical effects of ozone and related photochemical oxidants (S.D. Lee, et al., Eds.) pp. 325-361. Princeton Scientific Publishers, Inc., Princeton, N.J., 1983.
4. C. Borek, M. Zaider, A. Ong, H. Mason, and G. Witz, Ozone acts directly and synergistically with ionizing radiation to induce in vitro neoplastic transformation. Carcinogenesis 7:1611-1613 (1986).
5. Reed, S. Glaser, and J. Kaldor, Ozone toxicity symptoms among flight attendants. Am. J. Ind. Med. 1:43:54 (1980).
6. W. A. Pryor, M.M. Dooley, and D.F. Church, Mechanisms for the reaction of ozone with biological molecules, the source of the toxic effects of ozone. In: Advances in Modern Toxicology. The biological effects of ozone and photochemical oxidants. (S.J. Lee et al., Eds.) pp. 7-19. Princeton Scientific Publishers, Inc., Princeton, N.J., 1983.
7. M.A. Mehlman and C. Borek, Toxicity and biochemical mechanisms of ozone. Environ. Res. 42:36-53 (1987).
8. D.A. Johnson, Ozone inactivation of human alpha 1-proteinase inhibitor. Am. Rev. Respir. Dis. 121:1031-1038 (1980).
9. C.K. Chow and A.L. Tappel, An enzymatic protective mechanism against lipid peroxidation damage to lungs of ozone-exposed rats. Lipids 7:518-524 (1972).
10. A.J. DeLucia, P.M. Hogue, M.G. Mustafa, and E.C. Cross, Ozone interaction with rodent lung, effect on sulfhydryls and sulfhydryl-containing enzyme activities. J. Lab. Clin. Med. 80:559-566 (1972).
11. M.G. Mustafa and S.D. Lee, Pulmonary biochemical alterations resulting from ozone exposure. Ann. Occup. Hyg. 19:17-26 (1976).
12. G. Witz, N.J. Lawrie, A. Zaccaria, H.E. Ferran Jr., and B.D. Goldstein, The reaction of 2-thiobarbituric acid with biologically active alpha, beta-unsaturated aldehydes. J. Free Radicals Biol. Med. 2:33-39 (1986).
13. W.A. Pryor, J.P. Stanley, and E. Blair, Autooxidation of polyunsaturated fatty acids: II. A suggested mechanism for the formation of TBA-reactive materials from prostaglandin-like endoperoxides. Lipids 2:370-379 (1976).
14. C. Borek, C.M. Fenoglio, and D.W. King, (Eds.) Membranes, Cancer and Aging, Advances in Pathobiology, Vol. 7. Thieme-Stratton, Inc., New York, 1980.

15. C. K. Chow, Influence of dietary vitamin E on susceptibility to ozone exposure. In: The Biochemical Effects of Ozone and Photochemical Oxidants (S.D. Lee, M.G. Mustafa and M.A. Mehlman, Eds.) pp. 75-93. Princeton Scientific Publishers, Inc., Princeton, N.J., 1983.
16. B. L. Fletcher and A.L. Tappel, Protective effects of dietary alpha-tocopherol in rats exposed to toxic levels of ozone and nitrogen dioxide. Environ. Res. 6:165-175 (1973).
17. C. Borek, A. Ong, H. Mason, L. Donahue, and J.E. Biaglow, Selenium and vitamin E inhibit radiogenic and chemically induced transformation in vitro via different mechanisms of action. Proc. Natl. Acad. Sci. U.S.A. 83:1490-1494 (1986).
18. C. Borek and W. Troll, Modifiers of free radicals in inhibit in vitro the oncogenic actions of x-rays, bleomycin and the tumor promoter 12-0-tetradecanoylphorbol-13-acetate. Proc. Natl. Acad. Sci. U.S.A. 80:1304-1307 (1983).
19. R. Zimmerman and P.A. Cerutti, Active oxygen acts as a promoter of transformation in mouse embryo C3H/10T-1/2/C18 fibroblasts. Proc. Natl. Acad. Sci. U.S.A. 81:2085-2087 (1984).
20. T.J. Slaga, A.J.P. Klein-Szanto, L.L. Triplett, L.P. Yotti, and J. E. Trosko, Skin tumor promoting activity of benzoylperoxide, a widely used free radical generating compound. Science 213:1023-1025 (1981).
21. B.N. Ames, Dietary carcinogens and anticarcinogens. Oxygen radicals and degenerative diseases. Science 221:1256-1263 (1983).
22. C. Borek, Radiation oncogenesis in culture. Adv. Cancer Res. 37:159-232 (1982).
23. O.F. Nygaard and M.G. Simic (Eds.) Radioprotectors and Anti-carcinogens, Academic Press, New York, 1983.
24. C. Borek, Radiation and chemically induced transformation: Free radicals, antioxidants and cancer. Br. J. Cancer. (in press, 1987).
25. L.W. Wattenberg, Inhibiting chemical carcinogens by minor dietary factors. In: Molecular Interrelations of Nutrition and Cancer (M.S. Arnot, et al., Eds.) pp. 43-56 Raven Press, New York, 1982.

EXERCISE TRAINING EFFECTS ON MITOCHONDRIAL ENZYME ACTIVITY, UBIQUINONES AND VITAMIN E

Johanna Lang, Kishorchandra Gohil, Laurie Rothfuss and Lester Packer

Membrane Bioenergetics Group, Lawrence Berkeley Laboratory and Department of Physiology/Anatomy, University of California Berkeley, CA 94720

INTRODUCTION

It is well-known that endurance exercise training leads to an overall increase in the oxidative capacity of oxidative type (red) skeletal muscles through increased biogenesis of mitochondria (1, 2). While overall muscle mitochondrial content increases, the mitochondrial lipid/protein ratio and the specific activities of mitochondrial enzymes remain constant (2). Besides their known physiological function as providers of ATP through oxygen reduction, mitochondria are also potential sites of oxygen free radical formation by erroneous, incomplete electron transfer to oxygen, most probably at the site of ubiquinone oxidoreduction (3). Mitochondrial free radical production is likely to increase with the massive increase in oxygen consumption during physical activity. This suggests that exercise should result in an increased demand on antioxidant systems to respond to the increased production of oxygen-derived and other free radicals.

This hypothesis has been tested in the present investigation. The mitochondrial content, based on total ubiquinone content and succinate cytochrome C reductase activity, and vitamin E concentrations were examined in various tissues of sedentary and endurance exercise-trained rats.

Modern analytical techniques make it possible to determine vitamin E (tocopherol isomers) and oxidized and reduced ubiquinone homologs in tissues with high sensitivity, accuracy and speed. To follow changes in tissue ubiquinone and vitamin E content with exercise-endurance training, we took advantage of a new HPLC assay, which uses in-line UV and electrochemical detection for the simultaneous determination of tocopherols, ubiquinols and ubiquinones from tissue lipid extracts (4). In addition, we determined succinate cytochrome C reductase as a mitochondrial marker enzyme.

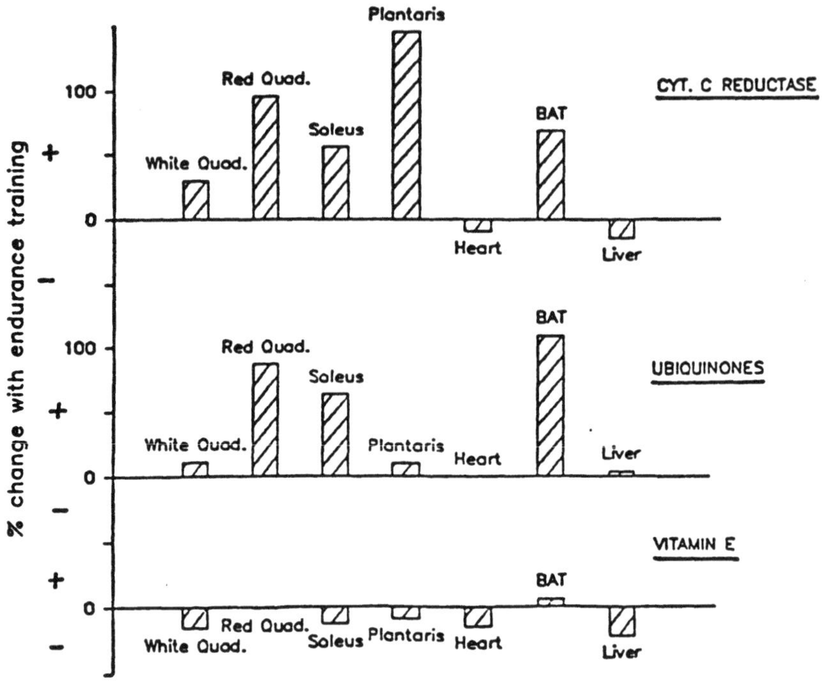

Fig. 1. Percentage change in succinate cytochrome C reductase activity and ubiquinone and vitamin E content with endurance training.

EXPERIMENTAL

Animals and training: 15 female, 9 weeks old Wistar rats were fed a normal diet. All animals were exercised for 10 - 30 min. on a treadmill at 15% gradient at an average speed of 10 m/min. for one week. Twelve animals which would run with little or no electrical stimulation were selected for the study. Six rats were randomly assigned to the sedentary group and were exercised for 5 min. on the treadmill at about 20 m/min., 3 days a week for 12 weeks. The remaining 6 rats were endurance trained for 12 weeks, 3 days per week, so that for the final 2 weeks of the study, the rats ran an average speed of 30 m/min. for 2 hours. During the week following the last week of training, 4 rats (2 sedentary and 2 trained) were sacrificed on alternate days and selected tissues were freeze-clamped and stored at $-70^{o}C$ for later analysis.

Assays: Succinate cytochrome C reductase was measured in whole homogenates by cytochrome C reduction in the presence of cyanide (5). Vitamin E, ubiquinone and ubiquinol contents of the tissues were analyzed from lipid extracts by HPLC with in-line UV and electrochemical detection (4). Significance of differences between sedentary and trained groups was analyzed by Student's t-test.

RESULTS AND DISCUSSION

Endurance training led to large increases in succinate cytochrome C reductase activity of red quadriceps (96%), plantaris (146%) and soleus (56%) muscles and brown adipose tissue (68%) but not in white quadriceps muscle, heart and liver (Table 1, Fig. 1).

Table 1. Succinate Cytochrome C Reductase, Total Ubiquinones and Vitamin E in Tissues of Sedentary and Endurance-Trained Rats

	Vitamin E (nmol/g)		Total Ubiquinones (nmol/g)		Succ. Cyt. C reductase (μmol.min x g)	
	Sed.	Trained	Sed.	Trained	Sed.	Trained
White quadriceps	19.6 ±2.3	16.4 ±1.5	9.9 ±0.5	11.0 ±0.8	0.6 ±0.1	0.7 ±0.1
Red quadriceps	29.3 ±3.4	29.0 ±2.3	52.0 ±3.6	97.9*** ±5.6	3.6 ±.4	7.1** ±0.6
Soleus	27.7 ±1.8	24.0 ±1.9	20.7 ±2.3	33.9** ±2.2	2.5 ±0.4	3.9*** ±0.1
Plantaris	26.6 ±3.0	24.3 ±1.4	46.6 ±3.8	51.4 ±4.6	1.3 ±0.1	3.2*** ±0.2
Heart	63.6 ±5.4	53.6 ±1.3	168.5 ±5.6	168.0 ±5.5	11.8 ±0.8	10.8 ±0.8
White adipose tissue	63.8 ±5.0	75.6 ±1.3	2.1 ±0.2	3.7*** ±0.3	n.d.	n.d.
Brown adipose tissue	330.1 ±32.3	348.7 ±41.3	235.6 ±26.8	492.9* ±68.4	8.1 ±1.1	13.6 ±1.9
Lung	64.5 ±4.7	59.2 ±2.2	9.2 ±0.7	8.9 ±0.5	n.d.	n.d.
Liver	64.0 ±7.3	49.8 ±2.7	95.6 ±7.3	99.0 ±3.7	7.9 ±0.8	6.9 ±1.1

n.d. = not determined

significantly different from sedentary group * $p < 0.05$
** $p < 0.01$
*** $p < 0.001$

Values are mean +/- S.E.M., n = 6

Ubiquinone content correlated well with mitochondrial enzyme activity in all tissues, except plantaris muscle, where the more than twofold increase in succinate cytochrome C reductase activity with endurance training was not paralleled by an increase in ubiquinone content. At present we have no explanation for this unexpected behavior.

Tissue vitamin E concentrations did not correlate with either ubiquinone concentrations (Figs. 1 and 2) or mitochondrial enzyme activities (Fig. 1).

Ubiquinone 9/10 ratios and ubiquinone redox states show distinct differences between different tissues, but are unaltered by endurance training (Table 2).

Table 2. Total Ubiquinone 9/10 Ratio and Ubiquinol/Ubiquinone Redox Equilibrium in Tissues of Sedentary and Endurance-Trained Rats

	Ubiquinone 9 / Ubiquinone 10		% Oxidized ubiquinones	
	Sed.	Trained	Sed.	Trained
White quadriceps	4.5	4.1	80.1 ±2.0	80.2 ±2.5
Red quadriceps	12.3	15.1	72.4 ±2.9	69.7 ±2.1
Soleus	12.0	11.8	64.3 ±7.8	63.5 ±1.8
Plantaris	14.8	15.8	75.9 ±2.1	69.8 ±3.1
Heart	9.4	9.9	69.1 ±2.3	66.7 ±2.4
White adipose tissue	a)	a)	<1	<1
Brown adipose tissue	16.9	15.3	15.9 ±1.4	16.0 ±0.7
Lung	7.1	7.7	93.0 ±0.9	88.8 ±1.6
Liver	b)	b)	30.6 ±2.9	24.6 ±1.7

a) Ubiquinone 10 below detection limit
b) Ubiquinol 10 not clearly separated from adjacent peak

CONCLUSIONS

The finding that increased mitochondrial content in skeletal muscle and adipose tissues in response to endurance training is not accompanied by paralleled increases in vitamin E suggests a net decrease of mitochondrial vitamin E content. This could be indicative of mitochondrial lipid peroxidation occurring in exercising animals.

The apparent increase in mitochondrial content is more likely explained by increased lipolysis which causes shrinkage of the adipose tissue and thus a relative increase in tissue concentration of cellular components (6). However, as vitamin E does not parallel increases in ubiquinones we can conclude that vitamin E is effectively lost from adipose tissues.

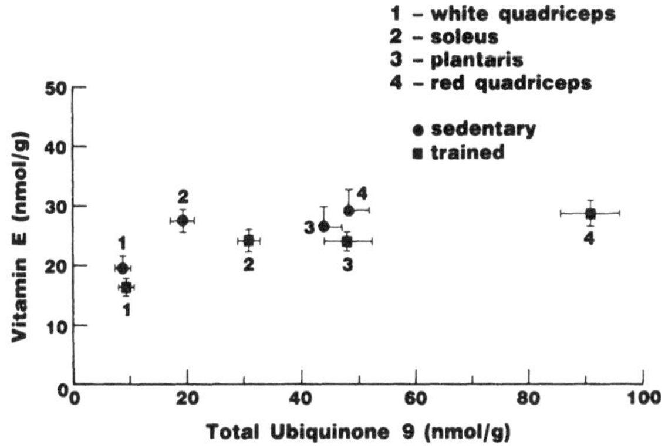

Fig. 2. Relationship between total ubiquinone 9 and vitamin E content of skeletal muscles.

ACKNOWLEDGEMENTS

The authors would like to acknowledge the technical assistance of Janet Volker. Research was supported by the National Foundation for Cancer Research, The Office of Naval Research, and Hoffman La Roche, Inc.

REFERENCES

1. J.O.Holloszy and F.W. Booth, Biochemical adaptation to endurance exercise in muscle, Annu. Rev. Physiol. 38, 237-291 (1976).
2. K.I.A. Davies, L. Packer, and G.A. Brooks, Biochemical adaptations of mitochondria, muscle, and whole animal respiration to endurance training, Arch. Biochem. Biophys. 209, 538-553 (1981).
3. A. Boveris and E. Cadenas, A.O.M. Stoppani, Role of ubiquinone in the mitochondrial generation of hydrogen peroxide, Biochem. J. 156, 435-444 (1976).
4. J. Lang, K. Gohil, and L. Packer, Simultaneous determination of tocopherols, ubiquinols, and ubiquinones in blood, plasma, tissue homogenates and subcellular fractions, Anal. Biochem. 157, 1016-116 (1986).
5. K. Cohil, L. Oacker, B. DeLumen, G.A. Brooks and S.E. Terblanche, Vitamin E deficiency and vitamin C supplements: exercise and mitochondrial oxidation, J. Appl. Physiol. 60(6), 1986-1991, (1986).
6. L. Bukowiecki, J. Lupien, N. Follea, A. Paradis, D. Richard, and J. Leblanc, Mechanism of enhanced lipolysis in adipose tissue of exercise-trained rats, Am. J. Physiol. (Endocrinol. Metabol. 2) E422-E429 (1980).

CANCER MORTALITY INVERSELY RELATED TO PLASMA LEVELS OF ANTIOXIDANT VITAMINS

K. Fred Gey[1], Georg B. Brubacher[1], and
Hannes B. Stähelin[2]

1 Vitamin Research Department, F. Hoffmann-La Roche & Co., Ltd., CH-4002 Basel, Switzerland
2 Geriatric Clinic, Kantonsspital Basel, University of Basel, Switzerland

Aggressive oxygen species (such as superoxide anion, hydroxyl radical, singlet oxygen and longer-lived reaction products (e.g. hydroperoxides, alkenals) have been implicated in cancerogenesis but it is very difficult to show their direct role in vivo. Indirect evidence can be obtained from the inspection of the body's multilevel defense system against oxygen radicals which includes essential antioxidants, i.e. ß-carotene and the vitamins A, C and E. Since the dietary supply of the latter can vary considerably cancerogenesis might, at least in part, be inversely related to the status of antioxidant vitamins. There is growing evidence in animals and humans in favour of this concept. In animals vitamin A deficiency results in metaplasia whereas experimentally induced tumors can be diminished by ß-carotene, vitamins A, C and E. In the human many dietary surveys have convincingly shown that the intake of fresh fruits and leafy green-yellow vegetables as well as the calculated consumption of the above mentioned essential antioxidants is inversely related to the mortality from cancers (1-4). Dietary surveys have, however, inherent weaknesses and thus require confirmation by the measurement of plasma antioxidants in prospective studies.

In males three prospective studies have consistently demonstrated that base-line plasma levels of ß-carotene were statistically, i.e., 14-38% lower in subjects with subsequent lung cancer (5-9). One study (9) described also an inverse correlation between plasma vitamin E and subsequent lung cancer. In females low base-line levels of vitamin E were significantly associated with subsequent breast cancer (10). In the prospective Hypertension Detection and Follow-up Program vitamin A was 28% lower in subsequent combined gastrointestinal cancers of both sexes (11). In some studies there were also some, although statistically unsignificant trends for vitamins A, C and E. Such uncertainties or even some inconsistencies may in part have been due to the fact that these results were obtained by comparison of deaths to a similarly low number of surviving controls matched for sex, age, smoking etc. One of the studies, i.e., Basel Study permitted, however, a reevaluation in

Table 1. Mean plasma values of antioxidant vitamins at base-line of 7-year follow-up of mortality of males in Prospective Basel Study. The values were age-standardized and their mean expressed in % of survivors. The absolute values of survivors (= 100%) are indicated in brackets. p for t-test.

State	N	ß-Carotene	Vit. A	Vit. C	Vit. E (α-Toc.)	CIAVIT+
Survivor	2'707	100 (0.35)	100 (2.83)	100 (47.0)	100 (35.5)	100 (581)
Cancer-All	102	80**	98	88*	95*	69**
-Lung	37	71**	101	93	94	62**
-Stomach	17	66**	88**	76*	88*	46**
-Colorect.	9	114	105	94	88	100
Lung Dis.	20	57**	94	89	92	49**
All Deaths	268	86**	100	92**	99	77**

* $p < 0.05$, ** $p < 0.01$ as compared to survivors, t-test
+Ciavit = µM ß-carotene x µM vitamin C x µM stand. α-tocopherol

more details (12). This was possible since in the Basel Study unlike all other ones the plasma vitamins were measured immediately after sampling in all subjects (5-7, 12). Thus in Basel 102 cancer deaths of males as registered in the 7-year follow-up could actually be compared to 2,707 male survivors. Clearly, this greater number of surviving controls can markedly increase the statistical power. In the reevaluation the levels of the antioxidants were age-standardized and their mean was expressed in % of surviving controls (Table 1).

The mean levels of subjects with subsequent lung cancer showed a lower ß-carotene at base-line as had to be expected . With the present improved statistics, however, low ß-carotene revealed a statistically significant association also with stomach cancer as well as with the overall mortality from cancer, and surprisingly with fatal non-malignant lung diseases as well. A lower mean value of vitamin A became in the new analysis a statistically significant predictor of subsequent stomach cancer, and low plasma vitamin C was now significantly associated with subsequent stomach cancer as well as with the overall mortality from cancer. The same was true for the mean value of vitamin E (α-tocopherol). CIAVIT, i.e., a normative cumulative index of antioxidant vitamins (actually arbitrarily calculated by multiplication of µmolar levels of ß-carotene, and vitamins C and E), was a statistically significant indicator of subsequent cancer of the lung, stomach, overall cancer mortality as well as of non-malignant lung diseases. Percentagewise the mean values of this normative index were substantially lower than those of individual antioxidants (Table 1).

The relative risk of different plasma levels is even more conclusive than differences in mean values, particularly regarding levels at risk and target levels respectively. The relative risk was calculated according to a procedure suggested by Peto (14). Briefly, the cases were categorized into subgroups of different plasma levels but of similar size, practically corresponding to tertiles or quintiles. So far as risk levels have been described before, e.g. for ß-carotene (8) and vitamin C (15), they were used as cut-off points. The relative risk of each antioxidant range was expressed by the ratio of the observed mortality over the mortality expected for the Basel Study population after adjustment for age and smoking. When the observed and expected mortality obviously did not differ in neighbouring categories they were combined in order to increase the statistical power, e.g. for vitamin E the critical bottom tertile was calculated as such whereas tertiles No. 2 and 3 were combined since they showed no difference.

Lung cancer (Fig. 1) showed an increased risk at ß-carotene levels below 0.2 µM, similar to a prospective study in Hawaii with men of Japanese ancestry (8). In the Basel population high ß-carotene levels were associated with a lower relative risk of lung cancer. The trend analysis was statistically significant and the relative risk of lung cancer at low ß-carotene levels was 2.6 times higher than that at high levels.

In the prospective study in Washington County, Md. (9) about twice as great risk of lung cancer was found for low levels of vitamin E as well. A corresponding tendency was also obvious in the Basel Study although it lacked statistical significance (Fig. 1). This discrepancy may be related to the fact that the mean vitamin E level in the Basel population was about 20% higher than that in Washington County. In Basel some tendency of increased risk for lung cancer was also observed at low levels of vitamin A and at a low to medium index of CIAVIT (Fig. 1).

Since in the Basel population all vitamin parameters varied almost independently the relative risk of discriminatory combinations was tentatively estimated in a few Basel subjects (Fig. 2). In the case of ß-carotene and vitamin E the risk of subsequent lung cancer was greatest, i.e. 1.94, when the level of both antioxidants was low, and vice versa minimal, i.e. 0.76, when the level of both plasma antioxidants was high. The risk of these two extremes differed by the factor 2.6. The cases with one antioxidant high and the other low showed a relative risk in between. For the few cases no statistical significance could be expected. Nevertheless synergistic links between ß-carotene and vitamin E are conceivable since the trend is regular and consistent with the study in Washington County, Md. (9).

The mean values of the non-malignant lung diseases (Table 1) indicated that they behave similar to those of lung cancer. This is confirmed by the calculation of the relative risk of death from the common combination of emphysema and bronchopneumonia without cancer (Fig. 1). An increased risk was predicted with statistical significance by low plasma levels of ß-carotene, and by low to medium values of CIAVIT. A corresponding although statistically unsignificant trend existed for vitamins A and E as well. These similarities between malignant and non-malignant fatal lung diseases (Fig. 1) suggest that an optimal status of various antioxidants is required for vital functions of the lung system. Whether their impairment by a marginal antioxidant status will then lead to cancer or not might depend on confounding risk factors.

The mean values of the Basel Study (Table 1) indicated an inverse association of stomach cancer with all vitamin antioxidants tested.

ß-Carotene levels below 0.2 µM revealed a relative risk of 2.01 but medium and high ß-carotene levels only of 0.69 (p=0.051). These risks of stomach cancer differed thus by the factor 3.0. A corresponding although more striking situation existed for vitamin A. Vitamin A levels below 2.09 µM were associated with a relative risk of 4.25, higher levels with a risk of 0.76 (p=0.002). The relative risk between low and high levels of vitamin A differed even by the factor 5.6. This finding with vitamin A may be related to the previous report (11) that low vitamin A is associated with an increased risk of combined gastro-intestinal cancers. The discriminatory combinations (Fig. 2) revealed that the relative risk of stomach cancer was highest at low plasma levels of both isoprene derivatives, i.e., the risk was then 6.84, but

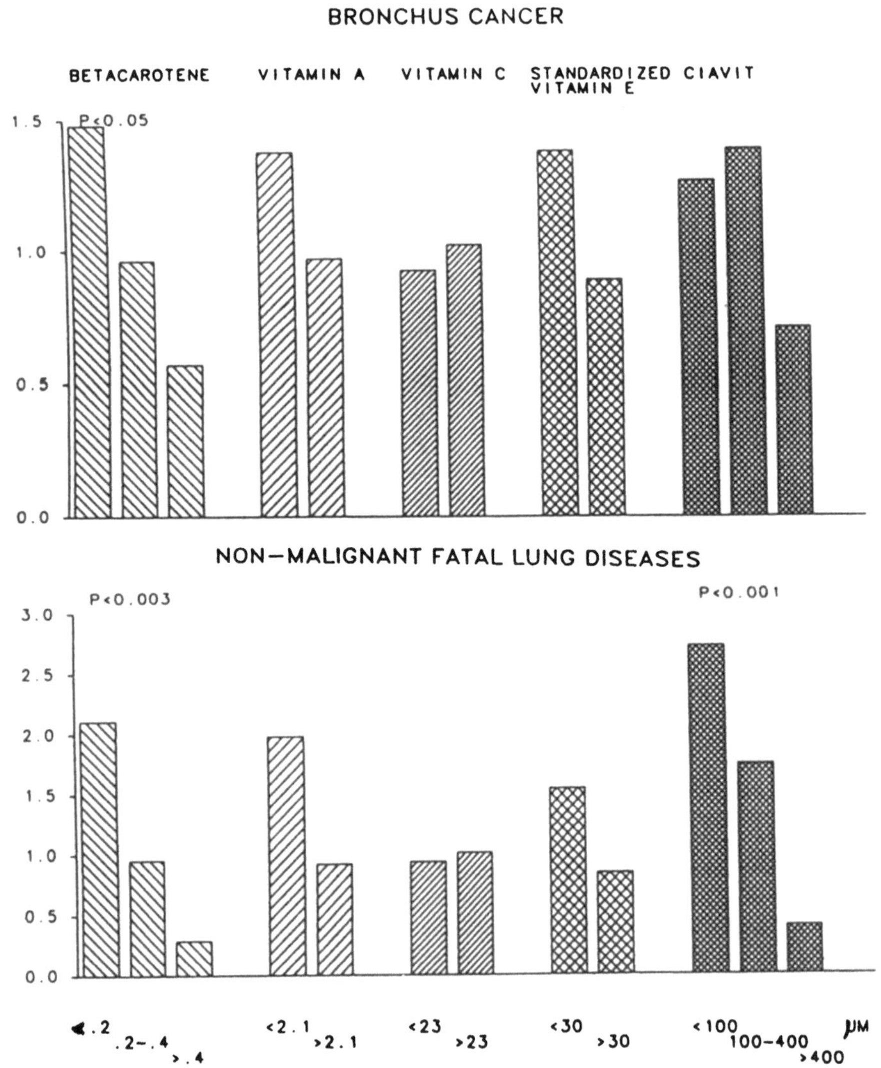

Fig. 1. Relative risk (observed mortality/expected mortality after adjustment for age and smoking; ordinate) of bronchus cancer (N = 37) and of non-malignant fatal lung diseases (N = 20) at different base-line categories of plasma oxidants in the 7-year follow-up of the Prospective Basel Study. P for trend analysis.

the risk was minimal at optimal plasma levels, i.e., only 0.57. The risk of these extremes differed by the factor 12. When only one isoprene derivatives was low and the other high an intermediate risk occurred. Vitamin A made a statistically significant contribution to stomach cancer when ß-carotene was accounted for in the logistic model. The plasma level of both isoprene derivatives has been accepted to behave completely independent when the diet contains sufficient amounts of preformed vitamin A as common in affluent westernized societies (16). Therefore, the present data suggest that ß-carotene can have actions independent from its provitamin A function. Although both isoprene derivatives might to some extent be able mutually to compensate, a potentially complete health prevention seems to depend on a simultaneously optimal level of both isoprene derivatives, i.e., the conceivable health preventive effect of ß-carotene may depend on an optimal status of vitamin A and vice versa.

Discriminatory combinations of vitamin A with vitamin E (Fig. 2) or of ß-carotene with vitamins C or E (not shown) gave corresponding results with statistical significance for stomach cancer. They suggest again that essential antioxidants aside from their specific functions can act as synergists.

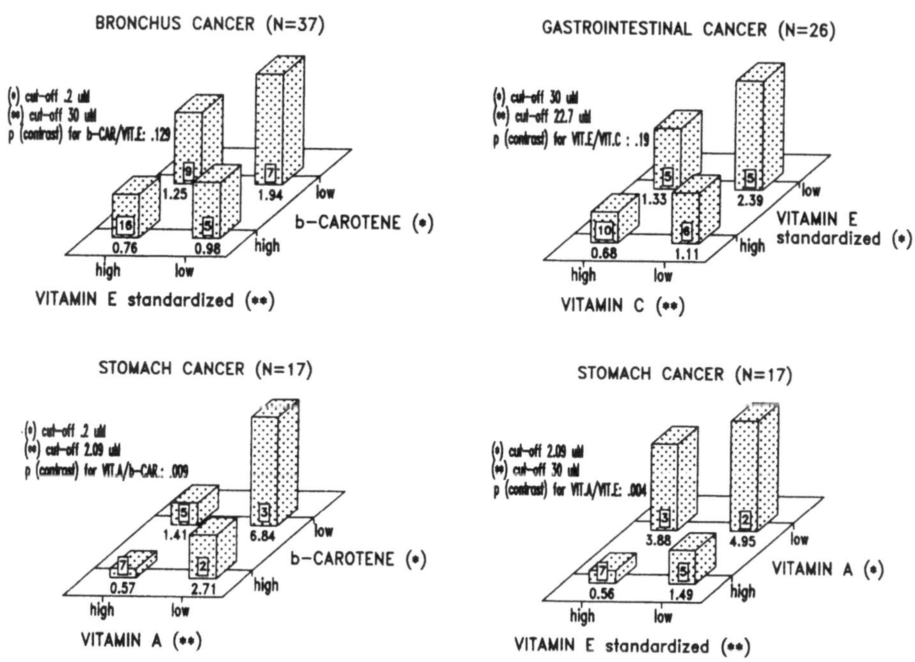

Fig. 2. Discriminatory combinations of low and high levels of various antioxidant vitamins in plasma at base-line of 7-year follow-up males in the Prospective Basel Study. The figures at the bottom of the bars indicate the number of cases, the values below the bars the relative risk (observed mortality/expected mortality after adjustment for age and smoking) for bronchus cancer (upper left), stomach cancer (lower left and right), and gastrointestinal cancers (upper right).

The mean values of the Basel Study (Table 1) indicated an association of both vitamins C and E with stomach cancer as well as with cancer of the large bowel. Therefore, all gastrointestinal tumors were combined for the calculation of the relative risk at different levels of vitamins C and E (Fig. 3). Vitamin C levels of accepted marginal deficiency, i.e., below 23 µM (or 0.4 mg/dl) were associated with a significantly increased risk. The risk of low and high levels of vitamin C differed by the factor 2.2. Since gastrointestinal cancers are known to be inversely related to plasma cholesterol as well, the vitamin E values of the Basel cases were considered without and with cholesterol-standardization. The latter had no principal effect on the result. The risks between low and high values of cholesterol-standardized vitamin E (Fig. 3) differed by 2.2. The CIAVIT figures behaved correspondingly (Fig. 3).

Discriminatory subgroups of vitamins C and E might also suggest synergistic links of these two antioxidants regarding the relative risk of combined gastrointestinal cancer (Fig. 2). When both compounds were simultaneously low, the risk was increased up to 2.39, when both levels were high, the relative risk dropped down to 0.68. The group with the greatest and lowest risk thus differed by the factor 3.5. Again the trend was obviously consistent with animal experiments, studies in tissue cultures as well as excellent physicochemical data which suggest the regeneration by vitamin C of vitamin E from its radical.

Synergism of all essential antioxidants has already been postulated before, at least for the initial states of stomach cancer (17). Vitamins C and E have been established to diminish the formation of mutagens such as nitrosamines as well as of endogenous fecal mutagens. Presumably vitamins C and E can simultaneously counteract tumor promotion through their synergistic protection against lipid peroxidation. Vitamin E is the major protector of the lipid-soluble phase and vitamin C interacts at the interphase and in the watersoluble compartment. Vitamins C and E are also able beneficially to modify immunoresponses as well as the eicosanoid system. Human intervention studies with the measurement of micronuclei in buccal mucosa of cancer-prone populations suggest that ß-carotene is involved in the prevention of DNA damage and/or DNA repair independently from the vitamin A status (18), possibly mediated by the particularly high radical scavenging properties of ß-carotene at low oxygen pressure. Vitamin A finally is essential for the cell differentiation. Vitamin C has been considered also to promote fibroplasia and thus to reduce tumor spreading. Clearly, this interpretation takes into account that tumorgenesis in general is a multifactorial and multistage process and that essential antioxidants have many, in part site-specific options to interfere with.

Since all major cancer types of the Basel population were somehow correlated with essential antioxidants this had to be reflected in the overall mortality from cancer. The present biostatistical procedure revealed for the first time also an inverse correlation with the overall mortality from cancer (Fig. 3). The relative risk of low and high levels of antioxidants differed by the factor 1.7 for ß-carotene, 2.0 for vitamin A, 1.3 for vitamin C, 1.6 for standardized vitamin E and 1.8 for the normative CIAVIT value. These trends were statistically significant for all parameters except vitamin C. This may be explained by the fact that vitamin C was only involved in gastrointestinal cancers.

Thus far only statistical associations between major causes of death and the plasma status of essential antioxidants have been demonstrated. Although these associations justify ongoing intervention studies, only the latter can finally prove the preventive potentials of these antioxidants in the human carcinogenesis (19).

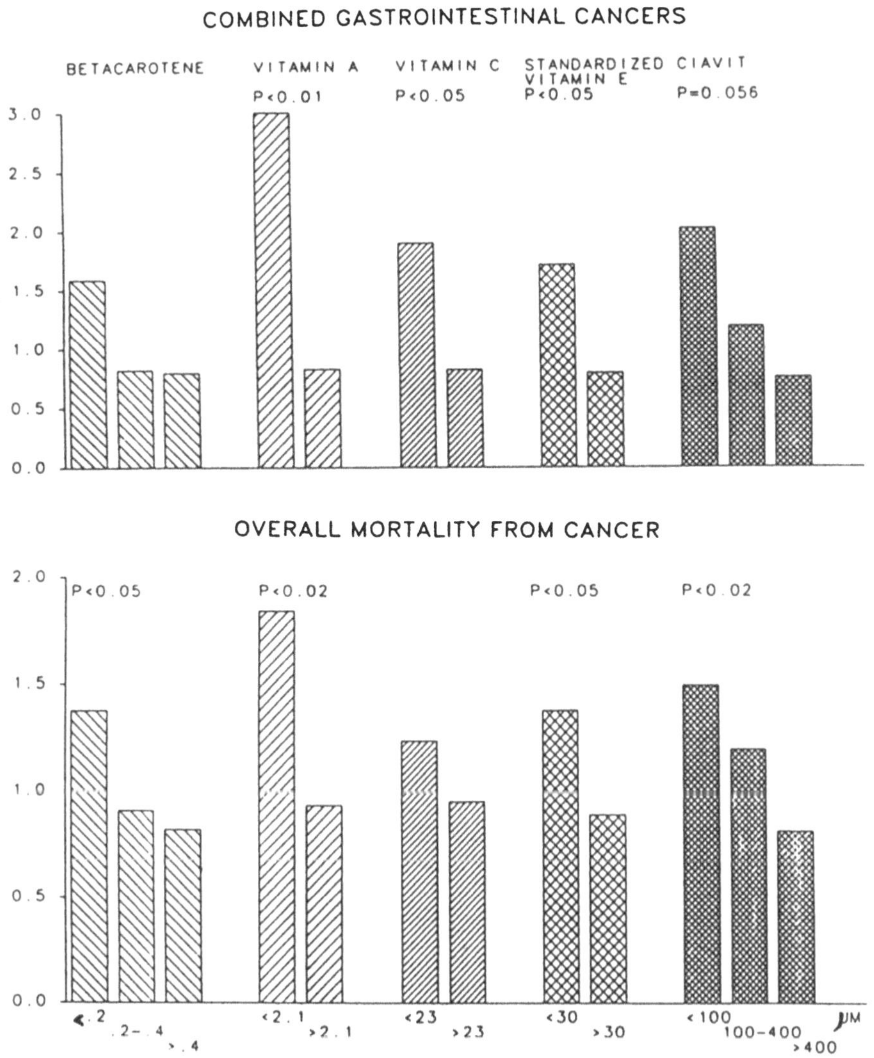

Fig. 3. Relative risk (observed mortality/expected mortality after adjustment for age and smoking; ordinate) of combined gastrointestinal cancers (N = 26) and of overall mortality from cancer (N = 102) at different base-line categories of plasma antioxidants in the 7-year follow-up of males in the Prospective Basel Study. P for trend analysis.

In summary, the reevaluated data of the Basel Study suggest that

(a) certain organs are associated with a specific risk profile of essential antioxidants, i.e.,

- the lung system primarily with ß-carotene and to some extent secondarily with vitamins A and E,

- the stomach with all major essential antioxidants,

- the large bowel with vitamins C and E.

(b) synergistic links exist between various essential antioxidants which would be in accordance with animal experiments. Therefore, a conceivable preventive effect of these antioxidants might depend on the simultaneous optimization of all interacting compounds rather than on the improvement of a single one.

REFERENCES

1. Committee on Diet, Nutrition and Cancer, Diet, nutrition and cancer. National Academy Press, Washington, DC., 1982.

2. S. Palmer, Diet, nutrition, and cancer. Progr. Food Nutr. Sci. 9, 283-341 (1985).

3. R. Doll and R. Peto, The causes of cancer: quantitative estimates of avoidable risks of cancer in the United States today. J. Natl. Cancer Inst. 66, 1191-1308 (1981).

4. R. Peto, R. Doll, J. D. Buckley, and M. B. Sporn, Can dietary ß-carotene materially reduce human cancer rates ? Nature 290, 201-208 (1982).

5. H. B. Stähelin, E. Buess, F. Rösel, L. K. Widmer, and G. Brubacher, Vitamin A, cardiovascular risk factors, and mortality. Lancet, 1, 394-395 (1982).

6. H. B. Stähelin, E. Rösel, E. Buess, and G. Brubacher, Cancer, vitamins, and plasma lipids: Prospective Basel Study. J. Natl. Cancer Inst. 73, 1463-1468 (1984).

7. H. B. Stähelin, F. Rösel, E. Buess, and G. Brubacher, Dietary risk factors for cancer in the Basel Study. Biblthca Nutr. Dieta 37, 144-153 (1986).

8. A. M. Y. Nomura, G. N. Stemmermann, L. K. Heilbrun, R. M. Salkeld, and J. P. Vuilleumier, Serum vitamin levels and the risk of cancer of specific sites in men of Japanese ancestry in Hawaii. Cancer Res. 45, 2369-2372 (1985).

9. M. S. Menkes, G. W. Comstock, J. P. Vuilleumier, K. J. Helsing, A. A. Rider, and R. Brookmeyer, Serum ß-carotene, vitamins A and E, selenium and the risk of lung cancer. N. Engl. J. Med. 315, 1250-1254 (1986).

10. N. J. Wald, J. Boreham, J. L. Hayward, and R. D. Bulbrook, Plasma retinol, ß-carotene and vitamin E levels in relation to the future risk of breast cancer. Br. J. Cancer 49, 321-324 (1984).

11. W. C. Willett, B. F. Polk, B. A. Underwood, et al., Relation of serum vitamins A and E and carotenoids to the risk of cancer. N. Engl. J. Med. 310, 430-434 (1985).

12. K. F. Gey, H. B. Stähelin, G. B. Brubacher, H. Riggenbach, F. Bernasconi, and L. K. Widmer, Inverse association of antioxidant vitamins in plasma with subsequent riks of cancer. Initial level of vitamins A, C, E and ß-carotene on 2,975 men followed-up for 7 years in the prospective Basel Study, (in preparation) 1987.

13. L. K. Widmer, H. B. Stähelin, C. Nissen , and A. da Silva, Eds., Venen-, Arterien-Krankheiten, koronare Herzkrankheiten bei Berufstätigen. Prospektiv-epidemiologische Untersuchung Basler Studie I-III. Huber, Bern, 1981.

14. R. Peto, M. C. Pike, N. E. Day et al., Guidelines for simple, sensitive significance tests for carcinogenic effects in long-term animal experiments. In Long-term and short-term screening assays for carcinogens: a critical appraisal. (WHO/IARC, Eds.). IARC Monographs, Suppl. 2, pp. 311-42,6 International Agency for Research on Cancer, Lyon, 1980.

15. Bureau Nutr. Sci., Food Directorate Health Protection Branch, Dept. Natl. Health and Welfare, Vitamins. In Recommended nutrient intakes for Canadians. pp. 48-106, Canadian Governm. Publ. Centre, 1983.

16. J. A. Olson, Serum levels of vitamin A and carotenoids as reflectors of nutritional status. J. Natl. Cancer Inst. 73, 1439-1444 (1984).

17. P.Correa, The gastric precancerous process. Cancer Survey 2, 437-450 (1984).

18. H. F. Stich, P. Hornby, and B. P. Dunn, A pilot ß-carotene intervention trial with inuits using smokeless tobacco. Int. J. Cancer 36, 321-327 (1985).

19. P. Greenwald and G. Schreiber, Perspectives on chemoprevention in men. In Modulation and mediation of cancer by vitamins (F. L. Meyskens and K. N. Prasad, Eds.), pp.40-46, Karger, Basel, 1983.

BETA-CAROTENE AND CHEMOPREVENTION OF CANCER

Charles H. Hennekens

Channing Laboratory
Department of Medicine
Department of Preventive Medicine and Clinical Epidemiology
Harvard Medical School
Brigham and Women's Hospital

INTRODUCTION

Cancer is the second leading cause of mortality in the U.S., responsible for over 20% of the approximately two million total annual deaths (1). Treatment of cancer has provided and will continue to provide enormous benefits to some categories of affected individuals. At the same time, however, interventions that may prevent cancer in healthy individuals could, at least in theory, also afford great benefits to society as a whole. For example, a complete cure for acute leukemia, which accounts for several thousand of the approximately 400,000 total cancer deaths in the U.S. each year, would certainly be a remarkable breakthrough in medical research. In terms of public health impact, however, even a small reduction, on the order of 30%, in the development of epithelial cell cancers, which account for 90% of U.S. cancer deaths, due to dietary supplementation with micronutrients could conceivably prevent over 100,000 cancer deaths annually. With respect to known etiologies, cigarette smoking is the leading identified avoidable cause of cancer as well as mortality from all causes, accounting for about 30% of all deaths due to malignancy in the United States (2). Heavy alcohol consumption, which ranks second, is responsible for about 3% of cancer deaths. Recently, promising, but unproven, hypotheses have suggested that a substantial percentage of cancers may be preventable through dietary changes. Although diet has been postulated to account for as many as 35% of the annual deaths from cancer, which characteristics of diet, if any, increase or decrease the risk of cancer remains unclear (2).

For several reasons, administration of supplements is an attractive design strategy for the testing of hypotheses concerning diet and cancer. First, unlike a study in which subjects are asked to make modifications in their diet, a trial using supplements can isolate a particular dietary component which may affect cancer risk. Moreover, the requirement that subjects take a daily pill is likely to result in a higher rate of compliance than any comparable dietary intervention. In addition, because the general public seems to find prescription of anti-cancer substances more acceptable than proscription of carcinogens, the identification of specific micronutrients which reduce cancer risk could have a greater and more immediate public health impact than the discovery of specific macronutrients which act as cancer promotors. The most direct means to evaluate whether dietary supplementation decreases cancer risk among healthy individuals is a randomized trial (3). In this report, the rationale for studying beta-carotene in cancer chemoprevention based on available epidemiologic evidence is reviewed, and the design features of an ongoing randomized trial are presented.

EPIDEMIOLOGIC EVIDENCE CONCERNING BETA-CAROTENE AND CANCER CHEMOPREVENTION

Since Peto and colleagues raised the possibility that a high intake of dietary beta-carotene might decrease the risk of cancer in humans (4), considerable interest has focused on this area of research. The evidence for a preventive role of beta-carotene derives primarily from observational epidemiologic studies. In most locations throughout the world, one or two foods account for the majority of beta-carotene consumed by that population, such as carrots in the U.S., yellow and green vegetables in Japan, red palm oil in West Africa and parts of Brazil, and dark green leafy vegetables among the Chinese in Singapore. Consequently, epidemiologic studies of diet and cancer have been able to examine specific relationship of beta-carotene with cancer risk simply because their general dietary questionnaires obtained information on the chief carotene containing foods consumed by the populations studied. The vast majority of questionnaire studies of cancer in relation to consumption of vegetables rich in beta-carotene have showed a protective effect. These studies were conducted in Europe, the U.S., the Middle East and the Orient, and looked at eight different sites of cancers, all derived from epithelial cells, the type of malignancy whose development beta-carotene may inhibit at late stages (5).

The strongest and most consistent evidence of a protective effect of high intake of carotene-rich foods comes from studies of persons with lung cancer. Eight investigations in a variety of geographic locations have all found a statistically significant protective effect of consuming above average amounts of beta-carotene (6-13). The relative risks in these studies ranged from to 0.77 in New Jersey and Japan (6,11) to 0.14 in Chicago (12). Among questionnaire studies of beta-carotene consumption and risk of various gastrointestinal cancers, 10 of 13 investigations reported statistically significant reductions in cancer risk of at least 29% among those consuming the largest amounts of carotene-rich foods (14-21). With respect to other cancer sites, there are fewer data on the possible protective effect of a diet rich in beta-carotene, although available evidence is similarly consistent. Investigators at Roswell Park in Buffalo have analyzed consumption of vegetables in relation to cancers of the larynx (22), bladder (23), breast (24) and cervix (25), and found an inverse association for each site. In addition, two studies of prostate cancer conducted in Japan (11) and Minnesota (26) found a protective effect of high levels of carotene consumption. Two other studies of prostate cancer reported positive associations, one in Buffalo (27) and the other conducted in Hawaii (28). Most recently, a protective effect of beta-carotene against prostate cancer was reported in persons with low fat intake (29).

An additional epidemiologic design strategy to provide information concerning the carotene and cancer hypothesis is a prospective cohort study in which blood samples are collected, frozen and stored at baseline for a large number of individuals without cancer. After many years of follow-up, specimens from those who develop cancer as well as a comparison group of those who remain free from the disease are analyzed for blood levels of particular micronutrients. Intake of carotene is highly correlated with blood concentrations (30), so serum levels reported in these studies are likely to be good indices of beta-carotene consumption. Of the published prospective blood-based studies, three that measured serum beta-carotene found an inverse relation with subsequent risk (31-33), particularly of lung cancer. A fourth, which found no association (34-35), measured total carotenoids, of which beta-carotene represents only a small fraction. In addition, the analyses of blood specimens collected among those screened for the Multiple Risk Factor Intervention Trial indicate a strong inverse relationship of baseline blood levels of beta-carotene and subsequent development of lung cancer (36).

THE NEED FOR RANDOMIZED TRIALS

Overall, currently available data are suggestive of a statistical association between eating above average amounts of vegetables rich in beta-carotene, as assessed by questionnaire or blood based approaches, and a lower than average risk of developing cancer. Their interpretability, however, is quite limited. Specifically, it is not known what aspect of vegetable consumption, if any, is truly protective. Although the beta-carotene content of the vegetables may be responsible for the observed decrease in risk seen in questionnaire studies, it could also be due to some other component, such as dietary fiber. Moreover, in questionnaire and blood-based population data the association might be explained by an effect of other confounding variables. For example, if those people who eat more than average amounts of vegetables also tend to eat less fat than people with below average vegetable intakes, it may be that eating fat increases cancer risk rather than that vegetable consumption decreases risk. In addition, non-dietary causal factors may be correlated with lower than average intake of vegetables. For these reasons, observational epidemiologic studies of dietary intake and assays of blood levels of beta-carotene have provided, at most, only promising leads.

The most reliable method to test the possible cancer chemoprevention role of beta-carotene is the conduct of a randomized, placebo-controlled trial. Because the investigators determine exposure status, by assigning participants at random to receive either the agent under study or a placebo, known and unknown confounding variables which might affect risk of the outcome under study will be present, on average, in equal proportions in all treatment groups. Thus, while observational studies can collect baseline information on known confounders and then adjust for their effects in the analysis, only properly conducted randomized trials allow control of both known and unknown confounding variables. As such, randomized trials are able to control confounding variables to an extent which is simply not possible using observational design strategies.

An additional unique feature of randomized trials is the ability to detect small to moderate effects. As with other preventive interventions, the most plausible hypotheses concerning beta-carotene postulate reductions in cancer risk on the order of 30% (37). Although an observational analytic study, whether case-control or cohort, could reliably detect larger effects, the magnitude of the effect of uncontrolled confounding variables could easily be as great as the small to moderate effects presumed to be most likely related to intake of beta-carotene. In contrast to observational designs, a clinical trial in which the treatments are assigned at random and the sample size is sufficiently large can reliably detect small to moderate effects and thus provide the most direct evidence upon which to make a judgement about whether an association is one of cause and effect (38).

Among trials of beta-carotene administration in humans, two (39-40) have involved Filipinos who chew betel nuts, a habit known to cause breakage in the chromosomes of the oral mucosal cells. Both trials indicate strong protective effects of beta-carotene against chromosomal breakage in the oral mucosal cells. In the second of two reports (40), the carotenoid canthaxanthin was also tested and showed no association, suggesting that it is beta-carotene itself which decreases oral chromosome breakage, rather than some other carotenoid. The finding of a protective effect against oral chromosome breakage is compatible with the possibility of a similar effect of beta-carotene on the initiation of carcinogenesis. The only way to assess this question directly, however, is the conduct of a randomized trial of beta-carotene and risk of development of cancer.

In planning a randomized trial of a possible chemopreventive agent such as beta-carotene, one of the most critical issues is timing. For both ethical and practical reasons, there must be sufficient doubt generally about the agent to be tested to allow withholding treatment from half the subjects, at the same time as there must be sufficient belief in its potential to justify exposing the other half.

An additional factor contributing to the timeliness of testing beta-carotene as a cancer chemopreventive agent is the already widespread use of multivitamins among the general U.S. public. For example, in a cohort study of over 120,000 middle-aged American nurses, 30-35% already take daily multivitamins (41). Further, even after adjusting for inflation, during the past decade there has been a threefold increase in dollar sales of vitamins nationally (42). Because excess vitamin intake is largely removed by the kidneys, what is known for certain today is that the U.S. excretes the richest urine in the world; what is not yet known is whether there are any health benefits attributable to this already widespread use of vitamin supplements. Although the observational data concerning beta-carotene and risk of cancer are suggestive, one fear is that the use of these supplements will continue to increase even in the absence of reliable evidence derived from controlled studies in human populations. A mistaken belief in the chemopreventive properties of beta-carotene could actually prove harmful if, for example, cigarette smokers used beta-carotene as a substitute for quitting smoking. Even if beta-carotene were effective in reducing deaths from lung cancer by as much as 50%, lifelong smokers would still have a risk 10 to 15 times greater than nonsmokers, rather than 20 to 30 times greater as they do now. Moreover, if beta-carotene is not effective, such a strategy would, of course, be completely worthless.

A RANDOMIZED TRIAL OF BETA-CAROTENE IN U.S. PHYSICIANS

From 1979 to 1981, several pilot studies were conducted among U.S. physicians to assess the feasibility of carrying out a large-scale trial of aspirin in decreasing cardiovascular mortality and beta-carotene in reducing risk of cancer. Based, in part, on the pilot study data, a grant application was submitted and funded by the National Institutes of Health for a large-scale trial. In 1982 invitational letters and questionnaires were sent to 261,248 U.S. male physicians, aged 40-84, identified from a master tape provided by the American Medical Association. A 2x2 factorial design was adopted to test the more immature hypothesis that beta-carotene (50 mg Lurotin on alternate days) reduces cancer risk in individuals with no such prior history as well as the more mature hypothesis that regular, low-dose aspirin (325 mg Bufferin on alternate days) consumption decreases cardiovascular mortality among those with no history of cardiovascular disease.

The decision to conduct the trial among physicians was made for several reasons. First, as regards the requirement that subjects in a study provide informed consent to participate, no group appeared more qualified than physicians. With respect to the need for accurate reporting of medical events and the ability to recognize possible side effects to the study medication, again, doctors appeared to be the population of choice. Finally, the experience of the pilot studies indicated that physicians were very conscientious participants, maintaining high rates of compliance with the regimen as well as completing and returning follow-up questionnaires as requested. This experience, coupled with the fact that doctors are less mobile than the general public (and easier to trace should they, in fact, move), allowed the trial to be conducted entirely by mail, at a small fraction of the usual cost for such investigations.

From the original enrollment mailing to 261,248 doctors, 123,135 returned their questionnaire, with 59,283 indicating that they were willing to participate in the trial. Of these, 33,211 were initially eligible to participate. Willing physicians were excluded if they had a personal history of cancer, myocardial infarction, transient ischemic attack or stroke, current liver or renal disease, peptic ulcer or gout; contraindications to aspirin consumption; current use of aspirin, other platelet-active drugs or non-steroidal anti-inflammatory agents; or current use of a vitamin A supplement.

The 33,211 initially willing and eligible participants were then enrolled into a run-in phase, where, for approximately 18 weeks, all physicians took their daily pills from calendar packs containing active aspirin and beta-carotene placebo. Since the postulated beneficial effects of aspirin are acute and side effects common, it was desirable to expose all willing and eligible subjects to active aspirin during the run-in prior to randomization. As regards beta-carotene, however, its possible beneficial effects are cumulative and side effects minimal, so that it was optimal to use beta-carotene placebo. At the end of the 18-week period, participants were sent questionnaires, and individuals who reported side effects or a desire to discontinue participation, as well as those who developed an exclusion criterion or even those who wished to continue but whose compliance was deemed inadequate, were excluded from the trial before randomization. Thus, a total of 22,071 physicians who were proven good compliers were randomized into the trial.

The decision to adopt a run-in was based on experience in earlier pilot studies, in which participants were randomized immediately and followed for two years. Virtually all the loss in compliance in the pilot investigations occurred during the first several months following enrollment. The chief reasons given for dropping out were willing spirits but weak flesh: inability to remember to take a daily pill or side effects due to aspirin. On the basis of this experience, it was concluded that if 30,000 physicians were enrolled and 30% dropped out before being randomized, the remaining 21,000 would represent a group of proven excellent compliers who would be much more valuable to a long-term trial than 30,000 who were randomized immediately upon enrollment and then failed to fulfill their commitments to the study. Thus, the implementation of a run-in period allowed for a substantial increase the power of the study, by yielding a group of committed compliers for long-term follow-up.

The 22,071 physicians who remained willing and eligible after the run-in were randomly assigned first to two groups, one taking aspirin and and the other, aspirin placebo. Each of these groups was further randomized into two subgroups taking either beta-carotene or beta-carotene placebo. Thus an individual participant could be receiving aspirin only, carotene only, both aspirin and carotene, or both placebos. The principal advantage of this 2x2 factorial design was the ability to answer these two unrelated questions in a single trial without loss of sensitivity or major increase in cost (43).

Brief follow-up questionnaires are mailed twice a year for the first 12 months and then annually, asking about compliance with the treatment regimens and the development of disease outcomes since the return of the last questionnaire. During the run-in period prior to randomization, each participating physician was also asked to send a blood specimen. The rationale for collecting these specimens and analyzing them for baseline levels of retinol, carotene and retinol-binding protein relates to increasing the sensitivity of the trial to identify which particular subgroup of doctors, if any, stands to benefit most from dietary supplementation with beta-carotene. Due to the excellent collaboration of participants, blood samples were received from 14,916 randomized physicians and are currently stored at -85°C in triplicate in each of three separate locations.

Each strategy in this study design was adopted with the aim of assembling a large cohort of committed participants so that valid and definitive answers to the two research questions could be obtained. The effectiveness of the various design strategies is reflected in the three-year compliance and follow-up figures for the trial. Among randomized subjects, 91.6% were still taking at least one of their two types of pills, and 88.3% were still taking both types of pill regularly. Thus, only 8.4% had stopped taking pills entirely. These compliance figures are based solely on self-reports, since the Physicians' Health Study is conducted entirely by mail and has participants located throughout the U.S. To evaluate the reliability of this self-reported compliance, blood specimens were collected and analyzed from a sample of Physicians' Health Study participants living in each of two separate geographic areas. Aspirin use was assessed by radioimmune assay of serum levels of thromboxane B2, the stable degradation product of thromboxane A2. There was nearly perfect correlation between aspirin use or non-use as reported on the questionnaire with the absence or presence of serum thromboxane B2. Further, as regards beta-carotene, blood levels of which were assessed using high pressure liquid chromotography, there was little overlap between the distribution of those assigned to active drug and those assigned to placebo (44). With respect to obtaining complete follow-up information from all participants, after three years of follow-up, 99.1% of the participating doctors had provided complete questionnaire data by either mail or telephone, leaving just 0.9% on whom only vital status was obtained. Thus, to date, three-year morbidity follow-up is 99.1% and mortality follow-up is 100%.

While compliance and follow-up rates are exceptionally high, the mortality rates of doctors in the trial have been far lower than expected. Because physicians are known to be, as a group, healthier than men in the general population and, moreover, only those physicians who had no history of myocardial infarction or cancer were included in the trial, it was anticipated that the mortality experience of the cohort would be relatively low. After further taking into account the healthy volunteer effect, it was projected that the study population would have a standardized mortality ratio (SMR) of 0.3 to 0.4, indicating that they would experience 30 to 40% the deaths of a comparable group of white men in the general U.S. population. In fact, after three years, the SMR was only approximately 0.15. In light of the unusually low SMR, it seemed important to attempt to extend the trial to ensure the accumulation of sufficient numbers of endpoints. A random sample of 300 participants was polled to ask if they would be willing to continue in the trial for an additional 3½ years beyond their original commitment. The response was overwhelmingly positive, with 298 of the 300 expressing willingness to continue their collaboration. On the basis of this response, as well as the continued high compliance and follow-up rates among the entire study population, the proposal to extend the trial's follow-up period was approved by the NIH. Because cancer and cardiovascular endpoints accumulate more than arithmetically with increasing age of the study population, this doubling of the duration of follow-up from 3½ to 7 years should result in approximately four times the number of events, which should allow for a valid test of each of these two research questions. It is crucial that randomized trials achieve either a definite positive result on which public policy can be based, or a null result which is truly informative (38).

Over the next several years, findings from this and other studies should provide clear evidence concerning the possible cancer chemopreventive role of micronutrients. In the case of the Physicians' Health Study, the trial's ability to definitively answer the beta-carotene and aspirin hypotheses derives from several aspects of the study design: the very large sample size, the use of physicians as the study population, the run-in period to eliminate poor compliers before randomization, the availability of pre-randomization blood samples and the continuing high compliance and follow-up rates. Ultimately, however, the ability to prove or refute the hypotheses being tested will be due, in large part, to the continued outstanding efforts of the 22,071 dedicated and conscientious physicians who are participating in the study, without whose collaboration the trial could not proceed.

ACKNOWLEDGEMENTS

I am particularly indebted to Michael Jonas for editorial assistance.

REFERENCES

1. E. Silverberg and J. Lubera, Cancer statistics, 1986. Ca 36, 9-25 (1986).
2. R. Doll and R. Peto, The causes of cancer. J Natl Cancer Inst 66, 1191-1308 (1981).
3. C.H. Hennekens and K.A. Eberlein, for the Physicians' Health Study Research Group, A randomized trial of aspirin and beta-carotene among U.S. physicians. Prev Med 14, 165-168 (1985).
4. R. Peto, R. Doll, J.D. Buckley, and M.B. Sporn, Can dietary beta-carotene materially reduce human cancer rates? Nature (London) 290, 201-208 (1981).
5. C.H. Hennekens, Vitamin A analogues in cancer chemoprevention. In Important Advances in Oncology (V.T. DeVita, Jr, S. Hellman, and S.A. Rosenberg, Eds.) pp. 23-55. J.B. Lippincott, Philadelphia (1986).
6. R.G. Ziegler, T.J. Mason, A. Stemhagen, R. Hoover, J.B. Schoenberg, G. Gridley, P.W. Virgo, R. Altman, and J.F. Fraumeni, Jr., Dietary carotene and vitamin A and risk of lung cancer among white men in New Jersey. J Natl Cancer Inst 73, 1429-1435 (1984).
7. E. Bjelke, Dietary vitamin A and lung cancer. Int J Cancer 15, 56-565 (1975).
8. G. Kvale, E. Bjelke, and J.J. Gart, Dietary habits and lung cancer risk. Int J Cancer 31, 397-405 (1983).
9. C. Mettlin, S. Graham, and M. Swanson, Vitamin A and lung cancer. J Natl Cancer Inst 62, 1435-1438 (1979).

10. R. MacLennan and J. DeCosta, Risk factors for lung cancer in Singapore Chinese, a population with high female incidence rates. Int J Cancer 20, 854-860 (1977).
11. T. Hirayama, Diet and cancer. Nutr Cancer 1, 67-81 (1979).
12. R.B. Shekelle, S. Liu, W.R. Raynor Jr., M. Lepper, C. Maliza, A.H. Rosof, O. Paul, A.M. Shyrock, and J. Stamler, Dietary vitamin A and risk of cancer in the Western Electric Study. Lancet 2, 1185-1189 (1981).
13. M.W. Hinds, L.N. Kolonel, J.E. Hankin, and J. Lee, Dietary vitamin A, carotene, vitamin C and risk of lung cancer in Hawaii. Am J Epidemiol 119, 227-237 (1984).
14. P. Cook-Mozaffari, F. Azordegan, N.E. Day, A. Ressicaud, C. Sabai, and B. Aramesh, Oesophageal cancer studies in the Caspian littoral of Iran: Results of a case-control study. Br J Cancer 39, 293-309 (1979).
15. C. Mettlin, S. Graham, R. Priore, J. Marshall, and M. Swanson, Diet and cancer of the esophagus. Nutr Cancer 2, 143-147 (1981).
16. A.J. Tuyns, G. Pequignot, and O.M. Jensen, Nutrition, alcool et cancer de l'oesophage. Bull Cancer 65, 58-64 (1978).
17. E. Bjelke, Dietary factors and the epidemiology of cancer of the stomach and large bowel. In: Aktuelle Probleme der Klinischen Diatetik (Suppl. to Aktuelle Ernahrungsmedizin), pp. 10-17. George Thieme Verlag, Stuttgart, 1978.
18. S. Graham, H. Dayal, M. Swanson, A. Mittelman, and G. Wilkinson, Diet in the epidemiology of cancer of the colon and rectum. J Natl Cancer Inst 61, 709-714 (1978).
19. R.L. Phillips, Role of life-style and dietary habits in risk of cancer among Seventh-Day Adventists. Cancer Res 35, 3513-3522 (1975).
20. B. Modan, H. Cudde, and F. Lubin, A note on the role of dietary retinol and carotene in human gastrointestinal cancer. Int J Cancer 28, 421-424 (1981).
21. P. Stocks, Cancer incidence in North Wales and Liverpool region in relation to habits and environment. In: British Empire Cancer Campaign 35th Annual Report, Part II, (Suppl.), pp. 111-113 (1958).
22. S. Graham, C. Mettlin, J. Marshall, R. Priore, T. Rzepka, and D. Shedd, Dietary factors in the epidemiology of cancer of the larynx. Am J Epidemiol 113, 675-680 (1981).
23. S. Graham and C. Mettlin, Dietary risk factors in human bladder cancer. Am J Epidemiol 110, 255-263 (1979).
24. S. Graham, J. Marshall, C. Mettlin, T. Rzepka, T. Nemotot, and T. Byers, Diet in the epidemiology of breast cancer. Am J Epidemiol 116, 68-75 (1982).
25. J.R. Marshall, S. Graham, T. Byers, M. Swanson, and J. Brasure, Diet and smoking in the epidemiology of cancer of the cervix. J Natl Cancer Inst 70, 847-851 (1983).
26. L.M. Schuman, J.S. Mandell, A. Radke, U. Seal, and F. Halberg, Some selected features of the epidemiology of prostatic cancer: Minneapolis-St.Paul, Minnesota case-control study, 1976-1979. In: Trends in Cancer Incidence: Causes and Practical Implications (K. Magnus, Ed.), pp. 345-354. Hemisphere Publishing, Washington, D.C. (1982).
27. S. Graham, B. Haughey, J. Marshall, R. Priore, T. Byers, T. Rzepka, M. Mettline, and J.E. Pontes, Diet in the epidemiology of carcinoma of the prostate gland. J Natl Cancer Inst 70, 687-692 (1983).
28. L. Kolonel, J. Hankin, and J. Lee, Diet and prostate cancer (abstract). Am J Epidemiol 118, 454 (1983).
29. B.E. Henderson, Summary report of the fifth symposium on cancer registries and epidemiology in the Pacific Basin. J Natl Cancer Inst in press (1987).

30. W.C. Willett, M.J. Stampfer, B.A. Underwood, J.O. Taylor, and C.H. Hennekens, Vitamins A, E and carotene: effect of supplementation on their plasma levels. Am J Clin Nutr 38, 559-566 (1983).
31. N. Wald, M. Idle, J. Boreham, and A. Bailey, Low serum-vitamin A and subsequent risk of cancer. Preliminary results of a prospective study. Lancet 2, 813-815 (1980).
32. H.B. Stahelin, R. Rosel, E. Buess, and G. Brubacher, Cancer, vitamins, and plasma lipids: Prospective Basel study. J Natl Cancer Inst 73, 1463-1468 (1984).
33. M.S. Menkes, G.W. Comstock, J.P. Vuilleumier, K.J. Helsing, A.A. Rider, and R. Brookmeyer, Serum beta-carotene, vitamins A and E, selenium, and the risk of lung cancer. N Engl J Med 315, 1250-1254 (1986).
34. W. Willett, B.F. Polk, B.A. Underwood, M.J. Stampfer, S. Pressel, B. Rosner, J.O. Taylor, K. Schneider, C.G. Hames, Relation of serum vitamins A and E and carotenoids to the risk of cancer. N Engl J Med 310, 430-434 (1984).
35. W.C. Willett, B.F. Polk, B.A. Underwood, and C.G. Hames, Hypertension Detection and Follow-up Program study of serum retinol, retinol-binding protein, total carotenoids, and cancer risk: A summary. J Natl Cancer Inst 73, 1459-1462 (1984).
36. J. Connett, M. Kjelsberg, and L. Kuller, for the MRFIT Group, A prospective study of the relationship between serum beta carotene, vitamin A, E, selenium and lung cancer (abstract). American Epidemiology Society (1987).
37. C.H. Hennekens, M.J. Stampfer, and W. Willett, Micronutrients and cancer chemoprevention. Cancer Detect Prev 7, 147-148 (1984).
38. C.H. Hennekens, Issues in the design and conduct of clinical trials. J Natl Cancer Inst 73, 1473-1476 (1984).
39. H.F. Stich, M.P. Rosin, and M.O. Vallejera, Reduction with vitamin A and beta-carotene administration of the proportion of micronucleated buccal mucosal cells in Asian betel nut and tobacco chewers, Lancet 2, 1204-1206 (1984).
40. H.F. Stich, W. Stich, M.P. Rosin, and M.O. Vallejera, Use of the micronucleus test to monitor the effect of vitamin A, beta-carotene and canthaxanthin on the buccal mucosa of betel nut/tobacco chewers. Int J Cancer 34, 745-750 (1984).
41. W. Willett, L. Sampson, C. Bain, B. Rosner, C.H. Hennekens, J. Witschie, and F.E. Speizer, Vitamin supplement use among registered nurses. Am J Clin Nutr 34, 1121-1125 (1981).
42. V. Herbert, The vitamin craze. Arch Intern Med 140, 173-176 (1980).
43. M.J. Stampfer, J. Buring, W. Willett, B. Rosner, K. Eberlein, and C.H. Hennekens, The 2x2 factorial design: its application to a randomized trial of aspirin and carotene among US physicians. Stat Med 4, 111-116 (1985).
44. S. Satterfield, for the Physicians' Health Study Research Group, Biochemical markers of compliance in the Physicians' Health Study. Controlled Clin Trials, in press (1987).

CANCER CHEMOPREVENTION WITH VITAMIN A AND BETA-CAROTENE IN POPULATIONS AT HIGH-RISK FOR LUNG CANCER

G.S. Omenn, G. Goodman, L. Rosenstock, S. Barnhart, B. Lund, M. Thornquist, and P. Feigl

Cancer Prevention Research Unit, Fred Hutchinson Cancer Research Center; University of Washington School of Public Health and Community Medicine, Seattle, WA

Vitamin A, beta-carotene, and synthetic retinoids have attracted wide interest as promising chemopreventive agents, based on work in rodents and in cultured cells and organs, and low dietary intake or low serum levels of vitamin A or beta-carotene have been associated with high risk for lung cancer (1-4). Even if the data were stronger, observational epidemiological studies cannot establish whether or not changing the vitamin A status of specific individuals or groups of individuals would alter their subsequent cancer risk. The hypothesis that increasing retinoid intake in humans would reduce the risk of lung cancer must be tested in humans. For both practical and ethical reasons, such studies should be undertaken initially among those at highest risk.

In the NCI-supported Cancer Prevention Research Unit at the University of Washington School of Public Health and Fred Hutchinson Cancer Research Center, we are engaged in a variety of intervention studies aimed at testing the hypothesis that specific dietary factors may be important in certain cancers by testing whether changes in those factors can reduce the incidence of those cancers. These modifications include retinoids in asbestos-exposed workers and in heavy smokers to prevent lung cancer, folate in women with cervical dysplasia to prevent cervical cancer, and low-fat diets for women at increased risk for breast cancer. Certain aspects of these studies have been reported previously (3,5).

Lung cancer is the leading cancer cause of death in the United States, responsible for more than 110,000 deaths per year (6). More than 80% of cases are due to cigarette smoking. About 4,000 cases may be due to irradiation from exposure to radon daughters in our natural environment. Occupational exposures to asbestos fibers are an important cause of lung cancers both through direct action of the asbestos fibers and through a synergistic effect with cigarette smoking, as summarized in the classic volume edited by Selikoff & Hammond (7). After decades of delay, major improvements in controlling or eliminating asbestos exposures in various workplaces have been achieved. Yet much more remains to be done. Meanwhile, such control measures must be complemented by bold efforts to prevent the development of cancers in individuals already exposed. Nicholson et al. (8) have predicted that already-exposed workers in the U.S. will suffer asbestos-related cancer deaths at a rate of 8,000 to 10,000 per year to the end of the century. Any interventions that would inhibit the late stages of the multi-stage carcinogenic process in these people could turn the time-bomb of the long latent period into a golden opportunity to prevent the occurrence of bronchogenic carcinomas or mesotheliomas, both of which are resistant to current therapies. The retinoids have been demonstrated to reduce the incidence of tumors in rodents well after exposure to the test carcinogen. Retinol is thought to act as a differentiation-maintaining agent, while beta-carotene is thought to act as an electron-scavenger (1,2).

THE SEATTLE CHEMOPREVENTION TRIAL IN ASBESTOS-EXPOSED WORKERS

The specific aims of this project are the following:

(1) to determine the efficacy of chemoprevention with oral retinol and beta-carotene in patients at high risk for bronchogenic carcinoma and for mesothelioma;
(2) to assess the toxicity and safety of such a chemoprevention regimen; and
(3) to demonstrate the feasibility and usefulness of recruiting subjects from populations defined through specific occupational exposures and located through employment and workers' compensation systems.

Our trial is a phase IV chemoprevention study of individuals with significant exposures to asbestos, based upon radiographic evidence of asbestos-related lung disease (parenchymal and/or pleural) or upon history of work in a known high-risk trade (plumbers, pipefitters, steamfitters, shipyard boilermakers, insulators, shipyard electricians, shipscalers, plasterboard workers and sheetmetal workers). Eligible subjects must be male, age 45 to 74, with a history of 15 years since first absestos work exposure; they must have a positive chest X-ray by ILO criteria (profusion of small opacities $\geq$ 1/0 and/or bilateral pleural thickening) or a history of 5 years work completed at least 10 years previously in a high-risk trade (see above).

Potential subjects have been identified through state, federal, and longshore and harbor workers' compensation systems; through medical specialists and lawyers with large numbers of cases; through unions in the high-risk trades; and through the U.S. Navy Asbestos Medical Surveillance program at Bremerton, WA. Naturally, these sources include a considerable overlap. Given our desire to protect the confidentiality and privacy of potential subjects, we investigators cannot know the degree of overlap until individuals respond to letters inviting an expression of interest, letters which are sent out for us by the various sources.

The study design can be summarized as follows:

1. Endpoints: incidences of lung cancer and of mesothelioma, determined by active and passive (SEER) follow-up.

2. Number of subjects required: minimum 2,500 (1,250 per arm), based upon expected incidence without treatment of 10 lung cancer cases per 1,000 person-years, 20% attrition over 5 years, and reduction in cancer incidence of 40-50% for a one-sided test or 50-55% for a two-sided test to obtain 80% statistical power (type I error 5%, type II error 20%) (5).

3. Choice of agent and dose: 25,000 units retinol plus 15 mg beta-carotene daily, seeking an adequate pharmacologic effect, but emphasizing the avoidance of adverse side effects.

4. Exclusions of potential subjects: based on a detailed protocol, including cardiopulmonary impairment, history of cirrhosis or hepatitis

within 12 months, diagnosis of malignant neoplasm (other than skin) within the past 5 years, those ingesting retinol or beta-carotene supplements, those with definite plans to leave the geographic area, and those with elevations of SGOT or alkaline phosphatase as signs of liver function impairment.

5. Randomization: after a two-month run-in period on placebo to evaluate compliance and baseline reports of possible toxicity, with stratification for smoking status (ever smokers vs. never smokers, less than 20 packs lifetime); length of time since first asbestos exposure (15-30 years vs. 30+ years); and chest X-ray results (presence of interstitial fibrosis).

6. Monitoring and toxicity: a regular, structured assessment, with oversight by an institutional toxicity review committee (see below).

7. Compliance: pill counts, patient self-reports, and annual assays of blood concentrations of alpha-carotene, beta-carotene, retinol, and retinyl palmitate, using a much-improved, isocratic, automated HPLC assay (Kalman et al., in preparation).

PROGRESS TO DATE

Recruitment began in May, 1985, after prolonged delays awaiting availability of the agents arranged by the National Cancer Institute. As of 20 February, 1987, more than 11,000 letters had been mailed from recruitment sources, with an estimated 20-30% duplication: only 9% had received a second mailing, which is standard procedure where there is no initial response. So far 2,061 responses have been received, including notification of patients' deaths, obvious ineligibility, or lack of interest. From these responses a regular flow chart is maintained of the process of evaluation of potential subjects and their progress into and on the study protocol. Figure 1 shows the status as of 20 February, 1987. Of the 1,187 men eligible for initial phone contact, based upon the simple postcard response to the recruitment letters, 1,086 have been reached thus far by telephone; 711 have been judged eligible for the evaluation appointment, of whom 676 have thus far completed that appointment. After various clinical and laboratory exclusions, 559 of the 676 were eligible for the enrollment appointment, of whom 500 thus far have completed that step after a two-month run-in on placebo. Among these, 34 are now off study for a variety of reasons: 14 declined to continue due to logistical difficulties or mild sid effects attributed to the pills that did not reach our critical toxicity levels; 3 went on our toxicity monitoring protocol and dropped out even though the signs of possible toxicity resolved; 3 were found to be ineligible after further

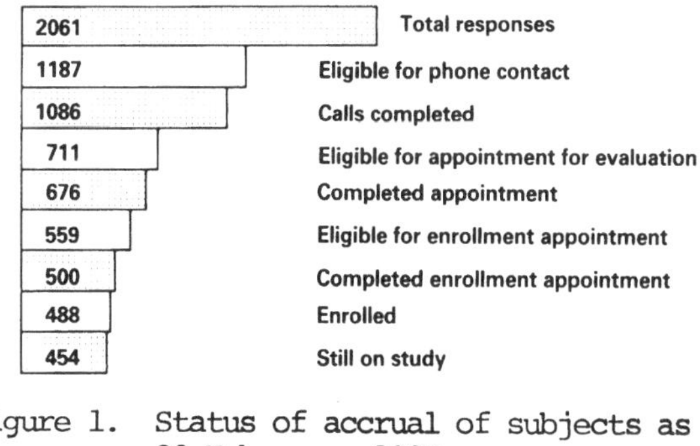

Figure 1. Status of accrual of subjects as of 20 February, 1987

review of X-rays and work histories; 6 had unrelated major medical problems; 3 developed cancers other than lung cancers; 4 died; and 1 was lost to follow-up.

Compliance based on both self-reports and pill counts has been excellent. For 458 individuals completing the two-month run-in on placebo by December 31, 1986, 95% missed 5 or fewer days during the two months; during the 0-4, 4-8, and 8-12 month periods on the placebo or retinol/beta-carotene double-blind regimen, 91%, 88% and 87% missed 5 or fewer days during the respective 4-month periods, with 226, 136 and 62 individuals, respectively. Pill counts and self-reports gave similar estimates.

Toxicity was evaluated according to a detailed protocol for eight groups of symptoms: blindness/nose bleeds; coloration of the skin; headaches; loss of weight or appetite; psychological symptoms; bone pain; nausea/vomiting; diarrhea. Liver function tests, total cholesterol, and triglycerides were measured, and mortality was monitored. Through 12/31/86 there were 4 deaths, 1 in the treatment arm and 3 in the placebo arm. Extensive analysis of mean toxicity scores for each symptom group and of percentages exceeding "critical toxicity scores" for each toxicity group at each follow-up period shows no differences thus far. We report periodically to the Institutional Safety Monitoring Committee, as well as to the National Cancer Institute. Thus far there is also no difference between the two treatment arms for elevation of SGOT, SGPT, alkaline phosphatase, cholesterol, or triglycerides.

Plans for the study consist of completion of recruitment and enrollment in the Seattle area during the next year, reaching at least 600 and possibly 800 or more subjects, with continued accumulation of data on compliance and toxicity. Interest in the study and willingness to continue seems very high among those who are actually enrolled, and we have established very good cooperation with all sources for the ongoing recruitment. Meanwhile, we are exploring the feasibility of recruiting collaborators in several additional geographic areas in order to reach the total sample size required.

An extensive team of faculty and staff has been assembled in our Cancer Prevention Rearch Unit to develop, direct, and carry out the many facets of this major chemoprevention trial, as well as a companion trail headed by Goodman pilot-testing alternative protocols in heavy smokers. My colleagues and I also have mounted some related studies, including the development of high resolution HPLC methods suitable for large-volume assays for retinol, retinyl palmitate, alpha-carotene, and beta-carotene; analysis of dietary intake from food frequency questionnaires and baseline blood values for alpha-carotene, beta-carotene and retinol; and studies of the responses of blood retinoids to supplementation, the relationship between smoking history and extent of fibrosis, and the limitations in applicability of ILO profusion categories in case identification among asbestos-exposed cohorts.

PROBLEMS AND LESSONS ABOUT PREVENTION TRIALS

There are dramatic differences between treatment trials and prevention trials. The presumption that experience from randomized clinical trials of treatments for cancer patients can guide the chemoprevention trials needs to be challenged and re-evaluated in light of the many striking differences in study design, logistics, and long-term relationships with subjects, summarized in Table 1.

Table 1. Contrasts Between Prevention Trials and Treatment Trials

	Prevention Trials	Treatment Trials
Endpoints	Lack of occurrence of infrequent event	Clinical improvement
Observation Period	Years	Days/weeks/months
Number of Subjects	Thousands	Tens
Dosage Strategy	Minimize dose/ emphasize safety	Maximize dose/ emphasize efficacy
Number of Drugs	Huge number (2500 Ss x 2 capsules/day x 5 yrs = 10 million capsules)	Few
Patient Relationships	Complex — not really patients	Direct — patients often eager for a trial
Data Management	Huge challenge	Relatively simple
Toxicity Assessment	Non-specific symptoms, difficult	High dose/higher risk, more specific, easier

CONCLUSION

It is urgent that we seek multiple means of preventing asbestos-related cancers. Primary prevention that deserves high priority includes stopping cigarette smoking and eliminating or markedly reducing asbestos exposures. Complementary efforts to capitalize on our speculative knowledges about multi-stage carcinogenic processes and to develop preventive intervention with drugs or diet also deserve high priority in our efforts to help those whose exposures cannot be erased. Rational selection of specific chemopreventive agents, well-tested in animal and *in vitro* studies, seems a useful experimental approach, as in the trial described here.

REFERENCES

1. R. Peto, R. Doll, J.D. Buckley, and M.B. Sporn, Can dietary beta-carotene materially reduce human cancer rates? Nature (London) 290: 201-208 (1981).
2. M.B. Sporn and A.B. Roberts, Role of retinoids in differentiation and carcinogenesis. Cancer Res. 43:3034-3040 (1983).
3. R.L. Prentice, G.S. Omenn, G.E. Goodman, J. Chu, M.M. Henderson, P. Feigl, G.D. Kleinman, D.B. Thomas, M.L. Hutchinson, B. Lund, and R.W. Day, Rationale and design of cancer chemoprevention studies in Seattle. Natl. Cancer Inst. Monogr. 69:249-258 (1986).
4. J.S. Bertram, L.N. Kolonel, and F.L. Meyskens, Jr., Rationale and strategies for chemoprevention of cancer in humans. Cancer Res. 47:3012-3031 (1987).
5. G.S. Omenn, G.E. Goodman, G.D. Kleinman, L. Rosenstock, S. Barnhart, P. Feigl, D.B. Thomas, D. Kalman, B. Lund, R.L. Prentice, and M.M. Henderson, The role of intervention studies in ascertaining the contribution of dietary factors in lung cancer: the Seattle chemoprevention trial of retinoids in asbestos-related exposed workers. Ann. N.Y. Acad. Sci., (in press).
6. G.S. Omenn, J. Merchant, E. Boatman, J.M. Dement, M. Kuschner, W. Nicholson, J. Peto, and L. Rosenstock, Contribution of environmental fibers to respiratory cancer. Environ. Health Perspect. 70:51-56 (1986).
7. I.J. Selikoff and E.C. Hammond (eds). Health Hazards of Asbestos Exposure. Ann. N.Y. Acad. Sci. 330:1-814 (1979).
8. W.J. Nicholson, G. Perkel, and I.J. Selikoff, Occupational exposure to asbestos: population at risk and projected mortality--1980-2030. Am. J. Ind. Med. 3:259-311 (1982).

ANTICARCINOGENIC ACTIONS OF PROTEASE INHIBITORS

Ann R. Kennedy and Paul C. Billings

Department of Cancer Biology
Harvard School of Public Health
665 Huntington Avenue
Boston, MA 02115

INTRODUCTION

There is a great variation in cancer incidence with diet, as has been recently reviewed (1,2). Epidemiological data suggest that environmental, specifically nutritional, factors play a major role in the etiology of cancer at many different sites (1-3). There are now several epidemiologic studies which suggest that components of vegetables might play a beneficial role in lowering the incidence of cancer (some examples of such studies are given in references 1-4). Although many compounds with anticarcinogenic potential are present in vegetables, it is possible that anticarcinogenic protease inhibitors contribute to the low cancer rates observed in certain human populations with high levels of vegetables in the diet. For example, the low cancer incidence rates in the Japanese and Seventh-Day adventists could be due to high levels of dietary protease inhibitors; it has been estimated that individuals in these populations ingest, on the average, more than 330 mg of protease inhibitors per day (5). There are, however, many other hypotheses which have been presented to explain the low cancer rates in these human populations. So many different variables are present in the diet that the effect of any specific anticarcinogenic agent cannot be distinguished in such epidemiologic studies; however, it is possible to distinguish the effects of specific anticarcinogenic agents in laboratory experiments. In this report, our own laboratory studies on anticarcinogenic protease inhibitors will be summarized and discussed. Although laboratory studies can give much information about the effects of potential chemopreventive agents, ultimately epidemiologic intervention studies must be performed to determine whether candidate chemopreventive agents are truly capable of preventing cancer in human populations. The current evidence that dietary protease inhibitors do have a role in lowering the cancer incidence in human populations has recently been reviewed (6 and 7).

IN VITRO STUDIES OF ANTICARCINOGENIC PROTEASE INHIBITORS

A large body of experimental data suggests that protease inhibitors from various sources have strong anti-carcinogenic activity *in vivo* (6 and 7). We (8-20) and now several other investigators (reviewed in ref. 6) have reported that protease inhibitors can suppress the malignant transformation of cells *in vitro*, as illustrated in Fig. 1. The results showing that protease inhibitors suppress transformation *in vitro* have utilized several different carcinogens, including X radiation, UV light, chemical carcinogens and steroid hormones as the inducing agents (reviewed in reference 6). These results suggest that protease inhibitors

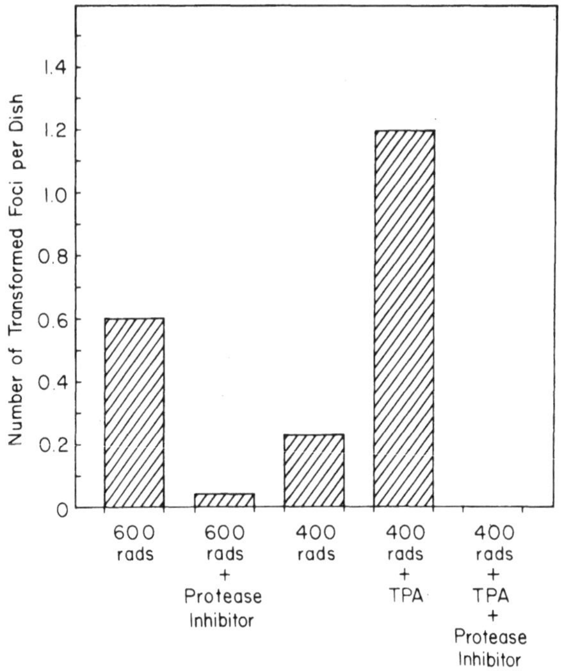

Fig. 1. Effects of a protease inhibitor (antipain) on radiation induced transformation in vitro (data are given in reference 12). Protease inhibitors are capable of suppressing the TPA enhancement of radiation transformation as well as the carcinogenic process brought about by a high dose of a complete carcinogen such as X irradiation, as discussed in references 8-20.

are capable of suppressing similar processes induced by different carcinogens in several different cell systems. Protease inhibitors have an unusual ability to suppress transformation in vitro in a highly significant manner, as opposed to the marginal effects observed for the other classes of possible human cancer chemopreventive agents which we have studied (see, for examples, refs. 6,15,19, 21-25).

We have reported that several different types of protease inhibitors suppress carcinogen-induced malignant transformation (12,14,17). We have observed that inhibitors of a specific protease, chymotrypsin, are the most effective for the suppression of malignant transformation (16,17). Treatment of cells with low concentrations of chymotrypsin inhibitiors, for short periods of time, will suppress carcinogen-induced transformation; effective concentrations of protease inhibitors which do not inhibit chymotrypsin are many orders of magnitude higher than this (16,17). Effective concentrations of some anticarcinogenic protease inhibitors are shown in Table 1; it can be observed that the inhibitors of chymotrypsin, such as chymostatin and the Bowman-Birk inhibitor, are effective at considerably lower molar concentrations than is antipain, which does not inhibit chymotrypsin. (Inhibitors of chymotrypsin are also more effective than other protease inhibitors at the suppression of several other phenomena related to carcinogenesis, as has been described elsewhere (17)). Of particular interest to us have been our studies with chymotrypsin inhibitors naturally present in vegetables--specifically, the soybean-derived Bowman-Birk inhibitor (BBI) (14, 16-19), the chick pea inhibitor (16) and the potato chymotrypsin inhibitor (20).

Table 1. Comparison of Protease Inhibitors (on a Molar Basis) in their Ability to Suppress Radiation Transformation In Vitro

(Lowest amount of each protease inhibitor to reduce radiation transformation in vitro to a comparable degree; i.e., to 10% of the dishes containing transformed foci).

Protease Inhibitor	Molar Concentration
Antipain	1.7×10^{-6}M
Chymostatin	1.7×10^{-12}M
Bowman-Birk Inhibitor	1.3×10^{-10}M

Our early studies showed BBI to be so effective in its ability to suppress malignant transformation in vitro that we wished to determine its ability to suppress carcinogenesis in vivo. As a crude extract of the inhibitor would potentially be of greater applicability in our future animal and human cancer chemopreventive studies than pure BBI-- due to the cost of its isolation and purification, etc- we developed a method to produce an extract of soybeans with the same ability as pure BBI to suppress malignant transformation in vitro; the anticarcinogenic activity of this extract is thought to be due solely to the presence of BBI (16). The extract contains 5 separate soybean protease inhibitors including BBI, but not including soybean trypsin inhibitor (SBTI), all of which are similar to BBI in molecular weight and trypsin inhibitory activity; these other trypsin inhibitors have been only partially characterized (26). We have observed that the BBI extract, at a low concentration (0.01 µg/ml) in the medium, has essentially the same ability as pure BBI to suppress malignant transformation in vitro induced by radiation in C3H/10T½ cells (16) and induced by radiation, benzo(a)pyrene and β-propiolactone (carcinogen treatments being given both with and without the cocarcinogen, pyrene) in 3T3 cells (18).

IN VIVO STUDIES OF ANTICARCINOGENIC PROTEASE INHIBITORS

In our first in vivo study with BBI, we studied its fate when ingested by mice and found that it was not inactivated by the stomach digestive process and over 90% of that ingested was fully active as a protease inhibitor in the small intestine (14). Thus, it seemed reasonable to determine whether dietary BBI could suppress colon carcinogenesis.

Our pilot study with BBI was designed to determine whether a dietary supplement of the BBI-containing extract described above could suppress dimethylhydrazine (DMH)-induced colon cancer in mice, and whether it had adverse effects on the health of treated mice. These studies have been described in detail (27). We observed a significant suppressive effect of BBI on DMH-induced colon carcinogenesis in mice and no apparent effects on the general health of animals maintained for essentially their entire lifespan on a diet high in protease inhibitor activity (i.e., containing 0.5% of the BBI extract). The results of these studies are summarized in Fig. 2. We have performed similar studies with a higher dose of DMH and lower levels of BBI in the diet. In these studies, we observed a significant suppressive effect of 0.1%, 0.01% and 0.0005% dietary BBI on DMH-induced colon carcinogenesis in mice, while 0.1% autoclaved BBI (in which the protease inhibitor activity was destroyed by heat) did not have a significant suppressive effect on DMH induced colon carcinogenesis in mice (unpublished data). One finding of great interest to us in this study was the fact that different dietary levels of BBI extract (varying in concentration over several orders of magnitude) had the same suppressive effect on DMH-induced colon carcinogenesis. Our in vitro studies have also demonstrated this effect-- purified BBI (16), the BBI extract (16) and several other different types of protease inhibitors (17) have essentially the same suppressive effect on carcinogen-induced transformation in vitro when

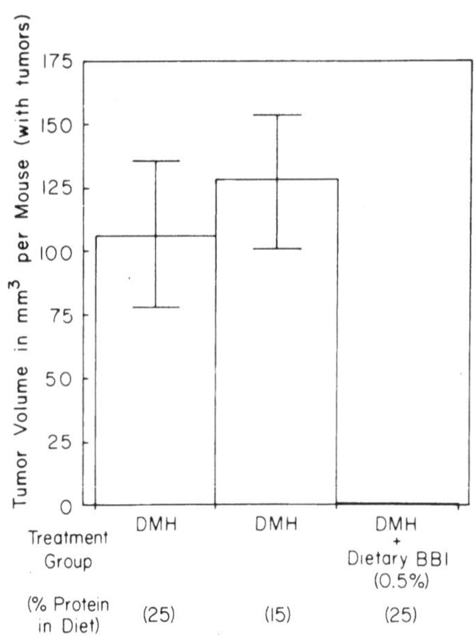

Fig. 2. Results of study performed to determine the effect of the Bowman-Birk inhibitor (BBI), given in the diet, on dimethylhydrazine (DMH) induced adenomatous tumors of the colon and rectum.
Details of this study are given in reference 27.

their concentrations in media have varied over several orders of magnitude. The "no effect" level is at an extremely low concentration for BBI and several other inhibitors of chymotrypsin (17). Our results discussed above suggest that very low dietary concentrations of BBI-type protease inhibitor preparations will be effective as chemopreventive agents for colon cancer.

We have also performed studies on the effect on BBI on dimethylbenz(a)anthracene (DMBA)-induced oral carcinogenesis in hamsters (28). The protease inhibitor preparations studied were: BBI, Autoclaved BBI (a control group in which all protease inhibitor activity was destroyed), and SBTI, another soybean-derived protease (trypsin) inhibitor which does not have the ability to suppress carcinogen-induced transformation in vitro (28). Gross and microscopic evaluations revealed a statistically significant reduction in the number of invasive carcinomas, the total number of tumors and the tumor mass for the DMBA + BBI treatment group, but not for the DMBA + Autoclaved BBI or SBTI treatment groups, compared to DMBA treatment alone. Some of these results are summarized in Fig. 3. In this study, the DMBA + Autoclaved BBI group served as an isocaloric diet (control) group for the DMBA-treated groups.

In these studies, the magnitude of the suppressive effect of BBI on oral carcinogenesis was not as great as that which we have observed in our colon carcinogenesis studies. Whether prevention of tumors, or merely a reduction in the number of tumors, is observed for any chemopreventive agent in the oral carcinogenesis system depends on the dose of carcinogen used. As examples, β-carotene (29) and Vitamin E (30) can reduce the number of tumors when a dose of DMBA is utilized which results in tumors in all of the animals, while these same chemopreventive agents can prevent tumor formation when a lower dose of DMBA is utilized. Our study with protease inhibitors, which is summarized in Fig. 3, utilized a sufficiently high dose of DMBA to result in a 100% tumor incidence in the animals (28). It is expected that BBI would be able to prevent oral

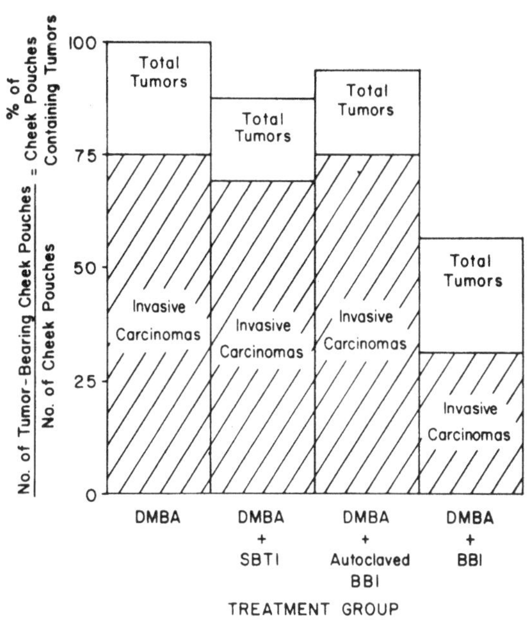

Fig. 3. Results of our study on the effects of protease inhibitors on dimethylbenz(a)anthracene (DMBA) induced oral carcinogenesis in hamsters. Abbreviations: SBTI - soybean trypsin inhibitor; BBI - Bowman Birk inhibitor; autoclaved BBI - Bowman-Birk inhibitor preparation in which protease inhibitor activity has been destroyed by heat. Details of these studies are given in reference 28.

carcinogenesis induced by a lower dose of DMBA, as BBI (28), β-carotene (29) and vitamin E (30) are thought to be comparable in their ability to suppress or prevent DMBA-induced oral carcinogenesis in hamsters (personal communication, Dr. Gerald Shklar).

In none of the long-term animal studies with anticarcinogenic levels of dietary BBI described above have we observed any effects on the growth of the animals or the general health of the animals; thus, we have no evidence to suggest that the use of protease inhibitors as chemopreventive agents will be complicated by possible toxicity problems. Thus far, our animal experiments have shown BBI to be a highly promising cancer chemopreventive agent.

CURRENT THOUGHTS ON THE MECHANISM OF ACTION OF THE ANTICARCINOGENIC PROTEASE INHIBITORS

From our previous studies, we believe that the first event in carcinogen-induced malignant transformation is a high frequency event occurring in many cells treated with low doses of either chemical or physical carcinogens (reviewed in refs. 6 and 31). We believe that the hypothesized high-frequency initiating event is likely to be caused by a change in gene expression, as it is known that such changes in gene expression can be caused by carcinogens (32) and can occur in a high proportion (≈ 80%) of carcinogen-treated cells (see, for example, ref. 33). We are assuming that a major consequence of this (presumed) carcinogen-induced change in gene expression is that it confers on cells an altered probability that a subsequent event, malignant transformation (which leads directly to the malignant state), will occur (reviewed in refs. 6 and 31). Our past work has suggested that protease inhibitors are able to stop what appears to be an ongoing cellular process begun by the carcinogen exposure (13,17), presumably by reversing the carcinogen-induced change in gene expression. Thus, we believe that a cellular process in many cells is begun by carcinogens and that anticarcinogenic protease inhibitors can turn this process off before a rare, later, mutation-like event (leading directly to the malignant state) can occur (13,17).

From our in vitro studies, we believe that the expression of the c-myc gene is of importance in the first step of transformation ("activation" of the c-myc gene is often associated with higher levels of expression, or deregulation, of this gene) and that an event such as the activation of ras, known to be activated by a point mutation (reviewed in reference 34), is our "later" (hypothesized) genetic event leading directly to the transformed state. It has already been shown that myc and ras cooperate in the induction of transformation (35,36) and that the activation of ras occurs in both carcinogen-transformed C3H/10T½ cells (37) and carcinogen-exposed normal diploid rodent cells as a late step in the transformation process (38). We have already reported that anticarcinogenic protease inhibitors such as antipain reduce the levels of c-myc expression in carcinogen-treated cells (39), and that anticarcinogenic protease inhibitors affect the first step of the transformation process (17). Thus, we believe that specific cooperating oncogenes, myc and ras, may play central roles in carcinogen-induced transformation in vitro.

The most direct approach to determining the mechanism of action of the anticarcinogenic protease inhibitors is to find out what they are interacting with in cells. To identify potential intracellular targets of the anticarcinogenic protease inhibitors, we have studied their ability to inhibit the cleavage of specific substrates by enzymatic activities present in cell homogenates. In these types of studies, we have detected a peptidase activity in C3H/10T½ cells which cleaves the substrate Boc-Val-Pro-Arg-MCA (40). This activity is inhibited by a variety of protease inhibitors in a manner that directly corresponds to their ability to suppress carcinogen-induced transformation of C3H/10T½ cells in vitro. The potency of various protease inhibitors, in terms of their ability to inhibit the enzymatic activity described above, was: BBI ≥ chymostatin > antipain > SBTI, as illustrated in Fig 4. We have purified and characterized this enzyme activity from

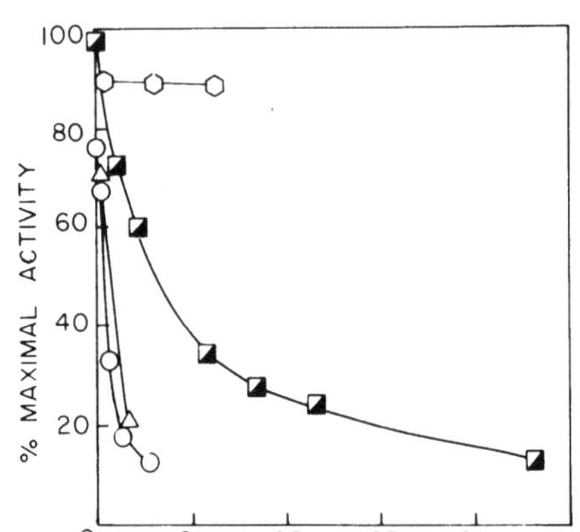

Fig. 4. The effect of protease inhibitors on Boc-Val-Pro-Arg-MCA hydrolyzing activity in C3H/10T½ cell homogenates. Aliquots of cell homogenates were incubated with the indicated protease inhibitor for 5 min. Boc-Val-Pro-Arg-MCA was then added and the reaction was directly monitored at room temperature in a spectrofluorometer, at excitation and emission wavelengths of 380 and 460 nm respectively. Details of these studies are given in reference 40.

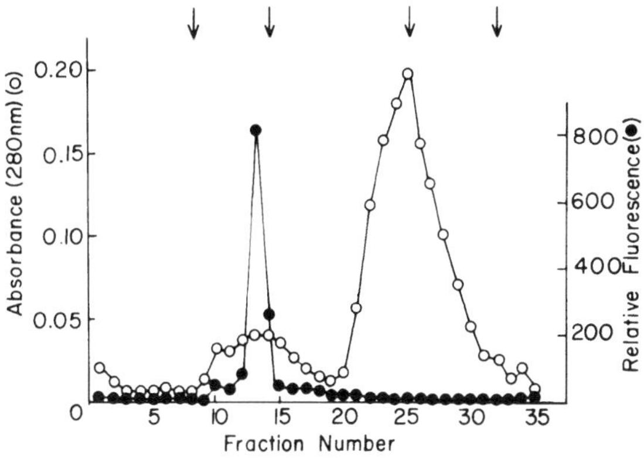

Fig. 5. Elution profile of Boc-Val-Pro-Arg-MCA hydrolyzing activity from a Sephadex G-200 column. Partially purified enzyme preparations were run over a G-200 column (0.7 x 120 cm) in 50 nM phosphate buffer (pH 7.0) at 4°C. Fractions were assayed for protease activity in 50 mM phosphate buffer (pH 7.0), 1 mM $MgCl_2$ with 16 uM Boc-Val-Pro-Arg-MCA as substrate. Arrows: position of elution of Blue Dextran (mw 2,000,000), Bovine Serum Albumin (mw 67,000) Chymotrypsinogen (mw 25,000) and Ribonuclease A (mw 13,700) respectively. Details are given in reference 40.

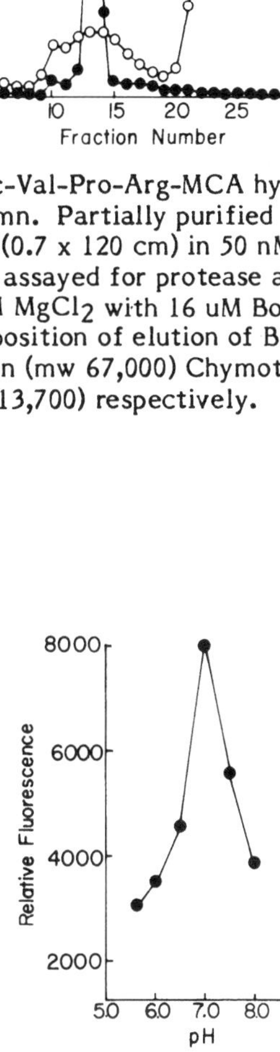

Fig. 6. Protease activity as a function of pH. Purified protease preparations were assayed for activity in 50 uM phosphate buffer, 1 mM $MgCl_2$ at the indicated pH with 16 uM Boc-Val-Pro-Arg-MCA as substrate. Samples were incubated for 60 min at 37°C and the fluorescense was determined. Details are given in reference 40.

C3H/10T½ cells. The enzyme elutes from a Sephadex G-200 column in a single peak with an apparent mass of 70-80 kilodaltons (Fig.5). The enzyme exhibits maximal activity at pH 7.0 (Fig. 6), is located in the cytosol, and is inhibited by diisopropylfluorophosphate (DFP), indicating that the enzyme is a serine protease (41). SDS-polyacrylamide gel analysis of our purified samples reveals a band of about 70 kilodaltons (40).

We have identified a similar protease in hamster cheek pouch epithelium (28) and mouse colonic epithelium (unpublished data). This proteolytic activity is elevated about 10-fold in DMBA-treated cheek pouch epithelium and can be detected at long time periods after carcinogen exposure, even in normal-appearing areas of the epithelium (28). In animals treated with DMBA and BBI, normal levels of Boc-Val-Pro-Arg-MCA hydrolyzing activity were present in the epithelium (28). Partial characterizations of the BBI-inhibitable proteolytic activities from the hamster cheek pouch and mouse colonic epithelia indicate that they are very much like the C3H/10T½ cell enzyme described above.

At the present time we do not know the intracellular function of this enzyme. One possible role it may play in vivo is that of protein processing. Many growth factors and peptide hormones are synthesized as inactive precursors which require post-translational modification to be active. Transformation of C3H/10T½ cells may require the presence of specific peptide hormones or growth factors. Although many of the pro-proteins that have been described contain pairs of basic amino acids at the processing site, processing sites containing a single basic residue are known to exist in several peptide hormones (42). The protease activity we have characterized cleaves a substrate with a basic amino acid at the site of cleavage and, thus, may be involved in peptide hormone or growth factor processing in C3H/10T½ cells. Inhibition of the Boc-Val-Pro-Arg-MCA hydrolyzing activity by anticarcinogenic protease inhibitors may suppress carcinogenesis by preventing the processing of a growth factor(s) or peptide hormone(s) which is necessary for the induction and/or expression of the transformed phenotype.

It is likely that an initial step in "multi-stage" carcinogenesis could involve the induction of a protease. We have already observed high levels of Boc-Val-Pro-Arg-MCA hydrolyzing activity in randomly chosen, normal-appearing areas of DMBA-treated hamster cheek pouch epithelium (28), as discussed above. If this proteolytic activity happens to be involved in the processing of a growth factor, one would expect more of the growth factor to be present than there normally is. Carcinogen-treated cells growing under the influence of abnormally large amounts of a growth factor would be likely to exhibit altered growth characteristics; such altered growth patterns are known to occur in a widespread fashion in carcinogen-treated tissues (see, for example, ref. 43) and have been characterized as "pre-neoplastic" changes. It is conceivable that a later step in carcinogenesis, occurring within these areas of altered growth characteristics, could lead directly to malignancy; the evidence for such a late step in the malignant transformation of C3H/10T½ cells and carcinogenesis in vivo has recently been reviewed (31).

There are numerous other possible roles for proteases in cancer cells. For example, proteases are considered essential for cancer cells to invade and destroy normal tissue (44). We are assuming that the proteolytic activity found in C3H/10T½ cells is involved in the initial stages of carcinogenesis primarily because it is inhibited by the anticarcinogenic protease inhibitors that suppress early stages of carcinogen-induced transformation in C3H/10T½ cells (12,13,17). In addition, we have observed that the protease inhibitors that suppress the transformation process do not have a significant effect on the growth or expression of C3H/10T½ cells that have already been transformed (6,12,17). The precise role of this protease in the malignant transformation of C3H/10T½ cells, however, remains to be determined.

Another mechanism of action for the anticarcinogenic protease inhibitors may be an "anti-promotional" effect. Many of the protease inhibitors we have studied are strongly antagonistic to the enhancing effects of promoting agents in the induction of transformation in vitro (12,15). It has been proposed that Bloom's syndrome is a human disease representing a defect in the area of promotion (see discussion in ref. 45). We have recently observed that protease inhibitors, including BBI (both the BBI extract and purified BBI), will greatly reduce the levels of spontaneously occurring chromosome aberrations and sister chromatid exchanges in cells of patients with Bloom's syndrome (45). Bloom's syndrome is an autosomal recessive genetic disease in which the high levels of chromosomal abnormalities are thought to be related to the high risk of cancer in people with the disease (46). Our results have suggested another hypothesis for the mechanism of the observed protease inhibitor effects. Radiation and chemical carcinogens may induce in normal cells a state or ongoing process resembling that which is always occurring in the Bloom's syndrome cells. Protease inhibitors may be capable of stopping that process, possibly by a mechanism discussed in reference 45. Much of our in vitro data suggests that such an ongoing cellular process can be initiated by carcinogen exposure (see references 31, 47-50) and that protease inhibitors appear to be able to stop this ongoing process (13, 17) even when added at late times to carcinogen-exposed cells. The mechanism for the observed anticarcinogenic actions of protease inhibitors is not yet clear, however; thus, more studies are needed to elucidate their mechanisms of action in both in vitro and in vivo systems.

REFERENCES

1. C. Grobstein, et al., Diet, Nutrition and Cancer. Committee on Diet, Nutrition and Cancer, Assembly of Life Sciences, National Research Council, National Academy of Sciences, Washington, D.C., (1982).
2. R. Doll and R. Peto, The causes of cancer: Quantitative estimates of avoidable risks of cancer in the United States today. J. Natl. Cancer Inst. 66, 1193-1308 (1981).
3. P. Correa. Epidemiologic correlations between diet and cancer frequency. Cancer Res. 41, 3685-3690 (1981).
4. R. L. Phillips, Role of lifestyle and dietary habits in risk of cancer among Seventh Day Adventists. Cancer Res. 35, 3513-3522 (1975).
5. B. H. Doell, C. J. Ebden, and C. A. Smith, Trypsin inhibitor activity of conventional foods which are part of the British diet and some soya products. Qual. Plant. Plant Foods Hum. Nutr. 31, 139-150 (1981).
6. A. R. Kennedy, Promotion and other interactions between agents in the induction of transformation in vitro in fibroblasts. In: Mechanisms of Tumor Promotion, Vol III, "Tumor Promotion and Carcinogenesis In Vitro", (T. J. Slaga, Ed.), Chapter 2, pp. 13-55. CRC Press, Boca Raton, 1984.
7. W. Troll, K. Frenkel and R. Wiesner, Protease inhibitors as anticarcinogens. J. Natl. Cancer Inst., 73, 1245-1250 (1984).
8. A. R. Kennedy and J. B. Little, Protease inhibitors suppress radiation induced malignant transformation in vitro. Nature (London) 276, 825-826 (1978).
9. A. R. Kennedy and J. B. Little, Radiation transformation in vitro: Modification by exposure to tumor promoters and protease inhibitors. In: Radiation Biology in Cancer Research, (R. E. Meyn and H. Rodney Withers, Eds.), pp. 295-307. Raven Press, New York, 1980.
10. A. R. Kennedy and R. R. Weichselbaum, Effects of 17 Bestradiol on radiation transformation in vitro; inhibition of effects by protease inhibitors. Carcinogenesis 2, 67-69 (1981).
11. J. B. Little and A. R. Kennedy, Promotion of X-ray transformation in vitro. In: Carcinogenesis, Vol. 7 (E. Hecker et al., Eds.), pp. 243-257. Raven Press, New York, 1982.
12. A. R. Kennedy and J. B. Little, Effects of protease inhibitors on radiation transformation in vitro. Cancer Res. 41. 2103-2108 (1981).
13. A. R. Kennedy, Antipain, but not cycloheximide, suppresses radiation transformation when present for only one day at five days post-irradiation. Carcinogenesis 3, 1093-1095 (1982).
14. J. Yavelow, T. H. Finlay, A. R. Kennedy, and W. Troll, Bowman-Birk soybean protease inhibitor as an anticarcinogen. Cancer Res. 43, 2454-2459 (1983).
15. A. R. Kennedy, Prevention of Radiation-Induced Transformation In Vitro. In: Vitamins, Nutrition and Cancer (K.N. Prasad, Ed.), pp. 166-179. S. Karger AG, Basel, 1984.

16. J. Yavelow, M. Collins, Y. Birk, W. Troll, and A. R. Kennedy, Nanomolar concentrations of Bowman-Birk soybean protease inhibitor suppress X-ray induced transformation in vitro. Proc. Natl. Acad. Sci. USA 82, 5395-5399 (1985).
17. A. R. Kennedy, The conditions for the modification of radiation transformation in vitro by a tumor promoter and protease inhibitors. Carcinogenesis 6, 1441-1446 (1985).
18. N. Z. Baturay and A. R. Kennedy, Pyrene acts as a cocarcinogen with the carcinogens, benzo(a)pyrene, B-propiolactone and radiation in the induction of malignant transformation of cultured mouse fibroblasts; soybean extract containing the Bowman-Birk inhibitor acts as an anticarcinogen. Cell Biol. Toxicol. 2, 21-32 (1986).
19. A. R. Kennedy, Implications for mechanisms of tumor promotion and its inhibition by various agents from studies of in vitro transformation. In: Tumor Promoters, Biological Approaches for Mechanistic Studies and Assay Systems (R. Langenbach, J. C. Barrett and E. Elmore, Eds.). Raven Press, New York (in press).
20. P. C. Billings, W. St. Clair, C. A. Ryan, and A. R. Kennedy, Inhibition of radiation-induced transformation of C3H/10T½ cells by chymotrypsin inhibitor 1 from potatoes. Carcinogenesis 8, 809-812 (1987)..
21. A. R. Kennedy, W. Troll, and J. B. Little, Role of free radicals in the initiation and promotion of radiation transformation in vitro. Carcinogenesis 5, 1213-1218 (1984).
22. A. R. Kennedy, Role of free radicals in the initiation and promotion of radiation-induced and chemical carcinogen induced cell transformation. In: Oxygen and Sulfur Radicals in Chemistry and Medicine (A. Breccia, M.A.J. Rodgers, and G. Semerano, Eds.), pp. 201-209. Edizioni Scientifiche, "Lo Scarabeo", Bologna, 1986.
23. A. R. Kennedy, Effects of antioxidants on the induction of malignant transformation in vitro. In: Vitamins and Cancer-Human Cancer Prevention by Vitamins and Micronutrients (F. L. Meyskens and K. N. Prasad, Eds.), pp. 51-64. Humana Press, Clifton, New Jersey, 1985.
24. B. S. Radner and A. R Kennedy, Suppression of x-ray induced transformation by Vitamin E in mouse C3H/10T½ cells. Cancer Lett 32, 25-32 (1986).
25. A. R. Kennedy and M. C. R. Symons. "Water Structure" vs. "Radical Scavenger" theories as explanations for the suppressive effects of DMSO and related compounds on radiation induced transformation in vitro. Carcinogenesis 8, 683-688 (1987).
26. D. L. R. Hwang, K. T. Davis-Lin, W. K. Yang and D. T. Foard, Purification, partial characterization and immunological relationships of multiple low molecular weight proteinase inhibitors of soybean. Biochim. Biophys. Acta. 495, 369-382 (1977).
27. H. Weed, R. B. McGandy and A. R. Kennedy, Protection against dimethylhydrazine induced adenomatous tumors of the mouse colon by the dietary addition of an extract of soybeans containing the Bowman-Birk protease inhibitor. Carcinogenesis 6, 1239-1241 (1985).
28. D. V. Messadi, P. Billings, G. Shklar and A. R. Kennedy, Inhibition of oral carcinogenesis by a protease inhibitor. J. Natl. Cancer Inst. 76, 447-452 (1986).
29. D. Suda, J. Schwartz and G. Shklar, Inhibition of oral carcinogenesis by topical B-carotene. Carcinogenesis 7, 711-715 (1986).
30. D. Trickler and G. Shklar, Prevention of oral carcinogenesis following Vitamin E administration. J. Natl. Cancer Inst. (in press).
31. A. R. Kennedy, Evidence that the first step leading to carcinogen-induced malignant transformation is a high frequency, common event. In: Carcinogenesis: A Comprehensive Survey, Vol. 9: Mammalian Cell Transformation: Mechanisms of Carcinogenesis and Assays for Carcinogens (J. C. Barrett and R.W. Tennant, Eds.), pp. 355-364. Raven Press, New York, 1985.
32. M. J. Fahmy, and O. G. Fahmy, Intervening DNA insertions and the alteration of gene expression by carcinogens. Cancer Res. 40, 3374-3382 (1980).

33. R. E. Scott and P. B. Maercklein, An initiator of carcinogenesis selectively and stably inhibits stem cell differentiation: a concept that initiation of carcinogenesis involves multiple phases. Proc. Natl. Acad. Sci. USA 82, 2995-2999 (1985).
34. A. Balmain, Transforming ras oncogenes and multistage carcinogenesis. Br. J. Cancer 51, 1-7 (1985).
35. H. Land, L. F. Parada and R. A. Weinberg, Cellular oncogenes and multistep carcinogenesis. Science 222, 771-778 (1983).
36. H. Land, L. F. Parada and R. A. Weinberg, Tumorigenic conversion of primary embryo fibroblasts requires at least two cooperating oncogenes. Nature (London) 304, 596-602 (1983).
37. L. F. Parada and R. A. Weinberg, Presence of a Kirsten murine sarcoma virus ras oncogene in cells transformed by 3-methylcholanthrene. Mol. Cell. Biol. 3, 2298-2301 (1983).
38. S. Sukumar, S. Pulciani, J. Doniger, J. A. DiPaolo, C. H. Evans, B. Zbar and M. Barbacid, A transforming ras gene in tumorigenic guinea pig cell lines initiated by diverse chemical carcinogens. Science 223, 1197-1199 (1985).
39. J. D. Chang, P. Billings and A. R. Kennedy, Reduction in c-myc expression is associated with antipain treatment of proliferating C3H/10T½ cells. Biochem. Biophys. Res. Commun. 133, 830-835 (1985).
40. P. C. Billings, J. A. Carew, C. E. Keller-McGandy, A. Goldberg and A. R. Kennedy, A serine protease activity in C3H/10T½ cells which is inhibited by anticarcinogenic protease inhibitors. Proc. Natl. Acad. Sci. USA 84, No. 14 (July, 1987).
41. K. A. Walsh and P. E. Wilcox, Serine Proteases. Methods Enzymol 19, 31-41 (1970).
42. J. Douglass, O. Cirielli, E. Herbert, Polyprotein Gene Expression: Generation of Diversity of Neuroendocrine Peptides. Annu. Rev. Biochem. 53, 665-715 (1984).
43. S. P. Ethier and R. L. Ullrich, Detection of ductal dysplasia in mammary outgrowths derived from carcinogen-treated virgin female BALB/c mice. Cancer Res. 42, 1753-1760 (1982).
44. E. Reich, D. B. Rifkin and E. Shaw (Eds.), Proteases and Biological Control. Cold Spring Harbor Conferences on Cell Proliferation, Vol. 2, Cold Spring Harbor Laboratory, 1975.
45. A. R. Kennedy, B. Radner and H. Nagasawa, Protease inhibitors reduce the frequency of spontaneous chromosome abnormalities in cells from patients with Bloom syndrome. Proc. Natl. Acad. Sci. USA 81, 1827-1830 (1984).
46. J. German, Bloom's syndrome X. The cancer proneness points to chromosome mutation as the crucial event in human neoplasia. In: Chromosome Mutation and Neoplasia (J. German, Ed.), pp. 347-357. Alan R. Liss, New York, 1983.
47. A. R. Kennedy, M. Fox, G. Murphy and J. B. Little, Relationship between X-ray exposure and malignant transformation in C3H/10T½ cells. Proc. Natl. Acad. Sci. USA 77, 7262-7266 (1980).
48. A. R. Kennedy and J. B. Little, An investigation of the mechanism for the enhancement of radiation transformation in vitro by TPA. Carcinogenesis 1, 1039-1047 (1980).
49. A. R. Kennedy and J. B. Little, Evidence that a second event in x-ray induced oncogenic transformation in vitro occurs during cellular proliferation. Radiat. Res. 99, 228-248 (1984).
50. A. R. Kennedy, J. Cairns and J. B. Little, The timing of the steps in transformation of C3H/10T½ cells by X-irradiation. Nature (London) 307, 85-86 (1984).

POSSIBLE MECHANISMS OF ACTION OF THE ANTICARCINOGENIC PROTEASE INHIBITORS

Jonathan Yavelow, Judith H. Kraft and Lorraine Schepis

Department of Biology
Rider College
Lawrenceville, NJ 08648-3099

INTRODUCTION

Protease inhibitors have been shown to possess anticarcinogenic properties. Two synthetic protease inhibitors, tosyl phenylethyl chloromethyl ketone (TPCK) and tosylarginine methylester (a competitive trypsin substrate), have been demonstrated to block the formation of skin tumors promoted with TPA in mice (1). These studies were repeated using the actinomycete-derived protease inhibitor leupeptin (2), thus establishing an in vivo basis for the involvement of proteases in the mechanism of carcinogenesis.

In addition to the topical application of protease inhibitors, an ingested protease inhibitor, ε-aminocaproic acid, blocks the formation of colon tumors induced by dimethylhydrazine in mice (3). It seems plausible then that ingested protease inhibitors may also act as anticarcinogens. If synthetic protease inhibitors do act as anticarcinogens, is it possible that foods rich in protease inhibitors (e.g., legumes) represent a source of a nutritionally relevant anticarcinogen?

Studies have proven that soybean diets protect animals from developing tumors. For example, Sprague-Dawley rats were protected from x-ray induced breast tumors (4), and spontaneous liver tumors in C3H/HEN mice were decreased in a dose-responsive manner (5). Epidemiological studies have corroborated animal experiments and have demonstrated that legumes affect the protection against breast and colon cancers in humans (6,7). Two mechanisms have been discussed whereby protease inhibitors may act as anticarcinogens--the direct and the indirect mechanisms (8). This chapter is confined to defining the process(es) involved in the direct anticarcinogenic action of protease inhibitors. In vivo, this mechanism appears to be most relevant to the colon.

One of the best known proteases involved in the mechanism of carcinogenesis is plasminogen activator (PA) (9). The Rous Sarcoma Virus temperature-sensitive for transformation increases PA activity only at the temperature permissive for transformation (9). How precisely PA contributes to cell transformation is still unclear, but it is known that PA is capable of cleaving substrates other than plasminogen. Two of these substrates, fibronectin and procollagenase, are relevant in transformation (10). Recent studies using monoclonal antibodies to PA have proven that metastasis can also be blocked. Thus, PA is involved in both transformation and metastasis. The release of PA from a cell-associated form to an extracellular form involves a chymotrypsin-sensitive enzyme (11). This enzyme is associated with the crude membrane fraction of normal chick-embryo fibroblasts and increases approximately two-fold upon transformation (12). Legume-derived protease inhibitors of the Bowman-Birk family possess anticarcinogenic activity. The mechanism whereby this Bowman-Birk inhibitor suppresses x-ray induced transformation has been studied in vitro using C3H/10T½ cells (13, 14). These experiments have proven that the chymotrypsin-inhibitory activity of BBI is responsible for suppression of transformation. This suggests a chymotrypsin-like enzyme may be involved in the mechanism of malignant cell transformation. In fact, other chymotrypsin inhibitors also suppress transformation.

The role of proteases in cells is becoming much more appreciated. Cellular proteases involve processes such as preprocessing and zymogen activation (including enzymes involved in growth factor maturation) (15); intracellular protein turnover (16); receptor expression (17); sister chromatin exchanges (18). The role of proteases in the mechanism of transformation may thus be involved in "potentiating" the function of other proteins. In this regard, it has been reported that the synthetic chymotrypsin substrate, acetyl tyrosine ethyl ester, inhibits receptor-mediated endocytosis of the insulin receptor in adipocytes (19).

Thus far, we have identified two BBI-sensitive membrane-associated enzymes (20). Studies using fluorescent BBI reveal that it is internalized in a time- and temperature-dependent manner (21). These studies suggest that the membrane-associated protease may act as a receptor. If a protease receptor exists, the receptor must be an integral plasma membrane protein. The studies described below show that this is indeed the case.

MATERIALS AND METHODS

Methylcholanthrene-transformed C3H/10T½ cells (clone 16) were grown to confluence in 150 cm^2 flasks as previously described (20). The cells were scraped and homogenized in a Dounce homogenizer. Crude membranes were fractionated using a discontinuous sucrose density gradient as previously described (22). The resulting bands were washed and then sequentially extracted using a high ionic salt solution and a glucopyranoside detergent in the following manner. Each band was resuspended in 2.0 M KCl (in 25 mM Tris-25 mM Glycylglycine-1 mM $CaCl_2$, pH 7.8) and homogenized with 30 strokes. The resuspended bands were then centrifuged at 100,000 x g for 20 minutes. The supernatants were saved and assayed for enzyme activity. The pellets were resuspended in 1% N-octyl-B-D-glucopyranoside, homogenized with 30 strokes, and centrifuged as before. Again, the supernatants were saved and assayed. all operations were carried out at 4°C.

Enzyme activity was assessed using the synthetic fluorescent substrate succinyl-ala-ala-pro-phe-aminofluorocoumarin, a similar substrate to succ-ala-ala-pro-phe-AMC. The enzyme assays were performed as previously described (20). Protein was determined by the method of Bradford (23).

The experiments were performed in triplicate.

RESULTS/DISCUSSION

Previous studies have shown that the crude membrane fraction of a cell contains a chymotrypsin-like enzyme that is inhibited by the Bowman-Birk protease inhibitor found in legumes. The enzyme can be identified by its ability to hydrolyze succ-ala-ala-pro-phe-AMC (20). Using a similar substrate, discontinuous sucrose density gradient centrifugation, and sequential salt and detergent extractions, we have attempted to discern in which cellular membrane the protease resides and whether the protease is a peripheral or integral membrane protein.

In order to determine in which cellular membrane the protease is located, the cell homogenates were applied to a discontinuous sucrose density gradient. Components of the plasma membrane and the Golgi apparatus band at the 20% to 40% interface (Band 1), and components of the mitochondria and the endoplasmic reticulum band at the 40% to 60% interface (Band 2). The pellet was labelled Band 3 (22).

The salt and detergent extractions are utilized to determine whether the protease is peripheral or integral to the membrane. The 2.0 M KCl solution releases the peripheral proteins into the supernatant, while the 1% N-octyl-glucopyranoside destroys the lipid bilayer, therefore releasing the integral proteins.

Table 1 shows the enzyme activity found in each of these supernatants from the particular bands isolated from a discontinuous sucrose gradient. The supernatants from extraction with KCl have almost no activity. However, these fractions do contain protein. Therefore, this step serves as a purification step. In contrast, the supernatants from the N-octyl-glucopyranoside extractions contain significant enzyme activity. This suggests that the protease is an integral membrane protein.

Most of the enzyme activity in the N-octyl-glucopyranoside fraction is found in Band 1, which contains mostly plasma membrane and some Golgi apparatus, indicating that the protease is located in the plasma membrane. In addition, a fairly large amount of activity is found in Band 2. There may be two possible explanations for this. First, the second band may be contaminated with plasma membrane. A second possibility may be that lysosomal proteases are present in Band 2.

Table 1. Chymotrypsin-like enzyme activity in the KCl and N-octyl-glucopyranoside supernatants of the bands isolated from a discontinuous sucrose density gradient.

KCl Extraction of:	Specific Activity (nmol/min/ug)	Total Protein (ug)	Total Activity (nmol/min)
Band 1	8.47×10^{-6}	160.56	1.36×10^{-3}
Band 2	8.92×10^{-5}	135.09	1.20×10^{-2}
Band 3	5.88×10^{-5}	159.24	9.36×10^{-3}
Detergent Extraction of:			
Band 1	3.77×10^{-3}	112.86	4.26×10^{-1}
Band 2	2.37×10^{-3}	120.82	2.86×10^{-1}
Band 3	9.72×10^{-4}	82.27	8.00×10^{-2}

C16 cells were homogenized and crude membranes were isolated by centrifugation. The crude membranes were then applied to a discontinuous sucrose density gradient. The bands were then sequentially extracted with 2.0 M KCl and 1% N-octyl-glucopyranoside, and the supernatants were assayed for succinyl-ala-ala-pro-phe-AFC hydrolytic activity.

Previous studies (20, 21) have suggested that the protease may act as a receptor. This study has reinforced this possibility by showing that the protease is an integrally-bound plasma membrane protein. The following studies are needed to clarify the involvement of chymotrypsin-like proteases in the mechanism of carcinogenesis. What is the precise biochemical nature of the integral membrane receptor protease(s)? What are the normal cellular substrates for the(se) enzyme(s)?

Thanks to Mrs. Elizabeth Falkenstein for manuscript preparation. This work was supported by grants from Research Corporation and the NIH.

REFERENCES

1. W. Troll, A. Klassen, and A. Janoff, Tumorigenesis in mouse skin: Inhibition by synthetic inhibitors of proteases. Science, 169, 1211-1213 (1970).
2. M. Hozumi, M. Ogawa, T. Sugimura, T. Takenchi, and H. Umezawa, Inhibition of tumorigenesis in mouse skin by leupeptin, a protease inhibitor from actinomycetes. Cancer Res. 32, 1725-1728 (1972).
3. J. G. Corasanti, G. H. Hobika, and G. Markus, Interference with dimethylhydrazine induction of colon tumors in mice by ε-aminocaproic acid. Science, 216, 1020-1021 (1982).
4. W. Troll, R. Wiesner, C. J. Shellabarger, S. Holtzman, and J. P. Stone, Soybean diet lowers breast tumor incidence in irradiated rats. Carcinogenesis, 1, 469-472 (1980).
5. F. F. Becker, Inhibition of spontaneous hepatocarcinogenesis in C3H/HeN mice by Edi Pro A, an isolated soy protein. Carcinogenesis, 2, 1213-1214 (1981).
6. P. Correa, Epidemiological correlations between diet and cancer frequency. Cancer Res. 41, 3685-3690 (1981).
7. B. Armstrong and R. Doll, Environmental factors and cancer incidence and mortality in different countries with special reference to dietary practices. Int. J. Cancer, 15. 617-631 (1975).
8. J. Yavelow, T. Finlay, A. R. Kennedy, and W. Troll, Bowman-Birk soybean protease inhibitor as an anticarcinogen. Cancer Res. 43, 2454s-2459s (1983).
9. J. P. Quigley, Proteolytic enzymes of normal and malignant cells. In:Surfaces of Normal and Malignant Cells (R. O. Hynes, Ed.) pp. 247-285. Sussex, England: John Wiley and Sons (1979).
10. D. E. Mullins and S. T. Rohrlich, The role of proteinases in cellular invasiveness. Biochim. Biophys. Acta, 695, 177-214 (1983).
11. J. O'Donnell-Tormey and J. P. Quigley, Inhibition of plasminogen activatory release from transformed chicken fibroblasts by a protease inhibitor. Cell, 27, 85-95 (1981).
12. J. O'Donnell-Tormey and J. P. Quigley, Detection and partial characterization of a chymostatin-sensitive endopeptidase in transformed fibroblasts. Proc. Natl. Acad. Sci. USA, 80. 344-348 (1983).
13. J. Yavelow, M. Collins, Y. Birk, W. Troll and A. R. Kennedy, Nanomolar concentrations of Bowman-Birk soybean protease inhibitor suppress X-ray-induced transformation in vitro. Proc. Natl. Acad. Sci. USA, 82, 5395-5399 (1985).
14. A. R. Kennedy, The conditions for the modification of radiation transformation in vitro by a tumor promoter and protease inhibitors. Carcinogenesis (Lond.), 6, 1441-1445, (1985).
15. M. B. Sporn and A. B. Roberts, Autocrine growth factors and cancer. Nature 313, 745-747 (1985).
16. L. Waxman, J. M. Fagan, K. Tanaka and A. L. Goldberg, A soluble ATP-dependent system for protein degradation from murine erythroleukemia cells. J. Biol. Chem. 260. 11994-12000 (1985).

17. J. J. Baldassare, S. Bakshian, M. A. Knipp, and G. J. Fisher, Inhibition of fibrinogen receptor expression and serotonin release by leupeptin and antipain. J. Biol. Chem. 260. 10531-10535 (1985).
18. A. R. Kennedy, B. S. Radner, and H. Nagasawa, Protease inhibitors reduce the frequency of spontaneous chromosome abnormalities in cells from patients with Bloom's syndrome. Proc. Natl. Acad. Sci. USA. 81 1827-1830 (1984).
19. A. L. Jochen and P. Berhanu, Chymotrypsin substrate analogues inhibit endocytosis of insulin and insulin receptors in adipocytes. J. Cell Biol. 103, 1807-1816 (1986).
20. J. Yavelow, M. Caggana, and K. A. Beck, Proteases occurring in the cell membrane: A possible cell receptor for the Bowman-Birk type of protease inhibitors. Cancer Res. 47, 1598-1601 (1987).
21. J. Yavelow, C. Scott, and T. C. Mayer, Fluorescent visualization of binding and internalization of the anticarcinogenic Bowman-Birk type protease inhibitors in transformed fibroblasts. Cancer Res. 47, 1602-1607 (1987).
22. J. P. Quigley, Association of a protease (plasminogen activator) with a specific membrane fraction isolated from transformed cells. J. Cell Biol. 71, 472-486 (1976).
23. M. Bradford, A rapid and sensitive method for the quantitation of microgram quantities of protein utilizing the principle of protein-dye binding. Anal. Biochem. 72, 248-254 (1976).

DIETARY CHOLINE DEFICIENCY AS A NEW MODEL TO STUDY THE POSSIBLE ROLE OF FREE RADICALS IN ACUTE CELL INJURY AND IN CARCINOGENESIS

Amiya K. Ghoshal, T.H. Rushmore, D. Ghazarian,
Amit Ghoshal, V. Subrahmanyan and E. Farber

Departments of Pathology and of Biochemistry
Medical Sciences Building, University of Toronto
Toronto, Ontario, CANADA M5S 1A8

INTRODUCTION

Over the last few years, working with a diet that is devoid of choline and in methionine (CD) we have observed that when this diet is fed to rats, they develop not only fatty liver, but also hepatocellular necrosis and cancer. In the course of our investigation as to the mechanism of cancer development by a dietary deficiency without any added carcinogen, we found that early nuclear lipid peroxidation (1), DNA alteration (2) and cell proliferation are very common features in the liver. Free radical generation and DNA alteration in a proliferating organ has been proposed as the initiating event in the development of liver cancer (see Fig. 1). However, the nature of free radical generated and the nature of DNA alteration are not known.

In this communication we summarize the evidence to date that free radicals are generated by this dietary imbalance and describe some ways to modulate the phenomenon in short-term studies of free radicals in liver cell injury.

RESULTS AND DISCUSSION

Generation of Free Radicals

Our hypothesis that free radicals are generated in the nuclei has been demonstrated by detection of diene conjugates in the nuclear membranes. One known consequence of lipid peroxidation, the genesis of hydroxy alkenals (3,4), has also been demonstrated (Fig. 2). However, the zonal distribution of alkenals is different when generated by a CD diet as compared with the generated by CCl_4 intoxication (Fig. 3). It should be mentioned here that CCl_4 causes microsomal and not nuclear lipid peroxidation whereas CD diet causes nuclear but not microsomal lipid peroxidation.

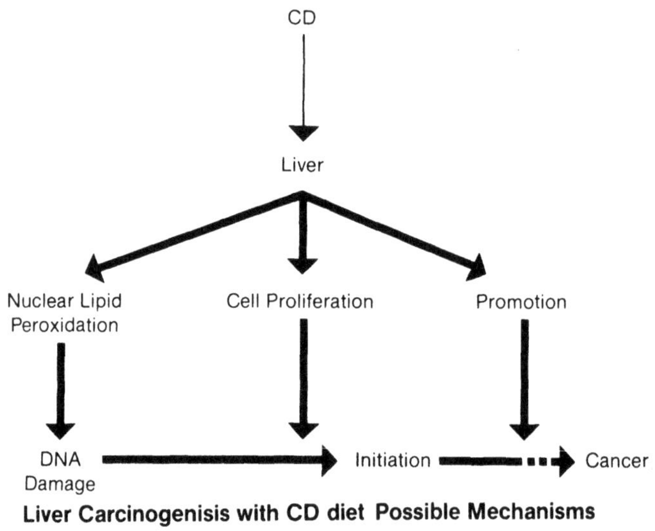

Fig. 1. A hypothesis for induction of liver cancer by choline deficient diet.

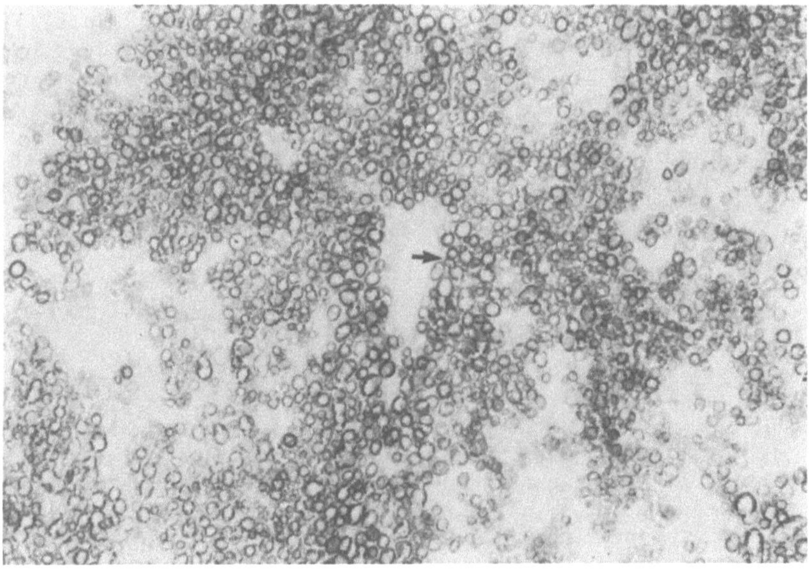

Fig. 2. Histochemical staining of hydroxyalkenals in liver section of rats on CD diet for 5 days. Note the presence of positive staining in zone 3 (→).

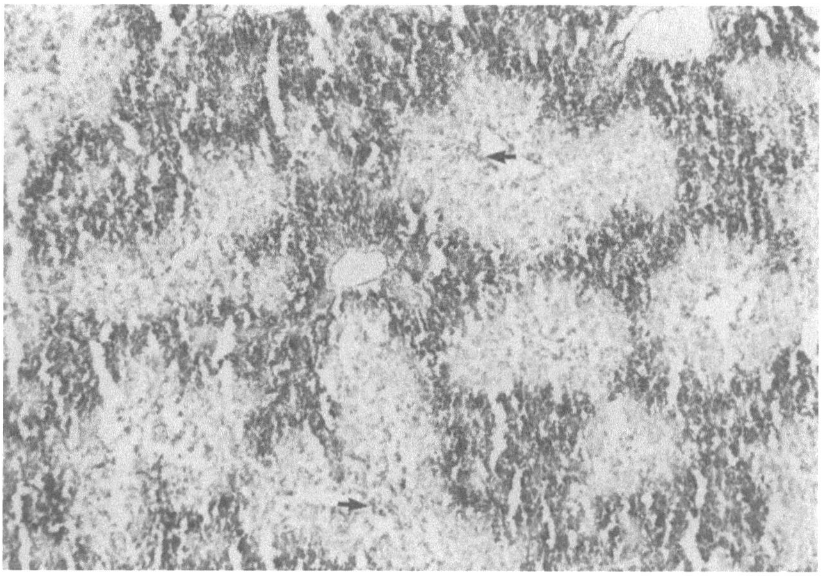

Fig. 3. Histochemical staining of hydroxyalkenals in liver section of rates 24 hours after CCl_4 intubation. Note the absence of staining in zone 3 (--) but presence of staining in zone 1 and 2.

Experiments With a Spin Trap Agent

Modulation of Free Radical Generation Affect the Biological Manifestation by CCl_4. As the free radical generated by the homolytic cleavage of CCl_4 is very short lived, its detection has been achieved by making the free radical more stable with the use of a spin trap N-tert-butyl-a-phenylnitrone (PBN) and detecting the signal by electron spin resonance spectroscopy (5). We gave PBN 2 hours before a large dose of CCl_4 (200 1/100 g b.w.), killed the animals 2 hours later and found that the diene conjugates in the microsomes were attenuated to almost 20% of that in the control animal with CCl_4 alone (Fig. 4).

Prevention by PBN of Lipid Peroxidation Induced by CD Diet. We do not know the nature of free radicals generated in the liver due to CD diet. We argued that if PBN can trap the free radical generated by CD diet, then it may be possible to prevent the deleterious effect of free radicals, e.g., lipid peroxidation, if rats are given PBN. Our expectation was fully realized. PBN, when given during the start of a CD diet, can not only prevent lipid peroxidation in the liver nuclei (not shown) but it can also stop a further generation of lipid peroxidation when rats are challenged with PBN while on the CD diet (Fig. 5). Work is in progress with Dr. Jazen of Guelph University, Ontario, Canada to detect possible stable free radicals by E.S.R. spectroscopy.

In this dietary model 100% of the rats develop preneoplastic nodules in their liver and more than 50% hepatocellular carcinoma. Very early and late alterations in the rat liver exposed to CD diet is shown in Table 1, in temporal sequence.

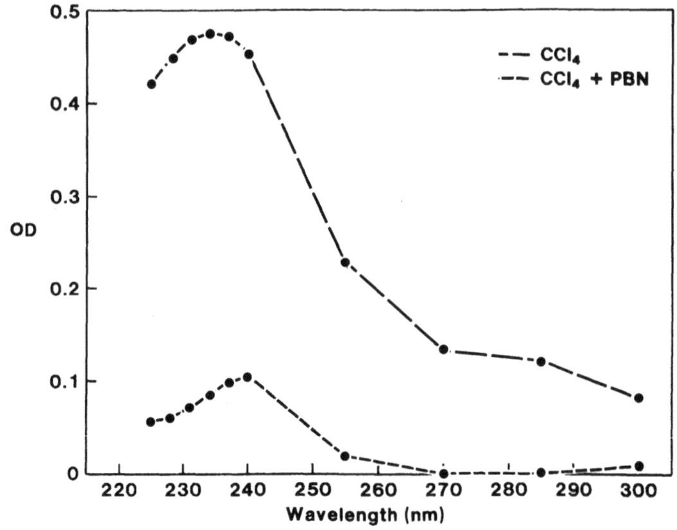

Fig. 4. Difference spectra. (— —)Diene conjugate peak in the microsomal lipid, 2 hours after CCl_4 intubation; (----) 10 mg PBN/100 g given 2 hours before CCl_4 inhibits more than 80% of diene conjugate peak.

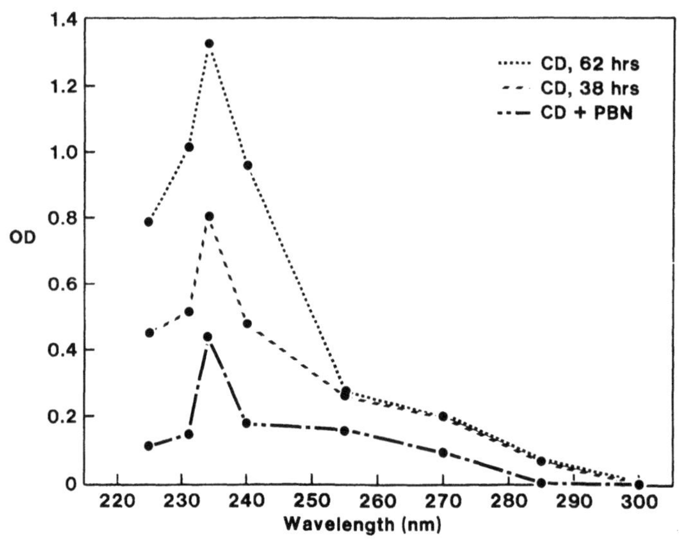

Fig. 5. Difference spectra. The upper curve (····) shows the peak of diene conjugates, 62 hours after initiation of CD diet without any additional treatment with PBN. Note the decrease in optical density at 235 nm (diene conjugates) in animals fed a CD diet and given PBN at 38 and 50 hours and examined at 62 hours (-·-·-). CD alone at 38 hours (----) without any PBN treatment. Not only was the increase in diene conjugate level from 38 to 62 hours after the initiation of CD diet prevented by treatment with PBN but the level seen at 38 hours was reduced.

Table 1. Temporal Sequence of Early and Late Events in the Rat Liver Exposed to CD Diet

Time after initiating the exposure to CD diet	Events
8 hours	Triglyceride (TG Accumulation)
1 day	Nuclear Lipid Peroxidation
2 days	DNA Alteration(s)
5 days	Mitochondrial Lipid Peroxidation
6 days	Cell Death and Cell Proliferation
10 weeks	Initiation
13 months (earliest)	Cancer Induction

Note: Microsomal lipid peroxidation could not be detected at any time up to 63 days.

In an attempt to find our whether the several changes observed in rats exposed to a CD diet are occurring sequentially or independently, we have studied the effects of several modulators, among them being extra dietary calcium.

Effect of Extra Calcium

Ca^{2+} has been shown to prevent lipid peroxidation in liver cells suspended in calcium-free medium (6). We have shown that when extra calcium is added to CD diet, some of the early changes due to choline deficiency, such as nuclear lipid peroxidation, DNA alteration, and cell death can be prevented, but little or no effect on the triglyceride accumulation in liver is observed. Strontium has a similar effect on the early events of CD diet. Similarly, extra Ca^{2+} also inhibited the promoting ability of CD diet when liver cells are initiated by diethylnitrosamine (7). Apparently, excessive accumulation of triglyceride in the liver could be separated from other pathological changes of a CD diet, such as nuclear lipid peroxidation, DNA alteration and cell death.

CONCLUSION

This dietary model, in which no carcinogen is used, gives us an excellent opportunity not only to study carcinogenesis without the interference of a xenobiotic but also enables one to study acute cell injury and the possible role of endogenous as well as exogenous free radicals in its development. The nature of the free radicals generated in the liver by a CD diet has not yet been indentified. The observations that (a) CCl_4 induces microsomal but not nuclear lipid peroxidation and CD diet induces nuclear but not microsomal lipid peroxidation, (b) PBN can inhibit both processes, and (c) extra dietary Ca^{++} can inhibit nuclear lipid peroxidation give us handles to begin to study the molecular pathogenesis of different membrane and other alterations in both acute cell death and cancer development as they may be related to the genesis of free radicals.

ACKNOWLEDGEMENTS

This work was supported by grants from the U.S. Public Health Service CA 41537, CA 21157, from the National Cancer Institute of Canada and Medical Research Council of Canada. We would like to express our sincere thanks to Lori Culter for her excellent secretarial help.

REFERENCES

1. T.H. Rushmore, Y.P. Lim, E. Farber and A.K. Ghoshal, Rapid lipid peroxidation in the nuclear fraction of rat liver induced by a diet deficient in choline and methionine. Cancer Lett., 24: 251-255 (1984).
2. T.H. Rushmore, E. Farber, A.K. Ghoshal, S. Parodi, M. Pala and M. Taningher, A choline devoid diet, carcinogenic in the rat, induces DNA damage and repair. Carcinogenesis, 7: 1677-1680 (1986).
3. A. Benedetti, M. Comporti and H. Esterbauer, Indentification of 4-hydroxynonenal as cytotoxic product originating from the peroxidation of liver microsomal lipids. Biochim. Biophys. Acta., 620: 281-296 (1980).
4. A. Benedetti, G. Malvaldi, R. Fulceri and M. Comporti. Loss of lipid peroxidation as a histochemical marker for preneoplastic hepatocellular foci of rats. Cancer Res., 44: 5712-5717 (1984).
5. E.K. Lai, P.B. McCay, T. Noguchi and K.L. Fong, In vivo spin-trapping of trichloromethyl radicals formed from CCl_4. Biochem. Pharmacol., 28: 2231-2235 (1979).
6. M.W. Fariss, K. Oltztsdottier and D.J. Reed, Extracellular calcium protects isolated hepatocytes from injury. Biochem. Biophys. Res. Commun., 121: 102-110 (1984).
7. A.K. Ghoshal, E. Laconi, F. Willemsen, A. Ghoshal, T. Rushmore and E. Farber, Modulation by calicum of the carcinogenic process in the liver induced by a choline-deficient diet. Can. J. Physiol. Pharmacol., 65: 478-482 (1987).

INHIBITION OF GENOTOXICITY BY DIALLYL SULFIDE AND STRUCTURAL ANALOGUES

Mark T. Goldberg

Department of Biomedical Sciences
University of Guelph,
Guelph, Ontario, Canada. N1G 2W1

INTRODUCTION

Higher plants contain an extensive array of biologically active chemicals, some of which are potent modifiers of chemical carcinogenesis (1). Specifically, some of these agents have been shown to be active in inhibiting the initiation stage of the carcinogenesis process (2). Members of the Allium genus, which include onions and garlic, are rich in sulfur-containing compounds. Among the major components of garlic oil are diallyl disulfide (DADS, 66%) and diallyl sulfide (DAS, 14%) (3). It has been previously shown that the latter agent, DAS, can inhibit 1,2-dimethylhydrazine-induced genotoxicity in the murine colonic epithelium (4) and cyclophosphamide-induced genotoxicity in the murine urothelium and hair follicles (5). Recently it has been shown that DAS can inhibit 1,2-dimethylhydrazine-induced colon tumorigenesis (6). The goal of this study was to determine the optimal time for the oral administration of DAS, prior to carcinogen treatment, and to determine the efficacy of other structurally analogous sulfur-containing compounds.

MATERIALS AND METHODS

Chemicals - Diallyl sulfide and the structural analogues allyl methyl sulfide, dipropyl sulfide, dimethyl sulfide, diallyl disulfide and dipropyl sulfone (Figure 1) were administered by gavage in medium chain triglyceride (MCT) oil (Mead-Johnson, Belleville, Ont.), a highly digestible hydrophobic vehicle. 1,2-dimethylhydrazine was administered i.p. (0.01 ml/g body weight) in 1mM EDTA at pH 6.5.

Animals - Male C57Bl/6J mice, 8-10 weeks old, were purchased from the Jackson Laboratory, Bar Harbor, ME and were randomly allocated into groups of 6 or 7. Mice were acclimatized for at least 1 week before treatments, during which time they were placed on a 12 hour light/dark cycle and given food and water *ad libitum*. Food was withheld from mice for 16 h before gavage with DAS or the structural analogues, but was made available after treatment.

Experimental Protocol - In the time course study, mice were given a single gavage (0.01 ml/g body weight) of DAS (200 mg/kg) or MCT vehicle alone at various times corresponding to 72, 48, 24, 18, 12, 5, 4, 3, 2, 1 or 0.5 h prior to DMH treatment, immediately before DMH and 1 h after

DADS	di–Allyl Disulfide	$CH_2{=}CHCH_2{-}S_2{-}CH_2CH{=}CH_2$
PSO	di–Propyl Sulfone	$CH_3CH_2CH_2{-}S{-}O_2CH_2CH_2CH_3$
MS	di–Methyl Sulfide	$CH_3{-}S{-}CH_3$
AMS	Allyl Methyl Sulfide	$CH_2{=}CHCH_2{-}S{-}CH_3$
PS	di–Propyl Sulfide	$CH_3CH_2CH_2{-}S{-}CH_2CH_2CH_3$
DAS	Diallyl Sulfide	$CH_2{=}CHCH_2{-}S{-}CH_2CH{=}CH_2$

Fig. 1. The structure of diallyl sulfide and sulfur-containing analogues.

DMH. In the study on the efficacy of sulfur-containing analogues of DAS, these agents were administered in doses which were the molar equivalents of 50 mg/kg DAS (low dose) or 200 mg/kg (high dose). Analogues were administered 4 h prior to DMH treatment. 1,2 Dimethylhydrazine was given at a dose of 20 mg/kg and always administered between 10:30 and 11:00 a.m. in both studies to avoid confounding of results by circadian rhythms. Also in both studies, one group of mice was treated with MCT vehicle and DMH (DMH control). Twenty four hours after DMH injection, mice were killed by cervical dislocation. Colons were removed, fixed and stained, then scored, blinded to treatment, for nuclear aberrations (7). The incidences of nuclear aberrations in 10 colonic crypts per animal were averaged within groups and expressed as a percentage of the response to the DMH control group. The statistical significance of results was assessed using Student's t-test. A p value <0.05 was considered significant.

RESULTS

The optimal time for the administration of DAS was determined to be 4 or 5 h prior to DMH treatment. A 4 h pretreatment gave more consistent inhibition (Figure 2) so this time of pretreatment was used in subsequent studies on sulfur-containing analogues of DAS. High doses of methyl sulfide (MS) or allyl methyl sulfide (AMS) significantly reduced the incidence of DMH-induced nuclear aberrations as did high and low doses of dipropyl sulfide (PS) (Figure 3). However, none of the sulfur-containing analogues of DAS were as effective as DAS in blocking DMH-induced nuclear aberrations (Figure 3).

DISCUSSION

Diallyl sulfide (DAS) has previously been shown to inhibit nuclear aberrations induced by DMH in the colon (4) and by cyclophosphamide in the bladder and hair follicle epithelial cells (5). It also prevents DMH-induced colon carcinogenesis (6). This study demonstrates that the most effective time to administer DAS is 4 to 5 hours prior to DMH treatment.

A pretreatment time of 4 hours was selected for studies on structurally related compounds, since this was the time when consistent inhibition was seen, as well as for practical reasons. Results indicated that pretreatment with allyl methyl sulfide and dipropyl sulfide

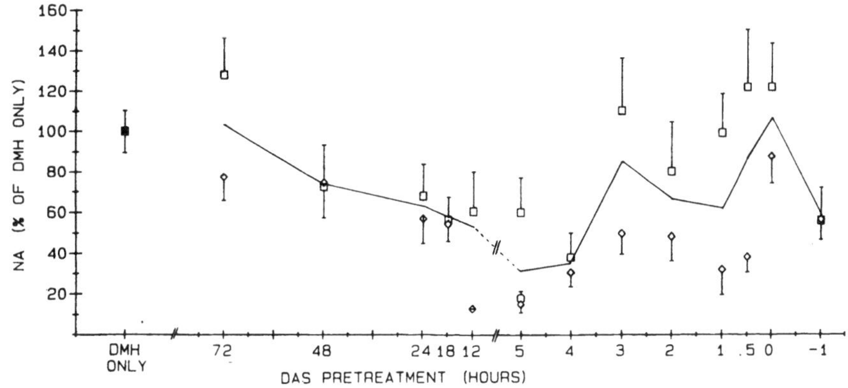

Fig. 2. Time course for the protective effect of diallyl sulfide on 1,2-dimethylhydrazine-induced nuclear aberration, expressed as a percentage of DMH only treated group. (N=6 or 7 except DMH group where N=25).

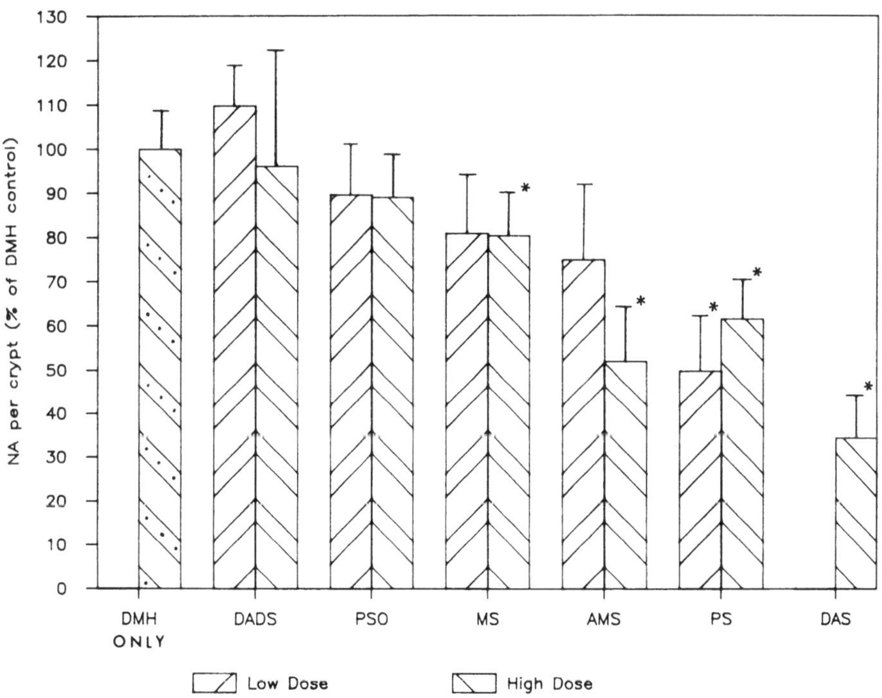

Fig. 3. Efficacy of sulfur-containing analogues of diallyl sulfide in inhibiting 1,2-dimethylhydrazine-induced nuclear aberrations in colonic epithelial cells. Abbreviations are as in Fig. 1. Low dose is the molar equivalent of 50 mg/kg DAS, high dose is the molar equivalent of 200 mg/kg (N=6 except DMH only group where N=25). $p<0.05$

significantly blocked DMH-induced nuclear aberrations in colonic crypts, although these agents were not as effective, on an equimolar basis, as diallysulfide.

ACKNOWLEDGEMENTS

The author wishes to thank Robert Braham and Dorothy Pharoah for technical assistance and help with the manuscript. Supported by the National Cancer Institute of Canada.

REFERENCES

1. National Academy of Sciences, Diet, Nutrition and Cancer, pp. 358-370. National Academy Press, Washington, D.C., 1982.
2. H.L. Newmark, A hypothesis for dietary components as a blocking agent of chemical carcinogenesis, plant phenolics and pyrrole pigments. Nutr. Cancer 6: 58-70 (1984).
3. M.H. Brodnitz, J.V. Pascale, and L.V. Derslice, Flavor components of garlic exract. J. Agric. Food Chem. 19: 273-275 (1971).
4. M.J.Wargovich and M.T. Goldberg, Diallyl sulfide: A naturally occurring thioether that inhibits carcinogen-induced nuclear damage to colon epthelial cells in vivo. Mutat. Res. 143: 127-129 (1985).
5. M.T. Goldberg and P.D. Josephy, Studies on the mechanism of action of diallyl sulfide, an inhibitor of the genotoxic effects of cyclophosphamide. Can. J. Physiol. Pharmacol. 65: 476-471 (1987).
6. M.J. Wargovich, Diallyl sulfide, a flavor components of garlic (Allium sativum) inhibits dimethylhydrazine-induced colon cancer. Carcinogenesis 8: 487-489 (1987)
7. M.J. Wargovich, M.T. Golderg, H.L. Newark, and W.R. Bruce, Nuclear aberrations as a short-term test for genotoxicity to the colon: evaluation of nineteen agents in mice. J. Natl. Cancer Inst. 71: 133-137 (1983).

PHYSIOLOGIC ANTICARCINOGENESIS: EFFECTS OF STABLE STRONTIUM ON METASTATIC BONE DISEASE

S.C. Skoryna, P. Koch, E. Yeghiayan, M. Fuskova,
A. Sauvé and J.F. Stara

St. Mary's Hospital Center, McGill University, Montreal, Que.
Department of Biology, Université de Montréal, Montreal, Que.
and Environmental Protection Agency, Cincinnati, Ohio

Metastatic inefficiency (1) presents a unique challenge for the investigation of local and systemic factors in physiologic anticarcinogenesis. It appears that a point has been reached whereby there is general agreement that the transfer of cancer cells from the primary tumor to secondary sites is a multi-step process (2) and likely also multifactorial since each step is affected by different factors (3,4). Purely mechanical factors seem to be insufficient to explain why 99.9% of circulating tumor cells are destroyed before they are able to produce a viable metastatic focus. Recent animal studies indicate that only specialized cancer cells are capable of initiating metastatic foci and that they are to a certain degree site specific (5). There is a need for further development of animal models, whereby experimental conditions are controlled, in order to explain the diversity of response to treatment (6).

A number of approaches to cancer treatment applicable to metastatic disease is outlined in this book (6,7,8,9). From the pathologic viewpoint, it seems useful to distinguish between the different properties of primary cancer. Growth is not the only property;invasiveness, metastability and anaplasia (10) are also likely to be affected by different local and systemic stimuli, not necessarily inherent to clonal differentiation (3,4). Although in animal models of carcinogenesis, the non-random spread of malignant cells was demonstrated (5), clinical observations indicate the omnipresence of randomness (3,4). The diversity of response to treatment is augmented by factors inherent to different autologous hosts as well as different environmental factors. The "seed and soil" hypothesis proposed 100 years ago by Paget (11) seems to be still valid, considering the number of factors which may affect either the seed or the soil or both.

Metastatic bone disease, when it occurs without soft tissue involvement, represents a specific group of patients. The survival of these patients is much longer than those with soft tissue metastases and this fact deserves further investigation of the possible role of physiologic anticarcinogenesis. Normal bone remodelling processes in the adult human skeleton (12) and in pre-existing conditions, such as osteoporosis, seem to contribute to the course of the disease by the presence of vertebral collapse and higher incidence of fractures. Galasko (13) demonstrated two types of osteolytic mechanisms which occur in both experimental metastases following injection of tumor cell suspension into medullary cavity of rabbits, as well as in human metastatic bone disease: 1) osteoclast-mediated osteolysis of enchondrial surface; 2) secretion of lytic tumor enzymes in later stages. In the experimental model of Galasko, the bone formation occurred on the periosteal surface, which resembles a bone repair process in fracture healing (13).

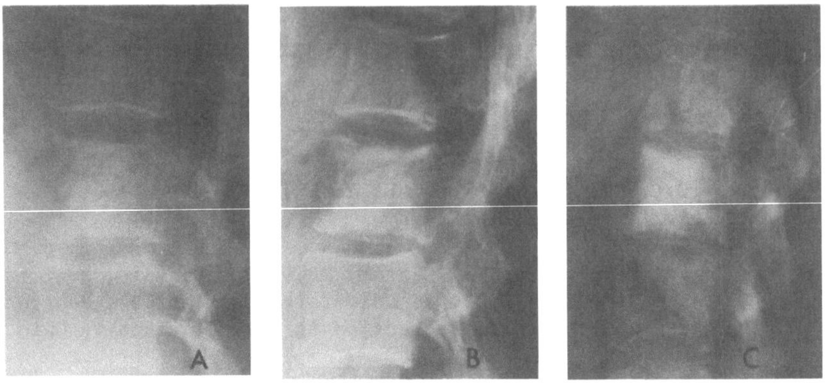

Fig. 1. Mixed osteoclastic and osteoblastic metastasis of the body of second lumbar vertebra, secondary to carcinoma of left breast. A) Radiograph taken prior to administration of stable strontium (April 1980). B) After 3 months of stable strontium administration (June 1980). C) After 9 months of administration of stable strontium (December 1980). Progressive sclerosis and mineralization is evident;an improvement of definition of anterior margin of body of the vertebra indicates new bone formation (4). Originally published in Canadian Medical Association Journal Vol. 125, October 1, 1981.

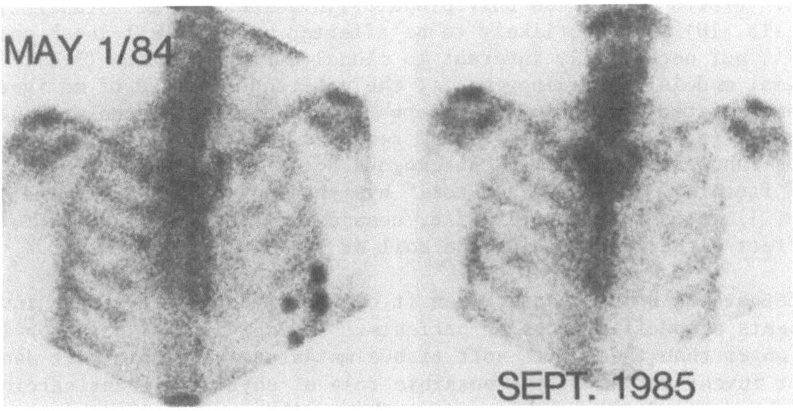

Fig. 2. Nuclear (Technecium MDP) scans. Patient with multiple bone metastasis, secondary to carcinoma of right breast, involving VI and VII ribs. A) Scans prior to administration of strontium and tamoxifen citrate. B) Scans after 16 months of continuous therapy.

Epithelial tumors of the prostate, breast, lung, thyroid, kidney and stomach frequently produce bone metastases. According to the review of Bhardwaj and Holland (14), in breast cancer the incidence varies between 49-84%; in prostatic cancer 47-84% of patients are affected. Blair and McAfee (15) pointed out that the diagnosis is frequently difficult: in 30% of metastatic bone disease, the radiographs are normal and in 2-8%, bone scans are negative.

The abundancy of current cancer protocols makes it difficult to collect a statistically significant number of cases with sufficient randomnization; multi-center studies have an inherent possibility of different conditions even if protocol is identical. The present paper is based on pre- and post-radiographic observations in 14 patients with malignant bone lesions, receiving low doses of stable strontium for periods ranging from 3 to 36 months;this includes 6 cases of primary breast cancer, 4 cases of prostatic cancer and 4 cases of multiple myeloma. Strontium carbonate compound (350-700 mg per day) has approximately a twice-higher content of elemental Sr^{2+} than strontium gluconate (188-290 mg per day); however, because of lower intestinal absorbability of Sr carbonate, the increase in serum level following Strontium administration is similar between the two salts, as determined by Atomic Absorption Spectrophotometry. Since this dosage of Strontium does not significantly affect serum Calcium concentration, the results were evaluated in terms of Sr:Ca ratio. Normal serum Sr:Ca ratio ranges between 1:1550 and 1:2400; following increased oral intake (on empty stomach), the serum Sr:Ca ratio ranged between 1:20 and 1:10 (16). Previous histomorphometric studies on iliac crest biopsy material indicate that bone Sr:Ca ratio ranges between 1:79 and 1:260, when measured by Atomic Absorption Spectrophotometry and between 1:40 and 1:170 when measured by microprobe analysis, following Strontium administration (17).

In all cases except one, evidence of partial or complete remineralization of metastatic bone lesions was observed (Fig.1); in one case (cancer of the breast in a male) no changes were observed after 3 months of Strontium administration. In cases where Strontium was administered simultaneously with anti-androgens (Fig.2) or anti-estrogens (Fig.3), the disappearance of metastatic lesions on Technecium DMP scans was particularly impressive. In cases of multiple myeloma, maintenance of metastatic bone status (Fig.4) over a period of 3 years was observed; in a case of myelomatous lesion of left humerus, an increase in cortical thickness is evident (Fig.5). Previous studies (Fig.6) indicate that in some cases, bone remodelling in osteoporotic skeleton can be maintained for a number of years. No adverse effects were observed except for some weight gain which was beneficial in cancer cases. All patients reported a significant decrease of bone pain (16).

DISCUSSION

Stable strontium is an old drug. Pharmacologic use of Strontium salts was incorporated into British,U.S.A. and other pharmacopeias, following the description in Squire's Companion in 1884 (18). The element was used in low dosage as Sr salicylate and Sr cinnamate in various diseases, including cancer and "rheumatism" on purely empirical basis. As late as 1955, Strontium compounds were listed in U.S.A. as prescription medication (19). Due to confusion in the public mind between the effects on bone of radioactive Sr^{90}, a product of atomic fission (20) and Stable Strontium, a naturally-occurring element (21), further clinical studies (22) were stopped.

Stable Strontium, a chemical analog of Calcium, differs basically from Calcium in physiologic properties (16): intestinal absorption of Strontium is not controlled homeostatically and cellular uptake appears to occur by passive diffusion; normal intracellular concentration of

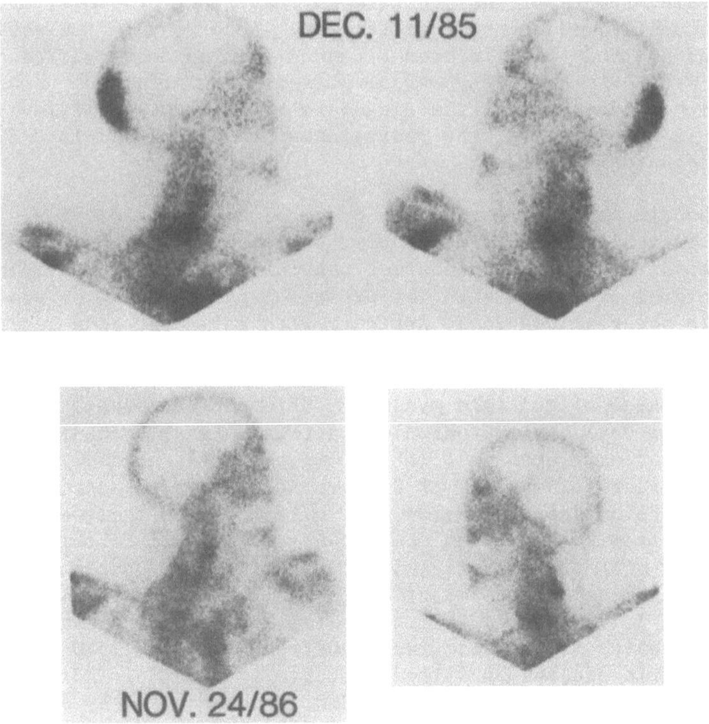

Fig. 3. Nuclear (Technecium MDP) scans of metastatic skull lesions, secondary to prostatic cancer. Dec.85: Occipital bone metastatic lesion prior to administration of stable strontium and anti-androgen; there is an increased uptake of radionucleotide related to metastasis. Nov.86: Complete disappearance of metastasis after one year of therapy.

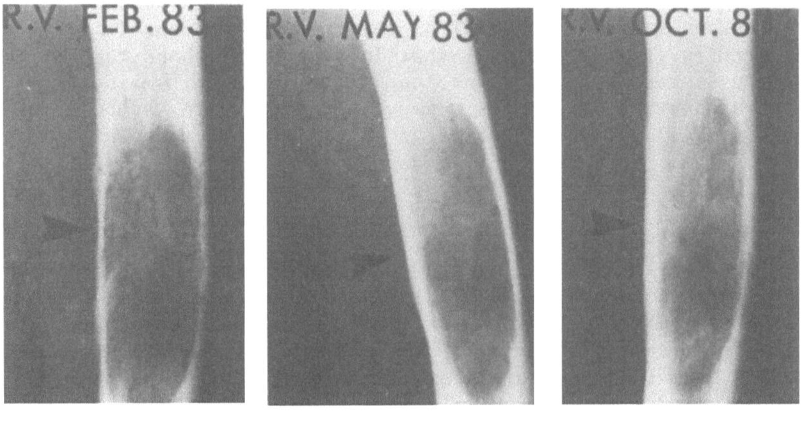

Fig. 4. Radiographs of lt. humerus of a patient with multiple myeloma lesion. A) Radiograph prior to administration of stable strontium. B) and C) Radiographs during continuous administration of stable strontium; progressive increase in cortical thickness is evident.

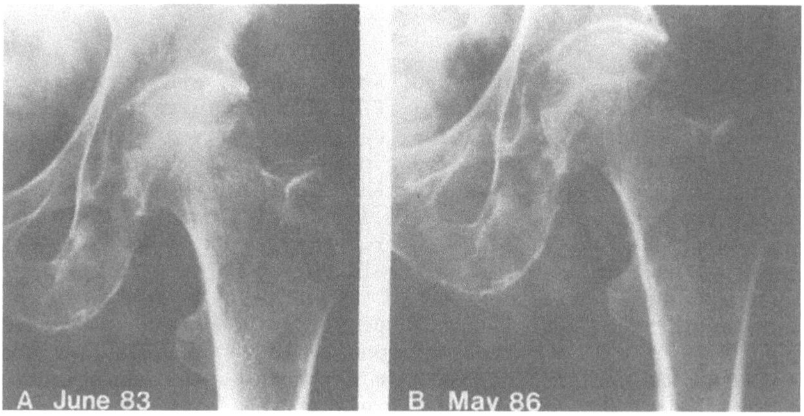

Fig. 5. AP radiographs of left hip of patient with multiple myeloma. A) Radiograph taken prior to Strontium administration; multiple osteolytic lucencies of left ischium. B) Radiograph taken after 3 years of continuous strontium administration. No progression of the disease and perhaps some improvement is apparent.

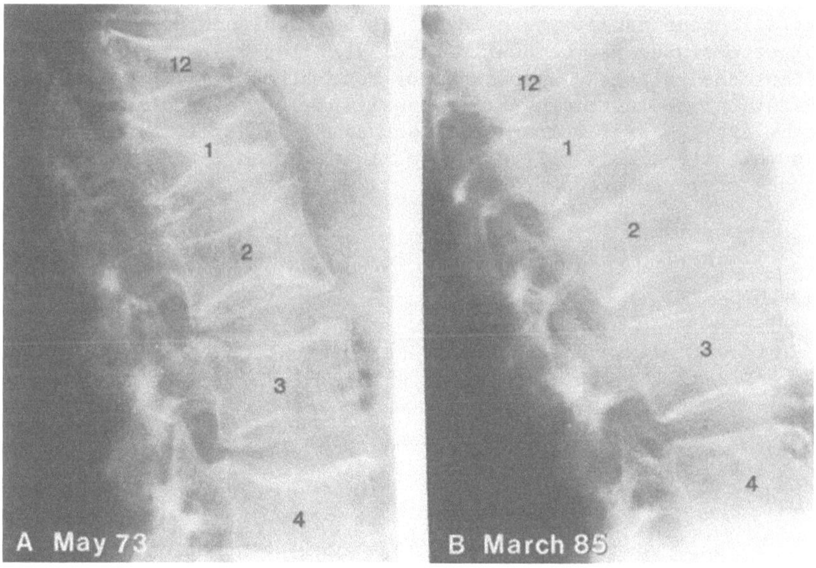

Fig. 6. Radiographs of the dorsolumbar spine of female patient with severe osteoporosis. A) There is collapse of bodies T_{12}, L_1, L_2, L_3 and L_4 at the age of 62, prior to strontium administration. B) No progression of osteoporosis is evident at the age of 74 following continuous administration of stable strontium.

Strontium on a standard diet is one order of magnitude higher than that of plasma (16).

Administration of low doses of Strontium may affect metastatic disease at two levels: 1) systemic effect due to increased cellular uptake, and 2) local effect on bone lesions due to affinity of Strontium to bone, resulting in remineralization of osteopenic areas. Systemic effect on metastases may be related to increased Strontium levels in PMN's (Pre-Strontium administration Sr:Ca ratio 1:8; Post-Strontium ratio 1:14) (23). If it is assumed that Calcium is excruded from PMN's in the presence of circulating tumor cells, this may contribute to prolonged survival of patients with bone metastases due to the effects of fluctuating plasma Calcium levels after administration of Strontium or physiologically by the release of Calcium from osteolytic bone lesions. In an animal model, the concentration of cationic proteins in neutrophils was demonstrated to affect exocytosis (24). On the other hand, significant hypercalcemia, resulting from osteolytic bone lesions, can be expected to decrease following Strontium administration, since according to studies of El Solh and Rousselot (25), there is a balance between Calcium and Strontium levels in the plasma.

Although the current report is based on a small number of cases, one cannot help but be impressed by the significant decrease in bone pain and remineralization of metastatic bone lesions following low dosage of stable strontium. The remineralization is probably due to secondary Calcium deposition, strontium serving only as an initiator of Ca^{2+} deposition in nucleation centers of the bone matrix, since the amount of Sr^{2+} is too small compared to Ca^{2+} content. Strontium affects only the bone metastatic remineralization process: the question to be answered is whether cancer cells in bone remain in a dormant state. Simultaneous administration of hormonal (or other) drugs, which exert cytotoxic effect on metastatic tumor cells in bone, is indicated.

Limits of methodology of our assessment of the effects of treatment of metastatic bone cancer are evident; as Bhardwaj and Holland pointed out "the neoplastic tissue may be nearly totally killed by chemotherapy without significant change in the appearance of bone films" (14). Obliteration of metastatic bone lesions by Sr^{2+} deposition occurs, remineralizing the space left by cytolysis of tumour cells and limiting tumour growth by proliferation of normal osteoblasts. Strontium administration is not indicated in the presence of sarcomatous growth.

ACKNOWLEDGEMENT

This work was supported by grants from St. Mary's Hospital Foundation and the Rideau Institute of Canada.

REFERENCES

1. L. Weiss, Dynamic aspects of cancer cell populations in metastasis. Am. J. Path. 97, 601-608 (1979).
2. P.A. Cerutti, Multistep carcinogenesis. In: Anticarcinogenesis and radiation protection 1987 (O. Nygaard, M. Simic and P. Cerutti, Eds.) Plenum Press, New York.
3. A. Horava and S.C. Skoryna, Observations on pathogenesis of neoplasia. C.M.A.J. 73, 630-638 (1955).
4. S.C. Skoryna, Systemic factors in carcinogenesis. C.M.A.J. 80, 689-697 (1959).

5. I.J. Fidler, Circumvention of biologic diversity of cancer metastases. In: Anticarcinogenesis and radiation protection 1987 (O. Nygaard, M. Simic and P. Cerutti, Eds.) Plenum Press, New York.
6. J.E. Talmadge, Preclinical studies with recombinant cytokines for the treatment of metastatic cancer. (These proceedings).
7. J.W. Smith and D.L. Longo, Clinical trials using monoclonal antibodies. (These proceedings).
8. A.R. Kennedy, Protease inhibitors as anticarcinogens. (These proceedings).
9. K.F. Gey, H.B. Stahelin and G.B. Brumbacker, Cancer mortality inversely related to plasma levels of antioxidant vitamins. (These proceedings).
10. S.C. Skoryna, Gastric cancer - An attempt to correlate experimental and clinical findings. Leber-Magen-Darm 6, 92-96 (1976).
11. J. Paget, The distribution of secondary growth in cancer of the throat. Lancet 1, 571-573 (1889).
12. H.M. Frost, Skeletal physiology and bone remodelling. An overview. In: Fundamental and Clinical Bone Physiology 1980 (M.R. Urist, Ed.) pp. 242-267, Lippincott, Philadelphia.
13. C.S.B. Galasko, Mechanisms of lytic and blastic metastatic disease of bone. In: Metastatic disease of bone 1982 (R.N. Levy, Ed.) pp. 20-27, Clinical Orthopedics 169, Lippincott, Philadelphia.
14. S. Bhardway and J.F. Holland, Chemotherapy of metastatic cancer in bone. In: Metastatic disease of bone 1987 (R.N. Levy, Ed.) pp. 28-37, Lippincott, Philadelphia.
15. R.J. Blair and J.C. McAfee, Radiological detection of skeletal metastases: radiographs versus scans. Int. J. Radiat. Oncol. Biol. Phys.1, 1201-1205 (1976).
16. S.C. Skoryna, Effects of oral supplementation with stable strontium. C.M.A.J. 125, 703-712 (1981).
17. P.J. Marie, R.J. Pivon, G. Chabot et al, Histomorphometry of bone changes in stable strontium therapy. In:Trace Subst. Env. Health 1985 (D.D. Hemphill, Ed.) Vol. 19, pp. 193-208, University of Missouri Press.
18. Squire's Companion to the British Pharmacopeia, 19th Ed., London, (1916).
19. A. Osol, G.E. Farrar Jr., K.H. Beyor Jr. et al, The dispensatory of the United States of America. pp. 1871-73, Lippincott Co., Philadelphia (1955).
20. S.C. Skoryna and D.S. Kahn, The late effects of radioactive strontium on bone. Cancer 12, 306-322 (1959).
21. H.T. Odum, The stability of the world strontium cycle. Science 114, 407-411 (1951).
22. F.E. McCaslin and J.M. Janes, The effects of strontium lactate in the treatment of osteoporosis. Proc. Mayo Clin. 34, 329-334 (1959).
23. S.C. Skoryna and P. Koch, Unpublished observations. (1987).
24. J.C. Foreman and L.M. Lichtenstein, Spontaneous histamine secretion from leucocytes in the presence of strontium. J. Pharm. Exp. Ther. 210, 75-81 (1979).
25. N. El Solh and F. Rousselet, Effects of Stable Strontium administration on Calcium metabolism. In: Handbook of Stable Strontium 1981 (S.C. Skoryna, Ed.) pp. 515-561, Plenum Press, New York.

Dr. J.F. Stara: Deceased prior to publication.

RADIATION VS CHEMICAL DAMAGE TO DNA

J.F. Ward, C.L. Limoli, P. Calabro-Jones and J.W. Evans

Division of Radiation Biology
School of Medicine, M-010
University of California, San Diego
La Jolla, California 92093

Ionizing radiation is considered to cause its damage to cellular DNA via OH free radicals and by direct ionization of the DNA (1). In attempting to simulate the OH portion of this damage we have used hydrogen peroxide (2). We showed that the probable mechanism of damage production by this agent was through OH radicals produced by its interaction with variable valency metal ions bound to DNA (in a Fenton type reaction). Previously it had been shown that hydrogen peroxide treatment causes single-strand breaks (SSB) (but not double-strand breaks, DSB) in cellular DNA (3). More recently the reaction of hydrogen peroxide with DNA was found to give the same range of base-damaged products as that formed by low-LET ionizing radiation (4). Therefore, the spectrum of all damage types and their relative yields produced in the individual moieties of the DNA (sugar and base) by OH radicals is the same for hydrogen peroxide as for low-LET ionizing radiation (5). When cells are treated with hydrogen peroxide at $0^{0}C$, killing does not occur until the number of SSB present (and the base damage concomitantly produced) is 2,600 times higher than that present when eqivalent cell kill is induced by ionizing radiation (2). Clearly although the same damage types are formed by hydrogen peroxide their presence is inconsequential to the cell survival. We conclude that it is not the chemical structure of the OH radical-damaged sites per se which is responsible for the ionizing radiation-induced cell killing. It is the concentration of these damaged sites in local regions of the DNA (producing what we have called locally multiply damaged sites, LMDS) which is responsible for the effectiveness of ionizing radiation in inducing cell killing.

We have suggested that the mechanism of energy deposition by ionizing radiation is responsible for its effectiveness in cell killing via the production of LMDS. Extrapolating from the hypotheses which radiation chemists developed to describe the radiation chemistry of water, we invoke the deposition of relatively high amounts of energy in small volumes, which have been called spurs, blobs and short tracks (6). Within these small volumes, multiple radicals are produced, which, when they overlap the DNA, react locally producing LMDS.

With a view to comparing the efficacy of ionizing radiation with chemical damage in cell killing, we collected data from a variety of agents on numbers of cellular DNA lesions necessary to kill mammalian cells (5).

Table 1. Yields of DNA Damage Necessary to Kill 63% of Cells.

Agent	DNA Lesion	Number of Lesions per cell per D_{37}	Reference
Ionizing Radiation	SSB DSB Total LMDS *	1000 40 440	14 15
Bleomycin A2	SSB DSB	150 30	16,17
UV light	$\hat{TT}$ dimer SSB	400,000 100	18,19,20
Hydrogen Peroxide $0^{\circ}C$ $37^{\circ}C$	SSB ?	<2,600,000	2
Benzo{α}pyrene-4,5-oxide	Adduct	100,000	21
Aflatoxin	Adduct	10,000	22
1-Nitropyrene	Adduct	400,000	23
Methyl Nitroso Urea **	7 Me-G † 0-6 Me-G 3 Me-G	800,000 130,000 30,000	24
N-Acetoxy Acetylaminofluorene Other similar aromatic amides behave equivalently.	Adduct	700,000	21

*Calculated
**Not from Survival Curve.
† 7-Methyl Guanine, 0-6 Methyl Guanine, 3-Methyl Adenine.
Table 1 includes the original collection (5) and data for other agents.

It is clear that ionizing radiation (and bleomycin) are different from other agents in their ability to kill cells. For these two noxae the number of DNA lesions necessary to kill 63% of the cells is in the range of 30-40 double-strand breaks (DSB) or 1000 single-strand breaks (SSB). These numbers are much lower than those for the other agents listed in Table 1. The straightforward explanation is that both ionizing radiation and bleomycin have the capability of producing damage in opposite strands of DNA in close proximity to each other (7). This LMDS damage, one type of which is measured as DNA double-strand breaks (DSB) presents a severe challenge to a cell's repair system. We have pointed out that if the individual SSBs of a DSB are exactly opposite in the DNA then genetic information is irretrievably lost, since a base is released at the site of every SSB (8). In addition to SSBs, base damage sites exist together with strand breaks forming lesions which also constitute LMDS. These sites as well as DSB constitute a loss of sequence integrity.

Such LMDS will not be formed by the other agents listed in Table 1 (with the exception of bleomycin). These other agents also differ from ionizing radiation in the time course of chromosomal aberration production (9): they require a "DNA synthesis-dependent repair (misrepair)", whereas following ionizing radiation,chromosomal aberrations are present at the first mitosis and their presence does not depend on "misrepair" as defined for these other lesions.

The repair of individual damaged sites on DNA can be a relatively straightforward process for the cell. The presence of the other undamaged strand acts as a template for repair synthesis. However, with LMDS, depending on the juxtaposition of the constituent lesions, there arises the potential for loss of genetic information either as initial damage or as a consequence of attempted short patch repair (10). Therefore, LMDS have a much greater lethal potential than the individually damaged sites.

In other work we have observed the occurrence of degradation of cellular DNA. This we monitor by the neutral elution assay: In this assay of DNA DSB the rate at which "test" DNA elutes under neutral conditions (pH 7.2) from a polycarbonate filter is compared to that of standard irradiated DNA (11). In the course of studies of the effects of hypotonic salt treatment on radiation response of Chinese Hamster V79 cells we found that after incubation of these cells under hypotonic conditions at 0°C for 20 min, DNA degradation begins to occur upon exposure to full medium at 24°C. Immediately after hypotonic exposure, no degradation is apparent (Figure 1). As the time of incubation in full

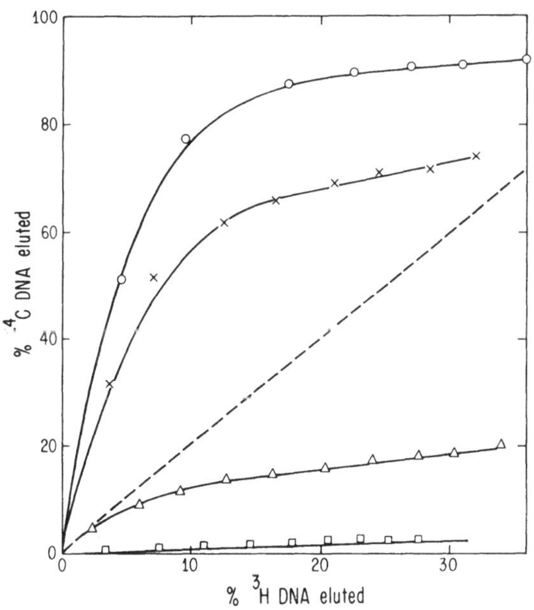

Fig. 1. Neutral elution profiles of the DNA (^{14}C-labelled) of V79 cells exposed to hypotonic salt (0.05 M phosphate-buffered saline at 0°C for 20 mins) after various times of subsequent incubation in full medium at 37°C. □:0 min, △:30 min, X:1 hr, O:3 hr. Comparison cells (^{3}H-labelled) given 60 gray radiation at 0°C. Dotted line represents elution curve of ^{14}C cells given 60 gray and no hypotonic treatment.

medium increases DNA degradation becomes apparent, i.e., the rate at which it elutes from the filter increases. In order to quantitate the extent of this degradation the rate of elution can be compared with the rate at which the DNA from cells irradiated with 60 gray elute (shown in Fig. 1). From such a comparision it is calculated that 3 hr after hypotonic treatment, 90% of the DNA elutes rapidly at a rate corresponding to a radiation dose of 200 gray. (8,000 DSB per cell).

We assayed the viability (clonogenic survival) of cells so treated and found it to be 28% of those which had not been exposed to the hypotonic salt treatment. Thus a significant fraction of the cells survive the presence of large numbers of DNA DSB which, had they been formed by ionizing radiation, would have been lethal. If the exposure to hypotonic salt was at 24°C no DNA fragmentation was observed. (other agents which lead to cellular DNA fragmentation, not caused directly by the agent, include: dimethyl sulfoxide (12) and polyethylene glycol (13)).

We have also examined the effect of the presence of these breaks on cell radiosensitivity. V79 cells were treated with hypotonic salt for 20 min at 0°C, transferred to full medium at 24°C and irradiated within 10 min of the transfer or 3 hr after the transfer, i.e., with and without the development of the double-strand cuts. Radiation survival curves are illustrated in Figure 2. The cell killing by the hypotonic treatment is

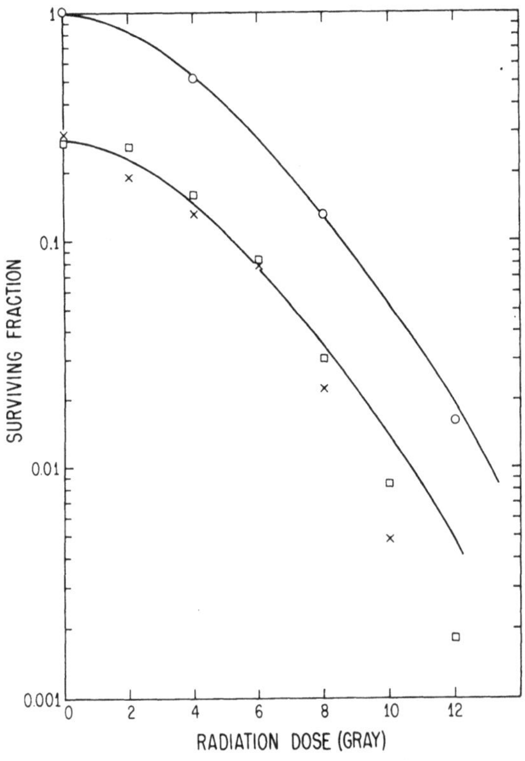

Fig. 2. Clonogenic survival curves of V79 cells irradiated after exposure to hypotonic salt (as in Fig. 1). □ After incubation in full medium for 20 min; X after 3 hr. O After exposure to 2-min hypotonic conditions at 0°C then 3 hr in full medium at 24°C. The solid lines are of the same shape, displaced vertically.

evident (0 dose), as is the lack of any major effects of the presence of the breaks on the survival parameters of the cells surviving the hypotonic treatment.

From these data we conclude that DNA double-strand cuts, probably induced enzymatically, are of lesser consequence than the DSB induced by ionizing radiation. The former represent clean sugar-phosphate bond cleavage, while in the latter case, each break is accompanied by a deoxynucleoside deletion. Hence repair of the enzyme-induced cuts only requires ligation, whereas the loss of sequence integrity in the latter case will not always be restored by closure of the DSB.

The term "misrepair" is used to describe the means by which errors are introduced into the genome following treatment by noxious agents. It is clear that "misrepair" can occur by mechanisms which are not the same for all types of damage. It is appropriate to consider the various possibilities which can lead to "misrepair", and perhaps to suggest which processes occur for the different damage types.

"Misrepair" can occur:

A. In replication of a singly damaged site (SDS) by:
   - The action of an error-prone enzymatic repair system.
   - The persistence of a damaged site through to S phase and its subsequent miscoding during replication.

B. If the damage is of the LMDS type where:
   - The sequence integrity has been lost from both DNA strands.
   - The sequence integrity will be lost during "normal" short-patch repair.

C. By the joining of a double-strand end with an incorrect partner.

It could be suggested that the process(es) of "misrepair" which are invoked in the conversion of SDS into damaging events may not be significant for ionizing radiation damage: The numbers of SDS necessary to cause death from chemical damage are several orders of magnitude greater than the yields of SDS present at doses of ionizing radiation necessary to kill. Therefore if "misrepair" is invoked for the lethal effects of chemically induced SDS, the frequency is relatively low. If this "misrepair" frequency is relatively lesion independent the yields of sites "misrepaired" would be low for ionizing radiation-induced damage. For ionizing radiation damage, errors in repaired DNA would be inherent in the damage itself (LMDS) rather than errors introduced by a repair system.

ACKNOWLEDGEMENT

This investigation was supported by PHS Grant number CA26279 awarded by the National Cancer Institute, DHHS. The authors thank Dr. J. J. McCormick for his assistance in tracking some of the information in Table 1.

REFERENCES

1. R. Roots and S. Okada, Protection of DNA molecules of cultured mammalian cells from radiation induced single-strand scissions by various alcohols and SH compounds. Int. J. Radiat. Biol. 21, 329-342 (1972).
2. J.F. Ward, W.F. Blakely, and E.I. Joner, Mammalian cells are not killed by DNA single-strand breaks caused by hydroxyl radicals from hydrogen peroxide. Radiat. Res. 103, 383-392 (1985).
3. M.O. Bradley and L.C. Erickson, Comparison of the effects of hydrogen peroxide and x-ray irradiation on toxicity and DNA damage/repair in mammalian cells (V79). Biochim. Biophys. Acta, 654, 135-141 (1981).
4. W.F. Blakely, D.P. Dodgen, E. Holwitt and M. Dizdaroglu, Chemical Characterization of DNA damaged by hydrogen peroxide. 34th Annual Meeting of the Radiat. Res. Soc., Atlanta, GA., Abstract Ek 2, (1987).
5. J.F. Ward, J.W. Evans, C.L. Limoli and P.M. Calabro-Jones, Radiation and hydrogen peroxide induced free radical damage to DNA. Br. J. Cancer 55, Suppl. VIII, in press (1987).
6. A. Mozumder and J.L. Magee, Models of tracks of ionizing radiations for radical reaction mechanisms. Radiat. Res. 28, 203-214 (1966).
7. J.F. Ward, Some biochemical consequences of the spatial distribution of ionizing radiation-produced free radicals. Radiat. Res. 86, 185-191 (1981).
8. J.F. Ward, Biochemistry of DNA lesions. Radiat. Res. 104, S103-S111 (1985).
9. A.T. Natarajan and G. Obe, Influence in DNA repair on radiation-induced chromosomal aberrations in human peripheral lymphocytes. In: Radiation Induced Chromosome Damage in Man (T. Ishihara and M.S. Sasaki, Eds.) pp. 127-140. Alan R. Liss, New York, 1983.
10. J.F. Ward, Mechanisms of DNA repair and their potential modification for radiotherapy. Int. J. Radiat. Oncol. Biol. Phys. 12, 1027-1032 (1986).
11. J.W. Evans, C.L. Limoli, and J.F. Ward, Does neutral elution measure intracellular levels of DNA double-strand breaks (DSB). 34th Meeting of the Radiat. Res. Soc., Las Vegas, Nevada, Abstract Cq 17 (1986).
12. J.W. Evans and J.F. Ward, Effects of dimethyl sulfoxide, a radioprotector on intracellular DNA. 33rd Annual Meeting of the Radiat. Res. Soc., Los Angeles, California, Abstract Eh 12 (1985).
13. J.F. Ward and C.L. Limoli, (unpublished).
14. M.M. Elkind and J.L. Redpath, Molecular and cell biology of radiation lethality. In: Cancer, A Comprehensive Treatise (F.F. Becker, Ed.), Vol. 6, pp. 51-99. Plenum Press, New York, 1977.
15. J.F. Ward, Ionizing radiation induced DNA damage: Identities and DNA repair. In: Mechanisms of DNA Damage and Repair (M.G. Simic, L. Grossman, and A.C. Upton, Eds.), pp. 135-138. Plenum Publishing, New York and London, 1986.
16. J.S. Lazo, I.D. Braun, B. Meandzija, K.A. Kennedy, E.T. Pham, and L.F. Smaldone, Lidocaine potentiation of bleomycin A2 cytotoxicity and DNA strand breakage in L1210 and human A-253 cells. Cancer Res. 45, 2103-2109 (1985).
17. M.O. Bradley and K.W. Kohn, X-ray induced DNA double-strand break production and repair in mammalian cells as measured by neutral filter elution. Nucleic Acids Res. 7, 793-804 (1979).
18. A.M. Rauth, Effects of UV light on mammalian cells in culture. Curr. Topics Radiat. Res. VI, 195-248 (1970).
19. R.B. Setlow, J.D. Regan, J. German, and W.L. Carrier, Evidence that xeroderma pigmentosum cells do not perform the first step in the repair of ultraviolet damage to their DNA. Proc. Natl. Acad. Sci. USA 64, 1035-1041 (1969).

20. J. Jagger, Ultraviolet inactivation of biological systems. In: Photochemistry and Photobiology of Nucleic Acids (S.Y. Wang, Ed.) Vol. II Biology, pp. 147-186. Academic Press, New York, 1976.
21. V.M. Maher, R.H. Heflich, and J.J. McCormick, Repair of DNA damage induced in human fibroblasts by N-substituted aryl compounds. In: NCI Monograph No. 58, "Carcinogenic and Mutagenic N-Substituted Aryl Compounds, 217-222 (1981).
22. E.M. Mahoney, J.C. Ball, D.H. Swenson, D. Richmond, V.M. Maher, and J.J. McCormick, Cytotoxicity and mutagenicity of aflatoxin dichloride in normal and repair deficient diploid human fibroblasts. Chem. Biol. Interact. 50, 59-76 (1984).
23. J.D. Patton, V.H. Maher, and J.J. McCormick, Cytotoxic and mutagenic effects of 1-Nitropyrene and 1-Nitrosopyrene in diploid human fibroblasts. Carcinogenesis 7, 89-93 (1986).
24. J.R. Connell and A.S.C. Metcalf, The induction of SCE with relation to specific base methylation of DNA in chinese hamster cells by N-Methyl-N-Nitrosourea and dimethylsulfate. In: Sister Chromatid Exchanges (R.R. Tice and A. Hollaender, Eds.) Part A, pp. 343-352. Plenum Press, New York and London, 1984.

RADON-INDUCED TRANSFORMATION

Eric J. Hall, Tom K. Hei and Gerhard Randers-Pehrson

Radiological Research Laboratory
College of Physicians & Surgeons, Columbia University
New York, NY 10032 USA

INTRODUCTION

The perception of radiation hazards in society must be completely re-evaluated as a result of the recent realization that radon represents the largest source of natural background radiation to the U.S. population (1). Studies designed to identify a link between lung cancer and radon in the home, by seeking a correlation between lung cancer incidence and variations in regional radon levels, have produced results that are equivocal. In epidemiological studies, uraninum miners exposed for many years to high radon levels in the mines show an enhanced incidence of lung cancer (2-4). The perceived risk to the public from breathing lower levels of radon in the home are based on an extrapolation from the high concentrations experienced by the miners. However, extrapolating from the uranium miner data involves complications because most miners studied were also heavy cigarette smokers. The number of non-smoking miners is too small to separate the effects of radon from those of the cigarette smoke.

The magnitude of the problem, and the complexity of interacting factors, emphasizes the urgent need for basic research on the biological effects of radon. The present report involves the use of an *in vitro* assay for oncogenic transformation to compare the effects of Helium-3 ions of high LET with X rays, and to investigate the interaction of this densely ionizing radiation with an environmental pollutant, namely asbestos fibers.

MATERIALS AND METHODS

The C3H 10T1/2 mouse embryo fibroblast cell line used for these studies was originally isolated and characterized by Reznikoff et al (5). Cells were maintained in Eagle's basal medium supplemented with 10% heat-inactivated fetal bovine serum (Hyclone Laboratories, Logan, UT.) and 25 µg/ml gentamycin (Gibco, Grand Island, N.Y.) Only cells from passage 8 through 13 were used for these studies. UICC standard reference samples of crocidolite were used and were prepared as described previously (6,7).

Helium-3 particles were accelerated in the 4-MeV Van de Graaff accelerator at the Radiological Research Accelerator Facility. These ions are virtually identical to the α-particles emitted by radon daughter products. A beam of helium-3 ions was directed vertically through the Mylar bottom of specially constructed irradiation dishes. Cells were grown as monolayers attached to the Mylar which had an average thickness of 6 μm. The details of the physical arrangement for track-segment irradiation has been published previously (8). Briefly, selected LETs were obtained by adjusting the accelerator energy such that charged particles of the appropriate energies were directed onto the irradiated cells. The specimen wheel had a capacity for 20 dishes and was remotely rotated by a stepping motor so that an individual dish was passed over a slot-shaped beam at a rate determined by both the dose to be delivered and the instantaneous dose rate of the beam. As a control low-LET radiation, X rays were used, generated by a Westinghouse X-ray machine operating at 250 kVp and 15 mA with a 0.5mm copper and a 1-mm Aluminum filter. Cells were irradiated at a distance of 50 cm where the dose rate was 125 rad per minute.

For the transformation assay, exponentially growing C3H 10T1/2 cells on Mylar dishes were irradiated with graded doses of helium-3 particles or X rays. Following irradiation, cells were trypsinized, counted and plated in 100-mm diameter petri dishes at a density such that approximately 300-400 viable cells would survive the treatments. In studies where the interaction between asbestos fibers and radiation were to be examined, the cultures were pretreated with 5 μg/ml crocidolite fibers for 24 hours followed by exposure to X rays or helium-3 ions. All treated and control cultures were maintained for a total of 6 weeks, with the medium changed every 10 days. The cultures were then fixed, stained and type II and III foci scored as transformants as described previously (5). All dishes were coded and scored independently twice. Previous studies from this laboratory have shown that cells from transformed foci were anchorage independent and tumorigenic upon inoculation into athymic nude mice. Corresponding dishes were plated at lower cell densities to determine the cytotoxicity of the two agents.

RESULTS AND DISCUSSIONS

The radiation hazard from exposure to radon in the atmosphere results from decay of its short-lived alpha-emitting daughters, the radio-nuclides Po-218 and Po-214. The initial energies of the alpha-particles from these radio-nuclides are 6 MeV and 7.7 Mev, with

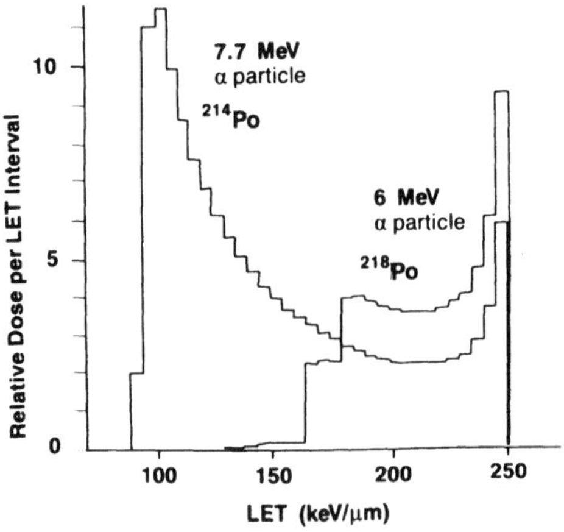

Figure 1. Spectra of LET values of the alpha particles emitted by Po-214 and Po-218.

corresponding Linear Energy Transfer (LETs) of about 80 and 65 keV/μm. Figure 1 shows the spectra of LET values as these α-particles are absorbed in tissue.

The critical cells for lung cancer induction are presumably the basal cells of the bronchial epithelium, the shallowest of which are 22 μm below the epithelial surface and are further protected by a mucus layer of about 15 μm thickness. Consequently, an alpha particle from a radioactive nucleus deposited on the mucosal surface must penetrate at least 37 μm before reaching the target cell (1). Greater distances must be travelled by particles that are not normal to the surface. As a consequence of these considerations, the critical cells in the bronchial epithelium will be exposed to very densely ionizing particles with LET values in the range 95 to 250 keV/μm. This is illustrated in Figure 2.

Figure 3 shows survival curves for C3H 10T1/2 cells irradiated with either X rays or helium-3 ions, chosen to simulate the α-particles from radon daughters.

The survival curve for X rays has a characteristic initial shoulder region before becoming steeper and approaching an exponential function of dose at higher doses. By contrast, the cell-survival curve for the high-LET helium ions, approximates closely to an exponential function of dose, indicating single-hit kinetics - i.e., the cell is killed by the traversal of a single particle. However, at a dose corresponding to a surviving fraction of 37%, about 14 particles traverse the nucleus of each cell killed (9). C3H 10T1/2 cells are quite large when stretched out and attached to a culture dish, with an average nuclear cross sectional area of about 240 μm^2 (10). This conclusion is quite interesting. While one particle kills the cell, on average about 13 particles traverse the cell without killing it for every one that does! This may be a clue to the high efficiency with which high LET-particles induce transformed foci.

Transformation data are most commonly plotted in terms of transformation incidence per surviving cell as a function of dose. This is convenient since this is the way in which various transformation data are compared. However, in applying these <u>in vitro</u> data to whole organisms, a more relevant quantity is transformation incidence per

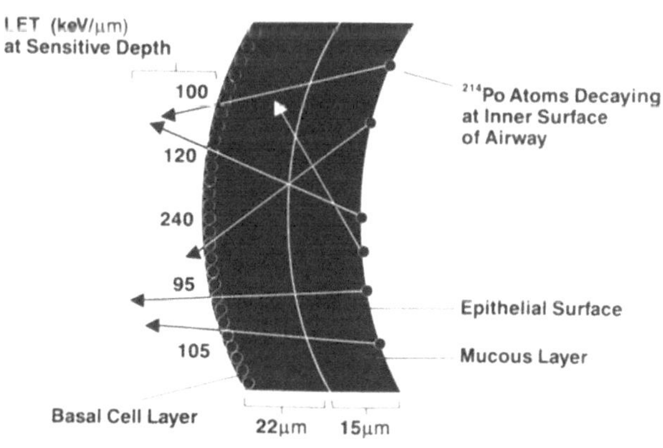

Figure 2. Illustrating that radon daughter products deposited on the surface of the lung will result in irradiation with very high LET particles of the basal cells which are at risk of producing cancer.

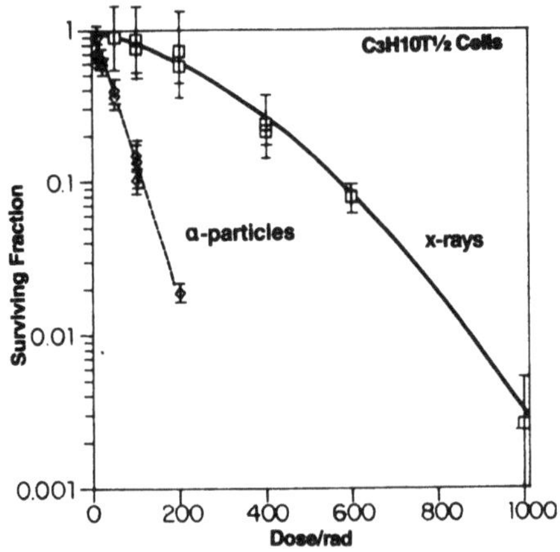

Figure 3. Survival curves for C3H T101/2 cells exposed to graded doses of either X ray, or helium-3 ions.

initial cell at risk. This is shown for X rays and radon in Figure 4. The overall curve shape reflects the combined effects of transformation incidence and cell lethality.

In the case of X rays, transformation incidence per initial cell at risk rises with dose at low doses, reaches a maximum at a few hundred rad, and subsequently falls at higher doses as cell killing becomes dominant, eventually becoming parallel with the cell survival curve. A dose-response relationship of this general shape has been observed for a number of tumors in experimental animals induced by whole-body irradiation. In the case of the high LET-particles, transformation incidence per initial cell at risk rises very sharply indeed at low doses, reaches a narrow maximum at about 50 rad, and subsequently falls rapidly to parallel the steep, shoulderless cell-killing curve.

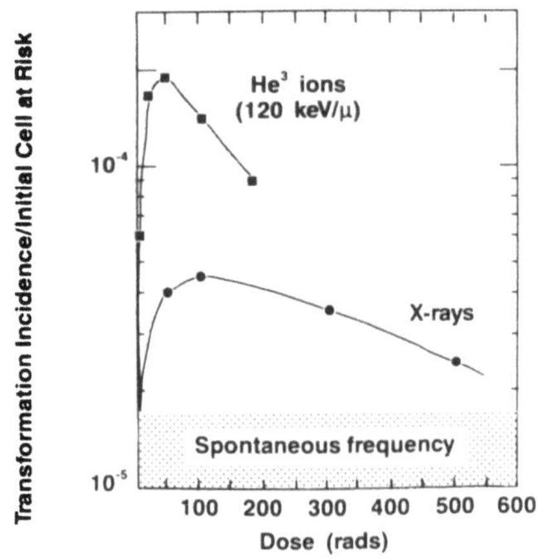

Figure 4. Transformation incidence expressed per initial cell at risk for graded doses of X rays and helium-3 ions.

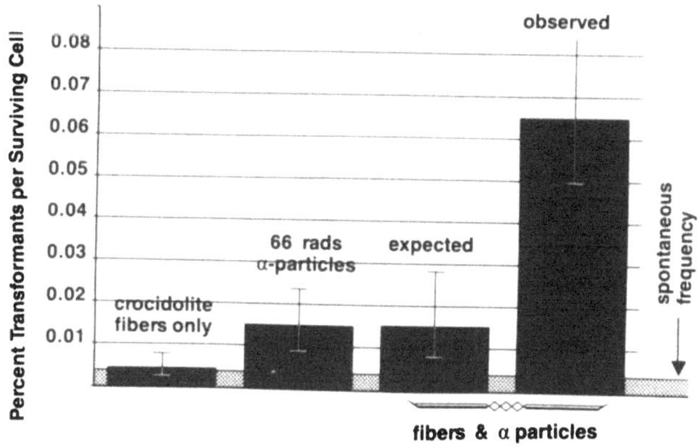

Figure 5. Transformation incidence in C3H 10T1/2 cells for asbestos fibers and helium-3 ions, delivered alone or in combination. Transformation appears to be supra-additive when asbestos and high-LET particles are delivered together.

However, the interesting point is that at a dose of about 50 rad, the transformation incidence per initial cell at risk for helium-3 ions is five-fold higher than can be achieved by any dose of X rays. This is presumably a consequence of a differing balance between the induction of transformed cells and the killing of transformed cells. High-LET particles are much more effective at cell killing than are X rays, but they are even more efficient at inducing transforming events.

Experiments have also been performed to investigate an interaction between densely ionizing particles and asbestos fibers. Figure 5 shows transformation data for 10T 1/2 cells resulting from a 24-hour pretreatment with 5 µg/ml crocidolite fibers, or a dose of 66 rad helium-3 ions or a combination of both. As shown previously, crocidolite fibers, at concentrations that resulted in only moderate cell killing, induced transformation frequencies indistinguishable from the spontaneous rate (6,7). However, cells pretreated with crocidolite fibers for 24 hours and subsdequently irradiated with α -particles exhibited a significantly higher transformation incidence than after radiation alone. The combination of asbestos fibers and high-LET particles produces a transformation incidence higher than the simple sum of the two agents alone. This apparent supra-additivity has implications for the possible interaction of radon and asbestos in the enviroment, and possibly for combinations of other environmental pollutants.

ACKNOWLEDGEMENTS

This investigation was supported by Contract No. DE-FG02-86ER60402 from the Department of Energy and Research Grant CA 12536 from the National Cancer Institute.

REFERENCES

1. NCRP Report No. 78, Evaluation of occupational and environmental exposures to radon and radon daughters in the U. S. National Council on Radiation Protection and Measurements, Washington, D.C., 1984.

2. W. J. Bair, Inhalation of radionuclides and carcinogenesis. In: Inhalation Carcinogenesis (M.G. Hanna, P. Nettesheim and J.R. Gilbert, Eds.), pp. 77-97. U.S. Atomic Energy Commission, Washington, D.C., 1970.
3. F. E. Lundin, J. W. Lloyd, E. M. Smith, V. E. Archer and D. A. Holaday, Mortality of uranium miners in relation to radiation exposure, hard-rock mining and cigarette smoking, 1950 through September 1967. Health Phys. 16, 571-578 (1969).
4. A. S. Whittemore and A. McMillan, Lung cancer mortality among U.S. uranium miners; A reappraisal. J. Natl. Cancer. Inst. 71, 489-493 (1983).
5. C. A. Reznikoff, J. S. Bertram, D. W. Brankow and C. Heidelberger, Quantitative and qualitative studies on chemical transformation of cloned C3H mouse embryo cells sensitive to post confluence inhibition of cell division. Cancer Res. 33, 3239-3249 (1973).
6. T. K. Hei, E. J. Hall and R. Osmak, Asbestos, radiation and oncogenic transformation. Br. J. Cancer 50, 717-720 (1984).
7. T. K. Hei, C. R. Geard, R.S. Osmak and M. Travisano, Correlation of <u>in vitro</u> genotoxicity and oncogenicity induced by radiation and asbestos fibers. Br. J. Cancer 52 591-597 (1985).
8. R. P. Bird, N. Rohrig, R. D. Colvett, C. R. Geard and S. A Marino, Inactivation of synchronized Chinese hamster V79 cells with charged particles track segments. Radiat. Res. 82, 277-289 (1980).
9. E. L. Lloyd, M. A. Giemmell, C. B. Henning, D. S. Giemmell and B. J. Zabransky, Cell survival following multiple track alpha particle irradiation. Int. J. Radiat. Biol. 35, 23-31 (1979).
10. C. R. Geard and T. Harding, Morphometric studies with attached mouse C3H 10T1/2 cells. In: Radiation Physcis, Biophysics and Radiation Biology. U.S. DOE Progress Report, 1982. Columbia University, pp. 274-280.

RADIATION-INDUCED NEOPLASTIC TRANSFORMATION OF HUMAN CELL HYBRIDS

J.L. Redpath

Department of Radiological Sciences
University of California, Irvine
Irvine, CA 92717

INTRODUCTION

In vitro cell systems are well established in the study of mechanisms of neoplastic transformation by a variety of agents including chemicals, ultraviolet light and ionizing radiation (for recent reviews see references 1-4). However, with few exceptions, systems that are used for quantitative studies are rodent-derived cell lines, usually of fibroblastic origin. This situation exists for the very good reason that human cells in culture are extremely difficult to transform, and where this has been achieved the frequency of transformation is three to four orders of magnitude less than that typically seen for rodent cells (2). There is an obvious need for a human cell system(s) for use in *in vitro* investigations of neoplastic transformation.

In this regard we have employed a human cell hybrid line (HeLa x skin fibroblast) as a model system for quantitative and mechanistic studies of transformation from the preneoplastic to the neoplastic state. These cells are well-established in studies of the genetic aspects of malignancy (5) and possess certain desirable properties. First, the cells are of human origin. Second, tumorigenic segregants of these cells possess characteristics of epithelial cells and produce undifferentiated carcinomas when injected into nude mice. Third, the target cells are capable of proliferation which allows for fixation and expression of carcinogen-induced damage. Fourth, and very importantly, the tumorigenic cells derived from the non-tumorigenic cell hybrids express a specific cell surface protein of molecular weight 75kD (6) for which monoclonal antibodies have been raised. Thus we can use immunofluorescence or immunoperoxidase staining techniques to identify colonies or foci of tumorigenic cells. Furthermore, the hybrid line we are using is extremely stable against spontaneous formation of tumorigenic segregants (frequency $<< 10^{-5}$). Finally, it is important to recognize that the non-tumorigenic cell hybrids possess many of the characteristics of tumorigenic cells and should be regarded as preneoplastic. Thus this system is useful for studying the transformation from the preneoplastic to the neoplastic state.

In this paper we report data showing that ionizing radiation can neoplastically transform these cells and that a reproducible dose-response relationship has been established. A preliminary report on some of the data presented below has been published elsewhere (7).

MATERIALS AND METHODS

Cell Line and Culture Conditions

The cell line used in this study is designated as CGL1 and was obtained from the third serial subclone of the hybrid line ESH5 in methylcellulose (8). This cell line does not form tumors when inoculated into nude mice and does not express the cell surface protein characteristic of tumorigenic cells. Culture conditions are described elsewhere (7).

Irradiations

Irradiations were carried out on a Shepherd and Associates Mark I self-shielded Cesium-137 gamma irradiator at a dose rate of 2 gray/min.

Morphology Assay

The morphology of the tumorigenic segregants is epithelial-like while that of the non-tumorigenic hybrids is more fibroblastic in nature (5). The tumorigenic cells form small, tight colonies which are generally round in appearance and stain darkly. The non-tumorigenic cells on the other hand form larger, loosely-packed colonies which often have spiky edges and stain lighter. It is thus possible to distinguish tumorigenic segregants based on morphologic appearance and staining density. However, this process requires skill which takes some time to acquire and is very time-consuming and labor intensive. Furthermore, it is certainly subject to error and may tend to underestimate yields if only glaringly obvious tumorigenic colonies are scored.

Immunoperoxidase Assay

A standard immunoperoxidase staining procedure was used and employed 3-amino-9-ethylcarbazole in dimethylformamide as the chromogenic substrate. Oxidation of the aminoethylcarbazole produces a reddish-brown water-insoluble precipitate. Colonies or foci containing cells positive for the 75kD protein appear brown against a clear background of unstained cells. This makes scoring the dishes for positive colonies extremely easy. Further details are described elsewhere (7).

Tumorigenicity Assay

Congenitally athymic nu/nu (nude) mice were used as the in vivo assay system with which to determine the neoplastic nature of the hybrid cell populations. The validity of this assay system has been demonstrated (9,10).

Experimental Protocol

The protocol for the experiments to be reported was as follows. Briefly, the cells were irradiated at or near confluence (late log phase) which represented densities of 2 to 5 x 10^6 cells per T-25 flask. They were then replated six hours later at densities typically in the range 600-1500 cells per T-25 flask. We often replated into T-75 flasks to save on labor and costs involved with immunoperoxidase staining. The flasks were then incubated for 10 to 25 days depending upon which assay (morphology or immunoperoxidase) was being used. However, some flasks were always stained at about 10 days in order to assay for survival.

RESULTS

Radiation-induced Tumorigenic Cells

Two separate colonies have been isolated from populations of irradiated cells which appeared to be tumorigenic on the basis of morphology. Populations derived from these colonies were subsequently shown to grow tumors in nude mice which upon histopathologic examination proved to be poorly differentiated carcinomas. Furthermore, these population of cells stained positively for the 75kD cell surface protein using both immunofluorescence and immunoperoxidase techniques.

Radiation Dose-Response Curves

Radiation dose-response data for both survival and transformants per cell exposed are shown in Figure 1. The incidence of transformation

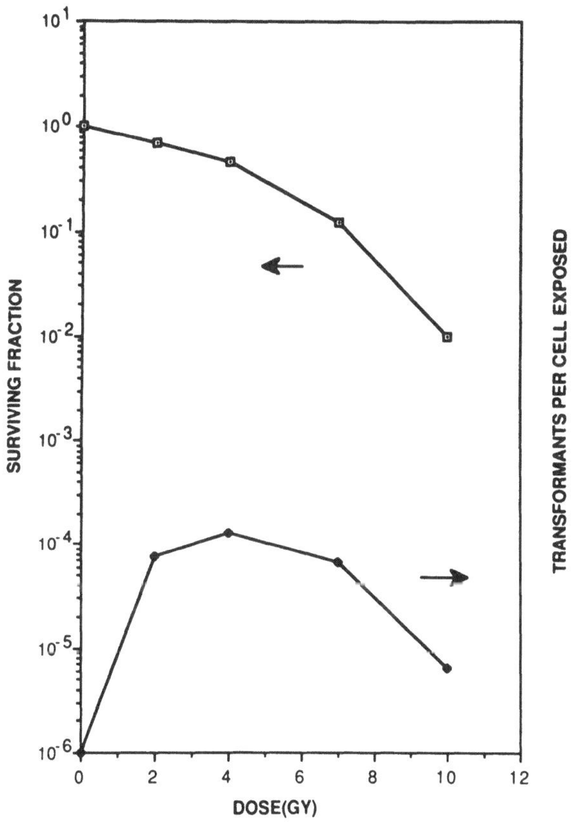

Fig. 1. Surviving fraction and transformants per cell exposed as a function gamma radiation dose to the human cell hybrid line CGL1.

increases at lower doses followed by a decrease at higher doses at the same rate as cell kill. This is a pattern typical of that seen with other systems, e.g. C3H10T-1/2 cells (11). In these experiments transformants were scored using immunoperoxidase staining at 17 to 19 days post-irradiation. At this time the cells are becoming confluent in the dishes so we are scoring foci of transformed cells.

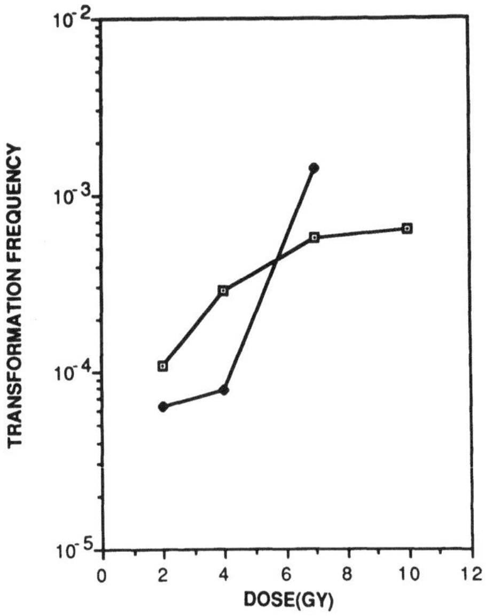

Fig. 2. Transformation frequency as a function of gamma radiation dose to the human cell hybrid line CGL1.
(◆) Morphology assay (□) Immunoperoxidase assay

We have also carried out a limited number of experiments using the morphology assay and the data obtained are shown in Figure 2 alongside those obtained using the immunoperoxidase assay for comparative purposes. Considering the uncertainty associated with the morphology assay (see Materials and Methods), these data are in reasonably good agreement.

Effect of Bicarbonate versus HEPES Buffer

Sodium bicarbonate (1.5 g/l, pH = 7.22) is the standard buffer solution that we use in our experiments. In an effort to avoid some problems we temporarily experienced with precipitation in our medium, we ran some experiments with HEPES buffer (5.9 g/l HEPES plus 1.5 g/l sodium bicarbonate, pH = 6.81). However, the transformation frequencies we observed with HEPES buffer were consistently lower than with bicarbonate (see Table 1). We do not have an explanation for this at present other than it may be pH-related since the two buffer conditions did result in different media pH values as indicated above and in Table 1.

Table 1. The Effect of Sodium Bicarbonate versus Hepes Buffer on the Gamma Irradiation-induced Transformation of Human Cell Hybrids

Buffer	pH	No. of cells at risk	No. of positively staining foci	Transformation frequency
Sodium Bicarbonate (1.5 G/1)	7.22	135900	118	8.7×10^{-4}
HEPES*	6.81	84416	8	9.5×10^{-5}

*N-2-Hydroxyethylpiperazine-N'-2-Ethanesulfonic Acid

DISCUSSION

We have demonstrated that we can quantitate the transformation of the human cell hybrid line CGL1 from its non-tumorigenic state to a state where it expresses a cell surface protein that correlates with tumorigenicity. We are in the process of establishing various aspects of the assay which allow for maximizing the yield of transformants. These aspects relate to the appropriate number of cells plated post-irradiation, the optimal time of post-irradiation subculture and the optimal time of assay. In addition, our data with HEPES buffer indicate that medium pH may be an important determinant of transformation frequency. We will further investigate this observation.

In summary, human cell hybrids such as those used in this study represent an addition to the library of cell systems which can be used to explore mechanisms of neoplastic transformation by radiation and other agents. It is stressed once more that the hybrid cell line we are using allows us to look at the transformation from the preneoplastic to the neoplastic state. In the future we plan to expand these studies to other hybrid lines including microcell hybrids, which contain a single human chromosome which is suppressing the expression of tumorigenicity (12). Studies with cell systems such as these should lead to a further understanding of mechanisms involved in the neoplastic transformation of human cells by radiation and other agents and hence to methods of preventing this process.

ACKNOWLEDGEMENTS

The technical assistance of M. Travisano, the experimental expertise and useful discussion of Dr. C. Sun, and the cooperation and support of Dr. M. Colman are gratefully acknowledged. We are indebted to Dr. E.J. Stanbridge for provision of the cell lines and expert advice. This work is supported by Grant No. CA39312 from NIH/U.S.D.H.H.S.

REFERENCES

1. J.C. Barrett and D.G. Thomassen, Use of quantitative cell transformation assays in risk estimation. In: Methods for Estimating Risk of Chemical Injury: Human and Non-human Biota & Ecosystems (V.B. Vouk, G.C. Butler, D.G. Hoel, and D.B. Peakall, Eds.), pp. 201-234. Scope, 1985.
2. C. Borek, The induction and control of radiogenic transformation in vitro: Cellular and molecular mechanisms, Pharmacol. Ther. 27, 99-142 (1985).
3. G.L. Chan and J.B. Little, Neoplastic transformation in vitro. In: Radiation Carcinogenesis (A.C. Upton, R.E. Albert, F.J. Burns, and R.E. Shore, Eds.), pp. 107-136. Elsevier, New York, 1986.
4. E.J. Hall and T.K. Hei, Oncogenic transformation of cells in culture: Pragmatic comparisons of oncogenicity, cellular and molecular mechanisms, Int. J. Radiat. Oncol. Biol. Phys. 12, 1909-1921 (1986).
5. E.J. Stanbridge, C.J. Der, C-J Doersen, R.Y. Nishimi, D.M. Peehl, B.E. Weissman, and J.E. Wilkinson, Human cell hybrids: Analysis of transformation and tumorigenicity, Science 215, 252-259 (1982).
6. C.J. Der and E.J. Stanbridge, A tumor-specific membrane phosphoprotein marker in human cell hybrids, Cell 26, 429-438 (1981).
7. J.L. Redpath, C. Sun, M. Colman, and E.J. Stanbridge, Neoplastic transformation of human hybrid cells by gamma radiation: A quantitative assay. Radiat. Res. 110, 468-472 (1987).
8. E.J. Stanbridge and J. Wilkinson, Dissociation of anchorage independence from tumorigenicity in human cell hybrids, Int. J. Cancer 26, 1-8 (1980).

9. E.J. Stanbridge, L.R. Boulger, C.R. Franks, J.A. Garrett, D.E. Reeson, D. Bishop, and F.T. Perkins, Optimal conditions for the growth of malignant and human animal cell populations in immunosuppressed mice, Cancer Res. 35, 2203-2212 (1975).
10. E.J. Stanbridge, Suppression of malignancy in human cells, Nature (London) 260, 17-20 (1976).
11. A. Han and M.M. Elkind, Transformation of mouse C3H10T1/2 cells by single and fractionated doses of X-rays and fission spectrum neutrons, Cancer Res. 39, 123-130 (1979).
12. P.J. Saxon, E.S. Srivatsan, and E.J. Stanbridge, Introduction of human chromosome 11 via microcell transfer controls tumorigenic expression of the HeLa cells, EMBO J. 5, 3461-3466 (1986).

MULTIPLE ONCOGENE ACTIVATION IN A RADIATION CARCINOGENESIS MODEL*

Seymour J. Garte, Mary Jean Sawey, Fredric J. Burns, Mary Felber, and Tracy Ashkenazi-Kimmel

Institute of Environmental Medicine
New York University Medical Center
550 First Avenue
New York, New York 10016

Radiation is the most thoroughly studied environmental carcinogenic agent and cancer induction is the most significant late effect of radiation. Ionizing radiation creates a number of lesions in DNA including base modification, single and double strand breaks, and sugar damage. A significant proportion of the radiation-induced mutations in somatic cells are associated with chromosomal deletions and gene rearrangements. These alterations are the result of breakage of the chromosomes followed by loss of genetic material (deletions) or rejoining of broken chromosomes (rearrangements) (1). Radiation has also been found to be a point mutagen in certain model assay systems. The relationship of these genotoxic effects of radiation to molecular mechanisms of radiation carcinogenesis is not yet understood. We have decided to address this issue using current molecular concepts involving oncogene activation applied to the biologically well-defined experimental model of rat skin carcinogenesis by ionizing radiation. Several types of tumors develop in rat skin following localized exposure to single or fractionated doses of ionizing radiation (2) which include squamous cell carcinomas, basal cell carcinomas, sarcomas, clear cell carcinomas, and sebaceous cell tumors. The histologic type of the tumor presumably reflects the cell type of origin.

Carcinogenesis is undoubtedly a progressive process involving multiple, independent steps (3). At the cellular level, establishment of the transformed phenotype may also be a multi-step process and activation of several, independent genes appears to be required. Land et al. (4) have used primary rat embryo fibroblast cells to show that two activated cellular oncogenes, c-ras and c-myc, may collaborate to produce morphologically transformed cells with tumorigenic potential.

There is evidence from animal systems to suggest that certain oncogenes may be activated by the direct action of the initiating carcinogen. Consistent activation by a point mutation of a single member of the ras oncogene family in different tumors produced by a single agent has been demonstrated in several experimental models (5-8). In contrast to the ras family, c-myc and other oncogenes have been shown to be activated by a process involving chromosomal translocations, enhanced expression, and/or gene amplification (9).

* Supported by NIH Grant No. CA43199.

We have examined a panel of 12 late stage rat skin tumors for activation of oncogenes from the *ras* and *myc* complementation groups (10). These tumors were of the following histologic types: four squamous cell carcinomas, three poorly differentiated carcinomas (clear cell), one each of basal cell carcinoma, sebaceous carcinoma, sarcoma, fibroma, and mixed (largely squamous) histology carcinoma.

Tumor DNA was purified by phenol extraction and transfected into NIH3T3 cells by the calcium phosphate coprecipitate method (10). The transfection efficiency of the six positive radiation tumor DNAs ranged from 0.020 to 0.110 foci/µg DNA. The positive tumor DNAs were from three poorly differentiated clear cell carcinomas, a sebaceous carcinoma, a squamous cell carcinoma, and a sarcoma. Data from these experiments are shown in Table 1. DNA from one of the primary transfectants was positive in a second round of transfection. The transformed phenotype of the transfectants was confirmed by anchorage independent growth and tumorigenicity in nude mice. Southern blot analysis of DNA from primary and secondary transfectants, as well as from nude mouse tumors arising after injection of transfectant cells revealed the presence of rat derived restriction fragments homologous to the K-*ras* oncogene against the mouse background. Similar experiments using N- and H-*ras* probes, revealed only the endogenous mouse fragments in transfectant DNA.

Southern hybridization of the original radiation induced tumor DNAs to a third exon human c-*myc* probe revealed gene amplification and extensive restriction fragment polymorphisms in 10 of the 12 tumors after digestion with Bam HI, Eco RI, or Hind III. The c-*myc* gene was amplified 5- to 20-fold by densitometric analysis. When the tumor DNAs were hybridized to the first exon of c-*myc*, there were no differences in either band intensity or restriction fragment pattern between tumor and normal rat DNA. This result implies that an internal rearrangement of the c-*myc* gene had occurred in these tumors as has been seen in other tumor systems, such as Burkitts lymphoma. Hybridization of the tumor DNAs to v-*abl*, c-*myb*, K-*ras*, or the first and second exons of N-*myc* showed no differences in band intensity or number.

In an attempt to determine whether the extensive pattern of restriction polymorphisms seen with c-*myc* was due to activation of another related gene, a series of hybridizations were carried out at various stringencies. The results from these experiments indicate that all five polymorphic bands are closely homologous to the c-*myc* 3rd exon, and that the higher molecular weight bands may represent more altered forms of the normal *myc* gene than the smaller size bands.

Northern and dot blot hybridization of poly A^+ RNA isolated from normal rat epidermis and 6 rat skin tumors revealed evidence for enhanced c-*myc* expression in amplification positive tumors. The levels of c-*myc* gene expression in the tumors were 2- to 6-fold higher than control rat epidermal levels.

No correlation was seen between any parameters of oncogene activation and age of animals, tumor growth rate, or dose or dose rate of radiation.

A summary of *myc* and *ras* oncogene activation in radiation-induced rat skin tumors is presented in Table 1. There appears to be a tissue-specific component in oncogene activation in this system. Concurrent activation of K-*ras* and c-*myc* oncogenes was found in four of the twelve tumors examined. Each of the poorly differentiated clear cell carcinomas and a sebaceous carcinoma exhibited activation of both *ras* and *myc* oncogenes, whereas *myc* alterations alone were observed in 4 out of 5 differentiated squamous cell carcinomas. A sarcoma was found to have an activated K-ras, but not an activated c-myc gene, whereas a benign fibroma contained only an activated c-myc oncogene. As mentioned above, there was a positive correlation between myc oncogene amplification and myc gene expression.

The finding of concurrent multiple oncogene activation in a panel of primary tumors induced by a single etiological agent provides evidence in favor of the hypothesis that activation of oncogenes from two or more complementation groups may contribute to carcinogenesis in animals and man.

TABLE 1. Oncogene Activation in Radiation Induced Tumors

		Transfection of Tumor DNA in NIH3T3 Cells		
Tumor Number	Tumor Type	No. Foci/ ug DNA[a]	K-ras in Foci	myc Amplification
NIH3T3	----	<0.003		
T24	Human Bladder Carcinoma	0.102		
RAD 1	Poorly Differentiated Clear Cell Carcinoma	0.100	N.D.[b]	+
RAD 2	Well Differentiated Cornified Squamous Cell Carcinoma	<0.005	-	+
RAD 3	Basal Cell Carcinoma	<0.005	-	+
RAD 4	Sebaceous Carcinoma (necrotic)	0.019	+	+
RAD 5	Poorly Differentiated Clear Cell Carcinoma	0.051	+	+
RAD 6	Fibroma	0.005	-	+
RAD 7	Squamous Cell Carcinoma	<0.010	-	+
RAD 8	Clear Cell Carcinoma	0.050	+	+
RAD 9	Sarcoma	0.020	+	-
RAD 10	Squamous Cell Carcinoma	0.020	+	-
RAD 11	Mixed Histology Carcinoma	<0.010	-	+
RAD 12	Squamous Cell Carcinoma	<0.010	-	+

a Tumor DNA (30 μg) was transfected into NIH3T3 cells by calcium phosphate co-precipitation (37). At least 9 replicate plates were scored for each tumor DNA.

b Not determined.

In order to further characterize the molecular alterations associated with activation of c-myc in radiation-induced tumors, we have recently cloned a c-myc gene from one of the squamous cell carcinomas showing amplification and restriction polymorphisms in c-myc. A genomic DNA library (using EMBL 4 bacteriophage lambda) of over 200,000 recombinants was screened with a 3rd exon c-myc probe. DNA from 25 positive plaques was further analyzed by restriction enzyme mapping and Southern analysis. A complete restriction map of the cloned gene is being constructed.

The apparent nature of the mechanism of activation of the c-myc gene (rearrangements and amplification) is consistent with the expected consequences of the clastogenic action of ionizing radiation. We are currently examining the c-myc oncogene in panels of small tumors arising shortly after irradiation and in clones of surviving cells immediately after irradiation. Preliminary data indicate that some degree of c-myc gene alterations are present in early tumors, but that a role for later stages of tumor progression in activation of this gene cannot be ruled out.

REFERENCES

1. H.P. Leenhouts and K.H. Chadwick, Radiation induced DNA double strand breaks and chromosome aberrations. Theor. Appl. Genet. 44, 167-172 (1974).
2. R.E. Albert, P. Phillips, P. Bennett, F. Burns, and R. Heimbach, The morphology and growth characteristics of radiation-induced epithelial skin tumors in the rat. Cancer Res. 29, 658-668 (1969).
3. R.K. Boutwell, The function and mechanism of promoters of carcinogenesis. Crit. Rev. Toxicol. 2, 419-443 (1974).
4. H. Land, L.F. Parada, and R.A. Weinburg, Tumorigenic conversion of primary embryo fibroblasts requires at least two cooperating oncogenes. Nature 304, 596-602 (1983).
5. H. Zarbl, S. Sukumar, A.V. Arthurs, D. Martin-Zanca, and M. Barbacid, Direct mutagenesis of Ha-ras-1 oncogenes by N-nitroso-N-methylurea during initiation of mammary carcinogenesis in rats. Nature 315, 382-385 (1985).
6. M. Quintanilla, K. Brown, M. Ramsden, and A. Balmain, Carcinogen-specific mutation and amplification of Ha-ras during mouse skin carcinogenesis. Nature (London) 322, 78-80 (1986).
7. I. Guerrero, A. Villasante, V. Corces, and A. Pellicer, Activation of a c-K-ras oncogene by somatic mutation in mouse lymphomas induced by gamma radiation. Science 225, 1159-1162 (1984).
8. S.J. Garte, A.T. Hood, A.E. Hochwalt, P. D'Eustachio, C.A. Snyder, A. Segal, and R.E. Albert, Carcinogen specificity in activation of transforming genes by direct-acting alkylating agents. Carcinogenesis 6, 1709-1712 (1985).
9. R. Dalla-Favera, S. Martinotti, R.C. Gallo, J. Erikson, and C.M. Croce, Translocation and rearrangements of the c-myc oncogene locus in human undifferentiated B-cell lymphomas. Science 219, 963-967 (1983).
10. M.J. Sawey, A.T. Hood, F.J. Burns, and S.J. Garte, Activation of c-myc and K-ras oncogenes in priamry rat tumors induced by ionizing radiation. Moled. Cell. Biol. 7, 932-935 (1987).
11. M. Wigler, A. Pellicer, S. Silverstein, R. Axel, G. Urlaub, and L. Chasin, DNA-mediated transfer of the adenine phosphoribosyltransferase focus into mammalian cells. Proc. Natl. Acad. Sci. 76, 1373-1376 (1979).

RADIATION AND ASBESTOS FIBERS: INTERACTION AND POSSIBLE MECHANISM

Tom K. Hei and Steve Kushner

Radiological Research Laboratory
College of Physicians and Surgeons, Columbia University
New York, NY 10032

INTRODUCTION

The association between exposure to asbestos fibers and the development of bronchogenic carcinoma of the ~~lung~~ and diffuse mesothelioma of the pleura and peritoneum have been well established in both man (1,2) and experimental animals (3,4). Furthermore, cigarette smoking can enhance the incidence of lung cancers in asbestos workers in a synergistic fashion. (5). It is not known for certain whether specific chemicals or the high LET α-irradiation from deposited polonium is the important factor in carcinogenesis by cigarettes and its enhancement by asbestos. Previous studies from this laboratory have found that asbestos fibers, at a concentraction which itself was ineffective in inducing oncogenic transformation <u>in vitro</u> did potentiate the oncogenicity of low LET ionizing radiation in a supra-additive fashion (6,7). Furthermore, leaching the fibers with acid to remove any possible chemical contaminants did not diminish their ability to enhance the oncogenic transforming potential of radiation (7). These results suggest that the mere physical presence of the fibers is crucial for the observed radiation-fiber interaction.

The cellular and molecular basis of the apparent supra-additivity between radiation and asbestos fibers in the induction of transformation is not known. The correlation between fiber dimension and carcinogenicity however, suggests the importance of fiber-cell interactions (8). The ability of cells to phagocytize asbestos fibers both <u>in vivo</u> and <u>in vitro</u> has been well documented (9,10). Fibers longer than 25 μm in length are generally not completely phagocytized by cells and common channels have been demonstrated between the extracellular and intracellular spaces at sites of interaction (11). The involvement of superoxide anions in asbestos-induced cytotoxicity has been suggested (12) and the radical scavenging enzyme, superoxide dismutase (SOD) has been shown to protect against the toxicity of asbestos <u>in vitro</u> (13). The utilization of superoxide dismutase applied either extracellularly or intracellulary via carrier molecules will undoubtedly help to clarify the mechanism by which asbestos induces tumors at the cellular level. In the present studies, the possible protective effects of SOD in the transformation incidence induced by both asbestos fibers and radiation were examined in an attempt to elucidate the role of oxygen free-radicals in the interaction.

METHODS AND MATERIALS

C3H 10T1/2 mouse embryo fibroblasts isolated and characterized by Reznikoff were used (14). These cells exhibit contact inhibition of growth and anchorage dependence under normal culture conditions. When treated with physical or chemical carcinogens, isolated foci consisting of morphologically altered cells appear. These cells are anchorage independent and produce fibrosarcoma upon inoculation into syngeneic animals. Cells were maintained in Eagle's Basal medium supplemented with 10% heat-inactivated fetal bovine serum (Hyclone Laboratories) and 25μg/ml gentamycin (U.S. Biochemicals). Only cells from passage 8-13 were used for these studies.

UICC standard reference crocidolite fibers were used in these studies. The fibers were used at 5 μg/ml all through the experiments and their preparation has been described previously (6).

For the transformation assay, exponentially growing C3H 10T1/2 cells plated in 100mm diameter petri dishes were pretreated with crocidolite fibers for 24 hours. Superoxide dismutase (Sigma Chemical) was added to the culture medium concurrently with the fibers at 25 units/ml. This concentration of enzyme has been determined previously to be non-cytotoxic to the cells. The cultures were subsequently irradiated with 4Gy of γ-rays from a cesium 137 irradiator at an absorbed dose rate of 1.37 Gy per minute. The cultures were then washed twice with buffered salt solution, trypsinized and replated into replicate dishes at a density such that 300-400 cells would survive the various treatments. All treated and control cultures were maintained for a total of six weeks with medium changed every 10 days. Fresh SOD was added to the medium at every change such that the enzyme was present throughout the experiment. The cultures were subsequently fixed, stained and type II & III foci scored as transformants as described previously (14). All dishes were coded and scored independently twice. Previous studies from this laboratory have demonstrated that cells from transformed foci are anchorage independent and tumorigenic upon inoculation into athymic nude mice. Overall, three separate experiments involving a total of 1500 dishes were involved in these studies. Corresponding dishes were plated at lower cell densities to determine cytotoxicity.

Data was analyzed using the two-tailed Student's t test for unpaired data. Differences between means were regarded as significant for $p \leq 0.05$.

RESULTS AND DISCUSSION

Figure 1 shows the phagocytosis of crocidolite fibers by a binucleated C3H 10T 1/2 cell. Smaller fibers can be seen around the perinuclear and other cytoplasmic sites. The long fiber as seen, however, is only partly phagocytized by the cell. The physical and biochemical consequences of asbestos-cell interactions have been documented previously (9-12). The *in vitro* cytotoxicity of asbestos fibers has been shown in a variety of cell types, including macrophages, V-79 lung cells and mouse embryo cells. The cytotoxic response is thought to be due to the membrane effects of the fibers. Wade et. al. using P388 D_1 cells, has shown that cytotoxicity was related to the phagocytic property of cells since non-phagocytic fibroblasts survived under conditions lethal to the macrophage-like cells (15). Time lapse microcinematography demonstrated that normal mitosis occurred even in the presence of intracellular asbestos. Furthermore, incomplete phagocytosis has been shown to be associated with increased membrane permeability and elevated lactic dehydrogenase activity in the media. (11). The presence of superoxide dismutase either alone or in combination with catalase can reduce the cytotoxicity of crocidolite fibers (>50 μg /ml) to macrophage by as much as 90% (13) suggesting the role of oxygen free-radicals in mediating asbestos-induced lethal cell injury.

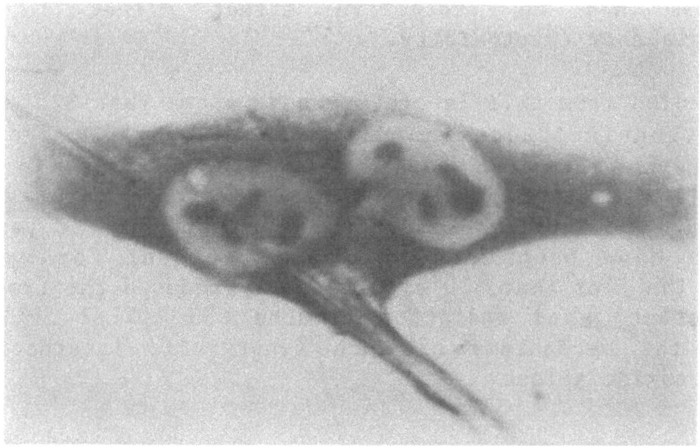

Fig. 1 Photomicrograph of a binucleated C3H 10T 1/2 cells showing partial phagocytosis of crocidolite fibers. Note the presence of small fibers around the perinuclear and other cytoplasmic sites (Giemsa stains, 400X)

Figure 2 shows the transformants per surviving cell resulting from the various treatments either in the presence or absence of SOD. As reported previously, crocidolite fibers alone, at concentrations used, induced transformation frequencies that do not differ significantly from the spontaneous rate. Addition of SOD to the culture medium has little or no effect on the cytotoxicity of the fibers as determined by the clonogenicity of treated cells. However, cells pretreated with fibers for 24 hours and subsequently irradiated with 4 Gy of γ-rays produced significantly higher transformation frequencies than radiation alone ($p \leq 0.005$). Addition of SOD obliterated the supra-additive response characteristic of the combination of radiation and asbestos ($p \leq 0.005$) and resulted in a transformation frequency similar to that due to radiation by treatment alone. In the present studies, addition of SOD to the

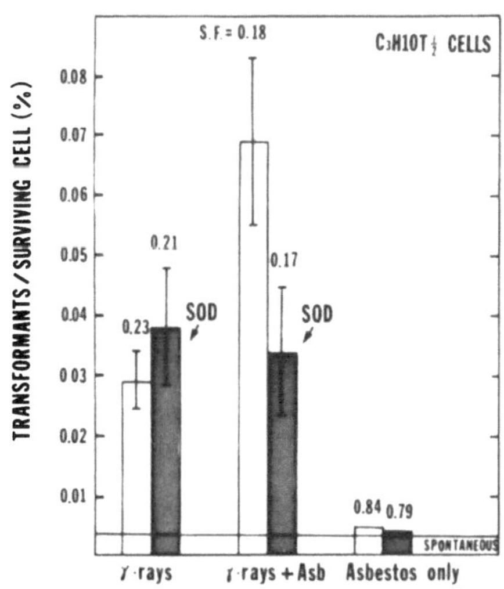

Fig. 2 Transformation incidence in C3H 10T 1/2 cells produced by the various treatment either in the presence (closed bar) or absence (open bar) of SOD. Results are pooled data from 3 experiments. Bars represent ± S.E. of means. S.F = surviving fraction.

irradiated cells had no protective effect either in terms of transformation yield or cytotoxicity.

Previous studies from this laboratory have shown that SOD can suppress the transformation incidence in x-irradiated C3H 10T1/2 cells (16). However, the present studies shows no protective effect by SOD. The reasons for this difference is not clear. Various serum batches may contain different antioxidant(s) levels and throughout the years several batches of SOD have been used which may account for some of the discrepencies. The fact that SOD significantly reduced the transformation incidence in asbestos and radiation treated C3H 10T1/2 cells strongly suggested that the mechanisms of such synergistic interaction may be mediated by superoxide anions.

REFERENCES

1. D.C. Barum, and R.D. Truan, An epidemiological study of lung cancer in asbestos miners. Arch. Ind. Hyg. Occup. Med. 17, 634-637. (1958).
2. J.C. Wagner, C.A. Sleggs, and P. Marchand, Diffuse pleural mesothelioma and asbestos exposure in the Northwestern Cape Province. Br. J. Ind. Med. 17, 260-271. (1960).
3. J.C. Wagner, and G. Berry, Mesothelioma in rats following inoculation with asbestos. Br. J. Cancer, 23, 567-581 (1969).
4. P. Gross, R.T.P. DcTredelle, E. Tolker, M. Kaschak, and M.A. Babyak, Experimental asbestosis: The development of lung cancer in rats with pulmonary deposits of chrysotile asbestos dust. Arch. Environ. Health 15, 343-355 (1967).
5. I.J. Selikoff, E.C. Hammond, and J. Churg, Asbestos exposure, smoking and neoplasia J. Am. Med. Assoc. 204, 104-110 (1968).
6. T.K. Hei, E.J. Hall, and R. Osmak, Asbestos, radiation and oncogenic transformation. Br. J. Cancer 50, 717-720 (1984).
7. T.K. Hei, C.R. Geard, R. Osmak and M. Travisano, Correlation of in vitro genotoxicity and oncogenicity induced by radiation and asbestos fibers. Br. J. Cancer 52, 591-597 (1985).
8. M.F. Stanton, M. Layard, and A. Tegeris, Carcinogenicity of fibrous glass: pleural response in the rat in relation to fiber dimension. J. Natl. Cancer Inst. 58, 587-604 (1977).
9. J.L. Kaw, D.H. and Zaidi, In vitro studies in the cytotoxic action of different varieties of asbestos dusts on macrophages. Acta. Pharmacol. Toxicol. 36, 225-235 (1975).
10. K. Miller, R.I.M. Handfield, and E. Kagan, E. The effects of different mineral dusts on the mechanism of phagocytosis: A scanning electron microscope study. Environ. Res. 15, 139-154 (1978).
11. E.J. Beck, Interaction between fibrous dust and cells in vitro. Ann. Anat. Pathol. 21, 227-236 (1976).
12. K. Hansen, and B.T. Mossman, Generation of superoxide from alveoler macrophages exposed to asbestiforms and non-fibrous particles. Cancer Res. 47, 1681-1686 (1987).
13. L.A. Goodglick, and A.B. Kane, Role of reactive oxygen metabolites in crocidolite asbestos toxicities to mouse macrophages. Cancer Res. 46 5558-5566 (1987).
14. C.A. Reznikoff, J.S. Bertram, D.W. Brankow, and C. Heidelberger, Quantitative & qualitative studies of chemical transformation of clonal C3H mouse embryo cells sensitive to post confluence inhibition of cell division. Cancer Res. 33, 3239-3249 (1973).
15. M.J. Wade, L.E. Lipkin, and A.L. Frank, Studies of in vitro asbestos cell interaction. J. Environ. Pathol. Toxicol. 2, 1029-1039 (1979).
16. R.C. Miller, R. Osmak, M. Zimmerman, and E.J. Hall, Sensitizers, protectors and oncogenic transformation in vitro. Int.J. Radiat. Oncol. Biol. Phys. 8, 771-776 (1982).

ROLE OF PEROXIDASE IN STIMULATION OF THE PENTOSE CYCLE OF A549 CELLS BY AEROBIC IRRADIATION

Marie E. Varnes[1] and John E. Biaglow[2]

Division of Biochemical Oncology, Department of Radiology, Case Western Reserve U., Cleveland, OH[1] and Division of Biochemical and Radiation Oncology Department of Radiation Therapy, U. of Pennsylvania Philadelphia, PA[2]

INTRODUCTION

The glutathione redox cycle is the major pathway for providing cells with reducing equivalents for bioreduction of drug- or radiation-induced superoxide and peroxide (1). GSH and NADPH, oxidized in the course of reductive metabolism, are regenerated <u>via</u> GSH reductase and the pentose cycle, respectively, with glucose as the ultimate hydrogen donor:

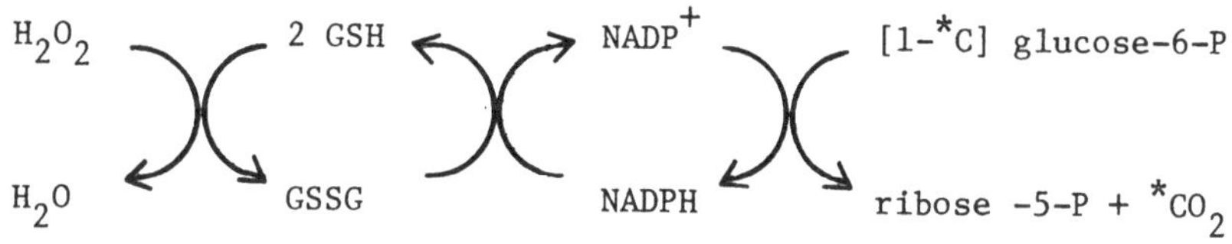

Biaglow <u>et al.</u> suggested that the pentose cycle (PC) pathway is important for radioprotection because it maintains intracellular GSH for direct hydrogen donation as well as for peroxidase-catalyzed detoxification of H_2O_2 and organic peroxides (2). This report shows that PC activity of human lung carcinoma cells is stimulated by both hypoxic and aerobic irradiation, and that, in the case of aerobic irradiation, activity of this pathway remains elevated over control levels for several hours after treatment. The hypothesis that this effect is largely due to utilization of NADPH as a reductant for GSH peroxidase has been tested by treating cells with various agents known to alter catalase or peroxidase activity.

MATERIALS AND METHODS

<u>Cell culture</u>. A549 human lung carcinoma cells (ATCC) were grown as monolayers in McCoy's 5A medium, supplemented with 10% calf serum and 5% fetal calf serum, and buffered with 2.0 g/l bicarbonate and 20 mM Hepes, pH 7.2. Cells were used in mid-log phase of growth.

<u>Irradiation protocol:</u> Cells were harvested by trypsinization and resuspended in Ca^{++}- and Mg^{++}-free phosphate-Hepes buffered saline, pH 6.7. One mM glucose (5.5-6.0 x 10^5 dpm/µmole) was added. Two ml aliquots of cell suspension were placed in 25 ml Erlenmeyer flasks, then the flasks

Table 1. Stimulation of Pentose Cycle Activity of A549 Cells by Aerobic and Hypoxic Irradiation

Gray	nmoles CO_2 Released / 10^6 Cells	
	Air	Nitrogen
0	3.5 +/- 0.2	2.0 +/- 0.3
10	5.6 +/- 0.1	3.2 +/- 0.4
20	7.0 +/- 0.3	3.9 +/- 0.4
30	8.0 +/- 0.2	4.5 +/- 0.4
40	9.1 +/- 0.3	5.5 +/- 0.6
50	10.5 +/- 0.5	7.6 +/- 1.0

Cells in suspension were irradiated, held for 30 min, then acidified to release CO_2. Results are averages from duplicate measurements of a representative experiment.

were capped and suspensions irradiated using a ^{60}Co gamma source, 12 Gy/min. Cell suspensions were maintained at 37°C throughout the experiments, except for the brief time of irradiation. For hypoxic conditions, cells were gassed with 100% N_2 at a flow rate of 1 l/min, for 15 min before irradiation.

Pentose cycle activity: PC activity in A549 cells was determined as previously described (3). Briefly, flasks containing suspended cells were sealed with serum stoppers through which a center well, containing 0.1 ml 5% KOH adsorbed onto a filter paper, had been inserted. Cell suspensions were incubated at 37°C for various times after irradiation or drug addition, then reactions were stopped and $^{14}CO_2$ collected by injection of 0.2 ml 6 N H_2SO_4 into the cell suspension. After 30 min, filters were removed, dried, and counted for radioactivity. The amount of CO_2 released was calculated from the specific activity of the added glucose.

RESULTS

Table 1 shows the dose-response for stimulation of PC activity of A549 cells by hypoxic and aerobic irradiation. In this experiment, CO_2 was collected during, and 30 min after, irradiation. The control rate of CO_2 release, reflecting NADPH turnover, is about two times faster under aerobic than hypoxic conditions. A radiation dose of 50 Gy increased PC activity by a factor of 3, within experimental error, in both air and hypoxia.

The time course for stimulation of PC activity by 50 Gy γ-irradiation, under aerobic conditions, is shown in Figure 1. The zero time point indicates cell suspensions which were acidified immediately after irradiation. For these experiments, the pH of the buffer was reduced from 7.3 to 6.7, in order to inhibit consumption of the labeled glucose via lactate formation (3). Lowering of the pH also resulted in a slight reduction of the PC activity (data not shown). For untreated cells, CO_2 release was linear with time, with a rate of about 0.15 nmoles/10^6 cells/min. Irradiation at 50 Gy stimulated an immediate increase in CO_2 release, and activity remained elevated over control levels for as long as three hours after irradiation. Stimulation of PC activity by misonidazole and t-butylhydroperoxide are shown for comparison. In the latter case, a rate of about 4.0 nmoles

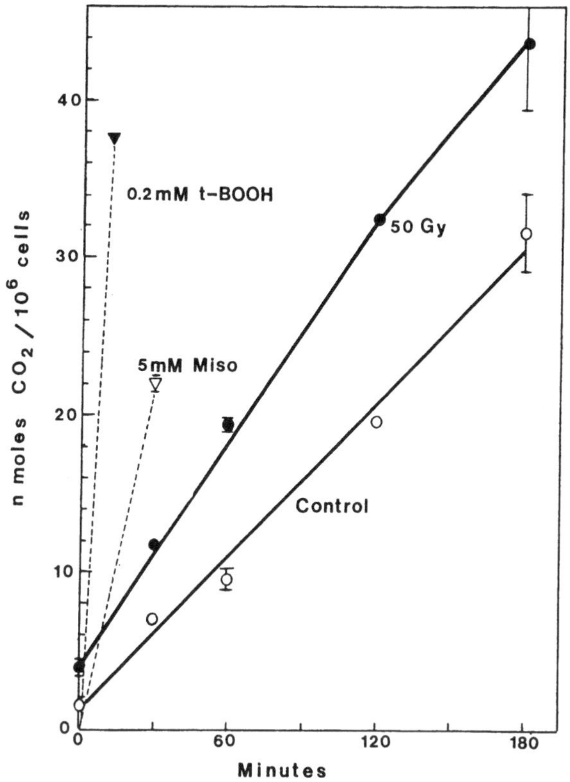

Fig. 1. Time course for stimulation of pentose cycle activity of A549 cells by 50 Gy irradiation. The effects of 5 mM misonidazole (Miso) and 0.2 mM t-butylhydroperoxide (t-BOOH), added at time 0, are shown for comparison. Results are averages, +/- s.e.m., for triplicate flasks from a typical experiment.

$CO_2/10^6$ cells/min represents the maximal rate of NADPH turnover observed for A549 cells.

Figure 2 shows the effects of various treatments which alter catalase or peroxidase activity on the ability of γ-irradiation to stimulate PC activity in A549 cells. Treatment conditions were selected on the basis of previous experience, as follows. Catalase, at 20 μg/ml, rapidly reduces exogenous H_2O_2 (4). Using a peroxidase electrode, Donahue et al. showed that addition of 30 μM sodium azide to A549 cells inhibits endogenous catalase activity (5). Biaglow et al. (6) demonstrated that treatment of log phase A549 cells for 72 hours with 0.2 mM buthionine sulfoximine (BSO) reduces intracellular GSH levels to less that 0.1% of controls, and Varnes et al. (4) showed that even this level of GSH depletion does not inhibit peroxidase activity of A549 cells. In contrast, treatment of cells with N-ethylmaleimide (NEM) removes GSH and completely blocks the ability of cells to reduce exogenous peroxide (5). Since control PC activity varied somewhat between experiments, the results of Figure 2 are reported as the difference in nmoles CO_2 released, at each time point, in irradiated vs unirradiated cell suspensions.

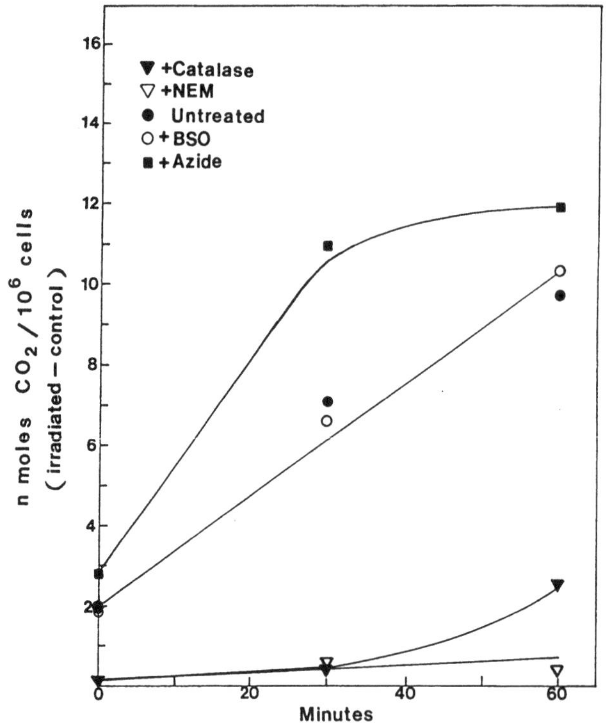

Fig . 2. Effects of various treatments on the ability of γ-irradiation to stimulate pentose cycle activity of A549 cells. Untreated cells (●); cells incubated, during and post irradiation, in the presence of 20 μg/ml catalase (▼) or 30 μM sodium azide (■); cells pretreated with 0.2 mM BSO for 72 hr (O) or 45 μM NEM for 5 min (▽), then washed and resuspended before irradiation. Results represent typical experiments, with nmoles CO_2 released from irradiated flasks subtracted from the respective controls at each time point.

Additional studies showed that the effect of irradiation on PC activity was cell density dependent, with greater release of $CO_2/10^6$ cells at a density of 6 x 10^5 than at 4 x 10^6 cells/ml. The possibility that radiation-induced PC stimulation might be due in part to a response to peroxide or organic peroxides formed primarily in the overlying buffer was tested. Glucose-containing buffer was irradiated at 50 Gy, then unirradiated cells were added, and incubations carried out as described above. A significant response to the irradiated buffer was observed, with a difference of 5 nmoles CO_2 released/10^6 cells after 10 min incubation, and 8 nmoles after 120 min incubation.

DISCUSSION

Our results demonstrate that NADPH, an important reducing species, is utilized by A549 cells in the response to ionizing radiation. Roberts _et al._ (7) previously observed stimulation of PC activity of human lymphocytes, but their data, in contrast to ours, showed a greater stimulation of the PC by hypoxic than by aerobic irradiation. Roberts _et al._ interpreted their results in terms of the competition model, with hydrogen atoms generated by the PC utilized either directly, or _via_ GSH, for repair of radiation-damaged DNA.

Figure 1 shows that γ-irradiation stimulates an immediate PC response, in addition to a prolonged response that continues for up to three hours after irradiation. This response is small, relative to the effects of misonidazole and t-BOOH, which stimulate PC activity _via_ utilization of NADPH as a cofactor for nitroreductase and the GSH peroxidase/reductase couple, respectively (2,3). It is not possible to precisely quantitate the amount of NADPH oxidized by irradiation from CO_2 release data; however, the observed PC stimulation corresponds within an order of magnitude with what would be expected from G values for H_2O_2 and organic peroxides (8, and personal communications of Drs. C. Koch and J. Raleigh).

The data of Figure 2 indicate that all the NADPH turned over by A549 cells during and post-irradiation is utilized in an indirect way, as a cofactor for GSH reductase/peroxidasae. Most indicative of this is the observation that treatment of cells with NEM, which completely blocks peroxidase activity (5), eliminates response of the PC to irradiation. In contrast, BSO treatment, which removes most of the GSH but does not block peroxidase activity (4), has no effect. H_2O_2 produced in the extracellular buffer during irradiation appears to account for most of the PC response, since addition of catalase, which cannot penetrate cells, largely eliminated the effect. In addition, irradiation of the buffer alone, followed by incubation with unirradiated cells, also produced a PC response, as described under "Results".

The observation that PC activity continues to be stimulated for a considerable time post-irradiation was somewhat surprising, and is not readily explained by the data of Fig. 2. Although the effects of catalase, and treatments with NEM and azide suggest H_2O_2 reduction, it is difficult to define how that H_2O_2 is formed. Raleigh and Shum have described radiation-generated peroxy radicals which can be long-lived and which can cause membrane damage (9). It is also possible that organic peroxides are formed from breakdown products of irradiated glucose in the incubation medium (10). Catalase might then inhibit post-irradiation stimulation of the PC by preventing the initiation of chain reactions which lead to formation of organic peroxides. These studies are continuing, with the objective of understanding the biochemical response of human tumor cells to radiation damage, and how that response ultimately influences cell survival.

ACKNOWLEDGEMENTS

These studies were supported by grants no. CA 40516 (MEV) and CA 13747 (JEB), awarded by the National Cancer Institute, DHHS. The authors wish to thank Drs. Cameron Koch and James Raleigh for very helpful discussions of radiation biology, and Mr. Keith Freeman and Mrs. Caroline Gray for their excellent technical assistance.

REFERENCES

1. D. J. Reed, Regulation of reductive processes by glutathione. Biochem. Pharmacol. 35, 7-13 (1986).
2. J. E. Biaglow, M. E. Varnes, E. P. Clark and E. R. Epp, The role of thiols in cellular response to radiation and drugs. Radiat. Res. 95, 437-455 (1983).

3. M. E. Varnes, S. W. Tuttle and J. E. Biaglow, Nitroheterocycle metabolism in mammalian cells: Stimulation of the hexose monophosphate shunt. Biochem. Pharmacol. 33: 1671-1677 (1984).
4. M. E. Varnes, J. E. Biaglow and S. W. Tuttle, Role of cellular non-protein thiols in oxygen consumption and peroxide reduction. In: Oxygen Transport to Tissue VIII, (I. S. Longmuir, Ed.), pp. 565-571. Plenum , New York, 1986.
5. L. Donahue, J. E. Biaglow and M. E. Varnes, Use of the oxidase electrode to determine tumor cell metabolism of peroxide. In: Oxygen Transport to Tissue VIII, (I. S. Longmuir, Ed.), pp. 573-581. Plenum , New York, 1986.
6. J. E. Biaglow et al., The effect of L-buthionine sulfoximine on the aerobic radiation response of A549 human lung carcinoma cells. Int. J. Radiat. Oncol. Biol. Phys. 12, 1139-1142 (1986).
7. W. Roberts, M. Kartha and A. Sagone, Effect of irradiation on the hexose monophosphate shunt pathway of human lymphocytes. Radiat. Res. 79, 601-610 (1979).
8. H. B. Michaels and J. W. Hunt, Determination of peroxides and hydroperoxides in irradiated solutions of nucleic acid constituents and DNA. Anal. Biochem. 87, 135-140 (1978).
9. J. A. Raleigh and F. Y. Shum, Radioprotection in model lipid membranes by hydroxyl radical scavengers: Supplementary role of α-tocopherol in scavenging secondary peroxy radicals. In: Radioprotectors and Anticarcinogens, (O. F.Nygaard and M. G. Simic, Eds.), pp. 87-102. Academic Press, New York, 1983.
10. G. O. Phillips, Chemical effects of ionizing radiations on aqueous solutions of aldohexoses. Radiat. Res. 18. 446-460 (1963).

TOXICITY AND RADIOPROTECTIVE EFFICACY OF BIS (3,5-DIISOPROPYLSALICYLATO) COPPER II AND $CUCL_2$*

Linda K. Steel[1], Sarath Seneviratne[1], and
William E. Jackson III[2]

[1]Radiation Biochemistry Department
[2]Computer and Electronics Department
Armed Forces Radiobiology Research Institute
Bethesda, MD 20814-5145

INTRODUCTION

Bis (3,5-diisopropylsalicylato) copper II (CuDIPs), a low molecular weight lipid-soluble complex, has been suggested to be an effective, non-toxic radioprotectant in mice exposed to gamma radiation from a ^{60}Co source (1). CuDIPs has also been shown to have antiinflammatory activity (2) and to catalyze the disproportionation of superoxide (3). In view of these observations, the present study examined the toxicity and radio-protective efficacy of CuDIPs, as well as those of diisopropylsalicylate (DIPs) and $CuCl_2$, in both male and female B6D2F1 mice.

MATERIAL AND METHODS

CuDIPs was the gift of Dr. John R. Sorenson (University of Arkansas). DIPs and $CuCl_2$ were purchased from Aldrich Chemical Co. (Milwaukee, WI) and Sigma Chemical Co. (St. Louis, MO), respectively. Drugs were dissolved or suspended by sonication in saline containing 0.25% Tween-80 and administered in a volume of 0.2 ml subcutaneously to the nape of the neck. B6D2F1 (C57B1/6J♀ x DBA/2J ♂) female and male mice obtained from Jackson Labs (Bar Harbor, ME) were housed in groups of nine for 2 weeks in a quarantine facility. Animals determined to be free of common murine diseases (random sampling) by serology and histology and whose sterile water bottle cultures were negative for *Pseudomonas* species were used. Female mice (20-24 g, avg. 22 g) and male mice (24-27 g, avg. 26 g), 13-16 weeks old, were housed in polycarbonate rodent cages with an air filter barrier and sterile wood-chip bedding, four animals per cage. Animals were kept on a 7 a.m. (light) to 7 p.m. (dark) cycle and provided standard rodent diet and water *ad libitum*.

Mice were bilaterally exposed to gamma radiation from a ^{60}Co source. Total dose was 8.0-12.5 Gy, at a dose rate of 0.4 Gy/min.

Calculations of dose-response relationships for animal survival (both drug toxicity and radiation) were performed by probit analysis (4). Appropriate linearizing transformations were made. Assessment of significant differences between probit fits were performed by the method of Finney (5). Comparison of two proportions were made with a Z-statistic (6).

RESULTS

DIPs was nontoxic at all doses tested (25-200 μg/g; data not shown). The toxic LD_{50} of CuDIPs was 87.3 μg/g and 222.1 μg/g in male and female mice, respectively (Fig. 1a). The toxic LD_{50} of $CuCl_2$ in females was 100 μg/g, corresponding to a Cu-equivalent dose of 376 μg/g CuDIPs (Fig. 1b). The toxic LD_{50} of $CuCl_2$ in males was not obtained with the doses examined (>106 μg/g, corresponding to a Cu-equivalent dose of >400 μg/g CuDIPs).

Animal mortality resulting from toxic doses of CuDIPs or $CuCl_2$ was associated with severe kidney damage (Fig. 2b,c). The kidney renal cortex region was characterized by enlarged vesicular nuclei, cast formation, sloughing of the proximal tubular epithelial cells into the lumen, flattening of the basement membrane, protein leakage, and mitotic figures. The renal medulla also exhibited extensive protein leakage and mineral-ization.

Mice were treated with nontoxic or minimally toxic doses of each drug or vehicle 24 h prior to bilateral gamma irradiation from a ^{60}Co source. The results are summarized in Table 1. DIPs administration did not improve survival compared to vehicle only. CuDIPs and $CuCl_2$ (at Cu-equivalent doses) were equally radioprotective when compared using dose modifying factors (DMF) as the end points.

CuDIPs or $CuCl_2$ administration 3 h prior to irradiation, or 3 or 24 h postexposure was less effective than when given 24 h prior to irradiation (Table 2).

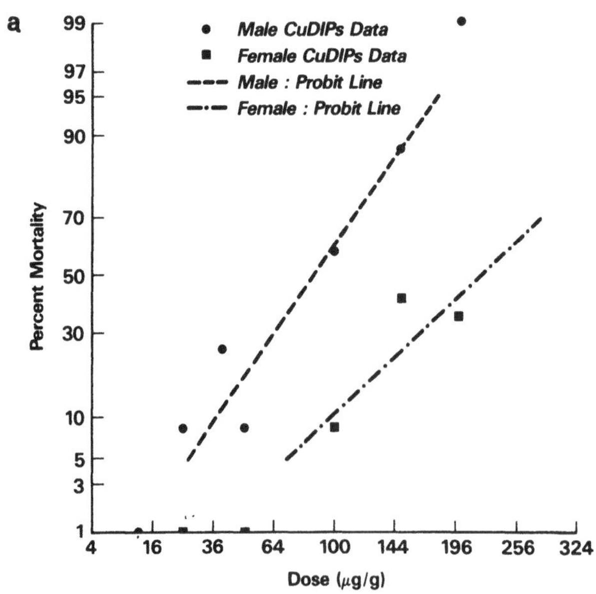

Fig. 1a. Toxicity of CuDIPs in male and female B6D2F1 mice

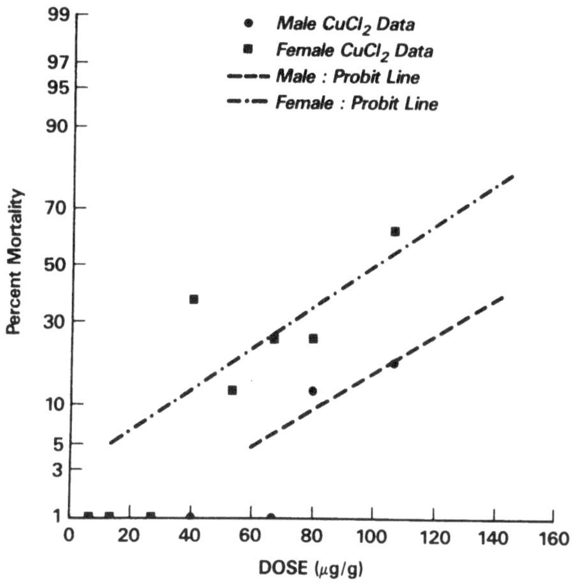

Fig. 1b. Toxicity of $CuCl_2$ in male and female B6D2F1 mice

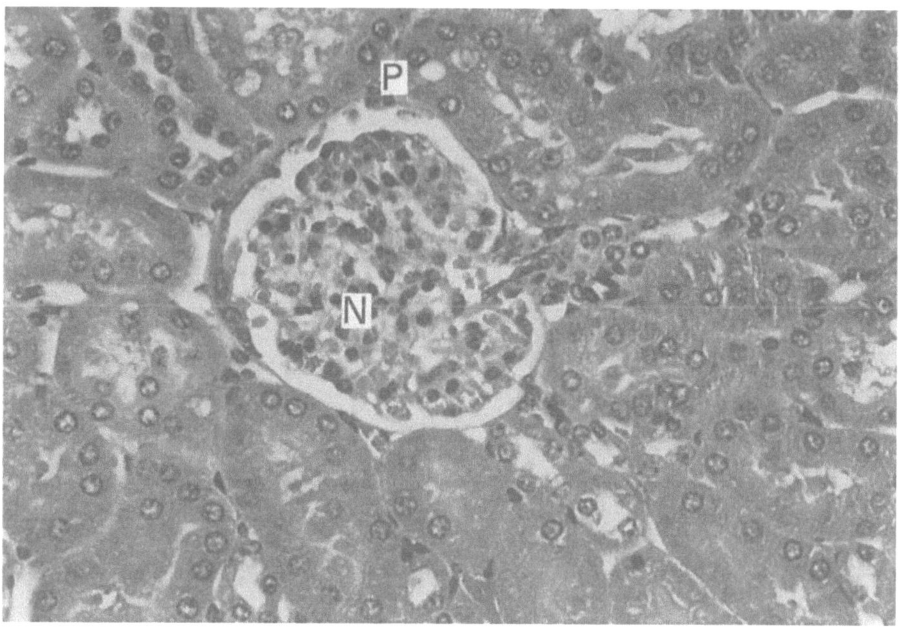

Fig. 2a. Cross section of a normal kidney cortex from a male mouse injected with vehicle 3 days prior to sacrifice. N = normal glomeruli; P = proximal tubules. (40x).

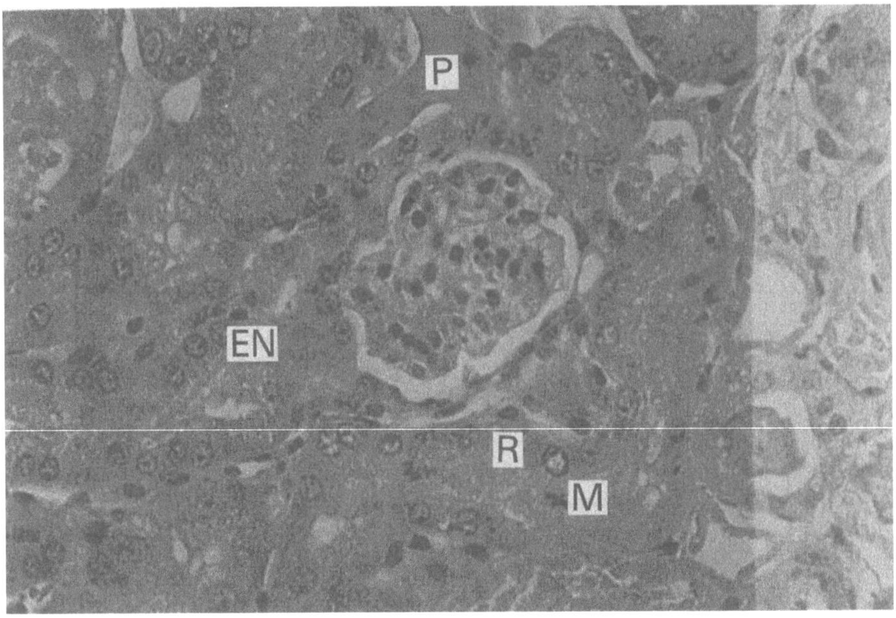

Fig. 2b. Cross section of the kidney renal cortex from a male mouse injected with $CuCl_2$ (79.8 μg/g) 3 days prior to sacrifice. P = protein leakage; M = mitotic figures; R = regenerating proximal tubules; EN = enlarged vesicular nuclei. (40x).

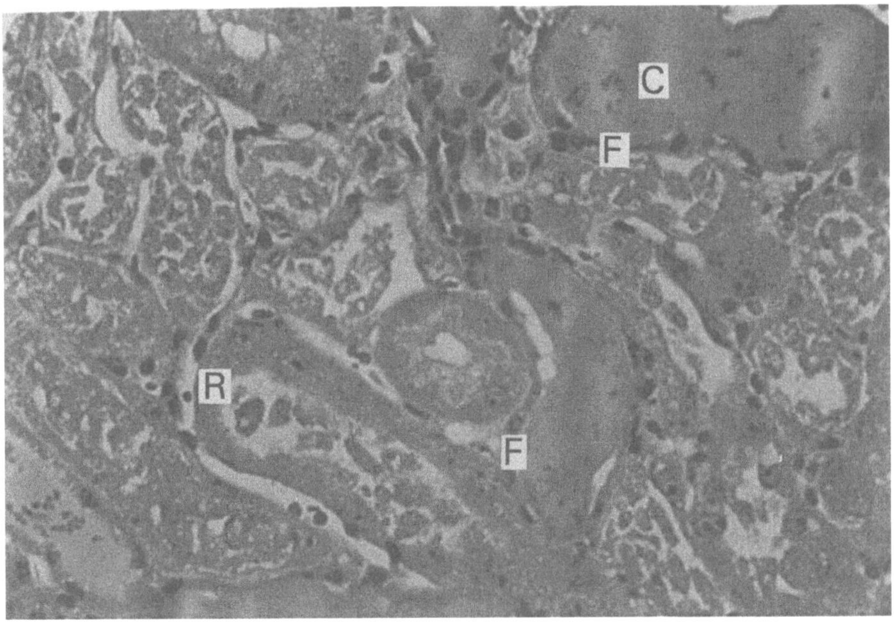

Fig. 2c. Cross section of kidney renal cortex from a male mouse injected with $CuCl_2$ (79.8 μg/g) 3 days prior to sacrifice. C = cast formation; F = flattening of the tubular epithelial cell along the basement membranes; R = regenerating proximal tubules. (40x).

Table 1. Radioprotective Efficacy of CuDIPs and $CuCl_2$, Administered 24 h Prior to Gamma Irradiation From a ^{60}Co Source

Sex	Drug	Dose ($\mu g/g$)[a]	LD50/60 (cGy)	Dose Modification Factor[b]
♂	vehicle		887	1.0
	CuDIPs	25	1043	1.18
		50	1031	1.16
	$CuCl_2$	6.6	1024	1.15
♀	vehicle		882	1.0
	CuDIPs	25	1032	1.17
		50	1059	1.20
		6.6	1090	1.23
	$CuCl_2$	13.3	1085	1.23
		26.6	1136	1.29

[a]6.6, 13.3 and 26.6 $\mu g/g$ $CuCl_2$ correspond to copper-equivalent doses of 25, 50 and 100 $\mu g/g$ CuDIPs.
[b]Ratio of vehicle treated and drug treated LD50/60 values.

Table 2. Survival Data for Mice Administered DIPs, CuDIPs, or $CuCl_2$ 24 h or 3 h Prior to, or 3 or 24 h Postexposure to Gamma Radiation From a ^{60}Co Source

Dose (cGy)	Sex	Percent Survival[a]					
		Drug Dose		Time of Drug Administration[b]			
		Drug	($\mu g/g$)[c]	−24 h	−3 h	+3 h	+24 h
1000		vehicle		0	37.5	0	25
		DIPs	25	8.3	0	12.5	0
		CuDIPs	12.5	ND	50	12.5	25
			25	58.3	50	25	25
			50	58.3	ND	ND	ND
		$CuCl_2$	13.3	75	ND	ND	ND
		vehicle	—	0	0	0	0
		DIPs	25	16.7	12.5	0	0
		CuDIPs	12.5	83.3	25	25	0
			25	41.5	37.5	12.5	37.5
		$CuCl_2$	6.6	50	ND	ND	ND
			13.3	25	ND	ND	ND
1100		vehicle	—	0	0	0	0
		DIPs	50	8.3	0	0	0
		CuDIPs	25	58.3	ND	ND	ND
			50	58.3	12.5	25	0
		$CuCl_2$	6.6	50	ND	ND	ND
			13.3	58.3	0	0	12.5

[a]Number of surviving mice divided by total; at least 8 mice were employed in each experiment.
[b]Relative to time of irradiation.
[c]$CuCl_2$ administered at doses of 6.6 or 13.3 $\mu g/g$ correspond to Cu-equivalent doses of 25 and 50 $\mu g/g$ CuDIPs, respectively.

CONCLUSION

Weight loss, kidney damage (Cu-toxicosis), and the modest DMFs obtained with CuDIPs or Cu-equivalent doses of $CuCl_2$ disfavors their use as antiradiation agents.

ACKNOWLEDGEMENT

Supported by the Armed Forces Radiobiology Research Institute, Defense Nuclear Agency, under Research Work Units MJ00064 and B2152. Views presented are those of the authors; no endorsement by Defense Nuclear Agency has been given or should be inferred. Research was conducted according to the principles enunciated in the "Guide for the Care and Use of Laboratory Animals," prepared by the Institute of Laboratory Animal Resources, National Research Council.

REFERENCES

1. J.R.J. Sorenson, Bis (3,5-diisopropylsalicytate) copper (II), a potent radioprotectant with superoxide dismutase mimetic activity. J. Med. Chem. 27: 1747-1749 (1984).
2. J.R.J. Sorenson, The anti-inflammatory activities of copper complexes. In: Metal Ions in Biological Systems (H.Siegel, Ed.), Vol. 14, pp. 77-124. Marcel Dekker, New York, 1982.
3. U. Weser, C. Richter, A. Wendel, and M. Younes, Reactivity of anti-inflammatory and superoxide dismutase active Cu(II)-salicylates. Bioinorg. Chem. 8: 201-213 (1978).
4. D.J. Finney, Probit Analysis, 3rd ed., pp. 1-99. Cambridge University Press, Cambridge, England, 1971.
5. D.J. Finney, Statistical Methods in Biological Assay, 3rd ed., pp. 370-379. MacMillan Publishing Company, Inc., New York, 1978.
6. G.W. Snedecor and W.G. Cochran, Statistical Methods, 7th ed., pp. 124-125. Iowa State University Press, Ames, Iowa, 1980.

MODULATION OF GLUTATHIONE LEVELS AND METABOLISM

Alton Meister

Cornell University Medical College
1300 York Avenue, New York, N.Y. 10021

INTRODUCTION

Glutathione (L-γ-glutamyl-L-cysteinyl-glycine), a tripeptide present in substantial concentrations in virtually all mammalian cells, plays a variety of roles in catalysis, metabolism, transport, and in cellular protection. A review of glutathione metabolism was presented at the first International Conference on Radioprotectors and Anticarcinogens (1); see also (2-6). A summary of glutathione metabolism is presented in Fig. 1. Understanding of the biochemistry of glutathione has led to procedures by which cellular levels of glutathione may be decreased or increased; glutathione metabolism may be modulated in other ways by selective inhibition of certain enzymes.

At the first conference, studies were described which showed that depletion of cellular glutathione by administration of an inhibitor of γ-glutamylcysteine synthetase led to a marked increase in the radiosensitivity of cells. Other effects of such glutathione depletion were summarized (1). Similar results have been found in other laboratories, and it has been found that increased protection of cells may be achieved by increasing their glutathione levels. It is now generally appreciated that modulation of glutathione metabolism may be an effective approach to the selective destruction of cells and also to their protection. This paper provides an overview of the methods that have been used for the modification of glutathione metabolism and of some of the results that have thus far been obtained.

METHODS FOR DECREASING CELLULAR GLUTATHIONE LEVELS

Glutathione is synthesized by the consecutive actions of γ-glutamylcysteine synthetase and glutathione synthetase, which catalyze reactions (1) and (2), respectively (see Fig. 1). Glutathione reacts with a variety of compounds of endogenous and exogenous origin to form S-substituted derivatives [reaction (7); Fig. 1.] When certain foreign compounds are administered to experimental animals, the levels of glutathione in certain tissues, including the liver, decreases. Thus, cellular levels of glutathione may be decreased by inhibition of either of the two enzymes required for its synthesis or by administration of compounds that react with glutathione. Another approach that can decrease glutathione levels is application of certain oxidants. This method is of

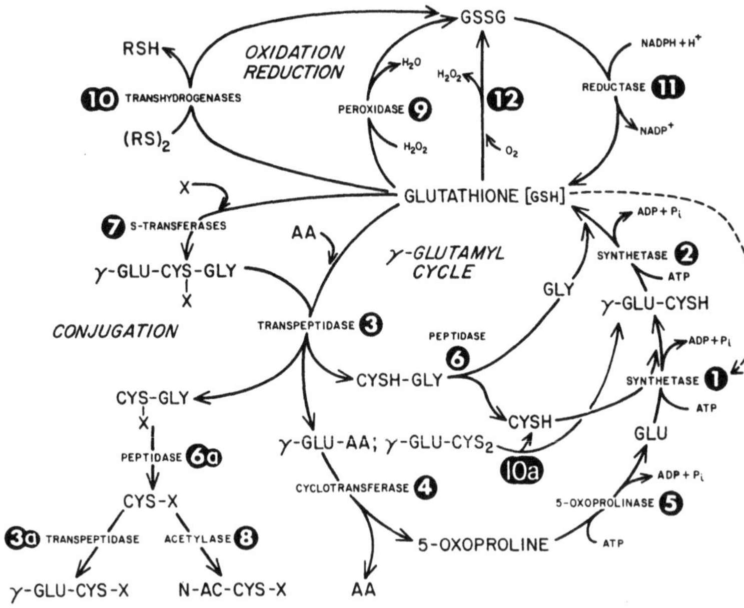

Fig. 1. An outline of glutathione metabolism (modified from Fig. 1 in reference (1)).

limited value inasmuch as it leads to accumulation of glutathione disulfide, which is toxic. A specific chemical oxidant that attacks only glutathione has not yet been found so that this approach leads to oxidation of other cellular components as well as of glutathione. Oxidation of glutathione is readily reversible because of the presence of widely distributed glutathione reductase activity. Although depletion of glutathione by administration of compounds such as diethylmaleate has been used in experimental work, this approach is limited because of lack of specificity, the occurrence of toxicity, and the fact that the depletion of glutathione is short-lived. Inhibition of glutathione synthetase would lead to glutathione depletion, but would be expected to produce other undesirable effects. Thus, patients who have a severe deficiency of glutathione synthetase experience marked, often life threatening, acidosis (7). These patients have decreased cellular glutathione levels which release the normal feedback inhibition of γ-glutamylcysteine synthetase (8). This results in overproduction of γ-glutamylcysteine which is converted by the action of γ-glutamyl cyclotransferase to 5-oxoproline. Formation of 5-oxoproline exceeds the capacity of 5-oxoprolinase and 5-oxoproline, an acid, therefore accumulates. These patients require continuous treatment with sodium bicarbonate (7).

The preferred approach to the production of a sustained decrease in cellular glutathione levles is inhibition of γ-glutamylcysteine synthetase. Compounds that are highly active in vitro and in vivo as inhibitors of γ-glutamylcysteine synthetase were studied (9). Their design was based on earlier studies on the inhibition of glutamine synthetase by methionine sulfoximine (10,11). The details of these investigations have been reviewed (12,13) and were summarized at the first conference (1). Thus, replacement of the S-methyl group of methionine sulfoximine by an S-butyl moiety led to a molecule (buthionine sulfoximine) which is several hundred times more effective than methionine sulfoximine in inhibiting γ-glutamylcysteine synthetase. Of much importance, buthionine sulfoximine does not inhibit glutamine synthetase significantly. It should be noted that other compounds including α-methylbuthionine sulfoximine (9), prothionine sulfoximine (14), and certain higher homologs of buthionine sulfoximine (15) are also effective *in vivo* inhibitors of γ-glutamylcysteine synthetase. When buthionine sulfoximine is given to rats or mice the levels of glutathione decrease in many tissues such as kidney, liver, pancreas, lung, and muscle (16,17). The rate at which tissue levels of glutathione decrease is closely related to the rate of glutathione turnover which is essentially equivalent to the rate of glutathione export from the cells. Cells that turn over glutathione slowly exhibit a slow decline in the level of glutathione when they are exposed to buthionine sulfoximine. A rapid decline in glutathione level occurs in liver and kidney. It is of interest that when buthionine sulfoximine is administered to mice or rats, the level of glutathione decreases rapidly in the liver and kidney reaching a level of about 15-20% of the control. Further decline in glutathione levels occurs much more slowly even after additional doses of buthionine sulfoximine are given. These findings are explained by the observation that glutathione is sequestered within the mitochondria. Mitochondria do not have enzyme activity capable of synthesizing glutathione and are thus supplied with glutathione by transport from the cytoplasm (17). During glutathione depletion, cytoplasmic glutathione disappears much more rapidly than does mitochondrial glutathione.

Mice depleted of glutathione by administration of buthionine sulfoximine exhibit a substantial increase in sensitivity to acetaminophen; the LD_{50} for acetaminophen decreases about 80% after administration of buthionine sulfoximine to mice (18). Suspension of human lymphoid cell lines in media containing buthionine sulfoximine led to depletion of glutathione to about 5% of the control levels, and this was associated with a marked increase in sensitivity to radiation (19). When mice bearing B16 melanomas were treated with buthionine sulfoximine, there was significant sensitization of the tumor to radiation (see (20)). These observations are in accord with the suggestion that treatment with sulfoximine inhibitors of γ-glutamylcysteine synthetase might make tumor cells more susceptible to irradiation and to certain chemotherapeutic agents (21). Many additional studies along these lines have led to similar effects. Thus, depletion of glutathione was reported to sensitize certain cells to radiation under hypoxic conditions and to decrease the oxygen enhancement ratio (22). Other studies showed sensitization of cells by buthionine sulfoximine under both aerobic and hypoxic conditions (23,24). Glutathione depletion by buthionine sulfoximine has been found to produce enhanced oxidative cytolysis of certain tumors (25) and to sensitize certain tumors to the effects of sulfhydryl reactive drugs (26). The very interesting observation was made that mouse L1210 leukemia cells that had become resistant to therapy with L-phenylalanine mustard developed glutathione levels that were about twice those of sensitive cells (27-29). L-Phenylalanine mustard appears to be detoxified by conversion to a non-toxic hydroxy derivative in a glutathione-dependent dehydrochlorination reaction. The mustard may also form a conjugate with glutathione. Notably, treatment of the resistant cells with buthionine sulfoximine led to re-sensitization of the tumor cells to L-phenylalanine mustard. Similar findings were later obtained in studies on human ovarian cancer cell lines (30,31); it was found that resistance to several drugs and to radiation was associated with increased levels of glutathione in the tumors. In these studies also, treatment with buthionine sulfoximine resulted in reversal of resistance. These observations have stimulated a program in which buthionine sulfoximine will be used as an adjuvant in the therapy of human ovarian cancer (31).

Although attention has been paid to the possibility of increasing the sensitivity of tumor cells to radiation and chemotherapeutic agents by decreasing tumor glutathione levels, it should be noted that several compounds are known that are toxified by glutathione. These include azathioprine (32) and bleomycin (33); treatment with buthionine sulfoximine would be expected to decrease such toxicity.

PROCEDURES THAT INCREASE GLUTATHIONE LEVELS

The upper level of cellular glutathione appears to be regulated by feedback inhibition of γ-glutamylcysteine synthetase by glutathione (8). The availability of substrates (i.e., glutamate, cysteine, glycine) are also important factors in glutathione synthesis; usually cysteine is the limiting amino acid.

Administration of glutathione and of cysteine may increase cellular levels of glutathione, but these effects are variable. Even moderate doses of L-cysteine are toxic (see (34)). Feeding of chemically defined diets containing excess L-cysteine to rats led to weight loss and death (35). Degenerative changes in the central nervous system have been found after injection of L-cysteine into young mice and rats (36,37). Cysteine is toxic to certain cells grown in culture (38). Glutathione is not effectively transported into cells; after it is given to animals, it is broken down extracellularly and the products are transported into cells where they may serve as substrates for intracellular glutathione synthesis. In general, administration of cysteine or of glutathione is not an efficient way to increase cellular glutathione levels.

Three methods for increasing cellular glutathione levels have been examined in our laboratory. (a) L-2-Oxothiazolidine-4-carboxylate (34, 39-42) is readily transported into cells and is converted intracellularly by 5-oxoprolinase to L-cysteine, which is utilized for glutathione synthesis. In this approach the toxic effects of L-cysteine (which seem to be largely extracellular) are avoided. After administration of L-2-oxothiazolidine-4-carboxylate, increased cysteine levels are found in many tissues (including the brain (34,43)), and increased glutathione levels are also observed. When ^{35}S-labeled L-2-oxothiazolidine-4-carboxylate was injected into mice, the highest levels of ^{35}S were found in liver and kidney but appreciable amounts of isotope were also found in other tissues (34). Administration of L-2-oxothiazolidine-4-carboxylate is highly effective in protecting mice against acetaminophen toxicity (39). Although N-acetyl-L-cysteine also promotes glutathione synthesis, it was less effective than L-2-oxothiazolidine-4-carboxylate in increasing the glutathione levels in the livers of mice treated with acetaminophen. Other ways in which cysteine might be delivered intracellularly have been considered (34); it would appear that the most effective approaches involve use of compounds that are readily transported into cells and which are converted intracellularly to cysteine by an enzyme-based mechanism.

(b) Since the upper level of cellular glutathione is limited by feedback inhibition of γ-glutamylcysteine synthetase by glutathione, procedures involving administration of cysteine precursors would not be expected to produce levels of glutathione that are higher than the normal feed-back regulated level. Since glutathione synthetase is not feedback inhibited by glutathione, it should be possible to obtain cellular levels of glutathione that are much higher than normal by supplying substrate for this enzyme. In this way, it would be possible to bypass the feedback regulated step of glutathione biosynthesis. The kidney has an effective system for transport of γ-glutamyl amino acids (44,45). When γ-glutamylcysteine, γ-glutamylcysteine disulfide, and γ-glutamylcystine were administered to mice, the levels of glutathione found in the kidney were much higher than those of untreated controls or of mice given equivalent amounts of glutamate plus cysteine or glutamate plus cystinyl-bis-glycine (46). In studies in which the sulfur atoms of γ-glutamylcystine were separately labeled with ^{35}S, the γ-glutamylcysteine moiety of γ-glutamylcystine was preferentially used for glutathione synthesis. There appears to be a pathway in kidney and possibly in other tissues in which γ-glutamylcystine formed by transpeptidation between glutathione and cystine is transported into the cell and reduced to cysteine and γ-glutamylcysteine. This constitutes an alternative pathway of glutathione synthesis.

(c) In a third approach to increasing cellular levels of glutathione, we examined several glutathione derivatives in the hope of finding one that would be readily transported into cells and which would be converted to glutathione intracellularly. Glutathione is not well transported into cells, but it was found that glutathione monoesters in which the glycine carboxyl group is esterified (i.e., L-γ-glutamyl-L-cysteinyl-glycyl esters) are very effectively transported. After transport, these are converted to glutathione intracellularly (47-49). In these studies we prepared a number of glutathione derivatives including the mono and diesters and the corresponding amides of glutathione (49). A promising derivative is the monoethyl ester of glutathione. This ester is more attractive than others because the other intracellular cleavage product, ethanol, is less toxic than methanol, higher alcohols, and ammonia. When glutathione monoethyl ester is administered to mice, there is a substantial increase in the glutathione levels of liver, kidney, and other tissues. Administration of glutathione esters increases glutathione levels of tissues of mice previously treated with buthionine sulfoximine. In our studies, we found that administration of glutathione monoethyl ester is highly effective in protecting against the letal effects of acetaminophen. Administration of glutathione monoethyl ester increases glutathione levels in the liver, kidney, spleen, pancreas, and heart, but there is no significant increase in the overall level of brain glutathione. Glutathione ester is transported into human erythrocytes and is slowly converted to glutathione (49). The ester is very effectively transported into human lymphoid cells, normal human skin fibroblasts, and fibroblasts from patients with genetic deficiencies of γ-glutamylcysteine and glutathione synthetases (48). Notably in some of these studies, cellular levels of glutathione were achieved that were about ten times higher than those normally found.

The use of glutathione monoesters has several advantages. Thus, there is no expenditure of cellular energy as is required for glutathione synthesis. The ester approach is not influenced by feedback inhibition by glutathione. The transport of glutathione monoethyl ester into cells is not significantly affected by the presence of high levels of glutathione, a result in accord with the observation that glutathione is not appreciably transported into cells.

Our studies on the protection of human lymphoid cells against radiation indicate that protection requires intracellular glutathione (50). Extracellular glutathione afforded little if any protection under the experimental conditions employed. We found that glutathione monoethyl ester is effective in protecting against radiation (48). In these studies (Fig. 2), human lymphoid cells of the CEM line were suspended in standard medium and irradiated with 500 rads. In the control **(Curve 1)**, the number of viable cells decreased to zero over

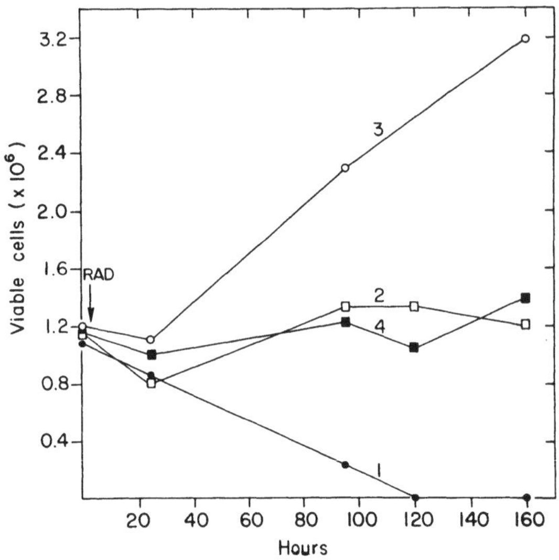

Fig. 2. Protection of human lymphoid cells against radiation damage; see text for identification of conditions (48).

a period of 120 hours. When cells were suspended in medium containing glutathione monoethyl ester prior to irradiation **(Curve 3)** , the number of viable cells increased at a rate that was similar to that found with untreated cells of this line. Some protection was noted when the cells were suspended in medium containing glutathione prior to irradiation **(Curve 4)**, and also when glutathione monoethyl ester was added after irradiation **Curve 2)**. That partial protection was observed when the cells were incubated with glutathione prior to irradiation probably reflects some uptake of glutathione degradation products and subsequent intracellular synthesis of glutathione. It is of particular interest that partial protection was found when glutathione monoethyl ester was added after irradiation. This indicates that some cellular damage is reversible and that glutathione probably functions in repair processes as well as in protection. Although the effectiveness of glutathione monoethyl ester is mainly associated with its intracellular conversion to glutathione, it is likely that the ester itself can interact with free radicals produced by radiation.

INHIBITION OF γ-GLUTAMYL TRANSPEPTIDASE

γ-Glutamyl transpeptidase catalyzes the initial step in the breakdown of glutathione. The enzyme is located mainly on the external surfaces of cells, whereas its major substrate, glutathione, is chiefly intracellular. Export of glutathione, a property of many cells, facilitates access of substrate to the enzyme. Although the functions of glutathione export require further study, it appears that this process is of significance in transport and in protection of the cell membrane. The kidney, which has a high level of γ-glutamyl transpeptidase activity and which is highly active in export of glutathione (from renal cell to tubule), has a system for transport of γ-glutamyl amino acids and related compounds . This pathway, as discussed above, mediates the transport of cystine as γ-glutamylcystine (46). Some tumors develop relatively high levels of γ-glutamyl transpeptidase (see (21)); this may facilitate transport of amino acids (e.g., cystine) into the tumor cells and thus provide a mechanism similar to that found in kidney for recovery of cysteine moieties.

Inhibition of γ-glutamyl transpeptidase, an essential catalytic component of this pathway, would be expected to interrupt this cysteine recovery mechanism. When inhibitors of γ-glutamyl transpeptidase are administered to mice, there is a substantial increase in the level of plasma glutathione and marked glutathionuria occurs. This is accompanied by a moderate decrease in tissue glutathione levels and in the rate of export of glutathione. These findings were made in short term studies; the effects of longer periods of inhibition of γ-glutamyl transpeptidase are not yet known.

Among the most potent of the known inhibitors of γ-glutamyl transpeptidase are the L- and D-isomers of γ-glutamyl-(o-carboxy)phenylhydrazide and L-(αS, 5S)-α-amino-3-chloro-4,5-dihydro-5-isoxazole acetic acid (AT-125) (51). The isomers of γ-glutamyl-(o-carboxy)phenylhydrazide are good inhibitors of the transpeptidase; however, they are split to a slight extent by this enzyme leading to formation of o-carboxyphenylhydrazine, which is toxic. Doses greater than about 2 mmol/Kg of body weight are often fatal to mice. These inhibitors also react with glutathione to form conjugates. Although AT-125 is a potent inhibitor, this enzyme is not specific for the transpeptidase; it also inhibits a number of other enzymes including glutamine amidotransferases.

A new approach to *in vivo* inhibition of γ-glutamyl transpeptidase arose from studies on the transport of γ-glutamylcysteine and related compounds into kidney (46). It was found that the urine of mice treated with γ-glutamylcystine and related compounds contains significant amounts of glutathione. It was subsequently discovered that administration of a variety of other γ-glutamyl compounds also induces glutathionuria, often of the same order of magnitude as found after administration of the hydrazide inhibitors or of AT-125 (52). Good correlation was found between the apparent affinity of γ-glutamyl transpeptidase *in vitro* for various γ-glutamyl amino acids and the extent of glutathionuria observed. The data indicate that administered γ-glutamyl compounds effectively compete with glutathione for the transpeptidase. Administration of certain γ-glutamyl amino acids thus appears to be a specific and non-toxic method for *in vivo* inhibition of transpeptidase that may be useful in experimental work on glutathione metabolism and function (52). It may also be useful for treatment of certain toxicities, and also for the modulation of the metabolism of endogenous glutathione conjugates, such as the leukotrienes.

MODELS FOR POTENTIALLY USEFUL MODULATIONS OF GLUTATHIONE

Most normal cells have high levels of glutathione, glutathione peroxidase, and catalase and thus have two systems for the destruction of hydrogen peroxide. A tumor that has no catalase or very low levels of this enzyme would have to depend upon glutathione peroxidase for destruction of peroxide formed in normal metabolism. Such a tumor might be expected to become less viable after therapy consisting solely of depletion of glutathione (13).

In another model, selectivity may be achieved or significantly increased through modulation of cellular glutathione levels (13,20). Some tumor cells seem to have a greater requirement for glutathione, i.e., a requirement for a higher glutathione level or for a higher capacity to synthesize glutathione, than do normal cells. Probably most normal cells have a very large excess of glutathione, whereas some tumors (and possibly certain parasites) have glutathione levels that are close to those required for survival. Such cells would be

Table I. Some Potentially Useful Modulations of Glutathione Metabolism

1. Treatment with buthionine sulfoximine (BSO) alone for tumors that have low or no catalase.
2. Treatment with BSO plus anti-cancer drug or radiation.
3. Treatment with BSO to decrease toxicity produced by drugs that are toxified by glutathione.
4. Treatment with L-2-oxothiazolidine-4-carboxylate (OTC) to increase tissue levels of cysteine and glutathione.
5. Treatment with OTC plus anti-cancer drug for tumors that have low or no 5-oxoprolinase.
6. Treatment with γ-glu-$(cys)_2$, to increase glutathione levels.
7. Treatment with glutathione mono esters to increase glutathione levels.
8. Treatment with γ-glut-derivative of an anti-cancer drug.
9. Administration of γ-glu amino acids to increase extracellular levels of glutathione and to decrease effective activity of γ-glutamyl transpeptidase.

expected to be more readily affected by chemotherapeutic agents and radiation, even after partial depletion of glutathione, than would normal cells. This is apparently true for certain tumor cells that have become resistant to radiation and to chemotherapeutic agents. For example, if doubling the glutathione level of a tumor is sufficient to protect it against a certain level of a chemotherapeutic agent, it would appear that simply decreasing the level of glutathione of such a tumor by 50% would be sufficient to restore sensitivity to the chemotherapeutic agent.

In another approach, advantage would be taken of the finding that certain tumors are lacking or markedly deficient in 5-oxoprolinase (13, 20). Such tumors would therefore not be able to convert L-2-oxothiazolidine-4-carboxylate to cysteine. A strategy involving the combined use of L-2-oxothiazolidine-4-carboxylate and chemotherapeutic agents might be useful. Normal tissues, which can convert L-2-oxothiazolidine-4-carboxylate to cysteine (which in turn can be used for glutathione synthesis) would therefore be protected, whereas the tumor would not. This would permit increased dosages of chemotherapeutic agents.

Another model may become feasible if it were found that glutathione monoesters are not transported or are poorly transported into tumors. This would provide a mechanism for increasing glutathione levels of normal tissues without appreciably affecting the levels in tumors; this would provide additional advantage in chemotherapy.

Modulation of other aspects of glutathione metabolism may also be therapeutically useful. Certain chemotherapeutic agents may serve as acceptor substrates of γ-glutamyl transpeptidase and thus may be transported as the corresponding γ-glutamyl derivatives. Inhibition of transpeptidase in such instances would lead to decreased transport of the anticancer agent into the cell and could therefore account for drug resistance. Observations consistent with such interpretations have been made in the study of azaserine-resistant cells which were found to be less efficient in concentrating azaserine than the sensitive parental lines (53). Wilms' tumor cells are sensitive to azaserine, but azaserine-resistant strains derived from these cells have greatly reduced transpeptidase activity suggesting that azaserine may be transported as a γ-glutamyl derivative. The possibility of administering certain anticancer drugs as the corresponding γ-glutamyl derivative should be considered. The action of γ-glutamyl transpeptidase might thus facilitate transport of an anticancer agent. It is possible that it might serve in detoxication of some agents. Inhibition of γ-glutamyl transpeptidase increases intracellular glutathione and produces marked glutathionuria. As noted above, this effect might be valuable for the treatment of renal toxicity. A list of some potentially useful strategies for modification of glutathione metabolism is given in Table I.

REFERENCES

1. A. Meister, Glutathione metabolism and transport. In: Radioprotectors and Anticarcinogens (O.F. Nygaard and M.G. Simic, Eds.), pp. 121-152. Academic Press, New York, 1983.
2. A. Larsson, S. Orrenius, A. Holmgren, and B. Mannervik (Eds.), Fifth Karolinska Institute Nobel Conference: Functions of Glutathione - Biochemical, Physiological, and Toxicological Aspects. Raven Press, New York, 1983.
3. I.M. Arias and W.B. Jakoby (Eds.), Glutathione, metabolism and function; Kroc Foundation Series, Vol. 6. Raven Press, New York, 1976.
4. A. Meister, New aspects of glutathione biochemistry and transport; Selective alteration of glutathione metabolism. Fed. Proc 43, 3031-3042 (1984).
5. A. Meister and M.E. Anderson, Glutathione. Annu. Rev. Biochem., 52, 711-760 (1983).
6. A. Meister, Glutathione: metabolism, transport and the effects of selective modifications of cellular glutathione levels. In: Thioredoxin and Glutaredoxin Systems: Structure and Function, Ninth Karolinska Institute Nobel Conference (A. Holmgren, C.I. Branden, H. Jornvall and B.-M. Sjoberg, Eds.), pp. 339-348. Raven Press, New York, 1986.
7. A. Meister, 5-Oxoprolinuria (pyroglutamic aciduria) and other disorders of the γ-glutamyl cycle. In: Metabolic Basis of Inherited Diseases, 5th ed. (J.B. Stanbury, J.B. Wyngaarden, D.S. Frederickson, J.L. Goldstein, and M.S. Brown, Eds.), Chapter 17, pp. 348-359. McGraw Hill, New York, 1983.
8. P. Richman and A. Meister, Regulation of γ-glutamylcysteine syntetase by nonallosteric feedback inhibition of glutathione. J. Biol. Chem., 250, 1422-1426 (1975).
9. O.W. Griffith and A. Meister, Potent and specific inhibition of glutathione synthesis by buthionine sulfoximine (S-n-butyl homocysteine sulfoximine). J. Biol. Chem., 254, 7558-7560, 1979.
10. A. Meister, On the synthesis and utilization of glutamine. Harvey Lect. Ser., 63, 139-178 (1969).
11. A. Meister, Glutamine synthetase of mammals. In: The Enzymes (P.D. Boyer, Ed.), Vol. 10, pp. 699-754. Academic Press, New York, 1974.
12. A. Meister, Inhibition of glutamine synthetase and γ-glutamylcysteine synthetase by methionine sulfoximine and related compounds, In: Enzyme-Activated Irreversible Inhibitors (N. Seiler, M.J. and Jung and J. Koch-Weser, Eds.), pp. 187-211. Elsevier-North Holland Biomedical Press, Amsterdam, The Netherlands, 1978.
13. A. Meister, Selective modification of glutathione metabolism. Science, 220, 471-477 (1983).
14. O.W. Griffith, M.E. Anderson and A. Meister, Inhibition of glutathione biosynthesis by prothionine sulfoximine (S-n-propyl-homocysteine sulfoximine): A selective inhibitor of γ-glutamylcysteine synthetase. J. Biol. Chem., 254, 1205-1210 (1979).
15. O.W. Griffith, Mechanism of action, metabolism and toxicity of buthionine sulfoximine and its higher homologs, potent inhibitors of glutathione synthesis. J. Biol. Chem., 257, 13704-13712 (1982).

16. O.W. Griffith and A. Meister, Interorgan translocation, turnover and metabolism. Proc. Natl. Acad. Sci. U.S.A., 76, 4932-4935 (1979).
17. O.W. Griffith and A. Meister, Origin and turnover of mitochondrial glutathione. Proc. Natl. Acad. Sci. U.S.A., 82, 4668-4672 (1985).
18. A.P. Seddon and A. Meister, unpublished data (1983).
19. J.K. Dethmers and A. Meister, Glutathione export by human lymphoid cells: Depletion of glutathione by inhibition of its synthesis decreases export and increases sensitivity to irradiation. Proc. Natl. Acad. Sci. U.S.A., 78, 7492-7496 (1985).
20. A. Meister, Modulation of intracellular levels of glutathione, In: Biochemical Modulation of Anticancer Agents: Experimental and Clinical Approaches (F. Valeriote and L. Baker, Eds.), pp. 245-275. Martinus Nijhaus, Boston, 1986.
21. A. Meister and O.W. Griffith, Effects of methionine sulfoximine analogs on the synthesis of glutamine and glutathione: Possible chemotherapeutic implications. Cancer Treat. Rep., 63, 1115-1121 (1979).
22. M. Guichard, G. Jensen, A. Meister and E.P. Clark and E.R. Epp, The role of thiols in cellular response to radiation and drugs. Radiat. Res., 94, 613 (1983).
23. J.E. Biaglow, M.E. Varnes, E.P. Clark and E.R. Epp, The role of thiols in cellular response to radiation and drugs. Radiat. Res., 95, 437-455 (1983).
24. J.B. Mitchell, A. Russo, J.E. Biaglow, and A. McPherson, Cellular glutathione depletion by diethyl maleate or buthionine sulfoximine: No effect of glutathione depletion on the oxygen enhancement ratio. Radiat. Res., 96., 422-428 (1983).
25. B.A. Arrick, C.F. Nathan, O.W. Griffith and Z.A. Cohn, Glutathione depletion sensitizes tumor cells to oxidative cytolysis. J. Biol. Chem., 257, 1231-1237 (1982).
26. B.A. Arrick, C.F. Nathan, and Z.A. Cohn, Inhibition of glutathione synthesis augments lysis of murine tumor cells by sulfhydryl-reactive antineoplastics. J. Clin. Invest., 71, 258-267 (1983).
27. K. Suzakae, B.J. Petro and D.T. Vistica, Reduction in glutathione content of L-PAM resistant L1210 cells confers drug sensitivity. Biochem. Pharmacol., 31, 121-124 (1982).
28. K. Suzukake, B.P. Vistica, and D.P. Vistica, Dechlorination of L-phenylalanine mustard by sensitive and resistant tumor cells and its relationship to intracellular glutathione content. Biochem. Pharmacol., 32, 165-167 (1983).
29. D.T. Vistica, S. Somfai-Relle, K. Suzukake and B. Petro, Inhibition of glutathione biosythesis by S-n-butyl homocysteine sulfoximine and sensitization of murine tumor cells resistant to L-phenylalanine mustard. J. Cell Biochem. Suppl., 6, 375 (1982).
30. K.G. Louie, B.C. Behrens, T.J. Kihsella, T.C. Hamilton, K.R. Grotzinger, W.M. McKoy, M.A. Winker and R.F. Ozols, Radiation survival parameters of antineoplastic drug-sensitive and resistant human ovarian cancer cell lines and their modification by buthionine sulfoximine. Cancer Res., 45, 2110-2115 (1985).
31. R. Ozols, T.C. Hamilton, K.G. Louie, B.C. Behrens and R.C. Young, Glutathiione depletion with buthionine sulfoximine: Potential clinical applications, In: Biochemical Modulation of Anticancer Agents: Experimental and Clinical Approaches (F. Valeriote and L. Baker, Eds.), pp. 277-294. Martinus Nijhaus, Boston, 1986.
32. P. deMiranda, L.M. Beacham III, T.H. Creagh and G.B. Elion, The Metabolic fate of methylnitroimidazole moiety of azanthioprine in the rat. J. Pharmacol. Exp. Ther., 187, 588-601 (1973).
33. W.G. DeGraff, A. Russo and J.B. Mitchell, Glutathione depletion greatly reduces neocarzinostatin cytoxicity in Chinese hamster V79 cells. J. Biol. Chem., 260, 8312-8315 (1985).

34. M.E. Anderson and A. Meister, Intracellular delivery of cysteine. Methods Enzymol., 143 (1987), in press.
35. S.M. Birnbaum, M. Winitz and J.P. Greenstein, Quantitative nutritional studies with water-soluble, chemically defined diets. III. Individual amino acids as sources of "non-essential" nitrogen. Arch. Biochem. Biophys., 72, 428-436 (1957).
36. J.W. Olney, O.L. Ho and V. Rhee, Cytotoxic effects of acidic and sulphur containing amino acids on the infant mouse central nervous system. Brain Res., 14, 61-76 (1971).
37. R.L. Karlsen, I. Grofova, D. Malthe-Sorenssen and F. Fonnum, Morphological changes in rat brain induced by L-cysteine and injection in newborn animals. Brain Res., 208, 167-180 (1981).
38. Y. Nishiuch, M. Sasaki, M. Nakayasu and A. Oikawa, Cytotoxicity of cysteine in culture media. In Vitro 12., 635-638 (1976).
39. J.M. Williamson, B. Boettcher and A. Meister, An intracellular cysteine delivery system that protects against toxicity by promoting glutathione synthesis. Proc. Natl. Acad. Sci. U.S.A., 79, 6246-6249 (1982).
40. J.M. Williamson and A. Meister, Stimulation of hepatic glutathione formation by administration of L-2-oxothiazolidine-4-carboxylate, a 5-oxo-L-prolinase substrate. Proc. Natl. Acad. Sci. U.S.A., 78, 936-939 (1981).
41. J.M. Williamson and A. Meister, New substrates of 5-oxo-L-prolinase. J. Biol. Chem., 257, 12039-12042 (1982).
42. A. Meister, M.E. Anderson and O. Hwang, Intracellular cysteine and glutathione delivery systems. J. Am. Coll. Nutr., 5, 137-151 (1986).
43. M.E. Anderson, M. Underwood, R.J. Bridges, and A. Meister, Glutathione transport and metabolism in the choroid plexus. Fed. Proc., 45, 1733 (1986).
44. O.W. Griffith, R.J. Bridges and A. Meister, Transport of γ-glutamyl amino acids; role of glutathione and γ-glutamyl transpeptidase. Proc. Natl. Acad. Sci. U.S.A., 76, 6319-6322 (1979).
45. R.J. Bridges and A. Meister, γ-Glutamyl amino acids; transport and conversion to 5-oxoproline in the kidney. J. Biol. Chem., 260, 7304-7308 (1985).
46. M.E. Anderson and A. Meister, Transport and direct utilization of γ-glutamylcyst(e)ine for glutathione synthesis. Proc. Natl. Acad. Sci. U.S.A., 83, 1246-1250 (1986).
47. R.N. Puri and A. Meister, Transport of glutathione as γ-glutamylcysteinylglycyl ester into liver and kidney. Proc. Natl. Acad. Sci. U.S.A., 80, 5258-5260 (1983).
48. V.P. Wellner, M.E. Anderson, R.N. Puri, G.L. Jensen and A. Meister, Radioprotection by glutathione ester: Transport of glutathione ester into human lymphoid cells and fibroblasts. Proc. Natl. Acad. Sci. U.S.A., 81, 4732-4735 (1984).
49. M.E. Anderson, F. Powrie, R.N. Puri and A. Meister, Glutathione monoethyl ester; preparation, uptake by tissues, and conversion to glutathione. Arch. Biochem. Biophys., 239, 538-548 (1985).
50. G.L. Jensen and A. Meister, Radioprotection of human lymphoid cells by exogenously-supplied glutathione is mediated by γ-glutamyl transpeptidase. Proc. Natl. Acad. Sci. U.S.A., 80, 4714-4714 (1983).
51. A. Meister, S.S. Tate and O.W. Griffith, γ-Glutamyl transpeptidase. Methods Enzymol., 77, 237-253 (1981).
52. M.E. Anderson and A. Meister, Inhibition of γ-glutamyl transpeptidase and induction of glutathionuria by γ-glutamyl amino acids. Proc. Natl. Acad. Sci. U.S.A., 83, 5029-5032 (1986a).
53. A. Perantoni, J.J. Berman and J.M. Rice, L. Azaserine toxicity in established cell lines. Correlation with γ-glutamyl transpeptidase activity, Exp. Cell Res., 122, 55-61 (1979).

INHIBITION OF MUTAGENESIS AND CARCINOGENESIS BY N-ACETYLCYSTEINE

Silvio De Flora, Carlo Bennicelli, Anna Camoirano, Domizio Serra, Cristina Basso, Patrizia Zanacchi and Carmelo F. Cesarone*

Institute of Hygiene and *Institute of General Physiology
University of Genoa
I-16132 Genoa, Italy

INTRODUCTION

N-acetylcysteine (NAC), a precursor of intracellular cysteine and glutathione (GSH)) (1), is extensively used in the treatment of patients suffering from respiratory diseases. Moreover, this synthetic aminothiol has been shown to possess a variety of antitoxic properties in humans, animals and in vitro test systems (2).

These features prompted us to investigate the inhibitory effects of NAC in mutagenesis and carcinogenesis. With this objective, we have been carrying out a series of studies, using various experimental models. The in vitro genotoxicity assays included the Ames reversion test, evaluating the reversion to prototrophy of his^- S. typhimurium strains (3), and a DNA-repair test, evaluating the differential lethality in E. coli WP2 (wild strain) and CM871 ($uvrA^-$ $lexA^-$ $recA^-$) (4). The in vivo studies aimed at assessing the modulation of biochemical parameters and the anticarcinogenic effets of NAC in rodents.

INHIBITION BY THIOLS OF THE "SPONTANEOUS" MUTABILITY IN S. TYPHIMURIUM

NAC did not affect the spontaneous mutations in a variety of S. typhimurium strains, i.e. TA1535, TA1538, TA98, TA100 and TA102. However, both NAC and GSH were consistently able to decrease the high spontaneous mutability of strain TA104. Dose-response curves with varying amounts of NAC are shown in Figure 1. Such phenomenon was observed under various experimental conditions, and irrespective of the addition of exogenous metabolic systems.

The differential genetic features of TA104, as compared to TA102 and other S. typhimurium strains (5), suggest that reactive oxygen species, producing a DNA damage reparable by the excision repair system, can be scavenged by thiols. This interpretation supports the view that oxygen radicals provide an important contribution to the spontaneous genetic instability (6).

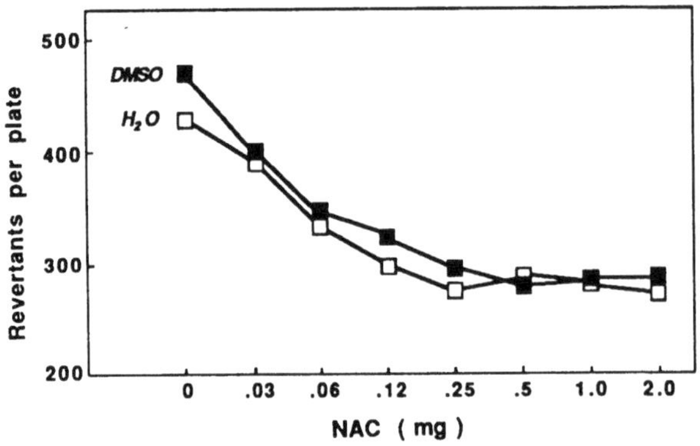

Figure 1. Inhibition by NAC of the spontaneous mutagenicity of strain TA104 of S. typhimurium.

EFFECT OF NAC ON DIRECT-ACTING GENOTOXIC COMPOUNDS

Previous studies demonstrated the ability of NAC to decrease, with dose-related effects, the direct mutagenicity in S. typhimurium strains of compounds of various chemical classes, including sodium dichromate, 4-nitroquinoline N-oxide (4NQO), epichlorohydrin and hydrogen peroxide (7), N-methyl-N'-nitro-N-nitrosoguanidine (MNNG) and nitrofluorene (8). Further assays provided evidence that the same trend also occurred with other compounds, such as formaldehyde (Figure 2) and glutaraldehyde (also under vapour form), the pesticides captan and folpet, the antitumor compounds ICR 191 and ICR 170, sodium nitrite and β-propiolactone (9).

Nucleophiles may be useful for distinguishing those mutagens which are bioactivated inside the target bacterial cells. This may be the case of nitrofurantoin, activated by bacterial nitroreductases, whose mutagenicity was poorly affected by GSH and NAC (Figure 3). Note however that the residual mutagenicity of nitrofurantoin in a nitrosoreductase-deficient strain (TA100NR) was decreased by NAC (Figure 3), which suggests a weak nitrosoreductase-independent and thiol-inhibible mutagenicity of this compound. Sodium azide is another compound activated by bacteria, which is poorly inhibited by thiols (8,10). Using the bacterial DNA-repair test, NAC decreased the genotoxic potency of hydrogen peroxide (11), 4NQO, MNNG and ICR 191 (9).

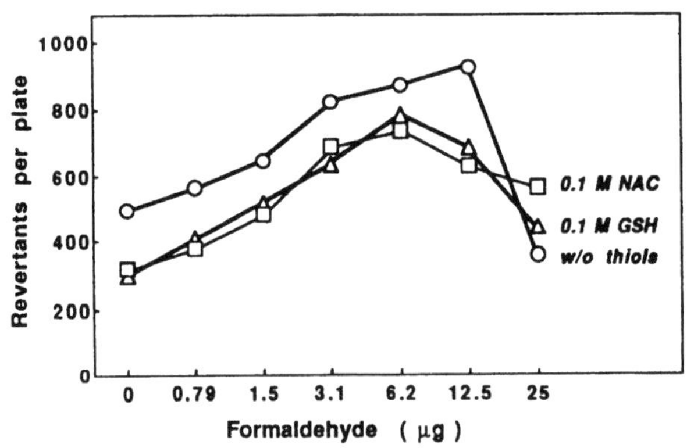

Figure 2. Protective effects of thiols on the spontaneous and formaldehyde-induced mutagenicity in S. typhimurium (TA104).

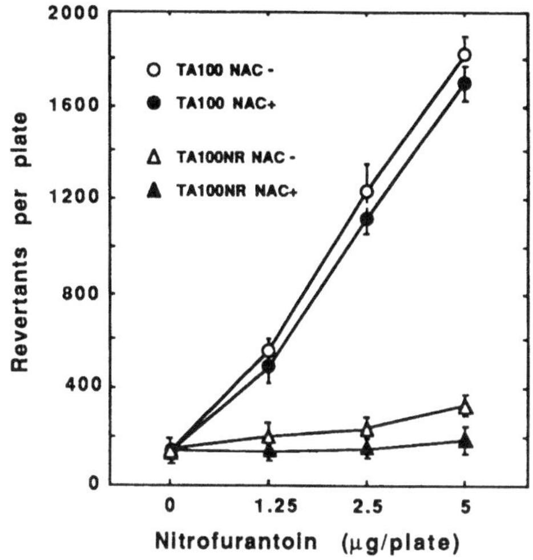

Figure 3. Effect of NAC (4 μmol) on nitrofurantoin mutagenicity in strains TA100 and TA100NR of S. typhimurium.

EFFECT OF NAC, GSH AND ASCORBIC ACID ON THE FORMATION OF NITROSODERIVATIVES AND ON THEIR GENOTOXICITY

The in vitro reaction of nitrite with the histamine H_2-receptor antagonists ranitidine (12) and famotidine (13), under acidic condition, led to formation of genotoxic derivatives. NAC and GSH lowered the mutagenicity of nitrite-drug mixtures to a similar extent as ascorbic acid, but with different mechanisms. In fact, while ascorbic acid inhibited the nitrosation reaction without affecting the mutagenicity of nitrosoderivatives, both thiols had poor effects on the reaction but decreased the mutagenicity of its products (9).

Ascorbic acid also failed to affect MNNG mutagenicity, which was conversely efficiently diminished by NAC and GSH (Figure 4). A similar

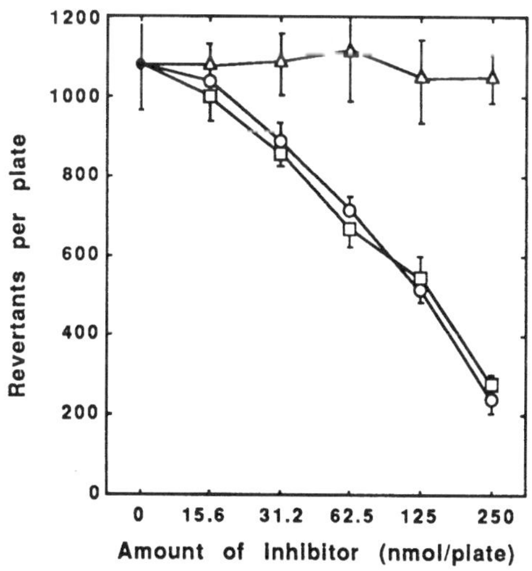

Figure 4. Effect of ascorbic acid (Δ), GSH (O) and NAC (□) on MNNG mutagenicity in strain TA100 of S. typhimurium.

trend was observed in the DNA-repair test (9). On the other hand, pretreatment of bacteria with thiols and with the GSH depletor diethyl maleate (DEM) resulted in an enhancement and in a decrease of genotoxicity, respectively (9). These data are consistent with the hypothesis that thiols play a dual role in MNNG genotoxicity, i.e. they exert an inhibitory effect when the reaction occurs outside cells, and an activating effect when the reaction occurs inside cells, in proximity of target DNA molecules (see ref. 10 for a thorough discussion).

EFFECT OF NAC ON GENOTOXIC COMPOUNDS REQUIRING METABOLIC ACTIVATION

The effect of thiols on the activity of promutagens is quite controversial (10). In vitro studies performed in our laboratory (7) showed that the highest nontoxic concentration of NAC abolished the S9-mediated mutagenicity of benzo(a)pyrene, 2-aminofluorene (2AF), cyclophosphamide, aflatoxin B1 (AFB1), a tryptophan pyrolysate product (Trp-P-2) and a cigarette smoke condensate. However, lower concentrations of NAC often resulted in some enhancement of mutagenicity. Further assays (14), also using liver or lung preparations of rats pretreated in vivo with either NAC, DEM, Aroclor, or their combinations, suggested that, in the Ames test, the effect of NAC on promutagens is the result of a balance between a stimulation of their activity and trapping of electrophilic metabolites. The final result also depends on the state of induction of animals (10,14). In the DNA-repair test, the genotoxicity of 2AF and AFB1 was consistently decreased by NAC at all tested doses (11).

IN VIVO METABOLIC STUDIES

In vivo studies were carried out by treating 10 groups of Sprague-Dawley rats (i.p. injections) with various combinations of NAC, diethyl maleate (DEM), buthionine sulfoximine (BSO) and Aroclor 1254 (14). In synthesis, NAC increased intracellular GSH in erythrocytes and in liver and lung cells, and generally replenished its stores following depletion (Figure 5). It did not affect cytochromes P-450 in hepatic and pulmonary microsomes (Figure 6), whereas it stimulated, especially in Aroclor-pretreated animals, cytosolic enzyme activities, such as glucose 6-phosphate dehydrogenase (G6PD), 6-phosphogluconate dehydrogenase (6PGD), GSSG reductase and NAD(P)H- dependent diaphorases (14). Moreover, treatment of rats with NAC enhanced GSH S-transferase, G6PD, 6PGD, diaphorases and Cr(VI) reduction in pulmonary alveolar macrophages (15). Thus, also in vivo NAC appears to exert protective effects by modulating GSH metabolism and the biotransformation of mutagenic/carcinogenic compounds.

ANTICARCINOGENICITY STUDIES

An anticarcinogenicity study (16) in Swiss albino mice provided evidence that the continuous supplement of NAC (0.2 %) to the diet, before (15 days) and after (4 months) a single i.p. injection of urethan (1 g/kg b.w.), efficiently prevented the induction of lung tumors (Figure 7). Irrespective of urethan administration, NAC also significantly enhanced GSH S-transferase activity in liver preparations of the same animals. Other treatment schedules - i.e. dietary NAC only before or at various times after injection of the carcinogen - did not significantly affect the frequency nor the multiplicity of lung tumors.

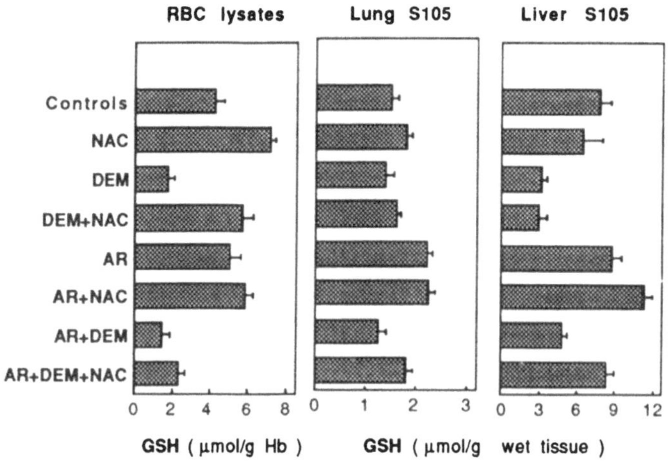

Figure 5. Concentration of GSH in red blood cell (RBC) lysates and lung or liver S105 fractions from variously treated rats.

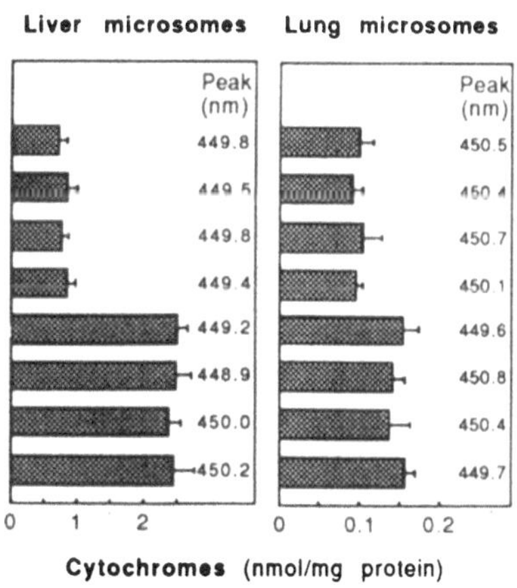

Figure 6. Properties of cytochromes P-450 in microsomes from variously treated rats.

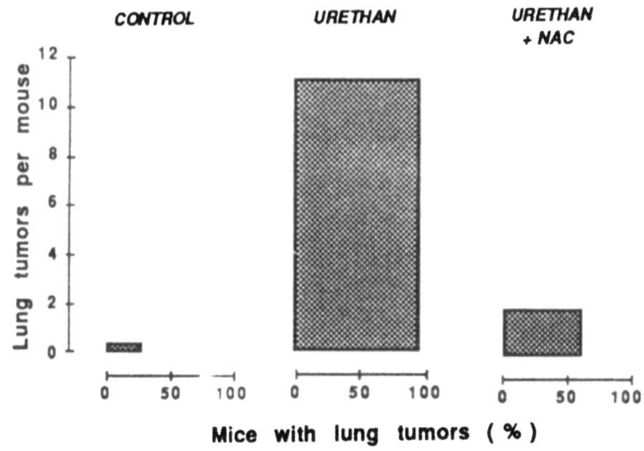

Figure 7. Inhibition by NAC of the induction of lung tumors by urethan

As reported separately (17), NAC also showed protective effects in a carcinogenesis model in the rat, evaluating the induction of Zymbal gland tumors and of liver preneoplastic lesions by 2-acetylaminofluorene.

ACKNOWLEDGEMENTS

These studies were supported by CNR (Special Projects "Oncology" and "Preventive and Rehabilitative Medicine", and National "Cardiorespiratory" Group).

REFERENCES

1. P. Moldéus, I. A. Cotgreave, and M. Berggren, Lung protection by a thiol-containing antioxidant: N-acetylcysteine. Respiration 50, 31-42 (1986).
2. F. M. Muggia, et al., N-acetylcysteune (NAC): a significant chemoprotective adjunct. Seminars in Oncology 10, 1-92 (1983).
3. D. M. Maron and B. N. Ames, Revised methods for the Salmonella mutagenicity test. Mutat. Res. 113, 173-215 (1983).
4. S. De Flora, P. Zanacchi, A. Camoirano, C. Bennicelli, and G. Badolati, Genotoxic activity and potency of 135 compounds in the Ames reversion test and in a bacterial DNA-repair test. Mutat. Res. 133, 161-198 (1984).
5. D. E. Levin, M. Hollstein, M. F. Christman, E. A. Schwiers, and B. N. Ames, A new Salmonella tester strain (TA102) with A-T base pairs at the site of mutation detects oxidative mutagens. Proc. Natl. Acad. Sci. USA 79, 7445-7449 (1982).
6. H. Joenje, Introduction: oxygen metabolism, DNA repair, and the origin of spontaneous genetic instability. In Antimutagenesis and Anticarcinogenesis Mechanisms, 1986 (D. M. Shankel, P. E. Hartman, T. Kada and A. Hollaender, Eds.) pp. 339-341. Plenum Press, New York and London.
7. S. De Flora, C. Bennicelli, P. Zanacchi, A. Camoirano, A. Morelli, and A. De Flora, In vitro effects of N-acetylcysteine on the mutagenicity of direct-acting compounds and procarcinogens. Carcinogenesis 5, 505-510 (1984).
8. M. Wilpart, P. Mainguet, D. Geeroms, and M. Roberfroid,

Desmutagenic effects of N-acetylcysteine on direct and indirect mutagens. Mutat. Res. 142, 169-177 (1985).
9. S. De Flora, et al., manuscripts in preparation.
10. S. De Flora, C. Bennicelli, D. Serra, A. Izzotti, and C. F. Cesarone, Role of glutathione and N-acetylcysteine as inhibitors of mutagenesis and carcinogenesis. In Absorption and Utilization of Amino Acids (M. Friedman, Ed.). CRC Press, Inc., Boca Raton, FL (in press).
11. S. De Flora, Detoxification of genotoxic compounds as a threshold mechanism limiting their carcinogenicity. Toxicol. Pathol. 12, 337-343 (1984).
12. S. De Flora, C. Bennicelli, A. Camoirano, and P. Zanacchi, Genotoxicity of nitrosated ranitidine. Carcinogenesis 4, 255-260 (1983).
13. S. De Flora, A. Camoirano, C. Basso, M. Astengo, P. Zanacchi, and C. Bennicelli, Bacterial genotoxicity of nitrosated famotidine. Mutagenesis 1, 125-130 (1986).
14. S. De Flora, C. Bennicelli, A. Camoirano, D. Serra, M. Romano, G. A. Rossi, A. Morelli, and A. De Flora, In vivo effects of N-acetylcysteine on glutathione metabolism and on the biotransformation of carcinogenic and/or mutagenic compounds. Carcinogenesis 6, 1735-1745 (1985).
15. S. De Flora, M. Romano, C. Basso, M. Bagnasco, C. F. Cesarone, G. A. Rossi, and A. Morelli, Modulation of detoxifying activities in alveolar macrophages of rats treated with acetylcysteine, diethyl maleate and/or Aroclor. Anticancer Res. 6, 1009-1012 (1986).
16. S. De Flora, M. Astengo, D. Serra, and C. Bennicelli, Inhibition of urethan-induced lung tumors in mice by dietary N-acetylcysteine. Cancer Lett. 32, 235-241 (1986).
17. C. F. Cesarone, L. Scarabelli, M. Orunesu, M. Bagnasco, A. Izzotti, and S. De Flora, Protective effects of thiols on carcinogenesis induced in rats by 2-acetylaminofluorene. This volume.

PROTECTIVE EFFECTS OF THIOLS ON CARCINOGENESIS

INDUCED IN RATS BY 2-ACETYLAMINOFLUORENE

Carmelo F. Cesarone, Linda Scarabelli, Mauro Orunesu,
Maria Bagnasco*, Alberto Izzotti* and Silvio De Flora*

Institute of General Physiology and *Institute of Hygiene
University of Genoa
I-16132 Genoa, Italy

INTRODUCTION

Aminothiols play a fundamental role in the physiology of the living cell. The most abundant and best studied compound of this class is glutathione (GSH). It is considered of paramount importance for many functions, such as amino acids transport, protein and DNA synthesis, membrane structure and activity (for thorough reviews see refs. 1 and 2). Moreover, GSH being a nucleophilic compound, it hampers alterations induced by reactive oxygen species and by endogenous or exogenous metabolites (for a review see ref. 3). Depletion of hepatic GSH levels affects a number of biochemical functions, as well as the efficiency of the cell response to physically or chemically induced injuries (3,4).

EXPERIMENTAL

The data obtained in previous works (5-10), showing antimutagenic and anticarcinogenic effects of N-acetylcysteine (NAC), prompted us to further study the effect of GSH or NAC administration in rats exposed to a discontinuous feeding regimen containing 0.05% 2-acetylamino-fluorene (2AAF). This experimental model, described by Teebor and Becker (11), was used to induce hyperplastic nodules in the liver. Briefly, six groups of male Wistar rats, weighing 80-100 g, received a standard Vogt-Möller diet (24.5% protein) supplemented as follows: no supplement (group A), 0.1% GSH (B), 0.1% NAC (C), 0.05% 2AAF (D), 0.05% 2AAF + 0.1% GSH (E), 0.05% 2AAF + 0.1% NAC (F). A complete exposure consisted of a 4-cycle period. Each cycle was composed by a 3-week treatment with the aforementioned diets, followed by 1 week of standard meal.

SURVIVAL RATE AND INHIBITION OF ZYMBAL GLAND CARCINOMAS

The survival rate was 100% in groups A (24/24), B (24/24), C (24/24) and E (33/33), 86% (37/43) in group D and 91% (30/33) in group F.

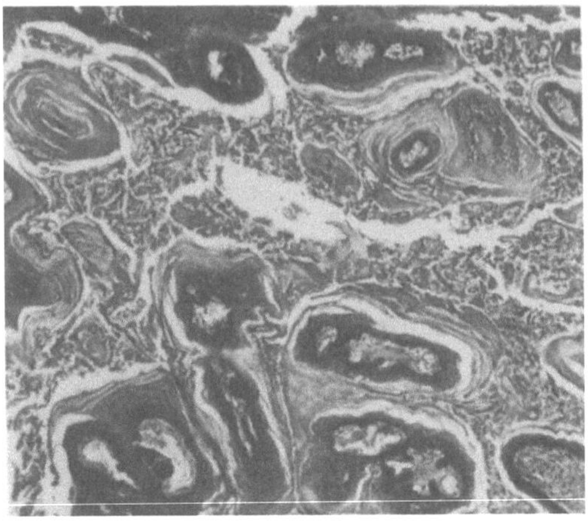

Figure 1. Histology of Zymbal gland carcinoma in 2AAF-treated rats.

After a complete exposure (4 cycles), a few animals received the standard diet for additional 4-8 weeks. All 5 remaining rats, exposed to 2AAF alone, developed deforming necrotic periauricular carcinomas of the sebaceous glands (Zymbal). Uncommon to occur spontaneously, these tumors, under histologic examination, exhibited clusters of squamous cells, containing epithelial beads and sebaceous material (Figure 1). No such tumors were detectable in control groups (A,B,C) nor in the 16 remaining rats fed with the carcinogen plus GSH or NAC. Protection by thiols was highly significant ($P < 0.001$).

EFFECT ON LIVER DNA DAMAGE AND REPAIR

Liver DNA damage was detected by a modification (12) of the alkaline elution technique (13). DNA was estimated by fluorometric assay (14), and the elution rate constant (K) was computed as described earlier (12). In the liver, on the 3rd week of each cycle, 2AAF treatment produced a marked increase in DNA fragmentation (Figure 2). It is noteworthy that, during the week of standard meal feeding (i.e. the 4th week of each cycle), although significantly repaired, the damage to DNA was still detectable. During the first two cycles,

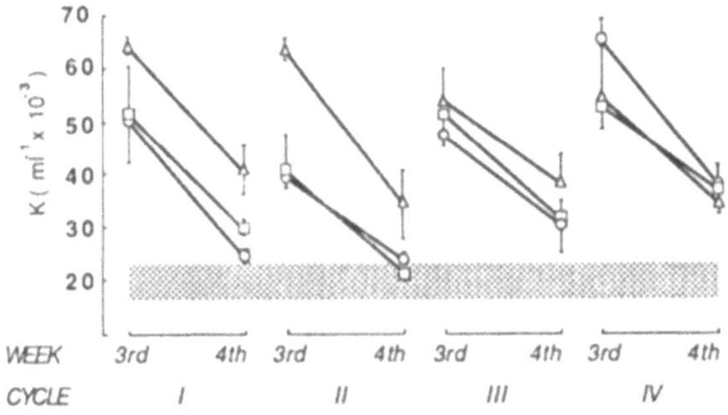

Figure 2. Elution rate constant for liver DNA from rats fed with 2AAF (△) or 2AAF plus GSH (○) or NAC (□) (dashed area = control values).

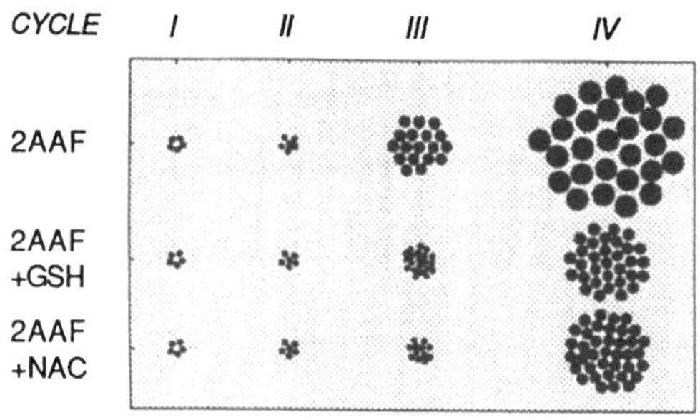

Figure 3. Number of GGT-positive foci per sq cm of liver in variously treated rats. Full circles represent foci of increasing diameter, i.e. < 0.05, 0.06-0.3, or > 0.3 mm.

both GSH and NAC protected DNA from the damage produced by 2AAF metabolites. Thiols were considerably less effective during the 3rd cycle of treatment and, during the 4th cycle, neither GSH nor NAC was able to reduce the residual damage.

EFFECT ON THE DEVELOPMENT OF LIVER GGT-POSITIVE FOCI

It is well known that the induction of liver hyperplastic nodules with chemical carcinogens stimulates the activity of gamma-glutamyl transpeptidase (GGT) (15). The treatment with 2AAF produced the appearance of a number of GGT-positive foci. The co-treatment with aminothiols did not affect their development during the first two cycles (Figure 3). The number and the size of GGT-positive foci was markedly increased in 2AAF-treated animals during the 3rd and the 4th cycle. As shown by the reduced size of lesions (Figure 3), the supplement of NAC or GSH to exposed rats inhibited the growth of GGT-positive foci.

MODULATION OF BIOCHEMICAL MECHANISMS

A number of biochemical parameters were also assayed, according to previously described procedures (16), in order to investigate the gamma-glutamyl cycle activities, along with other cytosolic pathways (Figure 4). A dramatic increase in membrane-bound GGT was induced, during the last two cycles, by 2AAF exposure. This enhancement was only slightly reduced by thiols that, moreover, had only a partial effect on the sharp decrease of another liver membrane-bound activity, i.e., G6Pase. The hepatic GSH level, as detected fluorometrically, was enhanced by 2AAF on the 4th cycle and further increased by GSH co-administration. Several catalytic activities, such as GSH S-transferase (GST), GSSG reductase, G6P dehydrogenase (G6PD) and NAD(P)H-dependent diaphorases were stimulated by 2AAF treatment, generally since the 1st cycle. The administration of GSH or NAC to rats fed with the standard diet resulted in a significant increase in GSH peroxidase, G6PD and diaphorases activity. In 2AAF-treated animals, both aminothiols produced a further stimulation in GST during the last two cycles.

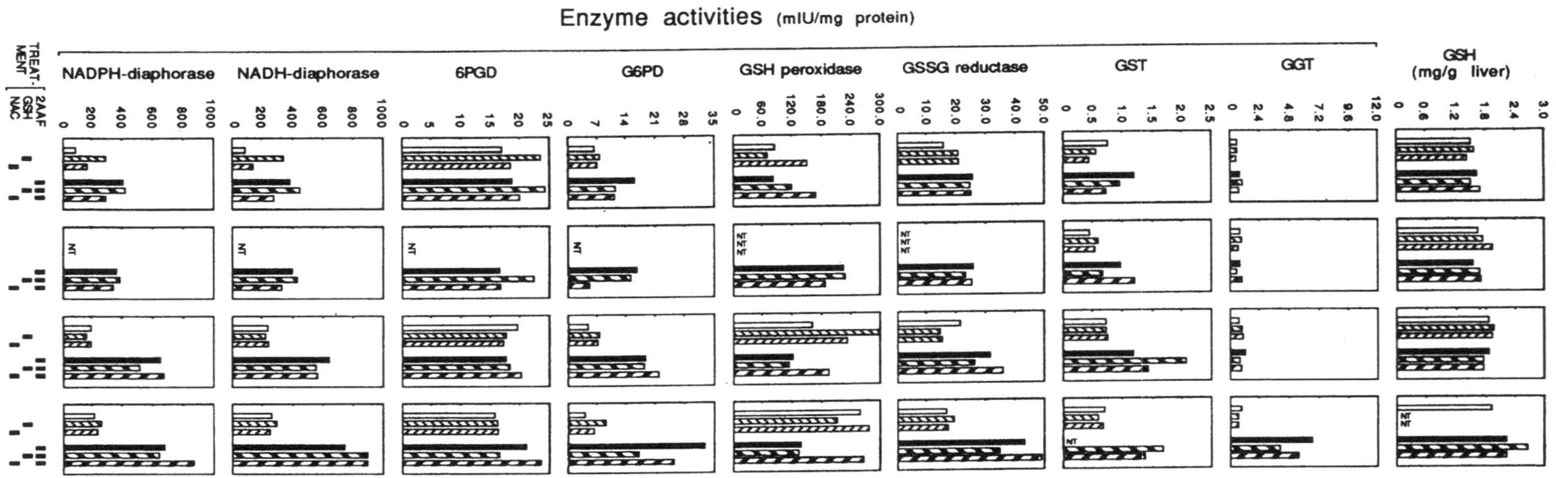

Figure 4. Measurement of hepatic GSH and of cytosolic enzyme activities in variously treated rats.

METABOLIC ACTIVATION OF ARYLAMINES

Liver S12 fractions were also assayed for their ability to activate 2AAF and other aromatic amines, either structurally related (i.e. 4AAF and 2AF) or unrelated (i.e. 2-naphtylamine and benzidine), to mutagenic metabolites in strain TA98 of S. typhimurium (17). In control rats, there was a consistent and marked trend to an age-dependent loss of metabolic activation of all tested arylamines during the 16 weeks of treatment. 2AAF feeding resulted in an evident autoinduction of metabolism, which was continuously amplified with time, even during the withdrawal weeks. In the same animals, the activation of the other amines was initially inhibited, but later on progressively shifted to a mild cross-induction which, in the case of the structurally related compounds, became significant at the end of the 4th cycle. The metabolic effects of GSH and NAC in this system were broadly variable, depending on the promutagen and thiol tested, on its co-administration with 2AAF, and on the week and cycle of treatment.

CONCLUSIONS

On the basis of the data obtained, it can be concluded that thiols exert a protective effect on liver cells during the early stages of 2AAF treatment, as assessed by evaluating the DNA damage and repair synthesis, as well as by the analysis of the development of GGT-positive foci. Moreover, thiols prevented Zymbal gland tumors in the same animals, which agrees with the protective effects produced by NAC in pulmonary (8) and colon (9) carcinogenesis and in various genotoxicity test systems (18).

ACKNOWLEDGEMENTS

This study was supported by grants 85.02090.44 and 85.02125.44 of CNR Special Project "Oncology". L. S. is recipient of a fellowship from the National Institute for Research on Cancer (IST, Genoa).

REFERENCES

1. A. Meister and M.E. Anderson, Glutathione, Annu. Rev. Biochem. 52, 711-760 (1983).
2. A. Larson, S. Orrenius, A. Holmgren and B. Mannervik (Eds.), Functions of Glutathione: Biochemical, Physiological, Toxicological and Clinical Aspects. Raven Press, New York, 1983.
3. O.F. Nygaard, M.G. Simic and J.N. Hauber (Eds.) Radioprotectors and Anticarcinogens. Academic Press, New York, 1983.
4. J.H.N. Meerman and R.B. Tijdens, Effect of glutathione depletion on the hepatotoxicity and covalent binding to the rat liver macromolecules of N-hydroxy-2-acetylaminofluorene, Cancer Res. 45, 1132-1139 (1985).

5. C.F. Cesarone, L. Scarabelli and M. Orunesu, Effect of glutathione on alterations of liver DNA structure and metabolic activities induced in vivo by 2-acetylaminofluorene, Anticancer Res 6, 1233-1286 (1986).
6. C.F. Cesarone, L. Scarabelli and M. Orunesu, Effect of glutathione and N-acetylcysteine on hepatocellular modifications induced by 2-acetylaminofluorene, Toxicol. Pathol. 14, 445-450 (1986).
7. S. De Flora C. Bennicelli, A. Camoirano, D. Serra, M. Romano, G.A. Rossi, A. Morelli and A. De Flora, In vivo effects of N-acetylcysteine on glutathione metabolism and on the biotransformation of carcinogenic and/or mutagenic compounds, Carcinogenesis 6, 1735-1745 (1985).
8. S. De Flora, M. Astengo, D. Serra and C. Bennicelli, Inhibition of urethan-induced lung tumors in mice by dietary N-acetylcysteine, Cancer Lett. 32, 135-241 (1986).
9. M. Wilpart, A. Speder and M. Roberfroid, Anti-initiation activity of N-acetylcysteine in experimental colonic carcinogenesis, Cancer Lett 31, 319-324 (1986).
10. J.Y.H. Chan, D.L. Stout and F.F. Becker, Protective role of thiols in carcinogen-induced DNA damage in rat liver, Carcinogenesis 7, 1621-1624 (1986).
11. G.W. Teebor and F.F. Becker, Regression and persistence of hyperplastic nodules induced by N-2-fluorenylacetamide and their relationship to hepatocarcinogenesis, Cancer Res. 31, 1-3 (1971).
12. C.F. Cesarone, C. Bolognesi and L. Santi, DNA-damage induced in vivo in various tissues by nitrochlorobenzene derivatives, Mutat. Res. 116, 238-246 (1983).
13. K.W. Kohn, L.C. Erickson, R.A. Grimek-Eqing and C.A. Friedman, Fractionation of DNA from mammalian cells by alkaline elution, Biochemistry 15, 4629-4637 (1976).
14. C.F. Cesarone, C. Bolognesi and L. Santi, Improved microfluorometric DNA determination in biological material using 33258 Hoechst, Anal. Biochem. 100, 188-197 (1979).
15. H. Pitot and M.H. Hanigan, Gamma-glutamyl transpeptidase: its role in hepatocarcinogenesis, Carcinogenesis 6, 165-172 (1985).
16. C.F. Cesarone, M. Romano, D. Serra, L. scarabelli and S. De Flora, Effects of aminothiols in 2-acetylaminofluorene-treated rats. II. Glutathione cycle and liver cytosolic activities, In Vivo 1, No. 2 (1987).
17. D.M. Maron and B.N. Ames, Revised methods for the Salmonella mutagenicity test, Mutat. Res. 113, 173-215 (1983).
18. S. De Flora, C. Bennicelli, A. Camoirano, D. Serra, C. Basso, P. Zanacchi and C. F. Cesarone, This volume.

THE ROLE OF THIOLS IN RESPONSE TO RADIATION AND DRUGS

John E. Biaglow, Marie E. Varnes,
Edward R. Epp*, and Edward P. Clark*

University of Pennsylvania, Philadelphia, PA
Case Western Reserve University, Cleveland, OH
*Harvard University, Boston, MA

Glutathione (GSH) has been implicated as the major intracellular nonprotein thiol involved in protecting hypoxic cells against radiation injury (1-10). In contrast, we have demonstrated increased radiosensitivity for GSH-depleted aerobic and hypoxic A549 human lung carcinoma cells (2-5). This parallel increase in radiosensitivity conflicts with the popular interpretation of existing data regarding the effect of GSH depletion on hypoxic sensitivity alone (5). Subsequently, ratios of (hypoxic slope)/(aerobic slope) may be misleading if GSH depletion alters both responses, but by different mechanisms.

We have performed extensive research to substantiate the role of GSH in the radiation response. In keeping with the Alper - Howard-Flanders model, we hypothesize that the reaction of radiation-induced radicals with oxygen occurs rapidly, yielding peroxyl radicals and hydroperoxides (1). Therefore, GSH depletion would reduce hydroxyl radicals produced by radiolysis of water under aerobic or hypoxic conditions. GSH depletion would selectively alter the aerobic radiation response, however, because peroxyl radicals and hydroperoxides are not detoxified (Fig. 1), and in agreement with the original model (11), cellular lethality increases (6). It follows then that the hypoxic radiation response, under certain conditions of GSH depletion, may only represent the increased sensitivity for the aerobic cells. Consequently, the greater effect of GSH depletion on irradiated aerobic A549 cells results in the increased value obtained from the ratio (i.e., 5) of (hypoxic slope)/-(aerobic slope) (12-14). These results suggest that GSH depletion more effectively sensitizes aerobic cells than hypoxic cells. We are, therefore, interested in the relationship between GSH depletion and subsequent inhibition of key enzymes, which catalyze the reduction of hydroperoxides (these peroxides are proposed as oxygen intermediates involved in cellular lethality) (14, 15).

We have used the GSH synthesis inhibitor L-buthionine sulfoximine (L-BSO) to cause a decrease in cellular GSH in order to determine its importance in the radiation response. We have determined the effects of both L-BSO concentration level and length of incubation time on the GSH content of A549 cells (Fig. 2). Concentration of L-BSO of 0.1 mM or less for 72 hr did not affect the growth rate of these cells. High cell densities of 3.2×10^6, provided greatest accuracy for GSH determinations. Incubation for 72 hr in 0.1 mM L-BSO depleted GSH 2000-fold. Concentrations of 0.1-0.25 mM L-BSO added to the growth medium produced no growth inhibition nor did it increase toxicity.

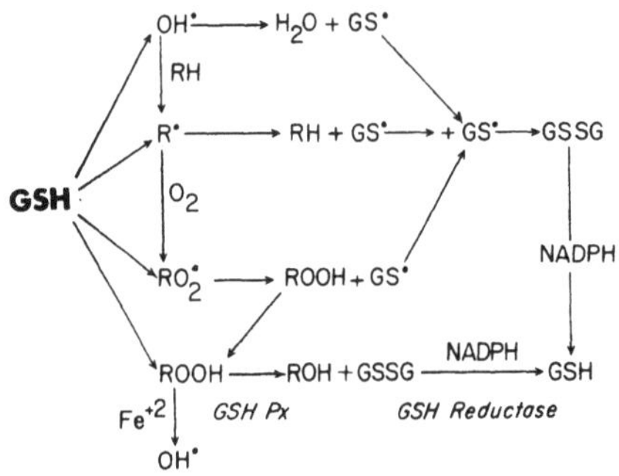

Fig. 1. Reaction of Glutathione with Hydroxyl Radical (•OH), Organic Radicals (R•), Peroxyl Radicals (ROO•), and Hydroperoxides (ROOH). GSH reacts with all radicals to produce the thiol radicals that self associate to produce oxidized glutathione (GSSG). GSH is regenerated by reduction with NADPH generated from the pentose cycle. It is believed that unbound Fe^{2+} reactions with peroxide can lead to additional generation of hydroxyl radicals and increased production of malondialdehyde (MDA).

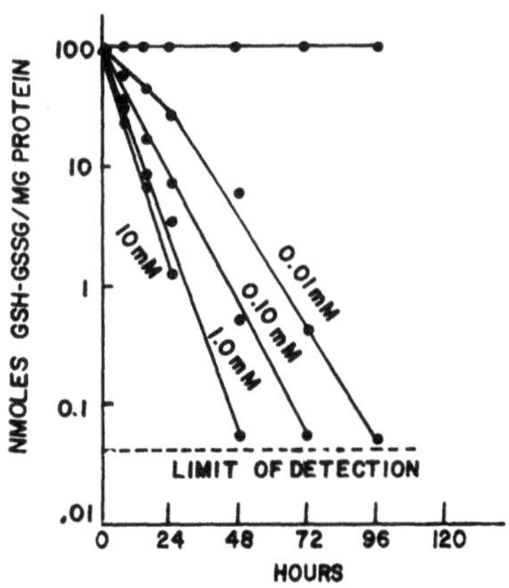

Fig. 2. Effect of L-Buthionine Sulfoximine Treatment on Cellular GSH Content. Most rapid depletion of cellular GSH occurs in the presence of 1-5 mM L-BSO. The rate of depletion was highest at 0.1 mM L-BSO concentration and least with 0.01 mM. The limit of detection is about 2 decades with cell densities at 10^5 and increases to greater than 4 decades with cell densities greater than 10^6. For all of our experiments we used 0.1 mM L-BSO for 72 hr or 0.25 mM for 52 hr to achieve the maximal GSH depletion.

And, although thiols were depleted more rapidly with yet higher concentrations of 1-5 mM L-BSO, toxicity increased and cell growth stopped.

Figure 3 shows the effects from time of exposure to 0.1 mM L-BSO on cellular response to 5 Gy (Panel A), 2 Gy (Panel B), and the oxygen enhancement ratio (Panel C). The estimated dose-modifying factor for the increased aerobic response is 2.4 after 120-hr incubation in 0.1 mM L-BSO. The oxygen enhancement ratio for control A549 cells at 0.01% SF is 2.9 to 5.0.

We have attempted to correlate enzyme inhibition with GSH depletion. Past results indicate that GSH depletion by L-BSO inhibits GSH-S-transferase but not GSH peroxidase (4, 5). However, GSH peroxidase inhibition occurs in cells incubated with GSH-S-transferase substrates diethylmaleate (DEM) and dimethylfumarate (DMF), or with thiol-reactive N-ethylmaleimide (NEM). These substrates form covalent bonds, which prevent oxidation of glucose-6-phosphate and subsequent electron transfer via the coupled enzyme system of glucose-6phosphate, 6-phosphogluconate dehydrogenase and GSH reductase to reform GSH. Figure 4 shows an aerobic radiosensitivity increase after 24 hr exposure to 0.5 mM L-BSO. We exposed the cells to increasing concentrations

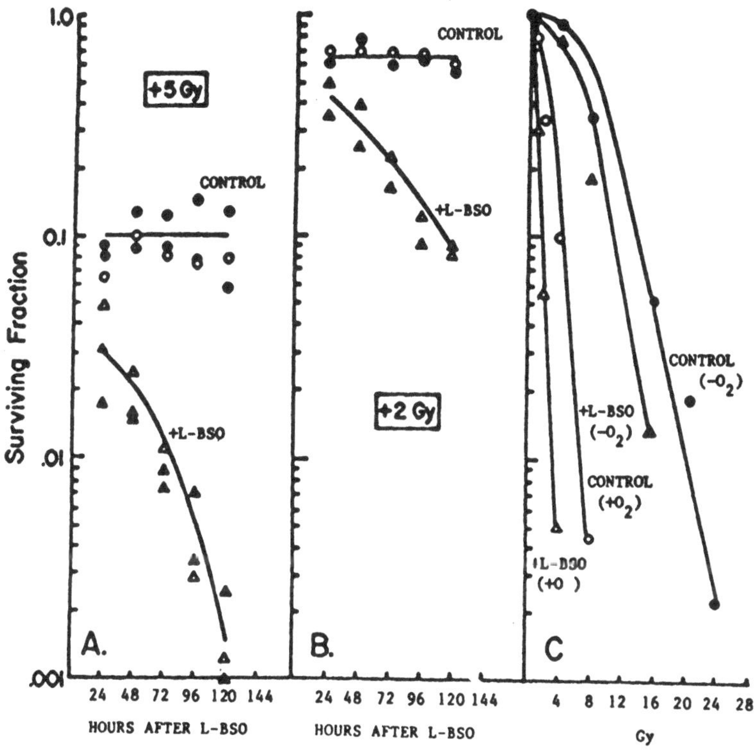

Fig. 3. Effect of GSH Depletion on the Aerobic and Hypoxic Radiation Response. A549 cells were exposed to 0.1 mM L-BSO for the indicated times. Cells were washed free of McCoy's 5a medium, trypsinized, and resuspended in a modified Hanks medium prior to irradiation. Cell density was 10^5, pH 7.4, 37°C during irradiation. The cell suspension was shaken during irradiation. Metabolic hypoxia was created by concentrating cells to 10^8 and cell slurry drawn into a glass syringe. Metabolic hypoxia occurred within 1 min at 37°C, measured by reduction of methylene blue to a colorless form. Cells were irradiated in the syringe, diluted, and plated for clonogenic assay. Production of hypoxia and subsequent irradiation required less than 10 min.

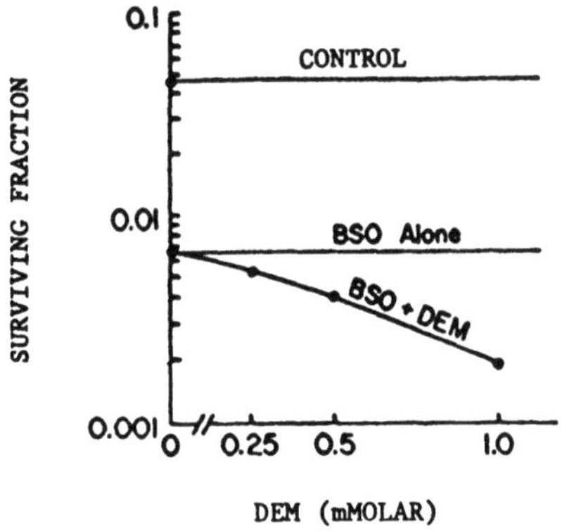

Fig. 4. Increased Aerobic Response of GSH-Deficient Cells after Treatment with DEM. Cells were made deficient of GSH by treatment with 0.5 mM L-BSO. Residual GSH was depleted by a second treatment with DEM. DEM is a substrate for GSH-S-transferase, and mitochondrial- or enzyme-bound GSH may be removed by this treatment. We found repeatedly that L-BSO treatment did not result in the inhibition of cellular capacity to reduce peroxides or hydroperoxides. However, treatment with substrates for GSH-S-transferase resulted in total inhibition of cellular reduction of t-butylhydroperoxide and peroxide (6, 7). Cellular viability was not influenced by the two-hour treatment with DEM or DMF (data not shown).

of DEM added 2 hr prior to irradiation. Cellular capacity to reduce peroxides and hydroperoxides was completely inhibited as a result (4) (data not shown).

We have not yet assessed the toxicity of radiation-induced hydroperoxide or subsequent breakdown products. However, we have attempted to understand these toxic mechanisms by investigating the ability of exogenously added hydroperoxides to alter cell biochemistry and ultimately to cause cell death. We have tested the effect of tertiary butylhydroperoxide (TBOOH) on radiation response and cell lethality. We have focused on the effect of GSH depletion, on the inhibition of GSH peroxidase, and the non-selenium containing enzyme GSH-S-transferase. We have also studied the role that the pentose cycle may have in reducing the products of the reactions with GSH (7) (cf. Fig. 1).

Intracellular GSH depletion should make cells more vulnerable to hydroperoxides, since both transferase and peroxidase activities are altered (4). Figure 5 represents lethality of A549 GSH-depleted cells after 10-min exposure to different amounts of TBOOH. The top curve shows toxicity results for control cells; however, the absence of GSH enhanced lethality by 70-fold with 0.1 mM TBOOH. Also, lesser concentrations of TBOOH increased toxicity in thiol-depleted cells (cf. above for thiol-containing cells). In fact, thiol-depleted cells were susceptible to virtually any amount of TBOOH. Lack of significant toxicity in the control cells may reflect the reduction of TBOOH to the more innocuous alcohol, tertiary butyl alcohol (TBOH), without appreciable oxidation of GSH. Therefore, to understand the increased TBOOH-induced toxicity toward L-BSO treated cells, we studied its effects on cellular NADPH oxidation and subsequent regeneration.

We have demonstrated that TBOOH added to cells results in oxidation of pyridine nucleotides. This occurs because GSH catalyzed TBOOH reduction (2GSH + TBOOH----->GSSG + TBOH + H_2O). GSSG in turn is reduced by NAD(P)H

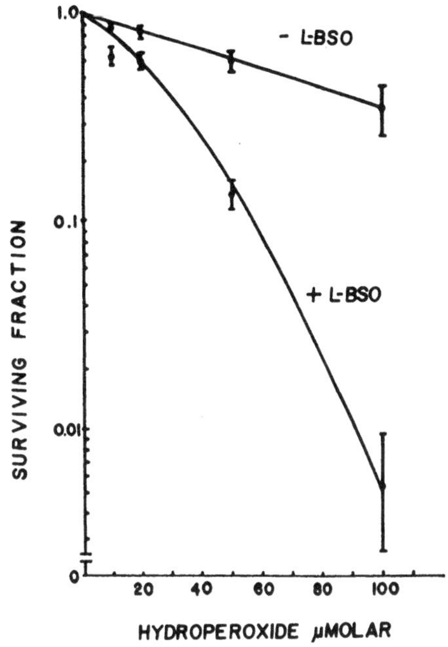

Fig. 5. Cytotoxicity of t-Butylhydroperoxide (TBOOH). A549 cell cultures were pretreated with 0.1 mM L-BSO for 72 hr. Cells were harvested and resuspended in Hank's salt solution (calcium and magnesium free) at a final density of 10^6/ml. TBOOH was added to the cell suspensions at the appropriate concentration for 10 min. Aliquots were removed and plated into fresh growth media to determine clonogenic survival. All incubations were performed at 37°C.

(GSH reductase) to produce $NADP^+$ and 2GSH. Figure 6 represents NAD(P)H changes for A549 cell suspensions following the addition of 100 nmoles TBOOH. The top curve shows oxidation and a return to control values within 3 min following TBOOH addition. Cells treated with L-BSO for 72 hr show a partial inhibition of return to control absorbancy. Bis-chloronitroso urea (BCNU) inhibits glutathione reductase (6). Therefore, among other things, it inhibits cellular reduction of GSSG, produced by TBOOH enzyme-linked oxidation of GSH. The addition of BCNU inhibits cyclic oxidation and reduction because it removes the substrate GSH from the reaction by forming a covalent bond with GSH.

Although we have not yet measured TBOOH reduction, we have developed methods for measuring the effect of various agents on cellular reduction of peroxide (Fig. 7). We found this reduction unaffected by GSH depletion, however, the amount of NEM required to inhibit peroxide reduction decreased 40-fold after L-BSO treatment. We found similar effects for NEM titration of NADPH (data not shown).

Hydroperoxides, if produced following radiation, also may have deleterious effects on cellular macromolecules because they can be a source of hydroxyl radicals (ROOH + Fe^{+2} -----> RO + OH + Fe^{+3}). Figure 8 shows that DNA breaks are caused by cells treated with 20 µM TBOOH for 10 min. These breaks are repaired in control cells, but not in GSH-depleted cells (upper curve).

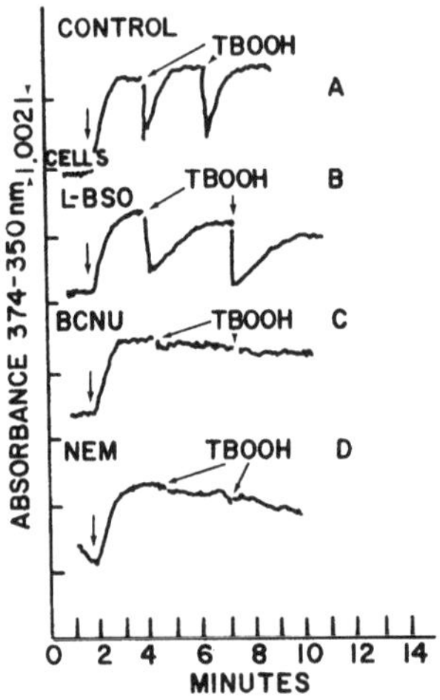

Fig. 6. Effect of TBOOH Additions on the Absorbancy Changes for Suspensions of A549 Cells. Cells were suspended after harvesting in 2 ml of 20 mM Hepes buffered saline with 10 mM KCl, pH 7.4, 37°C. At the arrow 200 nmoles TBOOH was added. Cyanide, when used, was 1 mM and NEM was 0.1 mM. Cells were treated with 0.1 mM L-BSO for 72 hr before use and, where indicated, with 0.1 mM BCNU for 1 hr.

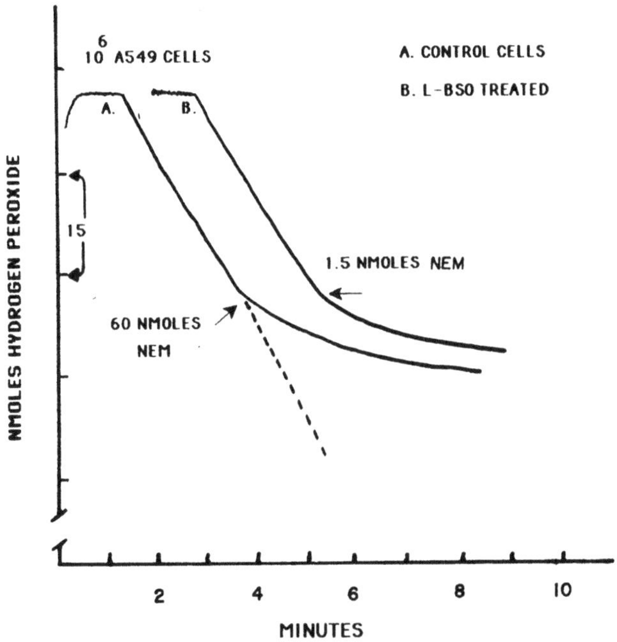

Fig. 7. Effect of NEM on Peroxide Reduction for Control Cells and for Cells Treated with 0.24 mM L-BSO for 52 hr. Catalase was inhibited by adding sodium azide to a final concentration of 0.02 mM. Cells were incubated in physiological saline buffered with 20 mM Hepes buffer, pH 7.4, with 10 mM KCl at 37°C. Peroxide was added prior to cell addition. Peroxide reduction is seen as a downward deflection.

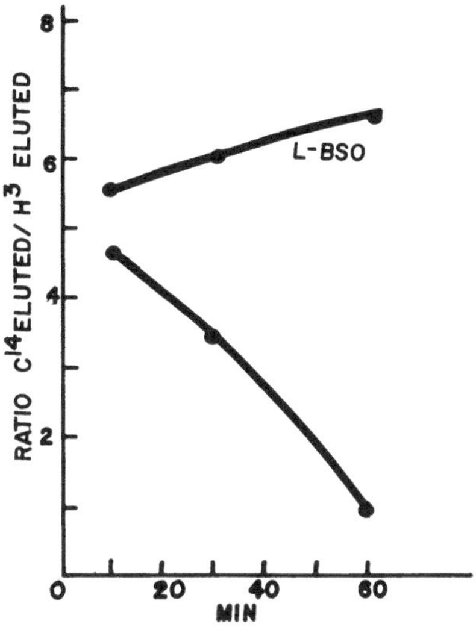

Fig. 8. Hydroperoxide Induced Lesions in Cellular DNA. The alkaline elution technique was used to determine the effect of TBOOH on the integrity of cellular DNA. Cells were pretreated for 52 hr with 0.25 mM L-BSO. One million cells were suspended in 10 ml Hank's with calcium and magnesium. TBOOH was added to the cell suspension to a final concentration of 50 μM.

According to our model and previously obtained data (4, 5, 12) the cellular response to radiation may be altered if hydroperoxides cause depletion of GSH (TBOOH + 2GSH----->TBOH + GSSG + H_2O). Therefore, we have tested the radiosensitizing capacity of TBOOH. Figure 9 shows TBOOH affects the radiation response of cells treated with BSO. The curve on the left indicates no measurable effect. However, the center curve shows TBOOH administered post-irradiation enhanced the cell's radiation response. Of major interest is the effect of hydroperoxides on the radiation response of GSH-depleted cells, represented by the curve on the right. These results suggest that TBOOH causes oxidation of "bound" GSH on the peroxidase enzyme resulting in GSSG formation. GSH disulfide will rapidly thiylate protein thiols or be excreted from the cells. GSSG thiylation will result in enzyme inhibition and an inability to repair either X-ray or TBOOH-induced damage. The major site of TBOOH damage has not been determined. It may be at the same site as radiation or at different sites. It may also involve redox changes to key repair enzymes. Figure 10 shows similar effects when cells are pretreated with BCNU, which did not remove GSH but did inhibit GSH reductase, thus affecting the cell's radiation response as much as TBOOH with L-BSO treated cells.

CONCLUSION

In conclusion, our work suggests that GSH depletion may occur first in the nucleus and cytosol, followed by the mitochondria (Fig. 11). Enzyme-bound GSH (i.e., transferase and peroxidase) may be more difficult to remove and may explain why we see reduction of peroxides and hydroperoxides following thiol depletion to values less than 0.04 nM/mg protein. Chemicals forming covalent

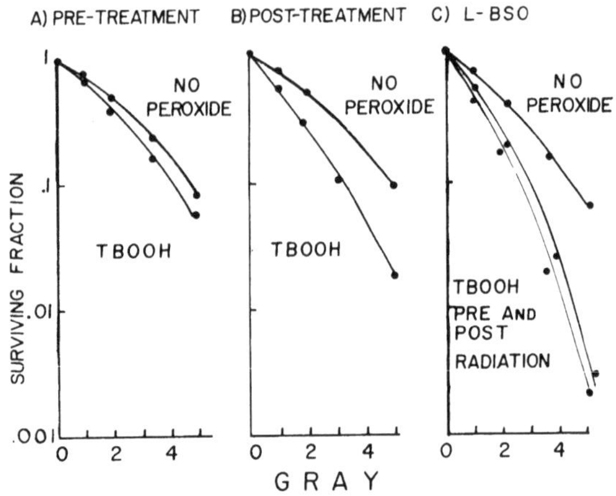

Fig. 9. Effect of TBOOH on the Radiation Response with Pre-treated (A), Post-treated (B), and GSH-Depleted Cells (C). In panel C, TBOOH was present for 5 min during and post-irradiation. 200 nmoles of TBOOH was added to 10 ml of 10^6 cells suspended in Hank's solution without magnesium and calcium but containing 20 mM Hepes, pH 7.4, 37°C.

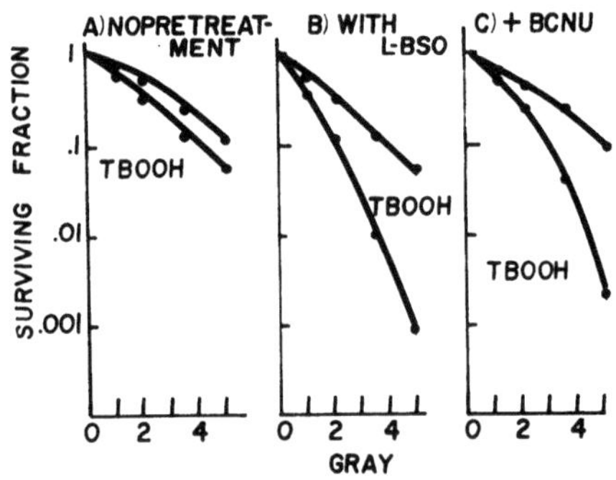

Fig. 10. Effect of L-BSO and BCNU on the Radiation Response of Cells Treated with TBOOH. The incubation conditions are the same as those listed in the legend of Fig. 7. Cells were treated with 0.1 mM BCNU prior to exposure to 20 µM TBOOH.

bonds with GSH (i.e., transferase catalyzed or chemically reactive) may become more efficient in removing enzyme-bound GSH, necessary for reaction with either organic peroxyl radicals and hydroperoxides via enzyme-catalyzed reduction. We agree with Selkoff and Reich (13), therefore, that electron-flow through the GSH system is as important as net GSH concentration during irradiation of cells. Small amounts of enzyme-bound GSH may afford significant protection via coupled enzyme systems. All enzymes generating NADPH can couple with the GSH reductase to maintain sufficient GSH to reduce radicals or hydroperoxides. Therefore, a small amount of GSH, enzyme bound, is capable of being turned over via the redox driving force of the pentose cycle, resulting in radical organic peroxyl and hydroperoxide reductions. The equations below illustrate the properties of the pentose cycle:

1. 2GSH + ROOH ------> ROH + GSSG + H_2O (GSH Px)
2. GSSG + NADPH ------> 2GSH + $NADP^+$ + H^+ (reductase)
3. $2NADP^+$ + Glucose-6-P ------> Ribulose-5-P + CO_2 (Pentose Cycle)

Readily, we see that removal of GSH by covalent bond formation (Eq. [1]), removal of glucose (Eq. [3]) (3, 6-7, 14), and inhibition of GSH reductase (BCNU, Eq. [2]) can block turnover of GSH and inhibit its role in the radiation response. Our results would agrée with the recent findings that transferases (with bound GSH) in the histone fraction within the nucleus are critical in reducing hydroperoxides produced by radiation (15).

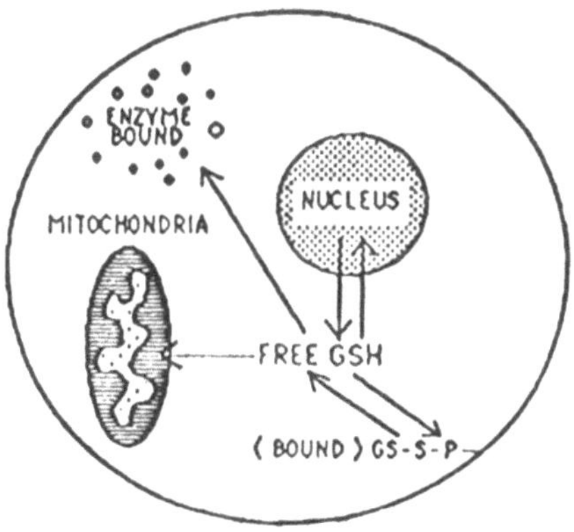

Fig. 11. Intracellular Location of Glutathione. GSH exists in nearly every compartment of the cell. It appears to be most rapidly removed from the cytoplasm after treatment with L-BSO. Depletion should follow in the nucleus, because it is believed that GSH is freely permeable into and out of the nucleus. GSH can be released from protein-bound stores, but its release requires reduction. GSH is not easily removed from the mitochondria, and its depletion from these cellular organelles appreciably lags behind cytoplasmic and nuclear depletion. We believe that enzyme-bound GSH is more difficult to remove, and requires agents such as NEM that are capable of forming covalent bonds. Conjugation of GSH with NEM and/or DEM or DFM results in an inability for the cell to "turn over" enzyme-bound GSH via coupled enzyme systems, consisting of GSH peroxidases, GSH transferases, GSH reductase, glucose-6-phosphate dehydrogenase, and GSSG.

Peroxide results show cellular toxicity and alterations in radiation response. In addition, toxicity is enhanced by either depleting thiols, or by blocking the reduction of GSSG by BCNU-linked inhibition of GSH reductase. Hydroperoxides also increase DNA breaks and alter the cellular radiation response when added post-irradiation. Our results indicate that cellular capacity to reduce peroxides or hydroperoxides is not directly related to effects on cellular lethality. Total inhibition of peroxide reduction does not occur following L-BSO treatment, but does occur when residual thiols are depleted and correlates with an increased radiation response.

Cellular redox effects caused by addition of TBOOH to either GSH-depleted cells post-irradiation or to thiol containing control cells may correlate with inhibition of enzymes (by thiylation as described above) involved in DNA repair. Inhibited TBOOH reduction by BCNU also suggests that TBOOH affects enzymes involved in DNA breaks and their subsequent repair. It is interesting to note that we have found that GSH depletion to very low levels influences cell lethality but not cellular repair capacity (data not shown), but that in the GSH-depleted state, TBOOH appears to sensitize cells post-irradiation.

Chronic L-BSO treatment results in GSH decreases to non-detectable values. However, there may still be enough enzyme-bound GSH to participate in coupled enzyme reactions and to afford protection against radiation damage.

We have tried to indicate in Figure 11 our concept of the cellular compartment of GSH. GSH is located in both the free and bound forms and may be equilibrated freely between the cytoplasm and the nucleus. GSH enters the mitochondria but is slowly depleted from this compartment in the presence of L-BSO. GSH may be bound by disulfide or molecular association with active sites on enzymes located throughout the cell. Our studies (12) indicate that GSH depletion inhibits transferase activity but GSH-peroxidase is less affected, perhaps because bound GSH is not easily removed. It would appear that this small pool of enzyme-bound GSH coupled with the pentose cycle is sufficient to afford protection against radiation-induced peroxide damage. We plan additional studies in the area of protein-bound GSH and its role in protecting cells.

ACKNOWLEDGEMENTS

This research was supported by National Cancer Institute Grants R01 CA 44982, R01 CA 40516, and R01 CA-18614.

REFERENCES

1. J. E. Biaglow, M. E. Varnes, E. P. Clark, and E. R. Epp, The role of thiols in cellular response to radiation and drugs. Radiat. Res. 95, 437-445 (1983).
2. J. E. Biaglow, M. E. Varnes, E. R. Epp, E. P. Clark, and M. Astor, Factors involved in depletion of glutathione from A549 human lung carcinoma cells: Implications for radiotherapy. Int. J. Radiat. Oncol. Biol. Phys. 10, 1221-1227 (1984).
3. J. E. Biaglow, E. P. Clark, E. R. Epp, M. E. Varnes, and J. B. Mitchell, Nonprotein thiols and the radiation response of A549 human lung carcinoma cells. Int. J. Radiat. Biol. 4, 489-495 (1983).
4. J. E. Biaglow, M. E. Varnes, E. R. Epp, E. P. Clark, and M. Astor, Redox enzymes and thiol radicals. In: Oxygen and Sulfur Radicals in Chemistry and Medicine (M.A.J. Rogers, A. Breccia, and G. Semerano, Eds.). Academic Press, Lo Scarabeo, Bologna, Italy, 1986.

5. J. E. Biaglow, M. E. Varnes, E. P. Clark, and E. R. Epp, The role of cellular glutathione in response of tumor cells to radiation and drugs. In: Biochemical Modulation of Anticancer Agents: Experimental and Clinical Approaches (F.A. Valeriote and L.H. Baker, Eds.). Martinus Nijhoff Publishing, Boston, 1986.
6. J. E. Biaglow, M. E. Varnes, E. R. Epp, and E. P. Clark, Factors influencing tertiary butylhydroperoxide metabolism. Submitted to J. Biol. Chem. (1987).
7. J. E. Biaglow, M. E. Varnes, J. B. Mitchell, and A. Russo, Role of the hexosemonophosphate shunt in the cellular response to radiation and drugs. Submitted to Cancer Res. (1987).
8. E.A. Bump, N.Y. Yu, and J.M. Brown, Radiosensitization of hypoxic tumor cells by depletion of intracellular glutathione. Science 217, 544-545 (1982).
9. J.B. Mitchell, A. Russo, J.E. Biaglow, and S. McPherson, Cellular glutathione depletion by diethyl maleate or buthionine sulfoximine: No effect of glutathione depletion on the oxygen enhancement ratio. Radiat. Res. 96, 422-428 (1983).
10. E.P. Clark, E.R. Epp, M. Morse-Gaudio, J.E. Biaglow, The role of glutathione in the aerobic radioresponse: I. Sensitization and recovery in the absence of intracellular glutathione. Radiat. Res. 108, 238-250 (1986).
11. T. Alper and P. Howard-Flanders, The role of oxygen in modifying the radiosensitivity of E.coli B. Nature, 178, 978-979 (1956).
12. J.E. Biaglow, M.E. Varnes, S.W. Tuttle, N.L. Oleinick, K. Glazier, E.P. Clark, E.R. Epp, and L. Dethlefsen, The effect of L-buthionine sulfoximine on the aerobic radiation response of A549 human lung carcinoma cells. Int. J. Radiat. Oncl. Biol. Phys. 12, 139-144 (1986).
13. I. Selkoff and J. Reich (Eds.), Cellular Bioenergetics. Academic Press, New York 1981.
14. J.E. Biaglow, M.E. Varnes, E.R. Epp, E.P. Clark, and K. Held, Role of glutathione in cellular response to drugs and radiation, IVth International Conference on Oxygen Radicals, San Diego, CA. Pergamon Press, New York 1987 (In press).
15. B. Ketterer, D. Meyer, and K.H. Tan, Intracellular factors affecting the formation and repair of lipid and DNA hydroperoxides, IVth International Conference on Oxygen Radicals, San Diego, CA. Pergamon Press, New York 1987 (In press).

THE ROLE OF MODIFICATION OF DNA DAMAGE IN THE RADIOPROTECTIVE ACTION OF AMINOTHIOLS

David Murray, Susanna vanAnkeren, Luka Milas, and Raymond Meyn

The University of Texas System Cancer Center
M.D. Anderson Hospital and Tumor Institute
Department of Experimental Radiotherapy
1515 Holcombe, Box 66
Houston, Texas 77030

The role of DNA damage modification in the protection of mammalian cells from the lethal effects of radiation by aminothiols has been the subject of much research over the last 20 years since the early demonstrations that cysteamine reduced the level of DNA single-strand breaks (SSBs) in irradiated cells (1, 2). Despite this considerable effort, the actual mechanisms of radioprotection at both the cellular and DNA level remain poorly defined. Perhaps one barrier to our understanding of these effects has been the failure to recognize that many chemical modifiers of radiosensitivity, including sensitizers such as oxygen and protectors such as cysteamine, may not simply dose-modify numbers of DNA lesions but may also change the "spectrum" of lesions induced by low-LET radiations. For example, lesions such as DNA-protein cross-links (DPCs) (3) and 8,5′-cycloadenosine (4) are actually preferentially induced under hypoxia despite the fact that this condition offers considerable radioprotection to the cell. Even these presumably minor lesions could contribute significantly to lethality under appropriate conditions, e.g. in cells exhibiting a particular repair defect (3). Evidence that aminothiols also cause a shift in the spectrum of radiation-induced lesions comes from the recent report by Radford (5) that cysteamine altered the levels of 4 types of DNA lesion--SSBs, double-strand breaks (DSBs), DPCs and base damage--in different proportions.

In the present study we examined the effect of two aminothiols derived from cysteamine--WR-1065 and WR-255591--and of the thiol dithiothreitol (DTT) on cell survival and on DNA damage in cultured Chinese hamster ovary (CHO) cells. DNA SSBs were measured using the alkaline elution method (6), which has the advantage that SSBs can be measured over the same low-dose range as cell killing. DNA DSBs were measured using the neutral elution method (7) adapted by using a pH of 7.0 for all solutions. Cell survival was assayed using a clonogenic assay. Correlation curves for these 3 drugs relating the degree of protection against cell killing and SSB induction are presented in Fig. 1, all measurements having been made on cells irradiated with either 5 or 10 Gy of γ-rays at 37°C. For both of the aminothiols, at all concentrations examined, the protection factors (PF) for SSBs were much lower than those for cell killing. On the other hand, for DTT the correlation between the modification of SSBs and killing was

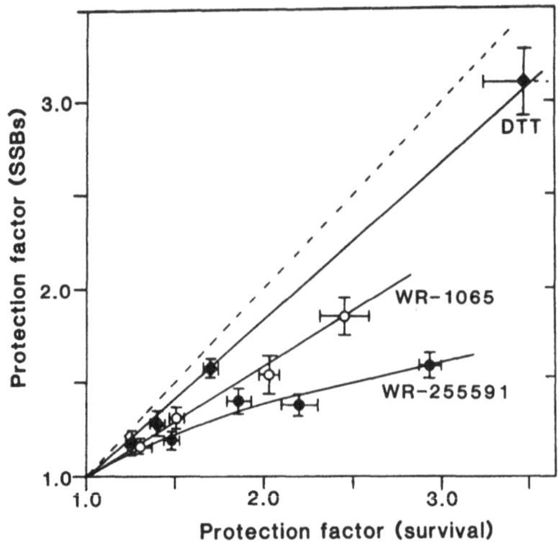

Figure 1. Correlation curves showing the relationship between the protection factors for cell survival and for DNA single-strand breakage for CHO cells treated with either WR-1065 (0-10 mM), WR-255591 (0-10 mM), or dithiothreitol (0-500 mM) at 37° for 30 min prior to irradiation. The dashed line represents the idealized one-to-one correlation between strand-breaks and cytotoxicity.

quite close, as reported recently by Held et al. (8). Table 1 shows that for the two aminothiols both SSB and DSB modification measured in chilled samples (to eliminate any repair) still underestimated the effect on survival for both drugs, although a somewhat greater PF was observed for DSBs than for SSBs.

This lack of correlation may reflect that aminothiols may be selectively modifying a particular subset of "lethal lesions"; in fact, Radford (5) has also shown that cysteamine preferentially protected against DSB induction compared with SSBs and concluded that DSB modification correlated strongly with the effect on cell killing. However, such a correlation may have been simply coincidental since the outcome of these lesions in terms of cell killing is not simply a consequence of the initial level of lesions induced in the cell. It is well established that subsequent competing biochemical pathways--for example, the action of repair enzymes--can influence the outcome in terms of survival. We therefore examined the effect of the two aminothiols on strand-break rejoining. DSB rejoining was unaffected by either pre- and/or postirradiation treatment with WR-1065. Both drugs appeared to retard SSB rejoining when present prior to irradiation, further suggesting that a major effect of the drug pretreatment is to alter the relative proportion of the different types of SSB induced, hence the apparent effect on their rate of rejoining. In addition to repair processes, cell progression (and its putative effect on lesion fixation processes that may compete with cellular repair mechanisms) may also be an important factor in cell death. It was therefore interesting to confirm previous observations with both cysteamine (2) and WR-1065 (9) that these aminothiols reversibly perturbed the progression of cells through their cycle, with their most prominent effect being an inhibition of DNA replication (data not shown).

TABLE I: Protection factors for cell killing and DNA strand-breakage in cultured CHO cells treated with aminothiol radioprotectors prior to γ-irradiation.

Drug	PF (survival)	PF (SSB)	PF (DSB)
WR-255591 (6 mM)	2.0	1.40 (37°C) 1.31 (0°C)	1.83 (0°C)
WR-1065 (4 mM)	2.0	1.54 (37°C) 1.23 (0°C)	1.60 (0°C)

In conclusion, based on current models for the radiobiological effect of γ-rays in which cell killing is determined by a competition between the repair and fixation of lesions, the relationship between initial damage induction and cytotoxicity ought only to exhibit a one-to-one correlation in cases where a true dose-modification of these initial lethal lesions occurs, and in the absence of a perturbing effect of the modifying agent on other cellular biochemical processes that can modify the ultimate biological effectiveness of the radiation. As reported by Held et al. (8), DTT appears to satisfy these criteria. On the other hand, in the case of aminothiols it is clear that these conditions are not met; this could well reflect the different structural properties conferred by the presence of the amino group, including DNA binding affinity, as suggested many years ago by Brown (10).

ACKNOWLEDGEMENTS

This investigation was supported by Contract Number DAMD17-86-C-6105 awarded by the U.S. Army Medical Research and Development Command. We would like to thank Donna Raye McGhee for her assistance in preparing this manuscript.

REFERENCES

1. P.H.M. Lohman, O. Vos, C.A. Van Sluis and J.A. Cohen, Chemical protection against breaks induced in DNA of human and bacterial cells by x-irradiation. Biochim. Biophys. Acta 224, 339-352 (1970).
2. S. Sawada and S. Okada, Cysteamine, cystamine, and single-strand breaks of DNA in cultured mammalian cells. Radiat. Res. 44, 116-132 (1970).
3. R.E. Meyn, S.C. vanAnkeren, and W.T. Jenkins, The induction of DNA-protein crosslinks in hypoxic cells and their possible contribution to cell lethality. Radiat. Res. 109, 419-429 (1987).
4. A.F. Fuciarelli, G.G. Miller, and J.A. Raleigh, An immunochemical probe for 8,5′-cycloadenosine-5′-monophosphate and its deoxy analog in irradiated nucleic acids. Radiat. Res. 104, 272-283 (1985).
5. I.R. Radford, Effect of radiomodifying agents on the ratios of X-ray-induced lesions in cellular DNA: use in lethal lesion determination. Int. J. Radiat. Biol. 49, 621-637 (1986).

6. K.W. Kohn, R.A.G. Ewig, L.C. Erickson, and L.A. Zwelling, Measurement of strand breaks and cross-links by alkaline elution. In DNA Repair: A Laboratory Manual of Research Procedures (E. Friedberg and P. Hanawalt, Eds.), pp 379-402. Marcel Dekker, Inc., New York, 1981.
7. M.O. Bradley and K.W. Kohn, X-ray induced DNA double strand break production and repair in mammalian cells as measured by neutral filter elution. Nucleic Acids Res. 7, 793-804 (1979).
8. K.D. Held, G.D. Bren, and D.C. Melder, Interactions of radioprotectors and oxygen in cultured mammalian cells. II. Effects of dithiothreitol on radiation-induced DNA damage and comparison with cell survival. Radiat. Res. 108, 296-306 (1986).
9. D.J. Grdina and B. Nagy, The effect of 2-[(aminopropyl)amino] ethanethiol (WR-1065) on radiation-induced DNA damage and repair and cell progression in V79 cells. Br. J. Cancer 54, 933-941 (1986).
10. P.E. Brown, Mechanism of action of aminothiol radioprotectors. Nature 363-364 (1967).

RADIOPROTECTION IN RAT SPINAL CORD WITH WR2721, WR77913 AND WR3689 FOLLOWING CEREBRAL LATERAL VENTRICULAR OR INTRACISTERNAL ADMINISTRATION

A.M. Spence, K.A. Krohn, J.E. Steele,
S.E. Edmondson and J.S. Rasey

Divisions of Neurology, Nuclear Medicine
and Experimental Biology, University of Washington
Seattle, Washington 98195

INTRODUCTION

The doses of radiotherapy that normal central nervous system (CNS) parenchyma and stroma can tolerate without injury are too low to provide cure for malignant astrocytic gliomas. One potential solution to this problem is to develop compounds that protect the normal CNS tissues more effectively than tumor cells against irradiation. The phosphorothioate radioprotectors such as WR2721, WR77913 and WR3689 are highly water-soluble and do not readily cross the blood-brain barrier. Injection of these drugs via routes that bypass the blood-brain barrier has allowed us to assess radioprotection in the rat cervical spinal cord.

MATERIALS AND METHODS

Adult female Fischer-344 rats (Simonsen Laboratories, Gilroy, CA) that weighed between 150 and 200 grams were used.

S-2-(3-aminopropylamino) ethylphosphorothioic acid (WR2721), S-(3-amino-2-hydroxypropyl) sodium phosphorothioate (WR77913) and S-2-[(3-methylaminopropyl)amino]ethyl sodium phosphorothioate (WR3689) were obtained from the Developmental Therapeutics Program, Division of Cancer Treatment, NCI. The purity of the drugs was confirmed by melting point and thin layer chromatographic analysis. Drugs were administered stereotaxically into the right lateral cerebral ventricle (WR2721 and WR77913) or the cisterna magna (WR3689) in 5 or 10 ul of phosphate-buffered saline as previously reported (1). The doses were: WR2721, 0.33 mg (0.60 X LD_{50}); WR77913, 2.75 mg (0.67 X LD_{50}); and WR3689, 2.0 mg (0.60 X LD_{50}) (1). Controls received an equivalent volume of PBS alone.

Gamma irradiations of the cervical spinal cord were delivered with a Picker ^{137}Cs teletherapy unit 45 minutes after drug administrations (2). The dose rate was 1.48 Gy/min; the radiation dose groups were 20, 26, 32 or 38 Gy; each group contained 7 to 13 rats. All irradiations and drug injections were performed with the rats lightly anesthetized with ketamine/xylazine.

Once weekly each rat was weighed and carefully observed for any signs of fore- or hindlimb paralysis, the principal endpoints for this study. Since forelimb paralysis always preceded hindlimb paralysis, animals were sacrificed after they showed definite hindlimb paralysis. The statistical significance of differences between animal groups in weeks to fore or hindlimb paralysis was assessed by means of the log rank sum test as previously reported (2).

Dose modifying factors (DMF) were calculated with the following formula:

$$\frac{D_{50}\ (X)\ \text{plus Radioprotector}}{D_{50}\ (X)\ \text{minus Radioprotector}}$$

D_{50} (X) = radiation dose to produce paralysis in 50% of rats at X weeks. For WR2721 the DMF for forelimb paralysis was calculated for each week's data between 20 and 30 weeks post-irradiation; similarly that for the hindlimb was calculated for each week's data between 26 and 36 weeks. DMF's for WR77913 and WR3689 were derived similarly (Table and Figures 1-3). The doses required to produce paralysis in 50% of rats for each week were calculated by probit analysis (2). The weekly DMF values were then averaged and the standard deviation and standard error were calculated. This approach was taken because there were large differences at some radiation doses between control and drug-treated groups in the time to either fore- or hindlimb paralysis.

RESULTS

The data on the median times to fore- or hindlimb paralysis (Table, left) show a statistically significant delay from WR2721 treatment in the onset of paralysis in all groups except for time to forelimb paralysis in the highest dose group. WR77913 delayed forelimb paralysis in the 26 Gy dose group and hindlimb paralysis in the 26 and 32 Gy dose groups. WR3689 delayed fore- and hindlimb paralysis in all the radiation dose groups.

The average DMF data for fore- and hindlimb paralysis (Table, right) show that the hindlimb values are higher than the forelimb and that WR2721 demonstrates the highest DMF's. Figures 1-3 are plots of the calculated DMF's vs. the weeks after irradiation. The forelimb DMF's do not vary over the time spans plotted whereas those for the hindlimb decrease with time.

DISCUSSION

The data show that the three drugs, WR2721 > WR3689 > WR77913, radioprotect the rat cervical spinal cord against ^{137}Cs at the radiation and drug doses we tested. From biodistribution studies of [S-35]-labeled WR2721, WR77913 and WR3689 following intraventricular or intracisternal injection we can estimate the cervical spinal cord drug concentrations at the time of irradiation (1). For the three drugs respectively these are 80, 280 and 720 ug of drug per gram of tissue.

The DMF for hindlimb paralysis is the more accurate indicator of the CNS radioprotective capacity of the drugs because this endpoint is totally due to radiation injury in the long tracts of the cervical spinal cord proper. The hindlimb DMF is not influenced by any radiation damage in nerves outside the subarachnoid space. In contrast, the DMF for forelimb paralysis is lower probably because the parallel opposed lateral cervical

TABLE

Protection Following Intraventricular Injection of WR2721 and WR77913 and Intracisternal Injection of WR3689 - Weeks to Forelimb or Hindlimb Paralysis		20 Gy	26 Gy	32 Gy	38 Gy	MEAN DMF Forelimb or Hindlimb (weeks analyzed)
Forelimb	Gamma only	49	23	19	20	
	Gamma + WR2721	60*	28*	31*	22	1.3 (20 - 30)
Hindlimb	Gamma only	58	23	20	20	
	Gamma + WR2721	65*	34*	35*	29*	1.6 (26 - 36)
Forelimb	Gamma only	30	19	17	16	
	Gamma + WR77913	29	25*	18	17	1.2 (15 - 28)
Hindlimb	Gamma only	40	22	19	19	
	Gamma + WR77913	37	34*	25*	20	1.3 (19 - 30)
Forelimb	Gamma only	47	43	22	19	
	Gamma + WR3689	55*	48*	30*	30*	1.2 (18 - 44)
Hindlimb	Gamma only	54	48	34	29	
	Gamma + WR3689	63*	57*	45*	34*	1.4 (27 - 48)

* p < .05 vs gamma only, log rank sum test.

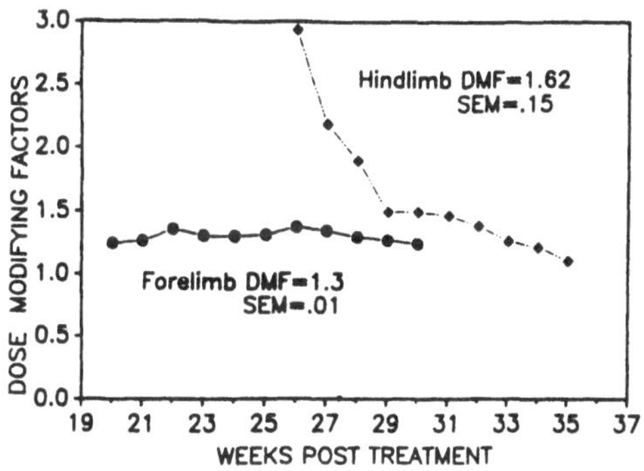

Fig. 1. WR2721

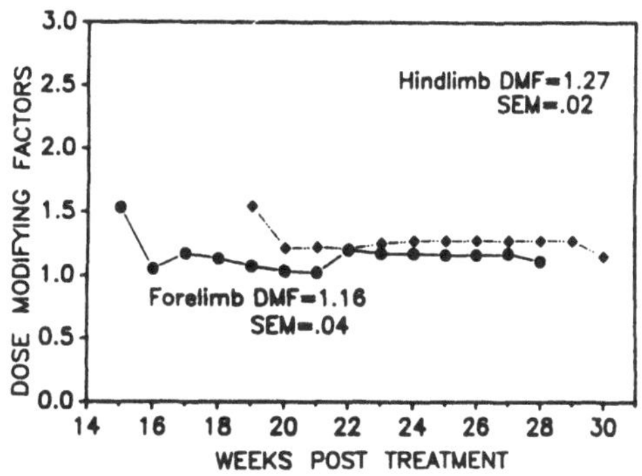

Fig. 2. WR77913

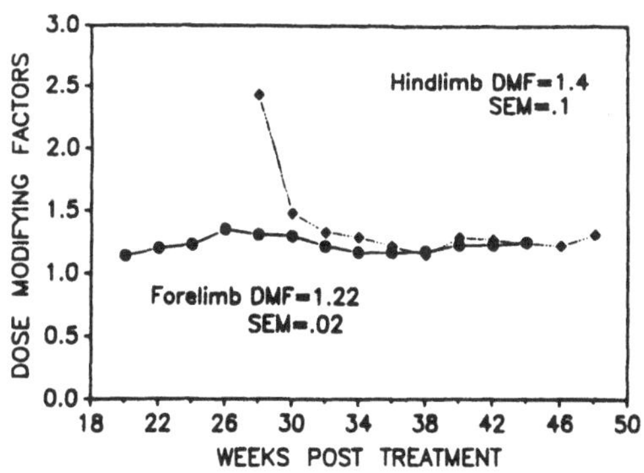

Fig. 3. WR3689

irradiation ports allowed irradiation of the nerve roots for the forelimbs that lay outside the subarachnoid space where they would not be accessible to and protected by the drugs.

The relatively early paralysis after high radiation doses indicates that the three phosphorothioates provide radioprotection predominantly in the oligodendroglial cellular elements of the spinal cord white matter. Delayed CNS radiation injury in the rat is thought to involve mainly two cell types, the oligodendroglial cells concentrated in white matter in which damage appears earlier but requires higher doses (>20 Gy for single dose treatments with photons) and the endothelial cells in which damage appears later but at lower doses (<20 Gy) (3). Because we explored a high dose range of 20-38 Gy that produces earlier manifestations of damage in the oligodendroglia, the radioprotection we observed very likely involved this earlier white matter oligodendroglial cell component of CNS radionecrosis. Had there been no radioprotection from the drugs in oligodendroglia, both the drug-treated and control rats would have shown early paralysis simultaneously. These data do not clarify whether there was any protection in the CNS vascular elements, although there may have been some. The delays to paralysis in drug-treated animals may have been due entirely to protection in white matter oligodendroglia and represented the time between the occurrence of white matter oligodendroglial damage in unprotected rats and the occurrence of vascular injury in drug-treated rats. The decline in the DMF values with time may also indicate little protection against vascular injury. Such an effect might be due to inadequate drug delivery to microvascular elements after injection into the lateral ventricle or the cisterna magna.

In conclusion, DMF's as high as 1.6 in the CNS are achievable with phosphorothioates when they are administered by routes that bypass the blood-brain barrier. These results need to be extended to establish whether brain tumor tissue derives greater or lesser radioprotection from phosphorothioates than normal brain. Simultaneously, further dose and fractionation protocols will have to be explored and less toxic, less water-soluble (more lipophilic) radioprotectors will have to be developed.

ACKNOWLEDGEMENT

This work was supported by NIH Grant, CA 36485.

REFERENCES

1. A.M. Spence, S.W. Edmondson, K.A. Krohn, Z. Grunbaum, J.S. Rasey, and J.E. Steele, Toxicity and biodistribution of the radioprotectors, WR2721, WR77913, and WR3689 in the CNS following intraventricular or intracisternal administration. Int. J. Radiat. Oncol. Biol. Phys. 12, 1653-1660 (1986).
2. A.M. Spence, K.A. Krohn, S.W. Edmondson, J.E. Steele, and J.S. Rasey, Radioprotection in rat spinal cord with WR2721 following cerebral lateral intraventricular injection. Int. J. Radiat. Oncol. Biol. Phys. 12, 1479-1482 (1986).
3. A.J. van der Kogel, The cellular basis of radiation-induced damage in the central nervous system. In Cytotoxic Insult to Tissues: Effects on Cell Lineage 1983 (C.S. Potten and J.H. Hendry, Eds.) pp. 329-352, Churchill-Livingstone, Edinburgh.

PRODUCTION AND CHARACTERIZATION OF MONOCLONAL ANTIBODIES TO THIOL-MODIFIED GLUTATHIONE

Joseph P. Messina, Joseph Mazurkiewicz, and David A. Lawrence

Departments of Microbiology-Immunology and Anatomy
Albany Medical College, Albany, N.Y. 12208

INTRODUCTION

Glutathione (GSH), a multifunctional tripeptide thiol found in virtually all mammalian cells, has a prominent role in the maintenance and execution of many homeostatic and regulatory processes of the cell (1). Of the many functions of GSH, perhaps the most studied, is its ability to protect the cell against the detrimental effects of ionizing radiation, reactive oxygen compounds, free radicals, and toxic xenobiotics (2). The GSH oxidation-reduction cycle is of primary importance when dealing with reactive oxygen intermediates as well as the detoxification of xenobiotics. Although the elimination of each is catalyzed by different enzymes, GSH peroxidase in the former and GSH S-transferases in the latter, the maintenance of a high ratio of GSH:GSSG by the cycle is of vital importance to both pathways.

Glutathione also influences lymphocyte reactivities (3). The evidence indicates that it may have a significant role in lectin-induced lymphocyte activation and proliferation. The involvement of GSH in carcinogenesis is suggested by the rise in intracellular GSH observed in transformed cells from various tissue sources (4). In addition the intracellular GSH concentration of tumor cells determines their sensitivities to various modalities of therapy.

Previously quantitation of GSH required fractionation procedures which precluded analysis of GSH compartmentalization and caused potential variance in the free vs. mixed-disulfide GSH pools. Monoclonal antibodies (mAb) known for their specificity and usefulness in a variety of immunoassays could alleviate the above problems. The production of antibodies capable of reacting with GSH or an easily derivatized form would be very useful in developing immunoassays (ELISA systems for quantitation and immunocytochemistry for subcellular distribution). We report here the production and characterization of monoclonal antibodies specific for thiol modified GSH.

MATERIALS AND METHODS

Immunization and Hybrydoma Production

KLH-GS-NEM was produced by conjugation (via an amino cross-linker) of keyhole limpet hemocyanin (KLH) with GSH which was previously alkylated with N-ethyl maleimide (NEM). Balb/c mice were immunized with KLH-GS-NEM in Freund's complete adjuvant. Splenocytes obtained from high titre mice were fused with P3X63Ag 8.653 plasmacytoma cells and cloned as previously described (5). The mAb's were propagated as ascites in Balb/c mice.

Elisa Assay

The wells of microtiter plates were coated with 10 μg/ml of GS-NEM conjugated to bovine serum albumin (BSA) by incubating at 4°C for 16 hrs, the plates were blocked with 0.2% BSA and were utilized for quantitation of anti-GS-NEM antibody with biotinylated goat anti-mouse IgG or (IgM) followed by avidin-alkaline phosphatase and the substrate paranitrophenyl phosphate Plates were read at 405 nm after a 30 min.incubation at 37°C.

Immunofluorescent Analysis of Cells

Immunofluorescent analysis of cells was performed to examine the intracellular distribution of GSH. Human peripheral blood mononuclear cells (PBMC) were obtained from volunteers and resuspended at a concentration of $5x10^6$ cells/ml in balanced salt solution (BSS) containing 10 μM NEM and incubated for 30 min at room temperature. The cells were washed three times with BSS and cytocentrifuged onto glass slides. The cells were fixed by incubating for 5 min in methanol at -20°C, rinsed in three changes of acetone followed by three two-minute rinses in PBS. Slides were stained with a 1:10 dilution of mAb 8.1-GSH or the appropriate control antibodies and incubated for 4 hrs at room temperature. Cells were washed three times with PBS and incubated for 2 hrs at room temperature with fluorescein conjugated goat anti-mouse IgG (gamma chain specific). Stained cells were washed three times with PBS, mounted, and photographed by using an epiluminescence microscope.

RESULTS

Development and Characterization of mAb to GS-NEM

The low molecular weight of GSH and its distribution throughout all mammalian tissues indicate that GSH has a very low antigenic potential. Its ability to act as an immunogen was increased by modifing the thiol group with NEM and subsequent conjugation to KLH. The immunogenicity of the KLH-GS-NEM complex was tested by injection into mice and rabbits, both species developed high antibody titers of 1:500 and 1:1000, respectively, which could be inhibited by preincubating the antisera with GS-NEM. To obtain monoclonal antibodies, splenocytes from mice with high antibody titers were fused to myeloma cells as described above. Several hybridomas secreting antibodies reactive with BSA-GS-NEM, but not BSA, were obtained. Two clones with high antibody titers in the culture supernatants were selected for further study. The isotypes of mAb's 8.1-GSH and 8.1-F8-G1 were determined by ELISA assay and found to be IgG_1 and IgM, respectively (Table 1). Although the binding of both antibodies could be inhibited to the same extent by low concentrations of GS-NEM (0.1 μM) only mAb 8.1-GSH was found to have the least cross-reactivity with Cysteine-NEM (CYS-NEM) and was purified further by protein A column chromatography.

Table 1. Analysis of Monoclonal Antibodies to GS-NEM[a]

Antibodies	Inhibitor[b]	OD 405/ well	CPM/ well	% Inhibition
8.1-GSH	None	0.827	ND	0
(IgG_1)	GS-NEM	0.608	ND	27
	CYS-NEM	0.810	ND	2
8.1-F8-G1	None	ND	9,647	0
(IgM)	GS-NEM	ND	7,444	22
	CYS-NEM	ND	8,249	15

[a]Analysis was by inhibition solid phase radioimmunoassay (cpm/well) or ELISA with alkaline phosphatase (OD405/well).
[b]Inhibitors were used at 0.1 μM.

Specificity and Sensitivity of mAb 8.1-GSH

Monoclonal antibody 8.1-GSH in ascites fluid had an ELISA titer of >1:50,000, with one-half maximal binding at a dilution of 1:10,000.

The sensitivity of mAb 8.1-GSH was determined by measuring the ability of GS-NEM to inhibit binding of the antibody to BSA-GS-NEM coated plates. Varying concentrations of GS-NEM were added to the mAb (1:20,000) and a competitive binding curve was generated (Fig. 1). The amount of GS-NEM required to block 50% (IC_{50}) of the antibody binding was 50 μM.

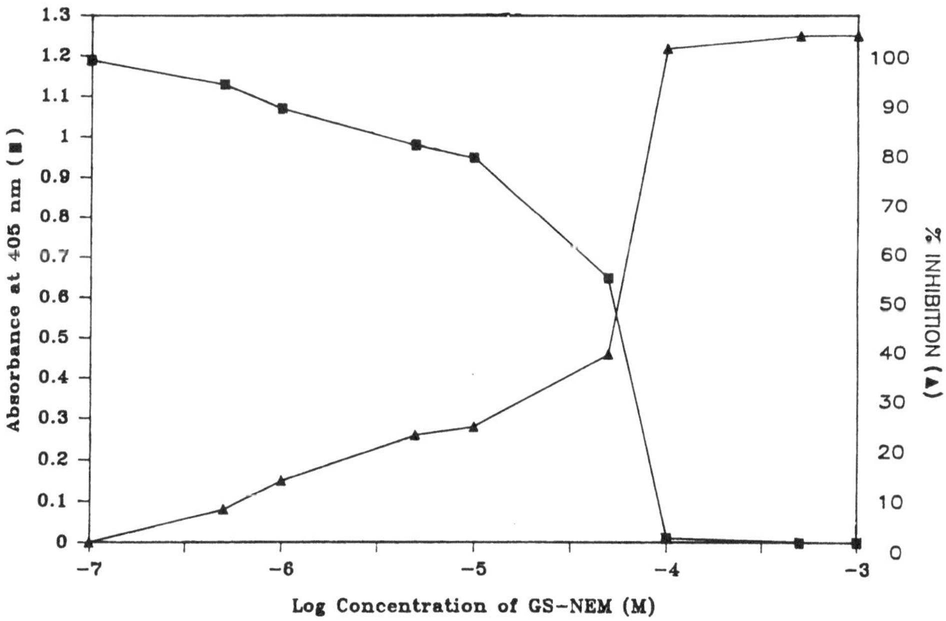

Fig. 1. Concentration of GS-NEM required to inhibit mAb binding to ELISA plates. MAb 8.1-GSH diluted 1:20,000 was incubated with various concentrations of GS-NEM in order to determine the concentration causing a 50% inhibition of the binding of the antibody to the GS-NEM antigen.

Table 2. Specificity of Monoclonal Antibody 8.1-GSH

Compound[a]	% Cross-Reactivity
GS-NEM	100
Cysteine	8.3
Glutathione (Reduced)	7.5
CYS-NEM	8.3
NEM	<0.01
Glutathione (Oxidized)	<0.01
S-Paranitrobenzyl-glutathione	<0.01
Ergothionine	<0.01
PGA_1-Glutathione	<0.01
Plasma Proteins-NEM	<0.01

[a]All inhibitors used at a concentration of 100 μM.
[b]Plasma Proteins were reacted with NEM dyalized against PBS and used at a final dilution of 1:2.

The specificity of this antibody was assessed further by examining the ability of closely related sulfhydryl compounds to inhibit its binding to BSA-GS-NEM. These results are shown in Table 2. Only GS-NEM was capable of 100% inhibition at a concentration of 100 μM. At a similar concentration Cys, Cys-NEM, and GSH produced <10% inhibition, while all other compounds tested showed no significant inhibition. This effect was also observed at higher concentrations of inhibitor and with the exception of GS-NEM, less than 50% inhibition was obtained at inhibitor concentrations of lmM. In addition to non-protein thiols, the cross- reactivity with protein thiols was determined. NEM-reacted plasma proteins, depleted of non-protein thiols, did not interfere with the binding of the mAb 8.1-GSH.

Fluorescent Analysis of Cells

The ability of mAb 8.1-GSH to bind GS-NEM adducts produced inside whole cells was determined by fluorescent microscopy. Human PBMC treated with 10 μM of NEM, reacted with a 1:10 dilution of mAb 8.1-GSH, and developed with fluoresceinated goat anti-mouse IgG demonstrated a high level of fluorescence (Fig. 2). Non-NEM treated cells exposed to mAb 8.1-GSH or NEM treated cells stained with an irrelevant antibody showed very little background fluorescence. The fluorescence intensity of the monocytes was much greater than the lymphocytes. This fluorescence was localized predominantly in the cytoplasm with very little staining of the plasma membrane and the nucleus (Fig. 3). Interestingly, patches with increased fluorescence were also observed in the cytoplasm.

DISCUSSION

The functional importance and wide distribution of GSH in tissues has led to the development of numerous assays for its quantitation. However, many of these assays are cumbersome, lack in sensitivity and in most cases quantitate only non-protein thiols or GSH plus GSSG. In an effort to develop an immunoassay we have produced polyclonal and monoclonal antibodies to thiol modified GSH. The modification was critical both for the promotion of antigenicity and for retaining some

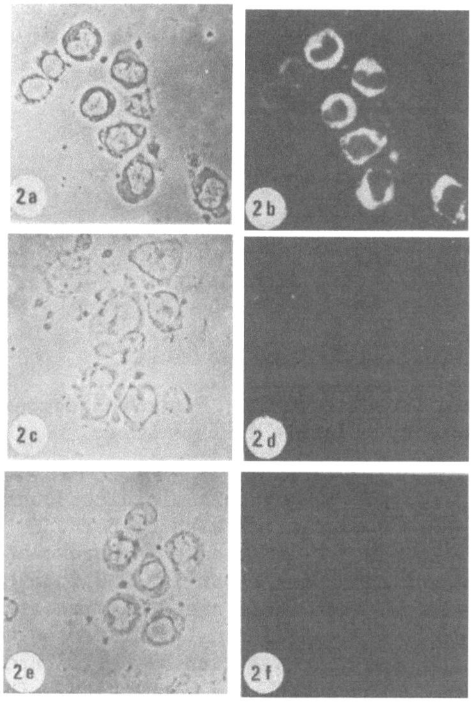

Figure 2. Immunofluorescent staining of human PBMC. Phase (a,c,e) and fluorescent (b,d,f) photomicrographs of fixed PBMC. Cells preincubated with 10 uM of NEM and stained with a 1:10 dilution of mAb 8.1-GSH (a,b); cells preincubated with 10 uM of NEM and stained with mouse IgG_1 (c,d); non-NEM treated cells stained with a 1:10 dilution of mAb 8.1-GSH.

specificity for the original GSH molecule. NEM was chosen because of its low molecular weight and its high sulfhydryl reactivity. Additional coupling of the GS-NEM adduct to the protein carrier KLH resulted in a highly antigenic complex as demonstrated by the high antibody titers produced in animals immunized with this complex.

The monoclonal antibodies produced bound to GS-NEM coupled to the protein carrier BSA but did not recognize the BSA alone. This binding was specific in that it could be inhibited in a linear fashion with GS-NEM, but not with other closely related sulfhydryl compounds. Of particular importance is the lack of inhibition by plasma sulfhydryl bearing proteins which had been pretreated with NEM. This suggests that the antibody does not cross-react with protein sulfhydryls and can be utilized in an immunoassay for the detection of free GSH.

Understanding the functions of intracellular GSH requires that we quantitate cellular GSH as well as detect its subcellular localization at various times during the cell cycle. Previously existing methodologies for quantitation of intracellular GSH precluded these possibilities since they required that the cellular integrity be disrupted in order to make the measurements. One of the attributes of antibodies is that they allow one to detect and quantitate intracellular components under conditions which minimally disrupt the cellular architecture. Using mAb 8.1-GSH we have found that GSH in human PBMC is localized mainly in the cytoplasm. The lack of nuclear GSH may be an artifact of the fixation procedures or indicative of a true low nuclear to cytoplasmic ratio of GSH in resting

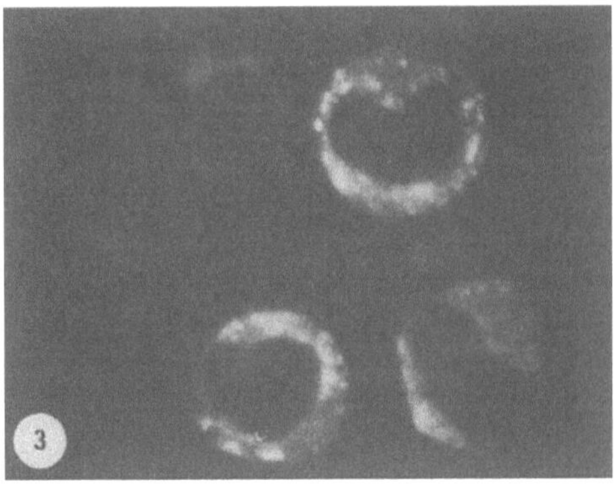

Figure 3. Detail of figure 2b. Monocytes demonstating patchy areas of intense cytoplasmic fluorescence.

(non-proliferating) cells. The cytoplasmic distribution of fluorescence suggests the presence of organelles with higher GSH concentration than the surrounding cytoplasm. The intense fluorescent vesicles in the monocytes may represent endosomes exporting GSH from the cell (6). The possibility also exists that these areas of intense fluorescence are mitochondrial GSH pools (7). Additional differences were also observed among the various cell populations. Much higher fluorescence is observed in the monocytes than in the lymphocyte population. This is in agreement with previous studies showing that monocytes have a higher GSH content than lymphocytes (8).

ACKNOWLEDGEMENT

This work supported by NIH grants CA 35889 and ES 03778.

REFERENCES

1. A. Meister and M. E. Anderson, Glutathione. Annu. Rev. Biochem. 52, 711-760 (1983).
2. B.A. Arrick and C.F. Nathan, Glutathione metabolism as a determinant of therapeutic efficacy: A review. Cancer. Res. 44, 4224-4232 (1984).
3. R.K. Fidelius, P. Ginouves, D. Lawrence, and Min-Fu Tsan, Modulation of intracellular glutathione concentrations alters lymphocyte activation and proliferation. Exp. Cell. Res. 170, 269-275 (1987).
4. S. Ahmad, A. Mulberg, J. Aljian, D.T. Vistica, Hepatic-mediated elevation and maintenance of metastatic tumor cell glutathione. Biochem. Pharmacol.. 35, 1697-1701 (1986).
5. R.H. Kennett, T.J. McKearn and K.B. Bechtol, eds., Methods for production and characterization of monoclonal antibody. In Monoclonal Antibodies, pp.363-419. Plenum Press, N.Y.
6. J. Swanson, B.Yiriner, E.Burke, A. Bushnell and S.C. Silverstein, Effect of alterations in the size of vacuolar compartment on pinocytosis in J-774.2 macrophages. J. Cell. Physiol. 128, 195-201 (1986).
7. D.J. Reed, and M.W. Fariss, Glutathione depletion and susceptibility. Pharmacol. Rev. 36, 25s-33s (1984).
8. D.A. Lawrence, S. Mudzinski, U. Rudosfsky, and G. Warner, Mechanisms of metal-induced immunotoxicity. In Immunotoxicology. (A. Berlin et.al. eds) pp.293-307. Martinus Nijhoff Publishers, Boston.

INDUCTION OF DIFFERENTIAL HYPOXIA IN TISSUES AND ITS IMPLICATIONS FOR THERAPY

G.E. Adams and I.J. Stratford

Medical Research Council, Radiobiology Unit
Chilton, Didcot
Oxon, OX11 ORD England

INTRODUCTION

The search for chemical substances that induce radiation protection in tissues has been in progress for many years. Interest has tended to concentrate on small molecules such as simple alcohols, dimethylsulphoxide and in particular, numerous thiols and thiophosphates. Major objectives have been two-fold. Firstly, the development of techniques for protecting individuals from the effects of exposure to whole-body radiation, accidental or otherwise and secondly, clinical strategies for reducing normal tissue morbidity in the radiation therapy of cancer. Several of the papers in these proceedings are concerned with various aspects of these topics. This paper, however, addresses another approach based on manipulation of oxygenation of normal tissues and tumours not only for inducing radiation resistance but also for obtaining increased therapeutic efficacy of some anti-tumour strategies.

EFFECT OF HAEMOGLOBIN STATUS ON TUMOUR RESPONSE TO RADIOTHERAPY

It has long been recognised by radiation therapists that anaemia adversely affects local control probability in radiotherapy. This has been recently reviewed (1). For example, the data of Bush et al. (2) reproduced in Figure 1a, clearly showed that the long-term control of patients treated with radical radiotherapy for stages IIB + III carcinoma of the cervix was significantly improved for patients with haemoglobin (Hb) levels greater than 12g%.

The criticality of haemoglobin status even within the normal range has been emphasized by the data of Overgaard et al. (3,4) from a large prospective randomised trial of radiotherapy in the treatment of pharyngeal tumours (Figure 1c). There is a clearly improved response rate for patients with haemoglobin values towards the upper end of the normal range compared with that for patients with lower values. None of the patients in either group were clinically anaemic at the time of treatment.

A particularly surprising finding in regard to Hb status comes from a re-examination of data (5) from a multi-centre trial of radiotherapy in hyperbaric oxygen for a large group of patients with advanced carcinoma of the cervix (6). The trial showed a significant improvement in local

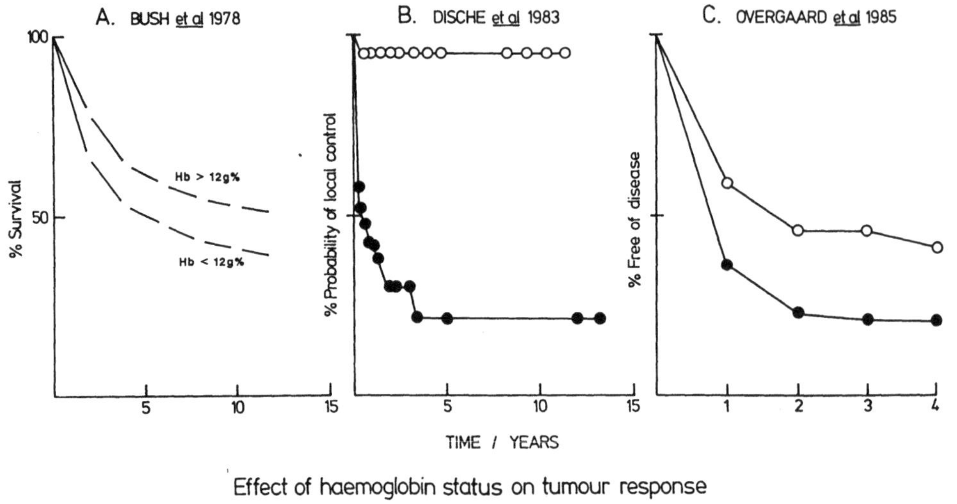

Figure 1. A, Actuarial survival curves for patients with STAGES IIb and III Ca cervix and haemoglobin levels >12 g% or <12 g% during radiotherapy. Patients treated and at risk at 12 years are, 696→149; and 429→131 respectively; at which time the difference between the two groups is significant at P = 0.001 from Bush (2).

B, Sub-group analysis of patients entered into the M.R.C. trial of hyperbaric oxygen with radiotherapy. Local tumour control in patients with stage III Ca cervix requiring blood transfusion, i.e. Hb <10 g%. Patients were divided according to treatment in air (●), 23 cases and hyperbaric oxygen (O) 15 cases. The difference between the two groups is significant at P = 0.0003 from Dische et al. (5).

C, Male patients, with supraglotic and pharynx tumours, free of local disease as a function of time after radiotherapy. O, Hb > 9 mmol/l (14.5 g%); ●, Hb < 9 mmol/l. Patients treated and at risk at 4 years are, O, 146→21; and ●, 120→9 from Overgaard et al. (3,4).

control rate for patients treated in oxygen. However, the analysis (5) of the outcome of patients who presented with anaemia and therefore required transfusion before radiotherapy treatment, produced the data shown in Figure 1b. A total of 39 patients were transfused of which, by chance, 23 were treated in air and 16 in oxygen. Despite transfusion, the former group showed a relatively poor response compared with the non-transfused patients who were treated in air. However, the group of transfused patients treated in oxygen showed a very good response, even better than the response rate in the total group of patients treated in oxygen.

The explanation of this behaviour is still not clear but it is suggested that prior anaemia could have a beneficial effect on the outcome of treatment with radiotherapy and oxygen provided the anaemia is corrected before treatment. This has prompted much laboratory research on the various effects of chronic and acute anaemia, with and without transfusion, on the radiation response of experimental tumours (7-11).

MODIFICATION OF HAEMOGLOBIN-OXYGEN ASSOCIATION

Oxygen delivery to all tissues is mediated in part by the equilibrium

$$HbO_2 \rightleftharpoons Hb + O_2$$

The equilibrium is shown diagrammatically in Figure 2a. Most normal tissues are adequately supplied with oxygen by the pool of haemoglobin, which is arterial blood, is almost completely saturated under normal conditions. This equilibrium can be displaced in a variety of ways which will be expressed by a change in the value of P50 (the partial pressure of oxygen at which half the haemoglobin exists in the oxy conformation). For example, small decreases in pH or temperature result in a decrease in P50, i.e., a displacement of the equilibrium to the left (left-shift). Similarly, decreases in erythrocyte concentrations of 2,3-diphosphoglycerate, an allosteric effector for the binding of O_2 to Hb can also decrease the value of P50. The effect of left-shifting will be to reduce the level of free oxygen in tissues, and thus should induce radio protection; in contrast, a right shift could induce some radiosensitization.

The Burroughs-Wellcome compound BW 12C, under development as an anti-sickling agent, is a potent left-shifter (12). Figure 2b reproduces data on the left-shifting activity in CBA mice treated with BW 12C (13). The left-shift is considerable and is still evident for at least 2-3 hours in this strain of mice. A recent study has shown that the reduction of free oxygen levels in tissue caused by the left-shift produces significant radioprotection in experimental systems (13,14). BW 12C protects against the whole-body acute radiation syndrome in CBA mice, and radiation-induced skin damage in both experimental mice and pigs (13,14). It also induces hypoxia in lymph nodes infiltrated by an experimental mouse T-cell lymphoma model. Treatment of mice with multiple doses of BW 12C induces necrosis of tumour tissue in the infiltrated nodes. There is no evidence of necrosis induction in normal tissue and therefore the observation of tumour necrosis might suggest a substantial deprivation of available oxygen in or around the tumour tissue.

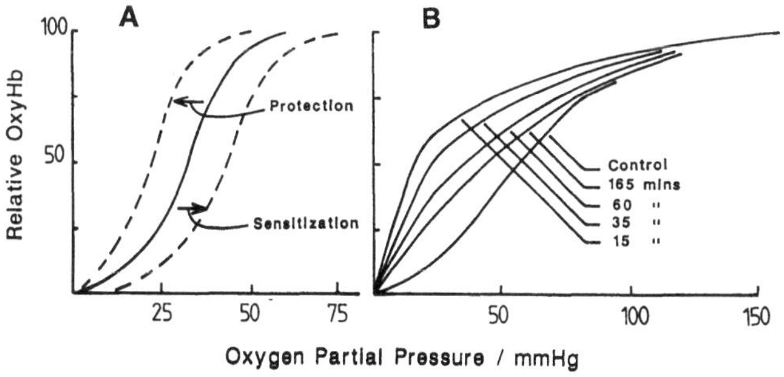

Figure 2. A, Diagrammatic representation of the relationship between oxygen association with Hb and the oxygen partial pressure.

B, Oxygen association curves for haemoglobin taken from Harwell CBA mice at various times after i.v. administration of 70 mg/kg BW12C.

There is much experimental evidence that even small tumours are considerably restricted in their oxygen supply compared with most normal tissues. Clinical evidence is accummulating that oxygen deficiency may be a feature of many human solid tumours also. The possibility, therefore, that further deprivation of oxygen could induce tumour necrosis and render tumours more susceptible to subsequent therapy, has prompted much investigation into ways of artificially increasing tumour hypoxia. A recent development and one that has possible promise for improved treatment for cancer has involved the study of the 'steal' phenomenon using vaso-active drugs such as the anti-hypertensive agent, hydralazine (15). These drugs induce vaso-dilation in normal tissue but do not exert this effect in the primitive vasculature in experimental tumours (15-18). The overall effect, therefore, is to divert blood away from the tumour and thereby substantially increase the level of tumour hypoxia. This method of inducing oxygen deprivation in tumours has been shown to cause substantial necrosis in experimental tumours and enhance the activity of some anti-cancer agents by possibly altering the pharmacokinetics of the drug in tumour tissue (Adams et al., Stratford et al. unpublished results). Further, it can increase the anti-tumour activity of bioreductive drugs and some electron-affinic radiosensitizers (see below).

An example of the ability of hydralazine to induce severe hypoxia in an experimental tumour, is shown in Figure 3 for the KHT tumour. Normally, about 10% of the clonogenic cells in this tumour are radiobiologically hypoxic when the tumour is grown subcutaneously on the backs of C3H mice. Tumour response in vivo is measured by in vitro clonogenic assay of surviving tumour cells obtained from tumour tissue excised 24 hours after treatment. The solid circles in Fig. 3 shows cell survival following irradiation of the tumour in situ. When the blood supply to the tumour is occluded by clamping for 10 minutes before and during irradiation (solid squares), the tumour becomes 100% hypoxic and the survival curve

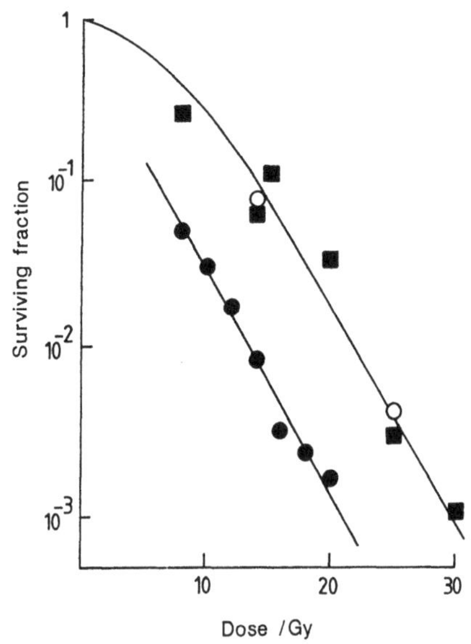

Figure 3. Response of the KHT sarcoma to single doses of X rays: ●, control tumours; ■, occlusion of tumour blood supply by clamping; O, tumours from mice treated iv with 5 mg/kg hydralazine 30 minutes prior to irradiation.

becomes identical with the survival curve for tumour cells irradiated in anoxia in vitro. The vertical displacement of the two survival curves is consistent with a 10% hypoxic fraction in unclamped tumours. The open circles in Figure 3 show survival data for irradiation of unclamped tumours in mice when 5 mg/kg hydralazine is administered i.v. 30 minutes before irradiation. The data points fall on the survival curve for clamped tumours indicating, therefore, that hydralazine induces complete radiobiological hypoxia in this tumour.

EXPLOITATION OF INDUCED TUMOUR HYPOXIA FOR INCREASING THE EFFICIENCY OF RADIOSENSITIZATION

Induction of additional tumour hypoxia which persists during radiation treatment will obviously be detrimental to treatment outcome. The data in Figure 4 indicate, however that induction of hypoxia post-irradiation appears to enhance considerably the efficacy of the electron-affinic sensitizer misonidazole in the KHT tumour. Figure 4a shows that in this tumour, a single dose of 100 mg/kg of misonidazole administered 60 minutes before a single dose of X radiation causes only a small increase in tumour sensitivity corresponding to an enhancement ratio of about 1.3 calculated from D_0 ratios. However, when hydralazine is given post-irradiation, there is a considerable increase in the amount of sensitization corresponding to an enhancement ratio of about 2.4. The data in Figure 4b show that a single dose of 10 gray reduces the surviving fraction of clonogenic cells in the KHT tumour to about 3×10^{-2}. This is steadily decreased to 5×10^{-3} as the administered dose of misonidazole given before irradiation is increased to 500 mg/kg. In contrast, when hydralazine is administered post-irradiation, the sensitization efficiency of misonidazole is enhanced by about one order of magnitude. Doses of misonidazole of 100 mg/kg or more, when combined with hydralazine give the maximum enhancement ratio achievable in this tumour system with misonidazole alone.

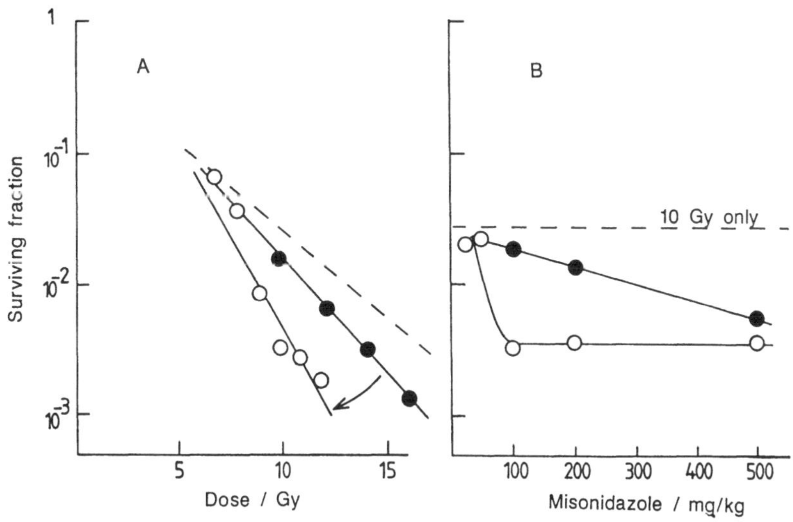

Figure 4. Misonidazole + Radiation ± Hydralazine. Response of the KHT sarcoma to single doses of X rays. A, dependence on radiation dose (± 100 mg/kg misonidazole); B, dependence on misonidazole concentration. Broken line, control tumours; ● tumours from mice treated with misonidazole only 60 minutes prior to irradiation; O, as above but with 5 mg/kg hydralazine administered iv immediately after irradiation.

Table 1. Effect of hydralazine (5 mg/kg) given immediately after 10 Gy X rays, on sensitization of tumour response by RSU 1069

Dose of RSU 1069 (mg/kg)	Surviving fraction x 10^3 (no hydralazine)	(hydralazine)
-	32	— *
5	32	8
10	20	4.2
20	12	2.0
40	8.5	1.9
60	3.4	2
80	1.4	1.1

* Comparison made at 14 Gy - no difference between hydralazine-treated and control.

Similar studies have been carried out with the dual-function bio-reductive radiation sensitizer RSU 1069 (19). This sensitizer is considerably more effective than misonidazole in various experimental tumour systems as is illustrated in Table 1.

As for misonidazole, the sensitizing efficiency of RSU 1069 is considerably enhanced by post-irradiation treatment with hydralazine. The mechanisms of this increase in sensitizing efficiency has not yet been fully characterised. Whether or not the effect is partly an expression of the hypoxic cytotoxic properties of the sensitizer requires further studies. These are in progress.

REFERENCES

1. R.S. Bush, The significance of anaemia in clinical radiation therapy. Int. J. Radiat. Oncol. Biol. Phys., 12, 2047-2050 (1986).
2. R.S. Bush, R.D.T. Jenkin, W.E.C. Allt, F.A. Beale, H.A. Bean, A.J. Dembo and J.F. Pringle, Definitive evidence for hypoxic cells influencing cure in cancer therapy. Br. J. Cancer 37 (III), 255-258 (1978).
3. J. Overgaard, S.H. Hansen, K. Jorgensen and M. Hjelm, Primary radiotherapy of larynx and pharynx carcinoma - An analysis of some factors affecting local control and survival. Int. J. Radiat. Oncol. Biol. Phys. 12, 515-521 (1986).
4. J. Overgaard, S.H. Hansen, A.P. Anderson, M. Hjelm-Handsen, K. Jorgensen, K. Sandberg, J. Rygard, R.H. Jensen and M. Petersen, Misonidazole as an adjuct to radiotherapy in the treatment of invasive carcinoma of the larynx and pharynx. Proceedings of the 3rd Int. Meeting on Progress in Radio-Oncology, March 1985, Vienna, Austria. Raven Press, New York, 1987, (in press).
5. S. Dische, P.J. Anderson, R. Sealy and E.R. Watson, Carcinoma of the cervix: Anaemia, radiotherapy and hyperbaric oxygen. Br. J. Radiol. 56, 251-255 (1983).

6. E.R. Watson, K.E. Halnan, S. Dische, M.I. Saunders, I.S. Cade, J.B. McEwen, G. Wiernik, D.J.D. Perrins and I. Sutherland, Hyperbaric oxygen and radiotherapy: A Medical Research Council trial in carcinoma of the cervix. Br. J. Radiol. 51, 879-887 (1978).
7. H.B. Hewitt and E. Blake, Effect of induced host anaemia on the viability and radiosensitivity of murine malignant cells in vivo. Br. J. Cancer 25, 323-336 (1971).
8. R.P. Hill, R.S. Bush and P. Yeung, The effect of anaemia on the fraction of hypoxic cells in an experimental tumour. Br. J. Radiol. 44, 299-304 (1971).
9. D.G. Hirst, J.L. Hazelhirst and J.M. Brown, The effect of haematocrit on tumour sensitivity to X-rays. Int. J. Radiat. Biol. 46, 345-354 (1984).
10. D.G. Hirst, Anaemia: A problem or opportunity in radiotherapy? Int. J. Radiat. Oncol. Biol. Phys. 12, 2009-2017 (1986).
11. D.G. Hirst, Oxygen delivery to tumours. Int. J. Radiat. Oncol. Biol. Phys. 12, 1271-1277 (1986).
12. C.R. Beddell, P.J. Goodford, G. Kneen, R.D. White, S. Wilkinson and R. Wootton, Substituted benzaldehydes designed to increase the oxygen affinity of human haemoglobin and inhibit the sickling of sickle erythrocytes. Br. J. Pharmacol. 82, 397-407 (1984).
13. G.E. Adams, I.J. Stratford, I. Ahmed and P.W. Sheldon, Some new developments in tumour radiosensitization. Proc "3rd Int. Meeting on Progress in Radio-Oncology", March27th-30th 1985; Vienna, Austria. Raven Press, New York, 1987 (in press).
14. G.E. Adams, D. Barnes, J. Loutit, S. Cole, P.W. Sheldon, I.J. Stratford, G.J.M.J. Van der Aardweg, J. Hopewell, R. White, G. Kneen, A. Nethersell and C. Edwards, Induction of hypoxia in normal and malignant tissues by changing the oxygen affinity of haemoglobin: Implications for therapy. Int. J. Radiat. Oncol. Biol. Phys. 12, 1299-1302 (1986).
15. D.J. Chaplin and B. Acker, Potentiation of RSU 1069 tumour cytotoxicity by hydralazine: A new approach to selective therapy. Int. J. Radiat. Oncol. Biol. Phys. 13, 1987 (in press).
16. D.B. Cater, C.M.B. Grigson and D.A. Watkinson, Changes of oxygen tension in tumours induced by vasocontrictor and vasodilator drugs. Acta. Radiol. 58, 401-408 (1962).
17. J.A. Kruuv, W.R. Inch and J.A. McCredie, Blood flow and oxygenation of tumours in mice II. Effect of vasodilator drugs. Cancer 20, 60-65 (1967).
18. W.H. Knapp, J. Debatin, K. Layer, F. Helus, A. Altmann, H.J. Sinn and H. Ostertag, Selective drug induced reduction of blood flow in tumour transplants. Int. J. Radiat. Oncol. Biol. Phys. 11, 1357-1366 (1985).
19. G.E. Adams, I. Ahmed, P.W. Sheldon and I.J. Stratford, Radiation sensitization and chemopotentiation: RSU 1069, a compound more efficient than misonidazole in vitro and in vivo. Br. J. Cancer 49, 571-578 (1984).

RADIOPROTECTION IN VIVO: CELLULAR HETEROGENEITY AND FRACTIONATION

J. Denekamp and A Rojas

Cancer Research Campaign
Gray Laboratory
Mount Vernon Hospital
Northwood, Middlesex HA6 2RN

INTRODUCTION

The curative treatment of cancer by radiotherapy is limited by the sensitivity of critical normal tissues surrounding the tumour that are inevitably included in the irradiation volume. In the last decade two new developments have progressed from the experimental stage in animals to clinical testing: Radiosensitizers that appear specific for tumour cells and radioprotectors that appear specific for normal tissues have both been developed. Of these the prototype compounds are nitroimidazole, misonidazole (miso), as a radiosensitizer (1) and the phosphorylated aminothiol WR-2721 (S-2-(3-amino-propylamino) ethyl phosphorothioic acid) as a radioprotector (2). The specificity of action of these agents does not relate to phenotypic alterations connected with the malignant transformation of cells, but rather to differences in regional oxygen tensions that result from the poor vascularity of tumours. The inadequate vascular array leads to regions of hypoxia which confer radioresistance (by a factor of 2.5 to 3.0). Miso can act as an oxygen-mimetic agent in such hypoxic regions and can repair damage from radiation-induced radicals, a process which already occurs efficiently if oxygen is present. Miso is ineffective in oxygen. Since hypoxic cells rarely exist in normal tissues the effect is tumour specific. Lesion repair by hydrogen donation by endogenous thiols is believed to compete with oxygen fixation of lesions, so it follows that the effect of thiols as radioprotectors is also influenced by local oxygen concentrations. Thiols are ineffective in anoxia. They are also relatively ineffective at extremely high oxygen concentrations, because the balance of the competition is loaded towards fixation. However, at oxygen tensions that are close to the transition region for radiosensitivity to radio-resistance (i.e., the K value), added thiols are maximally effective and the balance can most readily be tipped one way or the other. Since oxic cells predominate in normal tissues, the radioprotection by thiols seems to be specific for normal tissue.

The development of radioprotectors stemmed mainly from the military interest in protecting soldiers in the event of a nuclear attack. The concept of using these thiols in radiotherapy was introduced by John Yuhas. In 1969 Yuhas showed that there was much more protection of bone marrow and of intestine than there was of tumours (3). He later showed that uptake into tumours was less efficient than into normal tissues and reported a failure to achieve significant radioprotection in 16/17 tumours, the exception being a hepatoma (4).

Phillips et al. (5) studied six normal tissues and two types of tumour. They found very high protection factors (PF) for bone marrow (PF = 3.0), i.e., a three-fold increase in dose is needed for the same level of injury), but a range of values for the other normal tissues, falling as low as 1.2 for lung. In a leukaemia they measured a PF of 2.2, and in a solid tumour they saw protection at low doses but very little effect at high doses, which might be approaching the level of cell killing needed for local tumour control.

These animal data led to at least two clinical studies (6-8), and at the same time to a much wider interest in investigating the degree of protection that is observed in each study (see proceedings of conferences at Gaithersburg 1983 (9), and at Banff 1984 (10). Many of these results are summarised in reviews (e.g., Refs. 11 and 12). The expanded body of data shows that:

1. Precise oxygen tensions are critically important.
2. Endogenous thiol levels matter.
3. The ability to dephosphorylate and absorb the active product WR-1065 differs in different tissues.
4. No single protection factor can be assumed to apply to all normal tissues.
5. Because of heterogeneities of oxygenation the protection factor for one tissue differs at different dose levels.
6. The protection seen with a series of small X-ray treatments can be quite different from that seen with a large single dose.
7. Significant protection occurs in most tumours at clinically relevant X-ray doses.
8. Cumulative drug toxicity will limit the usefulness of WR-2721 in fractionated treatments.
9. The tolerated dose in man is, of course, much lower than that which can be given to mice.

The overall balance of these results is unfortunately not promising for widespread application of WR-2721 in clinical practice, but a greater scientific understanding of the factors influencing thiol radioprotection has emerged.

THIOL LEVELS IN CELLS

The effect of adding thiols to a tissue prior to irradiation will depend both on the uptake of thiols into that tissue, and the size of the endogenous pool of non-protein thiols (mainly glutathione) that is being supplemented. Both these parameters vary widely from tissue to tissue. Endogenous GSH levels are high in some tissues, e.g., kidney and liver, but are up to 10 times lower in most other tissues and in tumours. Averages for a whole tissue homogenate give no indication of the heterogeneity in thiol content in different cells (e.g., at different distances from the capillary supply, or in different phases of the intermitotic cycle), nor of the localised distribution of the GSH within each cell. Studies using BSO, a specific inhibitor of GSH synthesis, have shown that the cytosolic and nuclear GSH is more radpidly turned over than the mitochondial (13). Detailed micro-distribution studies are not available for most tissues but there is undoubtedly a heterogeneity of GSH levels in any cell population (14).

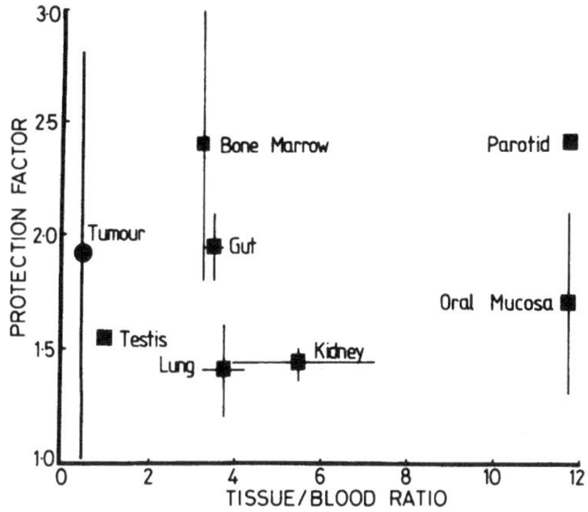

Fig. 1. There is no simple correlation between the radioprotection observed and the concentration of thiol (measured by ^{14}C or ^{35}S levels) in tumours and normal tissues. Uptake is low in tumours but a wide range of PF values (indicated by the bar) in seen at different levels of damage. More protection is observed in marrow than in oral mucosa, in spite of a 3-fold lower uptake of WR-2721 (see Ref. 11 for sources of data).

The uptake of the dephosphorylated form of WR-2721 has been studied using ^{35}S or ^{14}C labelled compound. It is quite variable from tissue to tissue. Figure 1 shows however that there is no simple relationship between the protection factor observed and the tissue/blood ratio of the labelled thiol. There is little uptake in tumours, and in normal tissues uptake appears to be passive in testis and the thiol to be concentrated by factors of 4- to 12-fold in various other organs. Factors influencing the rate and extent of WR-2721 uptake probably include blood flow, the microdistribution of capillaries, and the amount of phosphatase released by the endothelial cells in individual capillaries, which will locally convert the phosphorylated pro-drug into its active form, known as WR-1065. The active drug can then freely diffuse across the blood vessel wall into the surrounding tissues. The poor uptake into tumours may relate to the low levels of alkaline phosphatase that are present in immature tumour capillary endothelial cells. Only after dephosphorylation in other areas will the total blood levels of WR-1065 rise, providing a reservoir which can then diffuse at later times into the tumour.

However, local thiol content is only a part of the equation. Because thiols influence the damage inflicted by ionising radiation by redox competition with oxygen, the endogenous levels of oxygen also play a very important role.

HETEROGENEITY OF OXYGEN CONCENTRATION

The radiosensitivity of cells is a function of their oxygenation at the time of the ionising events. This has been extensively studied and the K curve for radiosensitivity has been characterised for many bacteria and mammalian cells in vitro (see Ref. 15, for summary). The K value is the oxygen tension at which half the maximum radiosensitivity is achieved. For cells in vitro this is achieved at a very low oxygen tension (5 mm Hg), i.e., when a monolayer is in equilibrium with a gas containing approximately 0.3% oxygen. In animals it is much more difficult to know

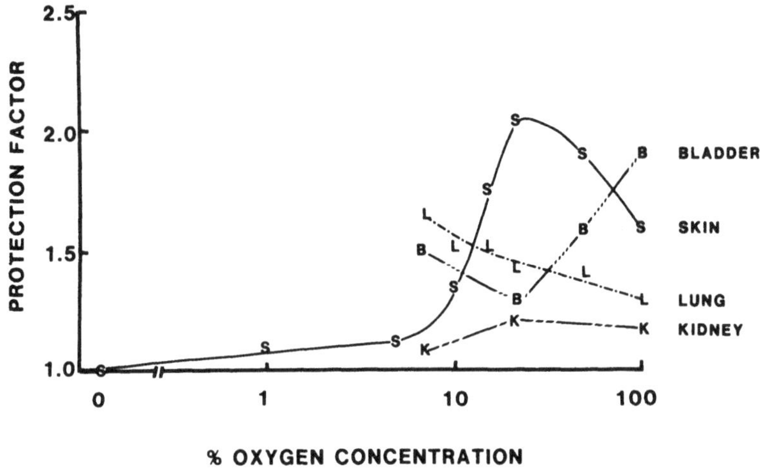

Fig. 2. Variation in the observed protection by WR-2721 when the oxygen concentration in the gas surrounding the animals at the time of irradiation is altered. The pattern is different in the four tissues, presumably reflecting differences in endogenous oxygen or thiol contents (see Ref. 16, 17 bladder: Stewart et al., unpublished; kidney: Stevens et al., unpublished).

what the precise oxygen tension in the critical clonogenic cells is. Studies in skin have shown that the K value is achieved when the animals are breathing 10-12% oxygen (11). This indicates that many skin cells exist at an oxygen tension very close to their natural K value: cells close to capillaries will be above that and cells most distant from any capillary will be below it.

The consequence of this is illustrated in Figure 2 for skin and for three other normal tissues. Pairs of dose-response curves have been obtained for mice irradiated in different ambient gases. As the oxygen supply is reduced the tissue becomes less sensitive. The protection factor (i.e., the dose needed with or without WR-2721 in that particular gas phase) also changes. In skin the maximum PF was seen in air, with less protection at lower or higher oxygen levels (16). In bladder, lung, and kidney the patterns are different, presumably reflecting the different natural levels of oxygenation in these tissues in air-breathing mice. In lung the PF increases as the inspired pO_2 falls (17), whereas the reverse is true for bladder (F. A. Stewart et al., unpublished). In kidney the PF was small at all pO_2 levels tested (G. Stevens et al., unpublished).

This microdistribution of oxygen will have a major impact on tissue responsiveness to a thiol radioprotector, even if it has little impact on the response to X rays alone. This is illustrated in Figures 3 and 4. The proportion of cells at high, moderate, low, or very low oxygen tensions will vary from one tissue to another (Fig. 3). The extreme example of tumours, which commonly contain 10-20% of hypoxic cells that are radiobiologically resistant is well recognised. The difference in degree of oxygenation in cells in different normal tissues, all of which are at least partially sensitized, is less well recognised, but is well-documented from micro-electrode studies (e.g., Ref. 18) and from studies with misonidazole (19,20).

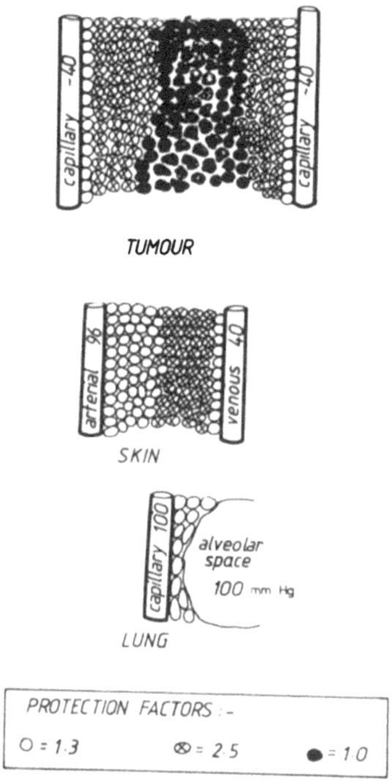

Fig. 3. Diagrammatic illustration of the subpopulations in tumours, skin, and lungs that will have different PF values because of their position in the oxygen gradients around blood vessels.

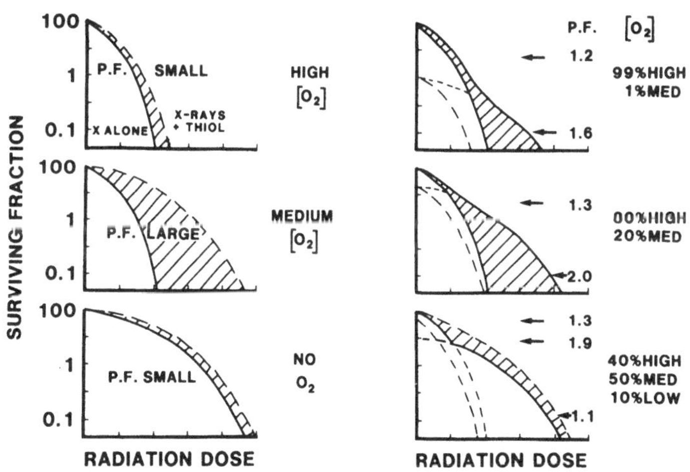

Fig. 4. Schematic survival curves for cells at different oxygen tensions treated with or without added thiol. Left-hand panels illustrate that the maximum thiol protection will be seen at oxygen tensions close to the K value. The right-hand panels show how the PF values will depend upon the mixture of subpopulations and the level of damage at which the isoeffective doses are derived.

If all cells are at a high oxygen tension the measured PF will be low; if all cells are close to hypoxia it will also be low (Fig. 4A). But if cells are just above the K value they will show a very large protection factor which can in principle approach the full OER of 3, if no mechanism other than oxygen-thiol competition is evoked. If a tissue (or tumour) contains a mixed population of cells at different oxygen tensions the response will be more complex and the measured protection factor will change depending upon the level of survival at which it is measured, i.e., upon the subpopulation which dominates the response at any particular dose (Fig. 4B).

This schematic example illustrates why it may be important to consider the dose level needed to achieve a measurable response in different tissues. This dose range is 5-10 Gy for bone marrow, 10-20 Gy for lung and kidney, and 20-50 Gy for skin. In assessing tumour response, clonogenic assays can be used at 5-15 Gy whereas local control may require doses of 40-60 Gy. Figure 4B demonstrates how misleading comparisons may be if they are not made over equivalent dose ranges.

Figure 5 shows a summary of PF values for all the published normal tissue and tumour data, plotted as a function of the X-ray dose used in each treatment, for the chosen level of tissue effect. Once again the very wide range of values for different normal tissues and tumours is apparent, and there appears to be a trend towards higher values at low doses, falling towards unity at very high doses. This would be consistent with a subpopulation of hypoxic cells dominating the response of both tumours and skin at doses above 35 Gy.

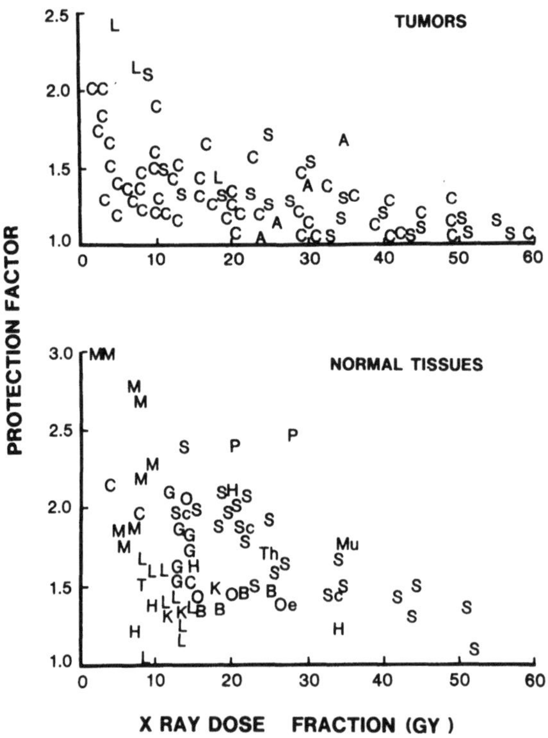

Fig. 5. Compilation of published PF values for tumours (A = adenoma, L = leukaemia, S = sarcoma, C = carcinoma) and a wide variety of normal tissues (see Ref. 11 for sources of data). There is a wide spread of PF values in both tumours and normal tissues, with higher values generally at low doses.

FRACTIONATION OF RADIATION DOSE

When repeated X-ray doses are used with WR-2721 it is not possible to use the same thiol dose as in single-dose studies because the drug shows a cumulative toxicity. Single doses of about 1000 mg/kg are needed to kill 50% of CBA Mice, and this increases to 2700 mg/kg for five daily or 4250 mg/kg for ten fractions(21). Although the total drug dose can be increased, the individual dose given with each irradiation progressively falls. Thus comparisons of single-dose and fractionated studies are difficult to make, because only those in which the same drug dose has been used with each fraction are valid.

Figure 6 shows in histogram format the change in PF that occurs when a fixed dose of WR-2721 is used with single X-ray doses or with each of several fractions. In three tissues the PF falls with fractionation, in three it is unaltered, and in lung there is a marginal increase. The lower panel summarises all the Gray Laboratory data for mouse tumours treated with either single doses or multiple fractions. In most cases there is a dramatic increase in PF as lower doses are used, whether in single or multiple treatments (21). It is this fact that is the single most worrysome contradiction to extrapolating the early WR-2721 studies to clinical application.

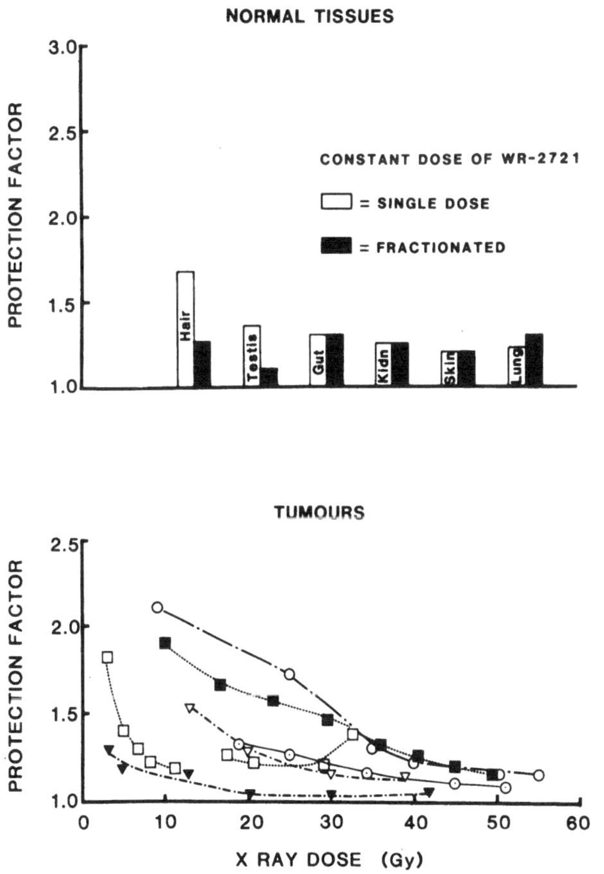

Fig. 6. Variation in PF with fractionation in normal tissues (upper panel) and with dose per fraction in tumours (lower panel) (see Ref. 11 and 21 for sources of data).

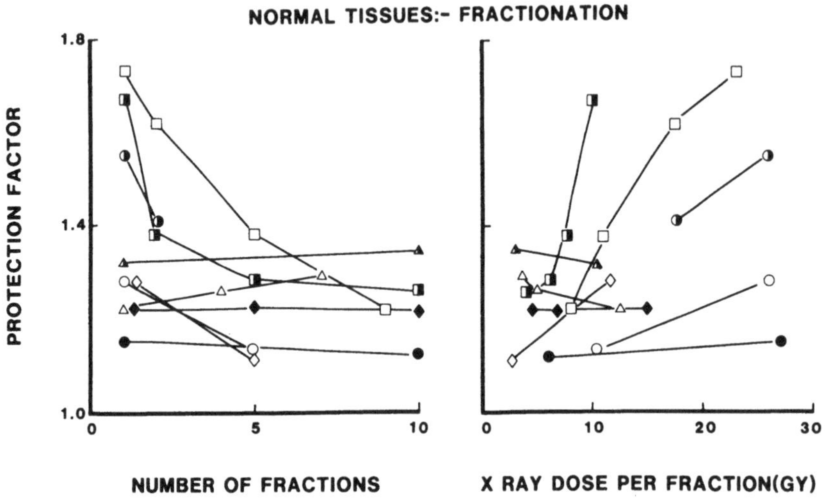

Fig. 7. A more detailed analysis of the change in PF with fraction number (left panel) and hence with dose per fraction (right panel) for a variety of normal tissues.

	Reference
Hair loss	□ ◧ (22)
Skin reactions	○ ◑ (21)
Skin reactions	● (23)
Lung	△ (24)
Lung	▲ (25)
Kidney	◆ (21)
Testis	◇ (26)

In Figure 7 the data for normal tissues are presented in more detail, as a function of fraction number (left-hand panel) and of the consequent dose per fraction (right-hand panel). Those studies which used only 1 or 2 fractions involved higher drug doses than those extending up to 10 fractions, because of the cumulative toxicity mentioned earlier, and the need to use the same tolerated dose with each fraction and with the single dose. Once again there is a variation from one tissue to another. The lower PF values with fractionation are seen for hair loss, skin reactions, and testis (where direct cytotoxicity of multiple WR-2721 complicates the interpretation). A virtually flat response is seen in another skin study and in lung and kidney.

For clinical application in radiotherapy, repeated (or fractionated) doses of 2-3 Gy are usually used, up to a total of 60-70 Gy in a period of 4-7 weeks. In such a situation the sensitive (oxygen-rich or thiol-poor) cells will be eliminated with the early fractions. The resistant (oxygen-poor or thiol-rich) cells may come to dominate the response with later fractions. Thus the influence of a thiol protector could vary depending on when it was administered during such a fractionated course of radiotherapy. However, many physiological and biochemical alterations may occur during the course of treatment and the heterogeniety of the population may be restored to its original balance, or indeed a new heterogenous set of subpopulations may result. In tumours, for example, it has been demonstrated that many of the hypoxic cells present before the first irradiation may "reoxygenate" between fractions, so that they become more radiosensitive in a fractionated treatment than they are with single doses. Thus for tumours the response to a series of small doses is better represented by considering the overall total dose. This may also be true for normal tissues.

The PF values all fall between 1.0 and 3.0, i.e., within the range of effect that is also achieved by changing from a high oxygen tension to anoxia. This is consistent with redox competition being the main mechanism of radioprotection by thiols. There is a wide variation in PF in individual tissues measured either by different laboratories, with different assays, or at different dose levels on a single pair of dose-response curves.

CONCLUSIONS

The clinical use of WR-2721 as a radioprotector of normal tissues is not justified for routine radiotherapy. The variability in normal tissues makes it difficult to choose a safe protection factor, by which the treatment dose can be increased. If too low a value is chosen the tumour may be undertreated since tumour radioprotection is often seen at clinically relevant doses. If too high a PF value is chosen the consequences from normal tissue morbidity will be disastrous. However, special circumstances that would allow local application of WR-2721 (e.g., for sparing buccal mucosa, salivary glands, rectum, or bladder) may be still worth further consideration.

Although the underlying mechanism of radioprotection by thiols may be simple (redox competition) its overall effect in tissues is difficult to predict. It is particularly important to realize that PF values cannot be extrapolated from one tissue to another, nor from single-dose studies to fractionated regimes.

ACKNOWLEDGMENTS

This work was performed at the Gray Laboratory and was entirely funded by the Cancer Research Campaign. We are grateful to Dr. Fiona A. Stewart and to Dr. Graham Stevens for allowing us to quote their unpublished data, and to Mrs. Jean Wilson for secretarial assistance.

REFERENCES

1. G.E. Adams, J.F. Fowler and P. Wardman (Eds), Hypoxic cell sensitizers in Radiobiology and Radiotherapy. Br. J. Cancer 37 (Suppl III). (1978).
2. D.E. Davidson, M.M. Grenan, and T.R. Sweeney, Biological characteristics of some improved radioprotectors. In: Radiation Sensitizers. Their use in the clinical management of Cancer. Vol. 5 (L.W. Brady, Ed.), pp. 309-210. Masson Publishing, New York, 1980.
3. J.M. Yuhas and J.B. Storer. Differential chemoprotection of normal and malignant tissues. J. Natl. Cancer Inst. 42: 331-335 (1969).
4. J.M. Yuhas, J.M. Spellman, and F. Culo, The role of WR-2721 in radiotherapy and/or chemotherapy. Cancer Clin. Trials 3: 211-216 (1980).
5. T.L. Phillips, L. Kane, and J.F. Utley, Radioprotection of tumor and normal tissues by thiophosphate compounds. Cancer 32: 528-535 (1973).
6. Y. Tanaka and T. Sugahara. Clinical Experiences of chemical radiation protection in tumor radiotherapy in Japan. In: Radiation Sensitizers: Their use in the clinical management of cancer (L.W. Brady, Ed.), pp. 421-425. Masson Publishing, New York, 1980.
7. M.M. Kligerman, M.T. Shaw, M. Slavik, and J.M. Yuhas, Phase 1 clinical studies with WR-2721. Cancer Clin. Trials 3: 217-221 (1980).

8. M.M. Kligerman, D.J. Glover, A.T. Turrisi, A.L. Norfleet, J.M. Yuhas, L.R. Coia, C. Simons, J.H. Glock, and R.L. Goodman, Toxicity of WR-2721 administered in single and multiple doses. Int. J. Radiat. Oncol. Biol. Phys. 10: 1773-1776 (1984).
9. O.F. Nygaard and M.G. Simic (Eds.), Radioprotectors and Anticarcinogens. Academic Press, London, 1983.
10. J.D. Chapman, G.F. Whitmore (Eds.). Chemical Modifiers of Cancer Treatment. Int. J. Radiat. Oncol. Biol. Phys. 10: 1161-1813 (1984).
11. J. Denekamp, A. Rojas, and F.A. Stewart, Is radioprotection by WR-2721 restricted to normal tissues? Radioprotectors and Anticarcinogens (O. Nygaard and M. Simic, Eds.), pp. 655-679. Academic Press, London, 1983.
12. J. Denekamp, Normal tissue radioprotection by WR-2721. In: Advances on Oxygen radicals and radiation protectors (A. Breccia, C.L. Greenstock, and M. Tamba, Eds.), pp. 153-171. La Scarabeo, Italy, 1984.
13. O.W. Griffith and A. Meister, Origin and turnover of mitochondrial glutathione. Proc. Natl. Acad. Sci. USA 82: 4668-4672 (1985).
14. H. O'Hara and T. Terasima, Variations of cellular sulfhydryl content during cell cycle of HeLa cells and its correlation to cyclic change of X-ray sensitivity. Exp. Cell. Res. 58: 182-185 (1969).
15. T. Alper, Cellular Radiology. Cambridge University Press, 1979.
16. J. Denekamp, B.D. Michael, A. Rojas, and F.A. Stewart, Thiol radioprotection in vivo: the critical role of tissue oxygen concentration. Br. J. Radiol. 54: 1112-1114 (1981).
17. C.S. Parkins, J.F. Fowler, and J. Denekamp, Low radioprotection by thiol in lung: the role of local tissue oxygenation. Eur. J. Cancer Clin. Oncol. 19: 1169-1172 (1983).
18. S. Schuchhardt, Comparative physiology of the oxygen supply. In: Oxygen Supply. Theoretical and Practical Aspects of Oxygen Supply & Microcirculation of Tissue (M. Kessler, D.F. Burley, L.C. Clark, D.W. Lubbers, I.A. Silver and J. Strauss, Eds.), pp. 223-229. Urban & Schwarzenberg, Munich, 1973.
19. F.A. Stewart, J. Denekamp, and V.S. Randhawa, Skin sensitization by misonidazole: A demonstration of uniform mild hypoxia. Br. J. Cancer 45: 869-877 (1982).
20. J.H. Hendry and M.L. Sutton. Care with radiosensitizers. Br. J. Radiol. 51: 927-928 (1978).
21. A. Rojas, F.A. Stewart, J.A. Soranson, D.A. Smith, and J. Denekamp, Fractionation studies with WR-2721: Normal tissues and tumor. Radiother. Oncol. 6: 51-60 (1986).
22. F.S. Echols and J.M. Yuhas. Chemoprotection against fractionated radiation exposures with WR-2721: skin injury. Radiat. Res. 66: 499-504, 1976.
23. F.A. Stewart and A. Rojas, Radioprotection of mouse skin by WR-2721 in single and fractionated treatments. Br. J. Radiol. 55: 42-27 (1982).
24. E.L. Travis, A.M. DeLuca, C.S. Parkins, and J.F. Fowler, Protection of early and late damage of mouse lungs by WR-2721 after single and fractionated doses of X-rays. Radiat. Res. 91: 399 (1982).
25. C.S. Parkins, J.F. Fowler, and J. Denekamp, Lung radioprotection by WR-2721 at low X-ray doses per fraction. Eur. J. Cancer Clin. Oncol. 20: 1445-50 (1984).
26. M.L. Meistrich, M.V. Finch, N. Hunter, and L. Milas, Protection of spermatogonial survival and testicular function by WR-2721 against high and low doses of radiation. Int. J. Radiat. Oncol. Biol. Phys. 10: 2099-2107 (1984).

PHARMACOLOGICAL INTERFERENCE WITH DNA REPAIR

Arthur B. Pardee, Robert Schlegel
and David A. Boothman

Dana-Farber Cancer Institute
44 Binney St., Boston, MA 02115

Many carcinogens and antineoplastic agents act primarily by damaging DNA. A very large portion of this damage is repaired by normal cells. Defects in these repair processes have serious consequences, as shown by xeroderma pigmentosum, ataxia telangiectasia and other cancer-prone genetic diseases. Drugs can also interfere with DNA repair pathways. They, like the genetic defects, increase the lethality of DNA damaging agents. Unlike these diseases, however, some drugs have been shown to decrease carcinogenicity. The basis for this paradoxical effect may be the ability of DNA repair inhibitors to convert normally sublethal misrepaired lesions into lethal ones. Drugs that interfere with DNA repair thus have promise not only for increasing the efficacy of chemotherapy, but also for decreasing carcinogenicity.

We have been investigating interactions of DNA-damaging agents and DNA repair modifiers in both cultured cells and animals over the past ten years. Our aims were to better understand the modes of action of repair inhibitors and to seek applications for cancer chemotherapy. Results obtained with three classes of drugs are described briefly here.

1. METHYLXANTHINES (MEX)

Of several classes of DNA repair inhibitors, the most extensively investigated are the methylxanthines (1-3). Caffeine is reported to decrease drug-induced carcinogenesis (4) and transformation (5). Other work has indicated increases in carcinogenicity (6). There have, similarly, been reports of both decreased and increased mutation frequencies (7). Caffeine can also potentiate cytotoxicity in cells that are resistant to a chemotherapeutic agent (8,9). We have proposed that caffeine greatly increases the lethality of alkylating agents by preventing cells (Syrian hamster fibroblasts) from having sufficient time to repair DNA lesions before mitosis. The still-damaged cells go through mitosis, develop shattered chromosomes, and emerge with fragmented nuclei. This is the basis of lethality for those cells that otherwise would have recovered (10).

Although caffeine-enhanced lethality is well documented in rodent cells, these effects were reported not to occur with human cells (11). We and others have shown that MEX strikingly increase the lethality of alkylating agents against human tumor cells in culture, particularly for polyfunctional agents such as thiotepa (12,13).

MEX also override mitotic cycle controls in undamaged normal rodent cells. Mitotic events are induced without the completion of DNA replication if MEX are provided (14). These events include breakdown of the nuclear envelope, condensation of chromosomes into a state very closely resembling premature chromosome condensation (PCC) (15), induction of mitosis-specific phosphoproteins, and rounding up of the cells. A temperature-sensitive mutant, tsBN2, also undergoes PCC at the restrictive temperature during S phase (16).

In essence, we have been able to short-circuit the cell cycle (Figure 1). The cells proceed around a mitotic cycle independent of the DNA synthetic cycle. Caffeine-induced PCC can only occur if cells have reached S phase, and are then prevented from completing DNA replication (14). Experiments with metabolic inhibitors (17) suggest that mitosis-related RNA accumulates during the inhibition of DNA replication, and this RNA is unstable when replication is allowed to resume. The proteins required for PCC are synthesized shortly after caffeine addition. Their activity is very labile in the absence of caffeine and is lost through an ATP-dependent mechanism.

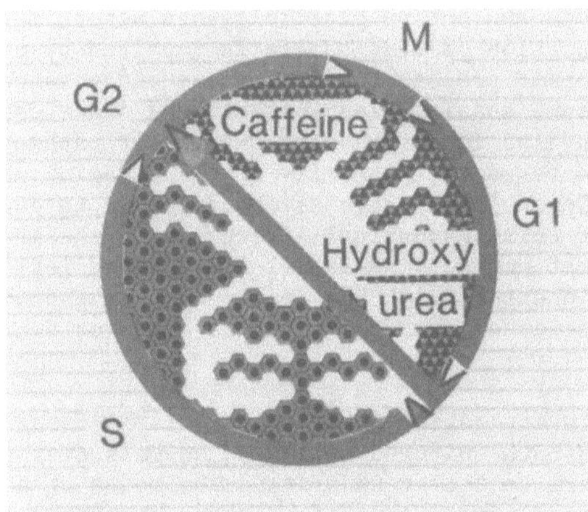

Figure 1. A short-circuited cell cycle

Caffeine (1-3-7-trimethylxanthine) is the most active compound we have examined (Table 1). Removal of the methyl group from the #7 position of caffeine (theophylline) greatly reduced the mitosis-promoting activity. Replacement of the #3 methyl, creating 3-isobutyl-1-methlyxanthine (IBMX), resulted in a further reduction. Modification or removal of the #1 methyl (pentoxifylline or theobromine, respectively) had much less effect. Of these compounds, IBMX and theophylline are the two most potent inhibitors of cAMP phosphodiesterase, yet they have little or no PCC-inducing activity. Thus, phosphodiesterase inhibition cannot be solely responsible for these effects.

TABLE 1. Premature Chromosome Condensation Induced by Methylxanthines (MEX)

MEX	mM	% PCC
Caffeine	2 (5)	41 (46)
Pentoxifylline	5	35
Theobromine	2	29
Theophylline	2	8
IBMX	2	1
Control	-	<1

Syrian hamster fibroblast (BHK) cells were plated at 1×10^5 per 60-mm dish in DME + 10% FCS and incubated for 24 hours. Synchrony in early S phase was achieved by incubation in isoleucine-deficient DME + 5% dialyzed FCS for 34 hours (27,28). Cells were then treated with DME + 10% FCS + 2.5 mM hydroxyurea for 16 hours, and finally were given the specified MEX along with Colcemid (0.3 μg/ml). Percentage of cells undergoing PCC was determined 8 hours after MEX addition, as previously described (14).

2. NICOTINAMIDE ANALOGS

Analogs of nicotinamide represent another class of DNA repair inhibitors. These agents block the synthesis of poly(ADP-ribose). Such compounds have been reported to inhibit malignant transformation of mouse cells (18). Various human tumor cells are differently sensitized by nicotinamide analogs (19) (for examples of reviews see 20,21).

3-Aminobenzamide (3-AB) acts quite differently from MEX (22). 3-AB inhibits repair of single-strand damage in both S-phase and quiescent cells. We propose that some alkylation-induced lesions are repaired during normal replicative DNA synthesis. When this repair is inhibited by 3-AB during S phase, sublethal damage is converted to lethal lesions which halt cells in G_2 phase.

3. β-LAPACHONE

β-lapachone is a naturally occurring tricyclic-O-naphtho-quinone. At low concentrations (<4 μM) it does not influence cell viability, growth or DNA synthesis in undamaged cultured human fibroblast or carcinoma cells (23,24). The viability of quiescent as well as log-phase growing human cells damaged with X irradiation or MMS was reduced by subsequent β-lapachone exposure. Thousand-fold higher concentrations of MEX or 3-AB were required to produce equivalent lethalities. DNA single-strand break repair was inhibited, but unscheduled DNA synthesis was stimulated. β-Lapachone inhibited the capacity to repair potentially lethal X-ray damage (PLDR) and caused other nonlethal lesions to be expressed (24). This agent also sensitized rodent cells to DNA-damaging agents (unpublished data).

We have recently investigated synergistic effects of β-lapachone upon nonhypoxic, halogenated pyrimidine radiosensitizers (24). Log-phase human laryngeal epidermoid carcinoma (HEp-2) cells that incorporated BrdUrd into their DNA for roughly 2 cell doublings were X irradiated and exposed to various post-treatments (Figure 2). β-Lapachone did not alter the viability of unirradiated, BrdUrd-sensitized cells (24). It severely reduced the surviving fraction of X irradiated, BrdUrd-sensitized cells. Dose enhancement ratios (DER) for BrdUrd-sensitized cells increased from 1.75 ± 0.04 for control cells receiving no post-treatment, to 3.03 ± 0.03 for cells given a 4

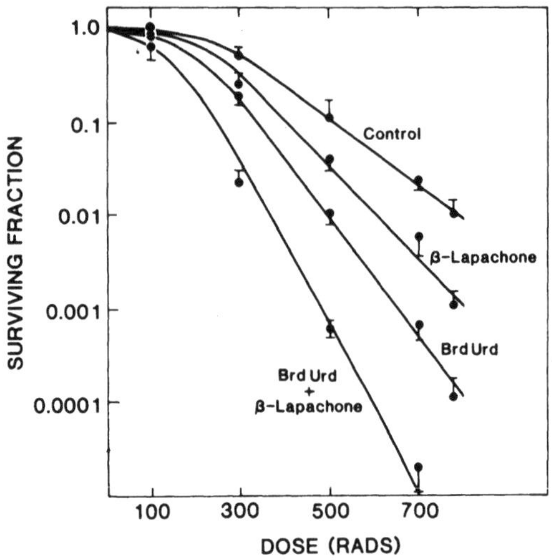

Figure 2. Lethal effects of X-rays on BrdUrd and β-lapachone treated cells.

5 x 10^5 human epidermoid laryngeal carcinoma (HEp-2) cells were seeded from subconfluent cultures onto 25-cm^2 tissue culture flasks, grown for 48 hours and exposed to 10 μM 5-bromo-2'-deoxyuridine (BrdUrd) for an additional 48 hours in a humidified 10% CO_2-90% air atmosphere. Following X irradiation, cells were exposed to DME containing 0.2% dialyzed fetal calf serum with or without 4 μM β-lapachone for 4 hours. The plating efficiency of control cells was 80 ± 2.4%.

hour post-X-irradiation exposure of 4 μM β-lapachone. Exposure to β-lapachone alone following X irradiation of unsensitized cells consistently resulted in DER values of 1.38 ± 0.05. Dose-response curves for BrdUrd-sensitized HEp-2 cells plated immediately or exposed to media containing low serum for 4 hours were statistically similar (data not shown), indicating no PLDR.

Another advantage of combining halopyrimidines and β-lapachone may be to decrease dangerous side effects of chemotherapy. Exposure of mammalian cells to BrdUrd is known to be very mutagenic, probably because analog incorporation (25) and imbalances in normal pool sizes following BrdUrd exposure (26) are thought to result in extensive misrepair. Thus, if β-lapachone inhibits misrepair it might reduce the number of mutagenic events. This effect is being investigated.

ACKNOWLEDGMENTS

We thank Susan Norris, Robert Curtin and Milton Datta for technical assistance, Marjorie Rider for preparing the manuscript, and Dr. Jean Gudas for reviewing the manuscript. This work was supported by USPHS Grant CA22427 to ABP.

REFERENCES

1. B.A. Kihlman, Caffeine and chromosomes. Elsevier Scientific Publishing Co., New York and Amsterdam, 1977.
2. J. Timson, Caffeine. Mutat. Res. 47:1-52, (1977).
3. J.J. Roberts, DNA repair. Adv. Radiat. Biol., 7:211-436, (1978).
4. T. Nomura, Comparative inhibiting effects of methylxanthines on urethan-induced tumors, malformations, and presumed somatic mutations in mice. Cancer Res. 43:1342-1346, (1983).
5. T. Kakunaga, Caffeine inhibits cell transformation by 4-nitroquinoline-1-oxide. Nature (London) 258:248-250, (1975).
6. P.J. Donovan, and J.A. DiPaolo, Caffeine enhancement of chemical carcinogen-induced transformation of cultured Syrian hamster cells. Cancer Res. 34:2720-2727, (1974).
7. M. Fox, The effect of post-treatment with caffeine on survival and UV-induced mutation frequencies in Chinese hamster and mouse lymphoma cells in vitro. Mutat. Res. 24:187-204, (1974).
8. D. Gaudin, and K.L. Yielding, Response of a "resistant" plasmacytoma to alkylating agents and X-ray in combination with the excision repair inhibitors caffeine and chloroquine. Proc. Soc. Exp. Biol. Med. 131:1413-1416, (1969).
9. T. Aida, and W.J. Bodell, Caffeine potentiates cytotoxicity and sister chromatid exchange induction in resistant rat brain tumor cells treated with 1,3-bis (2-chloroethyl)-1-nitrosourea. Cancer Res. 47:1361-1366, (1987).
10. C.C. Lau, and A.B. Pardee, Mechanism by which caffeine potentiates lethality of nitrogen mustard. Proc. Natl. Acad. Sci. USA, 79:2942-2946, (1982).
11. L.A. Zwelling, and K.W. Kohn, Platinum Complexes. In: Pharmacological Principles of Cancer Treatment (B. Chabner, Ed.), pp. 321. W.B. Saunders, Philadelphia, 1982.
12. H.J. Fingert, J.D. Chang, and A.B. Pardee, Cytotoxic, cell cycle, and chromosomal effects of methylxanthines in human tumor cells treated with alkylating agents. Cancer Res. 46:2463-2467, (1986).
13. J.E. Byfield, J. Murnane, J.F. Ward, P. Calabro-Jones, M. Lynch, and F. Kulhanian, Mice, men, mustards and methylxanthines: the potential role of caffeine and related drugs in the sensitization of human tumors to alkylating agents. Br. J. Cancer 43:669-683, (1981).
14. R. Schlegel, and A.B. Pardee, Caffeine-induced uncoupling of mitosis from the completion of DNA replication in mammalian cells. Science 232:1264-1266, (1986).
15. R.T. Johnson, and P.N. Rao, Mammalian cell fusion: induction of premature chromosome condensation in interphase nuclei. Nature (London), 226:717-722, (1970).
16. T. Nishimoto, E. Eilen, and C. Basilico, Premature chromosome condensation in a ts DNA⁻ mutant of BHK cells. Cell. 15:475-483, (1978).
17. R. Schlegel, R.G. Croy, and A.B. Pardee, Exposure to caffeine and suppression of DNA replication combine to stabilize the proteins and RNA required for premature mitotic events. J. Cell. Physiol., 131:85-91, (1987).
18. C. Borek, W.F. Morgan, A. Ong, and J.E. Cleaver, Inhibition of malignant transformation in vitro by inhibitors of poly(ADP-ribose) synthesis. Proc. Natl. Acad. Sci. USA, 81:243-247, (1984).
19. P.J. Thraves, K.L. Mossman, T. Brennan, and A. Dritschilo, Differential radiosensitization of human tumor cells by 3-aminobenzamide and benzamide: inhibitors of poly(ADP-ribosylation). Int. J. Radiat. Biol. 50:961-972, (1986).
20. J. Lunec, Introductory review: Involvement of ADP-ribosylation in cellular recovery from some forms of DNA damage. Br. J. Cancer, 49 (Suppl VI): 13-18, (1984).

21. D.M. Brown, J.W. Evans, and J.M. Brown, The influence of inhibitors of poly(ADP-ribose) polymerase on X-ray-induced potentially lethal damage repair. Br. J. Cancer, 40 (Suppl. VI): 27-31, (1984).
22. R.J. Boorstein, and A.B. Pardee, 3-Aminobenzamide is lethal to MMS-damaged human fibroblasts primarily during S phase. J. Cell. Physiol., 120:345-353, (1984).
23. R.J. Boorstein, and A.B. Pardee, β-Lapachone greatly enhances MMS lethality to human fibroblasts. Biochem. Biophys. Res. Commun., 118:828-834, (1984).
24. D.A. Boothman, S. Greer, and A.B. Pardee, Potentiation of halogenated pyrimidine radiosensitizers by β-lapachone, a novel DNA repair inhibitor. Cancer Res., (in press), (1987).
25. M.W. Heartlein, J.P. O'Neill, B.C. Pal, and R.J. Preston, The induction of specific-locus mutations and sister-chromatid exchanges by 5-bromo and 5-chlorodeoxyuridine. Mutat. Res. 92:411-414, (1982).
26. C.R. Ashman, G.P.V. Reddy, and R.L. Davidson, Bromodeoxyuridine mutagenesis, ribonucleotide reductase activity, and deoxyribonucleotide pools in hydroxyurea-resistant mutants. Somat. Cell Genet. 7:751-768, (1981).
27. R.A. Tobey, and H.A. Crissman, Preparation of large quantities of synchronized mammalian cells in late G_1 in the pre-DNA replicative phase of the cell cycle. Exp. Cell Res. 75:460-464, (1972).
28. J.L. Hamlin, and A.B. Pardee, S phase synchrony in monolayer CHO cultures. Exp. Cell Res. 100:265-275, (1976).

DNA STRAND SCISSION BY ACTIVATED BLEOMYCIN GROUP ANTIBIOTICS

Sidney M. Hecht

Departments of Chemistry and Biology
University of Virginia
Charlottesville, VA 22901 USA

The bleomycins (BLMs) are a family of antitumor antibiotics used clinically for the treatment of squamous cell carcinomas and malignant lymphomas (1-4). Bleomycin A_2 is the major constituent of the clinically used mixture of bleomycins. Although biochemical studies have indicated several loci that may contribute to the overall therapeutic effects obtained with bleomycin, the accumulated evidence suggests that the primary effect is at the level of

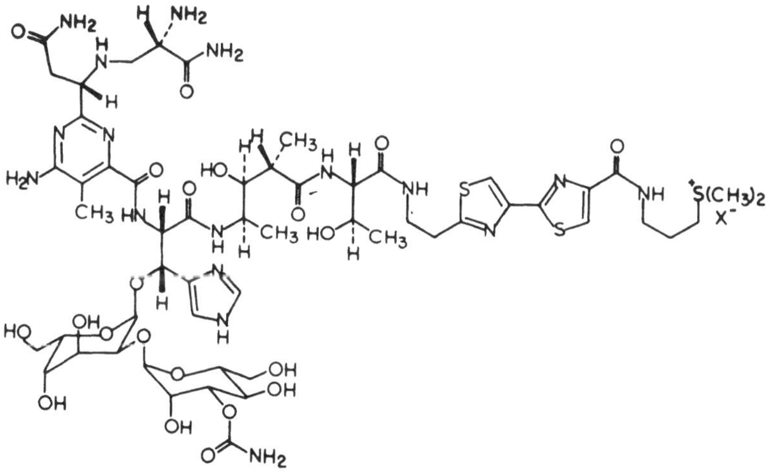

bleomycin A_2

oxidative DNA degradation (5). DNA degradation mediated by bleomycin is metal ion dependent; with the exception of Co•BLM (6), DNA degradation by metallobleomycins also requires oxygen (5,7). It is believed that the degradative event is actually mediated by a ternary complex consisting of bleomycin, a metal ion and oxygen. There are several mechanistically interesting facets of BLM-mediated DNA degradation, including the observed sequence selectivity for 5'-GT-3'

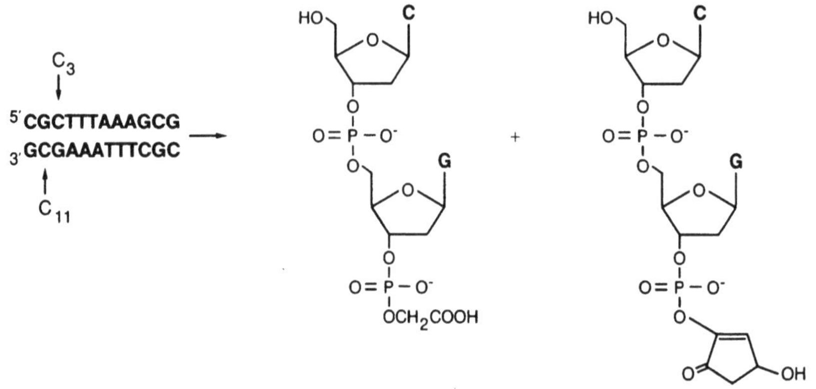

formed with Fe(II)•BLM

formed with Fe(II)•BLM, then alkali

sequences (7-9) and the fact that double-strand cleavage occurs at a frequency greater than that expected from the random accumulation of single-strand breaks (10,11).

In order to facilitate an understanding of the mechanism of DNA cleavage by bleomycin, we have employed a dodecanucleotide of defined structure as a substrate for degradation by Fe(II).BLM (vide infra). Bleomycin modified this oligonucleotide primarily at $cytidine_3$ and $cytidine_{11}$; strand scission at the former position was especially helpful in facilitating identification of the cleavage products as dinucleotide derivatives were released and found to be amenable to structure analysis. The reaction products from the oligomers were found to include an oligonucleotide terminating at the 3'-end with a phosphoroglycolate moiety. This species is believed to derive from a C-4' hydroperoxy derivative of deoxyribose, decomposition of which also results in the generation of base propenals having the trans-configuration (12,13). The other set of products formed included free bases and alkali-labile products, but no strand scission per se. However, subsequent treatment with alkali did effect strand scission, with concomitant formation of an oligonucleotide terminating at the 3'-end with a 2,4-dihydroxycyclopentenone moiety (14).

Further study of the alkali-labile lesion was carried out by mediating its degradation under controlled conditions with a number of reagents. These included hydrazine, which effected conversion to nucleotide pyridazine derivatives in quantitative yield, and alkylamines such as n-butylamine which provided oligonucleotides terminating with 3'-phosphate groups. The accumulated data suggests that the alkali-labile lesion results from C-4' hydroxylation of deoxyribose moieties in DNA. It is anticipated that these derivatives would lose free bases with facility, forming C-4'-hydroxyapurinic acids as a consequence. The C-4'-hydroxyapurinic acids are proposed to represent the alkali-labile lesions; the chemistry that this lesion is known to undergo at present is summarized in the following Scheme.

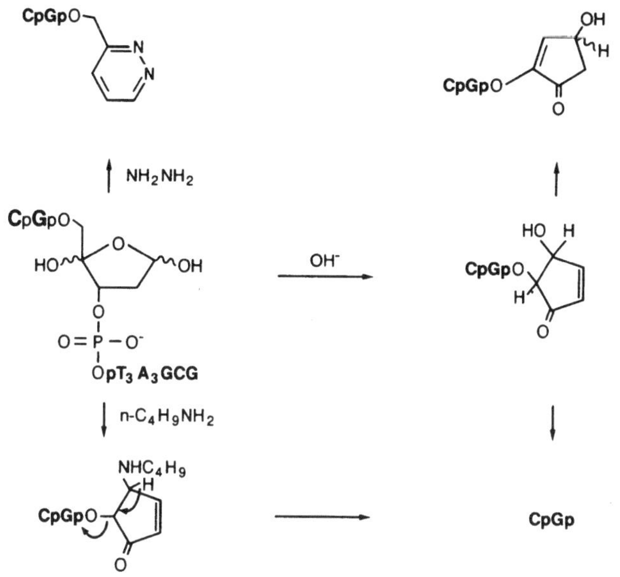

REFERENCES

1. S.K. Carter, The current role of bleomycin in cancer therapy. In:Bleomycin: Current Status and New Developments (S.K. Carter; S.T. Crooke; H. Umezawa, Eds.), pp. 9-14. Academic Press, New York, 1978.

2. S.M. Hecht, Summary of the bleomycin symposium. In Bleomycin: Chemical. Biochemical and Biological Aspects (S.M. Hecht, Ed.) pp. 1-23. Springer-Verlag, New York, 1979.

3. H. Umezawa, Recent studies on biochemistry and action of bleomycin. In:Bleomycin: Current Status and New Developments (S.K. Carter; S.T. Crooke; H. Umezawa, Eds.), pp. 15-19. Academic Press, New York, 1978.

4. H. Umezawa, Recent studies on bleomycin. Lloydia, 40, 67-81 (1977).

5. S.M. Hecht, DNA strand scission by activated bleomycin group antibiotics. Fed. Proc. 45, 2784-2791 (1986).

6. L.-H. Chang and C.F. Meares, Cobalt-bleomycins and deoxyribonucleic acid: sequence-dependent interactions, action spectrum for nicking, and indifference to oxygen. Biochemistry 23, 2268-2274 (1984).

7. A.D. D'Andrea and W.A. Haseltine, Sequence specific cleavage of DNA by the antitumor antibiotics neocarzinostatin and bleomycin. Proc. Natl. Acad. Sci. U.S.A. 75, 3608-3612 (1978).

8. M. Takeshita, A. Grollman; E. Ohtsubo and H. Ohtsubo, Interaction of bleomycin with DNA. Proc. Natl. Acad. Sci. U.S.A. 75, 5983-5987 (1978).

9. C.K. Mirabelli, A. Ting, C.-H. Huang, S. Mong, and S.T. Crooke, Bleomycin and talisomycin sequence-specific strand scission of DNA: A mechanism of double-strand cleavage. Cancer Res. 42, 2779-2785 (1982).

10. C.W. Haidle, R.S. Lloyd and D.L. Robberson. Molecular aspects of bleomycin-promoted damage of covalently closed circular DNA. In:Bleomycin: Chemical, Biochemical, and Biological Aspects; (S.M. Hecht, Ed.) pp. 222-243. Springer-Verlag, New York, 1979.

11. C.-H. Huang, C.K. Mirabelli, Y. Jan, and S.T. Crooke, Single-strand and double-strand deoxyribonucleic acid breaks produced by several bleomycin analogues. Biochemistry 20, 233-238 (1981).

12. L. Giloni, M. Takeshita, F. Johnson, C. Iden, and A.P. Grollman, Bleomycin-induced strand scission of DNA, J. Biol. Chem. 256, 8608-8615 (1981).

13. N. Murugesan, C. Xu, G.M. Ehrenfeld, H. Sugiyama, R.E. Kilkuskie, L.O. Rodriguez, L.-H. Chang, and S.M. Hecht, Analysis of products formed during bleomycin-mediated DNA degradation. Biochemistry 24, 5735-5744 (1985).

14. H. Sugiyama, C. Xu, N. Murugesan, and S.M. Hecht, Structure of the alkali-labile product formed during iron(II)-bleomycin-mediated DNA strand scission. J. Am. Chem. Soc. 107, 4104-4105 (1985).

PROSPECTS FOR LIPOSOMES AS DRUG CARRIERS

Gabriel Lopez-Berestein and Roman Perez-Soler

Immunobiology and Drug Carriers Section, Department of Clinical Immunology and Biological Therapy, The University of Texas M. D. Anderson Hospital and Tumor Institute at Houston, Houston, TX 77030-2603

Liposomes are vesicles composed of concentric phospholipid bilayers that are formed spontaneously when an aqueous solution is added to a dried lipid film (Figure 1). Several types of liposomes--multilamellar vesicles (MLV), small unilamellar vesicles (SUV), and reverse evaporation vesicles (REV)--have been described. MLV are used extensively as drug carriers for antineoplastic and antimicrobial drugs: lipophilic drugs can be incorporated in the MLV's large lipid compartment and hydrophilic drugs can be encapsulated between the vesicles' lipid bilayers. The liposomes' size,

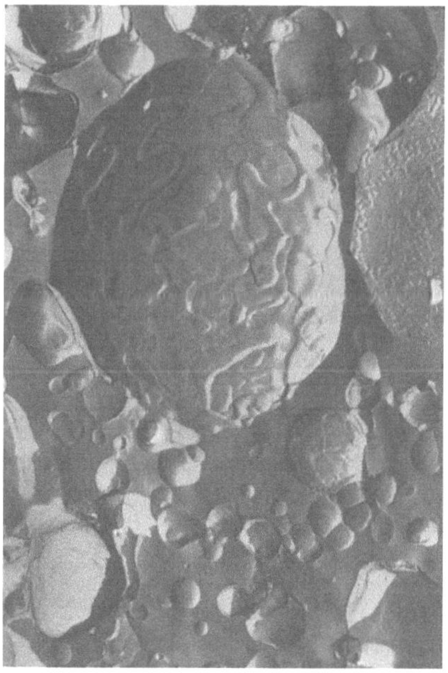

Figure 1. Freeze-fracture electron micrographs (magnification, 12,000) showing multilamellar liposomes of different diameters. Layered structure is evident in the vesicles observed.

TABLE 1. LIPOSOMES IN ANTIMICROBIAL THERAPY

Drug	Disease	Reference
Amphotericin	Leishmaniasis	1-3
	Histoplasmosis	4
	Cryptococcosis	5
	Candidiasis	6-12
Antimonials	Leishmaniasis	13-16
Cephalothin	Salmonellosis	17-18
Primaquine phosphate	Malaria	19-20
Immunomodulators		
Muramyl dipeptide	Herpes	21
	Candidiasis	22
Interferon	Hepatitis	23
Other antibiotics		

membrane charge, fluidity, and other characteristics may modify their *in vivo* behavior. The changes in drug bioavailability and biodistribution associated with liposome incorporation have been exploited to alter drug toxicity and enhance drug targeting. Design of liposomal drug carriers should be based on the rational selection of the drug or drug analogue; drug carriers; and the target disease, be it focused on a cell, an organ, or a tissue. We review here general pharmacologic concepts of the development of liposomal carriers for antimicrobial and antineoplastic agents.

LIPOSOMES AS CARRIERS OF ANTIFUNGAL DRUGS

Liposomes have been used extensively as carriers for a variety of antimicrobial drugs (Table 1) (1-27). We have, for several years, studied the use of liposomes as carriers of antifungal drugs in treatment of experimental and clinical systemic fungal infections (6-8, 10-12). These infections are a major cause of morbidity and mortality in immunocompromised patients and in patients with cancer, particularly those with hematologic malignancies. Nearly 20% of the fatal infections in patients with leukemia are caused by fungi, especially Candida and aspergillus spp. These prolonged infectious processes often lead to major delays in treatment of the patients primary disease, allowing the primary disease to recur. Amphotericin B (AmpB), a polyene antibiotic, is the mainstay of therapy for most systemic fungal infections, but its potential clinical effectiveness is hindered by its severe acute and chronic side effects. When we first studied the use of liposomes as carriers for AmpB in therapy for experimentally produced candidiasis (6-8) and leishmaniasis (3), we showed that liposomal AmpB (L-AmpB) is far less toxic and more effective than free AmpB in treatment and prophylaxis of systemic candidiasis in neutropenic and non-neutropenic mice (6-8) (Figure 2).

The effectiveness was also shown in treatment of leishmaniasis in hamsters and non-human primates (3). Leishmaniasis is widespread in the populations of Asia, Africa and South America, and it holds particular interest for us because this parasite selectively infects the reticuloendothelial system (RES), which is in turn the preferential target of liposomes. L-AmpB resulted in 98%-99% eradication of hepatosplenic parasites from infected monkeys; it was 170 to 750 times more active than antimony in leishmaniasis in hamsters and 60 times more active against the disease in monkeys (3).

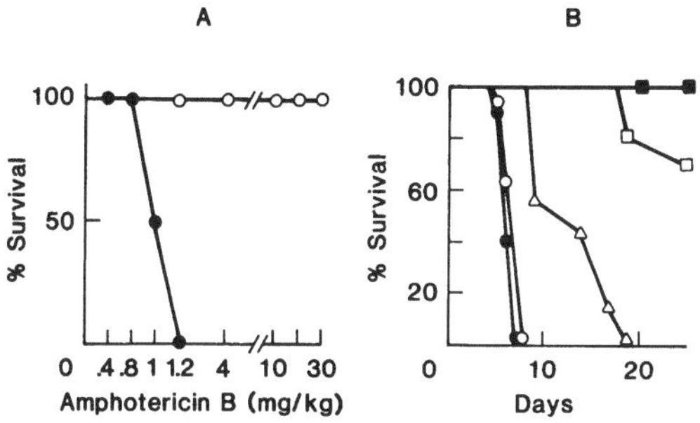

Figure 2. Toxicity and antifungal activity of free versus liposome-encapsulated AmpB in mice. All drugs were administered intravenously in 0.2 ml of 0.9% NaCl solution. A. Toxicity: (●) free AmpB, (o) L-AmpB. B. Antifungal activity: The drugs were injected two days after inoculation of yeasts. (●) untreated infected control, (o) "empty" liposomes, (△) single dose of 0.8 mg/kg free AmpB, (□) 0.8 mg/kg free AmpB daily for five days, (■) 4 mg/kg L-AmpB. Modified from Lopez-Berestein et al. (7).

CLINICAL EXPERIENCE WITH LIPOSOMAL AMPHOTERICIN B

These early experimental results encouraged us to study the clinical use of L-AmpB (11-12). Under a compassionate investigational new drug permit, we used L-AmpB to treat cancer patients for AmpB-refractory progressive systemic fungal infections, for which these patients had received lengthy treatment with the conventional form of AmpB (Fungizone); the persistent progressive nature of these fungal infections was documented by histologic tests or laboratory cultures, or both. L-AmpB (0.4 - 5 mg/kg) was administered intravenously for 10 to 45 minutes daily. Of 19 patients treated, 9 were cured of the fungal infections, 6 responded partially, and 4 patients did not respond to treatment. Acute side effects such as fever and chills occurred infrequently, and no long-term kidney or central nervous system side effects were observed. L-AmpB was in general less toxic than AmpB, easier to administer, and effective.

The mechanisms of L-AmpB's better effectiveness/toxicity ratio of are not well understood. Liposomes seem to alter the interactions of AmpB with cells, and although its antifungal activity is maintained by the incorporation in liposomes, its toxicity to normal cells--to lymphocytes and monocytes--is decreased. Altered drug distribution also seems to play a role. Higher concentrations of the drug were detected in the liver, spleen, and kidney of infected mice given L-AmpB than in infected mice given free AmpB. Factors such as leakage of liposomes through damaged endothelium, enhancement of immunity, and secondary transport of liposomes by peripheral phagocytes may also play a role.

LIPOSOMES AS CARRIERS OF ANTICANCER AGENTS

Liposomes as carriers of chemotherapeutic agents have been explored extensively in attempts to modulate drug toxicity and drug delivery to certain organs (Table 2) (28-50). These studies showed that liposome encapsulation alters the drugs' pharmacokinetics and distribution (28, 46)

TABLE 2. ENCAPSULATION EFFICIENCY OF LIPOSOMAL PREPARATIONS CONTAINING ANTICANCER AGENTS

Drug	Lipids	Type of vesicles	Percentage of Encapsulation	References
		Hydrophilic drugs		
Doxorubicin				
1	PC, Cho, Card, SA	SUV	55	28
2	PC:PG:Cho	SUV	50-65	29
3	PC:PG:Cho	SUV	50-60	30
4	PC:PG:Cho	LUV	95	31
Vincristine	PC:Cho:PS	SUV	10-11	32
Vinblastine	PC:Cho:SA	SUV	8-11	33
Methotrexate				
1	DPPC:Chol:SA	SUV	5	34
2	PC:Cho:DP	SUV	9.4	35
3	PC:Cho:SA	SUV	4.0	36
Ara C				
1	DMPC:Chol:SA	MLV	22	37
2	PC:PG:Cho	SUV	30-35	38
Act D	PC:Cho:SA	SUV	15.3	39
CDDP				
1	PC:Cho:DP	SUV	7.4	40
2	PC	SUV	1.5-2.5	41
3	DPPC:DSPC	SUV	3.9	42
Melphalan	PC:Cho	SUV	8.1	43
		Lipophilic drugs		
CPDP	DMPC:DMPG	MLV	85	44
NDDP	DMPC:DMPG	MLV	100	45
NDIDP	DMPC:DMPG	MLV	100	46
CDCPA	DPPC:Chol:PA	SUV	8.5-11.5	47
AraC-prodrug	PC:Cho:SA	SUV	85-97	48
DMPE:MTX	DOPC:CL:Cho:DP	SUV	100	49
NSC 251635	PC:Cho:SA	SUV	70-80	50

Abbreviations:

PC, phosphatidylcholine; Cho, cholesterol; Card, cardiolipin; SA, Stearylamine; SUV, small unilamellar vesicles; LUV, large unilamellar vesicles; PG, phosphatidyl serine; PS, phosphatidylserine; DPPC, dipalmitoyl phosphatidyl choline; DP, dicetyl phosphate; DMPC, dimyristoyl phosphatidyl choline; MLV, multilamellar vesicles; DSPC, distearoyl phospatidyl choline; DMPG, dimyristoyl phosphatidyl glycerol; PA, phosphatidic acid; DOPC, dioleoyl phosphatidyl choline; Ara C, cytosine Arabinoside; Act D, actinomycin D; CPDP, cis-bis-cyclopentenecarboxylato-1,2-diaminocyclohexane platinum II; NDDP, cis-bis-neodecanoate-1,2-diaminocyclohexane platinum II; NDIDP, cis-bis-N-decyliminodiacetato-1,2-diaminocyclohexane platinum II; CDCPA, cis-dichloro-bis-cyclopentylamine platinum II; DMPE:MTX, γ-diaminocyclophosphatidyl ethanolamine: methotrexate.

and that, as a result of these changes, such target organ toxicities as doxorubicin cardiotoxicity (51) can be reduced and antitumor activity against metastatic liver tumors increased (29, 30, 45, 52).

Several major problems must be overcome before anticancer agents can be successfully incorporated into liposomes. The rationale of drug selection has, in most instances, been based on the wide use of the chemotherapeutic agent, rather than on the drug's structural suitability to liposome incorporation. Generally, however, liposomal preparations containing agents specifically designed for liposome incorporation had high entrapment efficiencies.

To enhance liposome incorporation, changes in lipid composition were tested with some success in drugs such as doxorubicin, although the incorporation efficiency did not exceed 60% to 65%. Lipid composition did not seem to play an important role in determining the antitumor activity, except in cases--that of cytarabine (Ara C), for example--in which slow drug release is crucial (38).

Regardless of the drugs used and quite predictably, liposome incorporation altered the pharmacokinetics and organ distribution of all agents tested (28, 46). In general, liposome incorporation increased the peak drug blood levels, decreased distribution volume, and increased the Concentration x Time. As expected, the distribution studies showed a preferential drug localization in the organs rich in reticuloendothelial system (RES) cells and fenestrated capillaries. In the case of liposomal doxorubicin, cardiac drug levels, and therefore cardiotoxicity, were also significantly decreased by liposome encapsulation.

Toxicologic studies of liposomal doxorubicin showed that the subacute LD_{50} of doxorubicin was about two-fold higher after liposome entrapment (53, 54). Liposomal doxorubicin was reported to be less cardiotoxic, slightly less myelosuppressive and less vesicant than doxorubicin (51). Cardiotoxicity studies of equitoxic or maximum tolerated doses were not performed.

We have studied the use of liposomes as drug carriers of several hydrophobic cisplatin analogues specifically designed for liposome incorporation (44-46, 55). We observed the incorporation and liposome stability of some of the most promising analogues to be greater than 95%. Liposomal entrapment increased the lung, spleen, and liver levels of elemental platinum (46) several fold. Liposomal platinum was less toxic than cisplatin, non-nephrotoxic at a dose corresponding to the LD_{50}, and at least as effective as cisplatin against L1210 leukemia (55).

A concern ever-present with the use of liposome-incorporated cytotoxic agents is the potential toxicity to the reticuloendothelial system (RES), although macrophages are non-dividing cells and therefore relatively resistant to most chemotherapeutic agents. In our studies, liposomal platinum and cisplatin had similar effects on several in vitro functions of peritoneal macrophages (44). In vivo studies showed that liposomal platinum did not alter the phagocytic capacity of the RES (44).

Several studies have suggested that drugs incorporated in liposomes have access to different types of malignant tumors than do free drugs (56-58). Doxorubicin uptake by metastatic lymphoma cells in the liver, for example, was found to be higher for the liposome-incorporated drug (29). These differences may be related to the small vesicles used and abnormalities in tumor vasculature. In preliminary studies, however, we observed multilamellar liposomes 2 μm in diameter containing a platinum drug in the cytoplasm of malignant cells of macroscopic liver metastases of M5076 reticulum cell sarcoma and in normal hepatocytes. It is less evident, based on these studies, that the size of the capillary fenestrations is always a vital determinant in liposome distribution.

EXPERIMENTAL TUMORS

Experimental tumor models for testing antitumor activity of the liposomal antineoplastic drug should be selected carefully. Models of tumors metastasizing to the liver, spleen, lung and bone marrow are good choices because these organs are the natural target of liposomes. Both liposomal doxorubicin and liposomal platinum were shown to enhance the antimetastatic activity of the parent compound against phagocytic and nonphagocytic tumors (30, 45, 52) (Table 3).

TABLE 3. ANTITUMOR ACTIVITY OF LIPOSOME-ENCAPSULATED ANTITUMOR AGENTS AGAINST EXPERIMENTAL LIVER METASTASES

Drug	Tumor Model	Optimal Percentage of T/C
Free doxorubicin	M5076 reticulosarcoma	156
Liposomal doxorubicin		222
Free doxorubicin	Colon 26 carcinoma	100
Liposomal doxorubicin		134
Free doxorubicin	Colon 38 carcinoma	120
Liposomal doxorubicin		152
Free doxorubicin	J6456 lymphosarcoma	166
Liposomal doxorubicin		286
Cisplatin (optimal dose)	M5076 reticulosarcoma	144
Liposomal NDDP		196

Abbreviations:

NDDP, Cis-bis-neodecanoate-1,2-diaminocyclohexane platinum II; Percentage of T/C, (Median survival of treated animals: median survival of control animals) x 100.

The antitumor activity of the liposomal drug was found to be preserved against solid tumors and increased against regional lymph node metastases (41). Liposomal Ara C and liposomal cisplatin were also found to have prophylactic antitumor activity when administered up to 48 hours before tumor inoculation (45, 59).

CLINICAL TRIALS

Three clinical trials of liposome-incorporated antitumor drugs were reported or are now in progress. Sculier et al. reported the results of a phase I study of liposomal NSC 252635 (60), but these researchers could not establish the drug's dose-limiting toxicity because of drug preparation limitations. In preliminary results of a phase I study of liposomal doxorubicin Gabizon et al. (61) showed that bone marrow toxicity seems to be the dose-limiting toxicity, and that such side effects as nausea and vomiting were markedly reduced by liposome entrapment. A phase I study of liposomal doxorubicin is being conducted currently by Rahman and colleagues at Georgetown University (62).

CONCLUSIONS

Liposome incorporation of several drugs consistently changed their pharmacokinetics and distribution, correlated with decreased toxic effects such as cardiotoxicity and enhanced activity against liver metastases. Liposomal amphotericin B was shown in animal studies and clinical trials to be far less toxic than the free drug, in addition to being curative in situations in which the free drug was not effective. Liposomes are a good example of the potential application of drug carriers in clinical practice. New concepts in drug targeting, such as hormone- and virus-based carriers, need to be studied more extensively. The prospects are exciting, and clinical trials will give answers to some of our questions.

REFERENCES

1. R.R.C. New, M.L. Chance, and S. Heath, Antileishmanial activity of amphotericin and other antifungal agents entrapped in liposomes. J. Antimicrob. Chemother. 8 :371-381 (1981).
2. C.B. Panosian, M. Barza, R. Szoka, and D.J. Wyler, Treatment of experimental cutaneous leishmaniasis with liposome-intercalated amphotericin B. Antimicrob. Agents Chemother. 25 :655-656 (1984).
3. J. Berman, W.L. Hanson, W.L. Chapman, C.R. Alving, and G. Lopez-Berestein, Antileishmanial activity of liposome-encapsulated amphotericin B in hamsters and monkeys. Antimicrob. Agents Chemother. 30 6:847-851 (1986).
4. R.L. Taylor, D.M. Williams, P.C. Craven, J.R. Graybill, D.J. Drutz, and W.E. Magee, Amphotericin B in liposomes: a novel therapy for histoplasmosis. Am. Rev. Respir. Dis. 125 :610-611 (1982).
5. J.R. Graybill, P.C. Craven, L. Taylor, D.M. Williams, and W.E. Magee, Treatment of murine cryptococcosis with liposome-associated amphotericin B. J. Infect. Dis. 145 :748-752 (1982).
6. G. Lopez-Berestein, R.L. Hopfer, R. Mehta, K. Mehta, E.M. Hersh, and R.L. Juliano, Treatment of disseminated fungal infections in neutropenic mice with liposome-encapsulated amphotericin B. J. Infect. Dis. 150 :278-283 (1984).
7. G. Lopez-Berestein, R. Mehta, R.L. Hopfer, K. Mills, L. Kasi, K. Mehta, V. Fainstein, M. Luna, E.M. Hersh, and R.L. Juliano, R.L. Treatment and prophylaxis of disseminated infection due to Candida albicans in mice with liposome-encapsulated amphotericin B. J. Infect. Dis. 147 :939-944 (1983).
8. G. Lopez-Berestein, T. McQueen, and K. Mehta, Protective effect of liposomal-amphotericin B against C. albicans in mice. Cancer Drug Deliv. 2 :183-189 (1985).
9. F.C. Szoka, D. Milholland, and M. Barza, Effect of lipid composition and liposome size on toxicity and in vitro fungicidal activity of liposome-intercalated amphotericin B. Antimicrob. Agents Chemother. 311 :421-429 (1987).
10. G. Lopez-Berestein, M.G. Rosenblum, and R. Mehta, Altered tissue distribution of amphotericin B by liposomal-encapsulation: comparison of normal mice to mice infected with Candida albicans. Cancer Drug Deliv. 3 :199-205 (1984).

11. G. Lopez-Berestein, V. Fainstein, R.L. Hopfer, K. Mehta, M.P. Sullivan, M. Keating, M.G. Rosenblum, R. Mehta, M. Luna, E.M. Hersh, J. Reuben, R.L. Juliano, and G.P. Bodey, Liposomal-amphotericin B for the treatment of systemic fungal infections in patients with cancer: A preliminary report. J. Infect. Dis. 4 :704-710 (1985).
12. G. Lopez-Berestein, G.P. Bodey, L.A. Frankel, and K. Mehta, Treatment of hepatosplenic candidiasis with liposomal-amphotericin B. Antimicrob. Agents Chemother. 31 :421-429 (1987).
13. C.R. Alving, E.A. Steck, W.L. Chapman, Jr., V.B. Waits, L.D. Hendricks, G.M. Swartz, Jr., and W.L. Hanson, Liposomes in leishmaniasis: therapeutic effects of antimonial drugs, 8-aminoquinolines, and tetracycline. Life Sci. 26 :2231-2238 (1980).
14. C.R. Alving, E.A. Steck, W.L. Hanson, P.S. Loizeaux, W.L. Chapman, Jr., and V.B. Waits, Improved therapy of experimental leishmaniasis by use of liposome-encapsulated antimonial drug. Life Sci. 22 :1021-1026 (1978).
15. W.L. Chapman, Jr., W.L. Hanson, C.R. Alving, and L.D. Hendricks, Antileishmanial activity of liposome-encapsulated meglumine antimonate in the dog. Am. J. Vet. Res. 45 :1028-1030 (1984).
16. W.L. Chapman, Jr., V.B. Waits, and W.L. Hanson, Liposomes in leishmaniasis: effects of parasite virulence on treatment of experimental leishmaniasis in hamsters. Ann. Trop. Med. Parasitol. 78 :279-286 (1984).
17. J.V. Desiderio, and S.G. Campbell, Intraphagocytic killing of Salmonella typhimurium by liposome-encapsulated cephalothin. J. Infect. Dis. 148 :563-570, (1983).
18. J.V. Desiderio, and S.G. Campbell, Liposome-encapsulated cephalothin in the treatment of experimental murine salmonellosis. Res. J. Reticuloendothel. Soc. 34 :279-287, (1983).
19. P. Pirson, R. Steiger, and A. Trouet, The disposition of free and liposomally encapsulated antimalarial primaquine in mice. Biochem. Pharmacol. 31 :3501-3507 (1982).
20. P. Pirson, R.F. Steiger, A. Trouet, J. Gillet, and F. Herman, Primaquine liposomes in the chemotherapy of experimental murine malaria. Ann. Trop. Med. Parasitol. 74 :383-391 (1980).
21. W.C. Koff, S.D. Showalter, D.A. Seniff, and B. Hampar, Lysis of herpes virus-infected cells by macrophages activated with free or liposome-encapsulated lymphokine produced by a murine T cell hybridoma. Infect. Immun. 42 :1067-1072 (1983).
22. E.B. Fraser-Smith, D.A. Eppstein, M.A. Larsen, and T.R. Matthews, Protective effect of a muramyl dipeptide analog encapsulated in or mixed with liposomes against Candida albicans infection. Infect. Immun. 39 :172-178 (1983).
23. C. LaBonnardiere, Donnees preliminaires sur l'effet protecteur de interferon couple a des l'liposomes dans le modele souris-virus de l'hepatite murine. Ann. Microbiol. 129A :397-402 (1978).
24. I.A.J.M. Bakker-Woudenberg, A.F. Lokerse, J.C. Vink-van den Berg, F.H. Roerdink, and M.F. Michel, Effect of liposome-entrapped ampicillin on survival of Listeria monocytogenes in murine peritoneal macrophages. Antimicrob. Agents Chemother. 30: 295-300 (1986).
25. P.F. Bonventre, and G. Gregoriadis, Killing of intraphagocytic Staphylococcus aureus by dihydrostreptomycin entrapped within liposomes. Antimicrob. Agents Chemother. 13 :1049-1051 (1978).
26. T.I. Kozulitsina, O.A. Uvarova, V.I. Golyshevskaya, and M.P. Elshanskaya, Experimental study of chemotherapeutic activity of various antibacterial drug combinations. Probl. Tuberk. 7 :53-57 (1980).
27. J.R. Morgan, and K.E. Williams, Preparation and properties of liposome-associated gentamicin. Antimicrob. Agents Chemother. 17 :544-548 (1980).
28. A. Rahman, D. Carmichael, M. Harris, and J.K. Roh, Comparative pharmacokinetics of free doxorubicin and doxorubicin entrapped in cardiolipin liposomes. Cancer Res. 46 :2295-2299 (1986).

29. A. Gabizon, D. Goren, Z. Fuks, A. Meshorer, and Y. Barenholz, Superior therapeutic activity of liposome-associated Adriamycin in a murine metastatic tumor model. Br. J. Cancer 51 :681-689 (1985).
30. E. Mayhew, Y. Rustum, and W.J. Vail, Inhibition of liver metastases of M5076 tumor by liposome-entrapped adriamycin. Cancer Drug Deliv. 1 :43-58 (1983).
31. L.D. Mayer, M.B. Bally, M.J. Hope, and P.R. Cullis, Uptake of antineoplastic agents into large unilamellar vesicles in response to a membrane potential. Biochim. Biophys. Acta. 816 :294-302 (1985).
32. D. Layton, and A. Trouet, A comparison of the therapeutic effects of free and liposomally encapsulated vincristine in leukemic mice. Eur. J. Cancer 16 :945-950 (1980).
33. D. Layton, J. De Meyere, M.P. Collard, and A. Trouet, The accumulation by fibroblasts of liposomally encapsulated vinblastine. Eur. J. Cancer 15 :1475-1483 (1979).
34. K.R. Patel, and J.D. Baldeschwieler, Mouse lewis lung carcinoma and hepatoma ascites treatment by combination of liposome chemotherapy and non-specific immunotherapy. Int. J. Cancer 34 :717-723 (1984).
35. J. Freise, F.W. Schmidt, and P. Magerstedt, Effect of liposome-entrapped methotrexatate on Ehrlich ascites tumor cells and uptake in primary liver cell tumor. J. Cancer Res. Clin. Oncol. 94 :21-27, (1979).
36. H.K. Kimelberg, T.F. Tracy, S.M. Biddlecome, and R.S. Bourke, The effect of entrapment in liposomes on the in vivo distribution of H^3 methotrexate in a primate. Cancer Res. 36 :2949-2957 (1976).
37. K.R. Patel, and J.D. Baldeschwieler, Treatment of intravenously implanted lewis lung carcinoma with liposome-encapsulated cytosine arabinoside and non-specific immunotherapy. Int. J. Cancer 34 :415-420 (1984).
38. E. Mayhew, Y.M. Rustum, F. Szoka, and D. Papahadjopoulos, Role of cholesterol in enhancing the antitumor activity of cytosine arabinoside entrapped in liposomes. Cancer Treat. Rep. 63 :1923-1928 (1979).
39. S.B. Kaye, J.A. Boden, and B.E. Ryman, The effect of liposome (phospholipid vesicle) entrapment of actinomycin D and methotrexate on the in vivo treatment of sensitive and resistant solid murine tumors. Eur. J. Cancer 17 :279-289 (1981).
40. J. Freise, W.H. Mueller, P. Magerstedt, and H.F. Schmoll, Pharmacokinetics of liposome encapsulated Cisplatin in rats. Arch. Int. Pharmacodyn. Ther. 258 :180-192 (1982).
41. V.I. Kaledin, N.A. Matlenko, V.P. Nikolln, Y.V. Grutenko, and V.G. Budker, Intralymphatic administration of liposome-encapsulated drugs for mice: Possibility of suppression of the growth of tumor metastases in lymph nodes. J. Natl. Cancer Inst. 66 :881-887 (1981).
42. M.B. Yatvin, H. Muhlensiepen, W. Porshen, J.N. Weinstein, and L.E. Feinendegen, Selective delivery of liposome-associated cis-dichloro-diammine platinum (II) by heat and its influence on tumor drug uptake and growth. Cancer Res. 41 :1602-1607 (1981).
43. J. Khato, E.R. Priester, and S.M. Sieber, Enhanced lymph node uptake of melphalan following liposomal entrapment and effects on lymph node metastasis in rats. Cancer Treat. Rep. 66 :517-527 (1982).
44. R. Perez-Soler, A.R. Khokhar, P. Claringbold, L.P. Kasi, and G. Lopez-Berestein, Effects on the monocyte-macrophage system and antitumor activity against L1210 leukemia of cis-bis-cyclopentenecarboxylato-trans-R,R-1,2-diaminocyclohexane-platinum (II) encapsulated in multilamellar vesicles. J. Natl. Cancer Inst. 77 :1137-1143 (1986).
45. R. Perez-Soler, A.R. Khokhar, and G. Lopez-Berestein, Development of lipophilic cisplatin analogues incorporated in liposomes. In: Liposomes as Drug Carriers: Trends and Progress. (G. Gregoriadis Ed.). John Wiley & Sons, New York, In press.
46. J. Lautersztain, R. Perez-Soler, A.R. Khokhar, R.A. Newman, and G. Lopez-Berestein, Pharmacokinetics and tissue distribution of liposome-encapsulated cis-bis-N-decyliminodiacetato-1,2-diaminocyclohexane-

platinum (II). Cancer Chemother. Pharmacol. 18 :93-97 (1986).
47. G. Deliconstantinos, G. Gregoriadis, G. Abel, M. Jones, and D. Robertson, D. Incorporation of cis-dichlorobis-cyclopentylamineplatinum (II) into liposomes enhances its uptake by ADJ/PC6A tumours implanted subcutaneously into mice. Biochem. Soc. Trans. 5 :1326-1328 (1977).
48. W. Rubas, A. Supersaxo, H.G. Weder, H.R. Hartmann, H. Hengartner, H. Schott, and R. Schwendener Treatment of murine L1210 lymphoid leukemia and melanoma B16 with lipophilic cytosine arabinoside prodrugs incorporated into unilamellar liposomes. Int. J. Cancer 37 :149-154 (1986).
49. S.C. Kinsky, K. Hashimoto, J.E. Loader, M.S. Knight, and D.J. Fernandes, Effect of liposomes sensitized with methotrexate-γ-dimyristoyl phosphatidylethanolamine on cells that are resistant to methotrexate. Biochim. Biophys. Acta. 885 :129-135 (1986).
50. C. Brassinne, G. Atassi, J. Fruhling, W. Penasse, A. Coune, J. Hildebrand, J.M. Ruysschaert, and C. Laduron, Antitumor activity of a water-insoluble compound entrapped in liposomes on L1210 leukemia in mice. J. Natl. Cancer Inst. 70: 1081-1085 (1983).
51. E.H. Herman, A. Rahman, F.J. Ferrans, F.A. Vick, and P.S. Schein, Prevention of chronic doxorubicin cardiotoxicity in beagles by liposomal encapsulation. Cancer Res. 43 :5427-5432 (1983).
52. E. Mayhew, M. Goldrosen, J. Vaage, and Y. Rustum, Liposomal-adriamycin and survival of mice bearing liver metastases of colon carcinomas 26 or 38. Proc. Am. Assoc. Cancer Res. 27 :402 (1986).
53. A. Rahman, A. Fumagalli, B. Barbieri, P.S. Schein, And M. Casazza, Antitumor and toxicity evaluation of free doxorubicin and doxorubicin entrapped in cardiolipin liposomes. Cancer Chemother. Pharmacol. 16 :22-27 (1986).
54. F. Olson, E. Mayhew, D. Maslow, Y. Rustum, and F. Szoka, Characterization, toxicity and therapeutic efficacy of Adriamycin encapsulated in liposomes. Eur. J. Cancer Clin. Oncol. 18 :167-176 (1982).
55. R. Perez-Soler, A.R. Khokhar, M.P. Hacker, and G. Lopez-Berestein, Toxicity and antitumor activity of cis-bis-cyclopentenecarboxylato-1,2-diaminocyclohexane platinum (II) encapsulated in multilamellar vesicles. Cancer Res. 46 :6269-6273 (1986).
56. J.N. Weinstein, R.L. Magin, M.B. Yatvin, and D.S. Zaharko, Liposomes and local hyperthermia: Selective delivery of methotrexate to heated tumors. Science 204 :188-191 (1979).
57. J.N. Weinstein, R.L. Magin, R.L. Cysyk, and D.S. Zaharko, Treatment of solid L1210 murine tumors with local hyperthermia and temperature-sensitive liposomes containing methotrexate. Cancer Res. 40 :1388-1395 (1980).
58. A. Gabizon, D. Goren, Z. Fuks, Y. Barenholz, A. Dagan, and A. Meshorer, Enhancement of adriamycin delivery to liver metastatic cells with increased tumoricidal effect using liposomes as drug carriers. Cancer Res. 43 :4730-4735 (1983).
59. T. Kataoka, and T. Kobayashi, Enhancement of chemotherapeutic effect by entrapping 1-β-D-arabinofuranosyl cytosine in lipid vesicles and its mode of action. Ann. N. Y. Acad. Sci. 308 :387-394, (1978).
60. J.P. Sculier, A. Coune, C. Brassinne, C. Laduron, G. Atassi, J.M. Ruysschaert, and J. Fruhling, Intravenous infusion of high doses of liposomes containing NSC 251635, a water-insoluble cytostatic agent. A pilot study with pharmacokinetic data. J. Clin. Oncol. 4 :789-797 (1986).
61. T. Peretz, A. Gabizon, R. Catane, R. Ben-Yosef, S. Biran, S. Drukman, and Y. Barenholz, Clinical studies on liposome-associated doxorubicin clinical (L-DXR. Progress report. Proc. Am. Assoc. Clin. Oncol. 6 :43 (1987).

THE INHIBITORY EFFECT OF PARVOVIRUS H-1 ON CULTURED HUMAN CANCER CELLS OR TRANSFORMED CELLS

Zu-yu Luo, Zao-zhong Su, Lan-ping Guo,
Jun-zi Li and Ya-lun Liu

Department of Biology
Fudan University
Shanghai, Peoples Republic of China

INTRODUCTION

Hamster osteolytic virus H-1 is an autonomous parvovirus. Its genome, consisting of a linear single-stranded DNA molecule of 5176 nucleotides, carries 3 structural genes of capsid and non-capsid protein; the virion replication depends therefore almost completely on the enzyme system of the host cells (1). H-1 can inhibit the formation of tumours in experimental animals. For example, infection of Syrian hamster with H-1 reduced the incidence of "spontaneous" tumours from 0.05% to 0.0023%, the incidence of tumours induced by the chemical carcinogen DMBA from 95% to 38% (3), and the incidence of tumours induced by adenovirus from 67% to 25% (4), other parvoviruses possess similar functions. The reports from human epidemiological studies indicated that infection by parvovirus lowered significantly the incidence of several cancers (5, 6).

The mechanisms of the oncosuppressive activity of parvovirus remain unclear. It is believed that parvovirus infection stimulates the immune reaction of humans and animals to a higher state (3, 7, 8). However, AAV (adeno-associated virus) may reduce the oncogenic potential of adenovirus and HSV (herpes simplex virus) by way of preventing the expression of their oncogenes in the host cells (9); MVM (minute virus of mice) inhibits the transformation by SV_{40} of cultured mouse cells (10). These results indicate the possibility that parvoviruses inhibit the growth of tumour cells directly. Obviously, the establishment of an *in vitro* system is benefical to the exploration of the mechanisms of the antineoplastic activity of parvoviruses.

The present work established the *in vitro* system for studying the oncosuppressive activity of parvovirus and demonstrated that H-1 can be grown cytolytically in various human cancer cell lines, including 4 hepatoma and 3 gastric cells lines (Fig. 1). It was also grown cytolytically in primary cancer cell outgrowth from cancer tissue "clumps" removed surgically from two patients who suffered primary gastric cancer of differentiated or low-differentiated type, respectively (Fig. 2). However, a gastric cancer cell line of non-differentiated type (HGC-27) and a primary cancer cell culture from a lymphnode metastasis of a primary gastric cancer of very poorly differentiated type were not sensitive to the

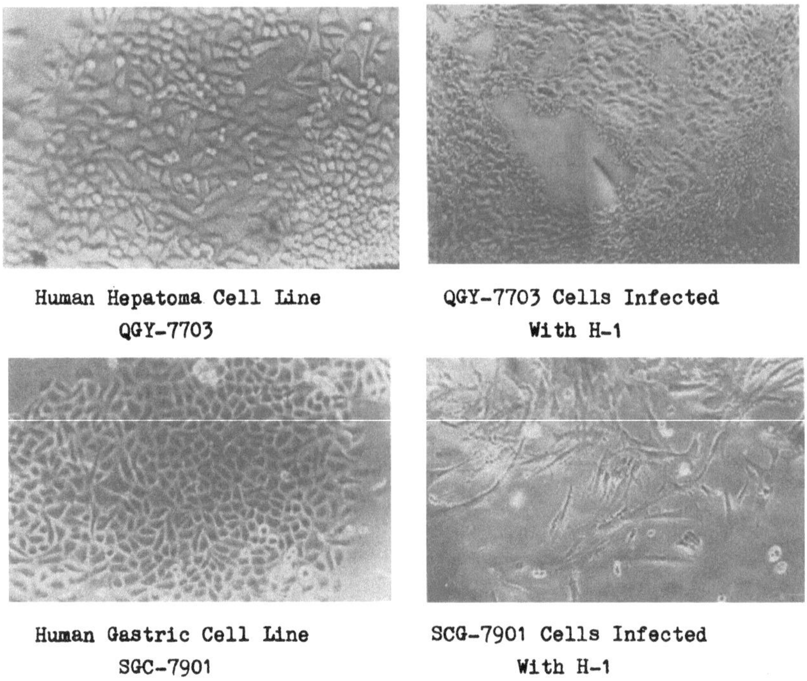

Fig. 1. Human hepatoma cell line QGY-7703 and gastric cell line SGC-7901 (left), as well as the appearance of CPE after infection with H-1 (right). 40X - 180X

infection H-1. H-1 did not block the growth of primary cells from normal gastric and hepatic tissues, as well as "normal" liver cell line HuL-1 (Fig. 3) (11). In addition, the growth of two skin fibroblast cell lines (XP 20S and XP 2YO), established from xeroderma pigmentosum patients, were not affected by H-1, however, their SV_{40}-transformed counterparts (XP 20S/SV and XP 2YO/SV) were sensitive to H-1 (Fig. 4).

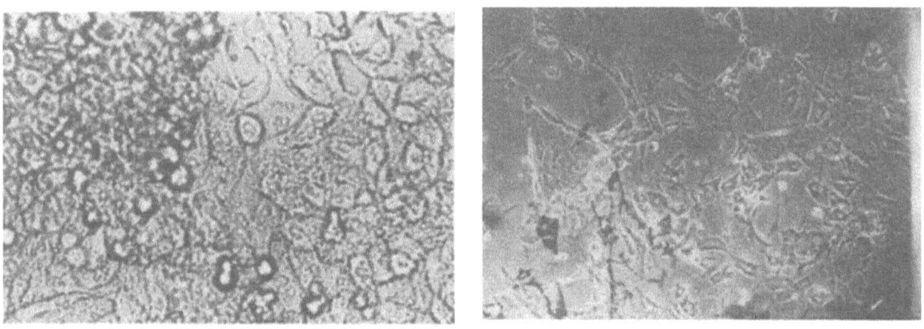

Fig. 2. Killing effect of H-1 on primary gastric cancer cells. Primary cell outgrowth from clumps of gastric cancer tissue (left); the dark area indicates a cancer clump. Right hand panel: 5 days after infection with H-1 (m.o.i. = 2) the cells were lysed. 250X

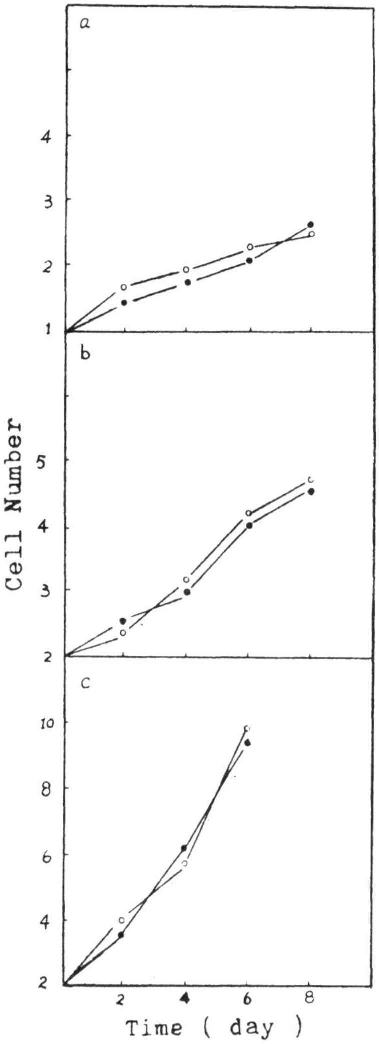

Fig. 3. Infection with H-1 did not affect the growth of primary hepatic (a), primary gastric (b) and HuL-1 (c) cells. Ordinate shows cell number X 10^{-5}.

Our results thus indicate that H-1 possesses selective inhibitory or killing effect on human cancer cells. This selectivity may relate to the state of differentiation of the malignant cells. The expression of oncoviral genes may activate the cellular sensitivity to parvovirus. The in vivo oncosuppressive activity of parvovirus may involve at least their direct inhibitory action on malignant cells. Certain human cancerous or transformed cells can support the cytolytic growth of H-1.

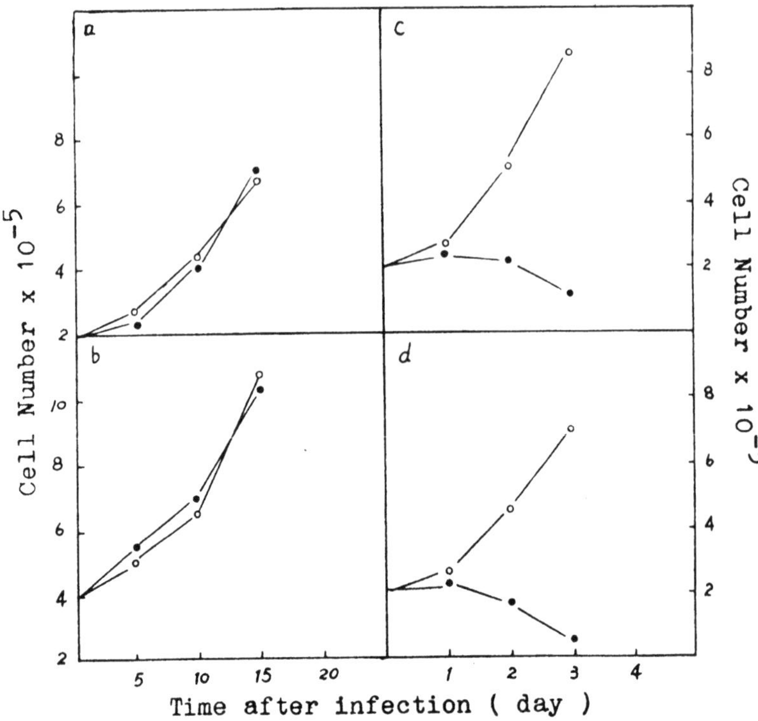

Fig. 4 Growth curves of XP 20S (a), XP2YO (b), XP 20S/SV (c), and XP 2YO/SV (d), Cells infected (·) or non-infected (o) with H-1 (m.o.i. 2).

ACKNOWLEDGEMENTS

This project was supported by National Nature Sciences Foundation of China. We thank Profs. M. Ikenaga, He Shaoqing, Shen Yu and Dong Rongchun for providng cell lines and Dr. Zhu Weiqi for surgical tissues. We are also grateful to Mr. Sun Guangrong for assistance with electron microscopy.

REFERENCES

1. S. L. Rhode, III and P. R. Paradiso, Parvovirces genome: Nucleotide sequence of H-1 and mapping of its genes by hybrid-arrested translation. J. Virol. 45:173-184 (1983).
2. H. W. Toolan, Lack of oncogenic effect of the H-viruses for hamsters. Nature (London) 214:1036 (1967).
3. H. W. Toolan, S. L. Rhode, III and J. F. Gierthy, Inhibition of 7,12-dimethylbenz(a)anthracene-induced tumors in Syrian hamsters by prior infection with H-1 parovirus. Cancer Res. 42:2552-2555 (1982).
4. H. W. Toolan and N. Ledinko, Inhibition by H-1 virus of the incidence of tumors produced by adenovirus 12 in hamsters. Virology 35:475=478 (1968).
5. S. Sprecher-Goldberger, L. Thiry, N. Lefebvre, D. Dekegel and F. de Halleus, Complement-fixation Cytomegaloviruses and herpes simples viruses in patients with tumors and in control individuals. Am. J. Epidemiol 94:351-358 (1971).
6. B. Georg-Fries, S. Biederlack, J. Wolf and H. Zur Hausen, Analysis of proteins, helper dependence, and seroepideniology of a new human parvovirus, Virology 134:64-71 (1984).

7. H. W. Toolan, E. L. Saunders, C. M. Southam, A. E. Moore and A. G. Levin, H-1 virus virmia in the human. Proc. Soc. Exp. Biol. Med. 119:711-715 (1965).
8. H. W. Toolan, H-1 virus viremia in the hamster. Proc. Soc. Exp. Biol. Med. 119:715-717 (1965).
9. K. I. Berns and W. W. Hauswirth, Adeno-associated viruses. Adv. Virus Res. 25:407-449 (1979).
10. S. Mousset and J. Rommelaere, Minute virus of mice inhibits cell transformation by simian virus 40. Nature(London)300:537-539 (1982).
11. H. Katsuta, T. Takaoka and N. Huh, Establishment of tissue culture cell strains from normal fatal human liver and kidney Jpn. J. Exp. Med. 50:329-337 (1980).

EVALUATION OF IMMUNOMODULATORY AND THERAPEUTIC PROPERTIES OF BIOLOGICAL RESPONSE MODIFIERS: A COMPARISON OF PRECLINICAL AND CLINICAL STUDIES

James E. Talmadge[1], Carl Pinsky[2], Ronald B. Herberman[3], Cedric Long[4], and Paul L. Black[1]

[1]Preclinical Screening Laboratory, Program Resources, Inc. NCI-Frederick Cancer Research Facility, Frederick, MD 21701 [2]Office of the Associate Director, Biological Response Modifiers Program, DCT, NCI-Frederick Cancer Research Facility Frederick, MD 21701. [3]Pittsburgh Cancer Institute, 230 Lothrop Street, Pittsburgh, PA 152. [4]Office of the Director NCI-Frederick Cancer Research Facility, Frederick, MD 21701

Biological response modifiers (BRMs) are those agents or approaches that modify the relationship between the tumor and host by modifying the host's biological response to the tumor cells with resultant therapeutic effects (1). BRMs may modify the host responses in several ways:

1. increase the host's antitumor responses through augmentation and/or restoration of effector mechanisms or mediators of the host's reaction which may be deleterious;
2. increase the host's defenses by the administration of natural biologicals (or the synthetic-recombinant derivatives thereof) as effectors or mediators of an antitumor response;
3. augment the sensitivity of the host's tumor cells to endogenous mechanisms for the control of tumor growth;
4. alter the transformed phenotype by increasing the differentiation/maturation of tumor cells;
5. increase the ability of the host to tolerate damage by cytotoxic modalities of cancer treatment.

Specifically, BRMs include immunoaugmenting, immunomodulating, and immunorestorative agents; cytokines (soluble protein mediators produced by the host, which include interferons, lymphokines, and monokines); cytokine inducers; inhibitors of tumor growth factors; thymic factors; tumor antigens; and modifiers of tumor antigen cell surface components, antitumor antibodies, antitumor effector cells [lymphokine activated killer cells (LAK) or tumor infiltrating lymphocytes (TIL)], and maturation and differentiation factors. A subset of BRMs which are termed biologics encompass those products of the immune system of the host, which may or may not be immunomodulatory agents but which have potential therapeutic activities (1). Thus, biologics include interferons, interleukins, colony stimulating factors, tumor necrosis factor, lymphotoxin, cytolysin, etc. At present, cancer immunotherapy appears to have relatively limited clinical value (2-6). This conclusion is in contradistinction to the optimistic and premature reports in the popular press and some of the research literature, which have inflated expectations and may ultimately discredit the field. Additional and more sophisticated preclinical studies are needed to optimize therapeutic protocols, understand the mechanism of action and probable toxicity prior to clinical entry of a BRM. In addition, a decrease in empirical clinical studies and an increased emphasis on hypothesis directed clinical trials are needed. Nontheless, several tumor histiotypes have responded to clinical immunotherapy under specific conditions. Our present understanding of immunoregulation and mechanisms of immunotherapy suggest that we may be able to develop therapeutic protocols that will be effective during the evolution of the tumor-host relationship.

The study of immunotherapy began with Cooley's toxins, and nonspecific immunotherapy began to gain prominence in the late 1960's when the clinical efficacy of Bacillus Calmette-Guerin (BCG) was reported (7, 8). In addition, these nonspecific crude bacterial adjuvants have potential lot-to-lot variations associated with microbial strain variants or differences in media or culture conditions used to grow the reagents. The emphasis switched from therapeutic regimes with BCG and other nonspecific and variable bacterial adjuvants to the development of clinically defined agents such as levamisole, isoprinosine, pyran copolymers, and double-stranded polyribonucleotides. Recently, the advent of recombinant DNA technology has produced highly potent immunoregulatory molecules such as the IFNs, IL-2, TNF, and most recently, CSFs which are suitable for testing in preclinical and clinical studies. The enthusiasm surrounding each compound appears to subside, however, after unanticipated toxicities are demonstrated clinically or as new and more exciting recombinant cytokines are developed.

In 1968, it was reported that BCG vaccination prolonged the survival of children with acute lymphocytic leukemia (7). Although these results could not be confirmed, other controlled studies in adults with acute myelogenous leukemia suggested similar benefits from BCG therapy as did preliminary reports of prospectively randomized trials (9-11). However, a final analysis of 25 such studies from the 1970's indicated that nonspecific BCG therapy does not improve the response rate or increase the duration of response in patients with acute myelogenous leukemia. For reasons that are not well understood, the major benefit with BCG in hematopoietic neoplasms has been the prolongation of survival in the subset of patients responding poorly to chemotherapy (12). Recently, the intravesicular instillation of BCG was evaluated with or without percutaneous administration in patients with transitional cell carcinoma of the bladder (13-16). It was found that the intracavitary use of BCG, in conjunction with appropriate surgical therapy, significantly decreased the number of recurrences, lengthened the disease-free interval, and reduced the number of patients who subsequently required cystectomy. Indeed, such protocols now appear to be the treatment of choice for transitional cell carcinoma of the bladder. Similarly, treatment protocols using BCG (17), C. parvum (18), or OK-432 (19) have demonstrated a palliative effect in the treatment of peritoneal and plural effusions. Unfortunately, the effusion studies have used a limited administration schedule (generally a single injection), and the therapeutic as opposed to a palliative efficacy of a prolonged therapeutic protocol remains to be determined.

The initial clinical studies with IFN-α were the subject of much enthusiasm when nonrandomized studies found therapeutic activity against malignant melanoma, osteosarcoma, and various lymphomas. However, subsequent trials with IFN-α have demonstrated significant therapeutic activity only against hairy cell leukemia, a few other types of lymphoma, and perhaps renal cell carcinoma (3, 4, 6, 20-25). Nonetheless, despite extensive clinical studies, IFN-α development is still in its early stages, and such basic parameters as optimal dose and therapeutic schedule remain to be determined. It is now axiomatic that IFN-α is not the "magic bullet" it was originally suggested to be, and additional studies will be required to realize its promise. Recombinant IL-2 has also undergone clinical trials, and the enthusiasm for this agent has been tempered by the recent disclosure of toxicity and mortality associated with its aggressive administration (26). Recombinant TNF has just entered clinical trials and has generated considerable enthusiasm because it is an agent that promises to necrotize tumors, although few therapeutic responses have been observed to date.

The concept that augmentation of host defense mechanisms may be therapeutically useful has been supported by preclinical data in various experimental therapy systems (27, 28). However, many of these experimental systems used transplantable tumors, and were based on therapy models that were not closely analogous to clinical situation. In many instances, the therapeutic efficacy of a BRM was tested against antigenic and well-vascularized, subcutaneous or intradermal tumors. Indeed, many of these studies have focused on tumor prophylaxis, where the hosts were treated prior to tumor challenge rather than the therapy of preexisting, well-established disease. Thus, based on such modeling approaches, it is not surprising that the impressive therapeutic or prophylactic activity of immunotherapy found with these rodent models has not been observed clinically.

Originally studies were undertaken using the intraperitoneal administration of a BRM to animals with a minimal tumor burden in the peritoneal cavity (ascites). This model has limited relevance to most human tumors because the drug gains immediate access to tumor cells without interference from anatomical barriers. This situation mimics a subset of patients with ovarian ascites or colorectal cancer, which represent the predominant tumors in the peritoneal cavity, and the few patients with breast pulmonary carcinomas, which form pleural effusions. Alternatively, antigenic and well-vascularized transplanted tumors growing subcutaneously, intradermally, or intramuscularly were treated. Although such solid tumors appear to be more difficult to treat than ascites tumors, all of these tumor models still fail to provide a demanding model for predicting the efficacy of agents against systemic neoplastic disease. Because metastasis is the major predisposing cause of death from cancer (29), approaches should stress the treatment of metastasis from transplantable tumors. Optimally this would include the study of spontaneous metastases in hosts which have been "tumor conditioned" for approximately a month prior to primary tumor resection and initiation of therapy. This model is more rigorous than "traditional" models and provides parallels to the clinical situation. However, it remains to be determined whether it will provide a correlation or prediction of the clinical activity of potential immunotherapeutic agents. It should be emphasized that the biggest clinical challenge is the development of therapeutic protocols for systemic metastases and that this represents the most important model for the development of antineoplastics.

The goal of the present manuscript is to provide an overview of immunotherapy and to discuss the preclinical approaches for the clinical utilization of BRMs and how the preclinical data correlates with clinical observations. In addition, approaches are discussed to facilitate preclinical-clinical interactions and data for correlation and predictive studies. This approach stresses concepts that may facilitate and coordinate clinical studies based on the preclinical development of clinical hypotheses. Limited clinical information is available for most BRMs and biologicals as compared to the data available preclinically, although in the case of IL-2, IFNs, and perhaps polyribonucleotides, a sufficient data base is available for preliminary comparisons.

A brief overview of one such systematic preclinical approach is outlined as follows: based on initial studies in normal or tumor-bearing animals to determine the cellular pharmacology as well as pharmacokinetics of the immunomodulators, one can address the therapeutic properties of the BRM with some insight into a therapeutically active protocol based on the pharmacokinetics and possible mechanism of activity. Further insight into protocol parameters may be provided by studies with similar agents. Based on these studies to determine the optimal therapeutic protocol (dose, route, schedule, sequence and timing and duration of BRM administration), a series of protocol variations are utilized to determine the effector cells or mechanisms responsible for the therapeutic activity. Then based on the optimal therapeutic protocol and the mechanism of therapeutic activity, clinical hypotheses can be formulated and subsequently tested with some assurance of a rational rather than empirical development and testing of clinical hypotheses.

One of the difficulties that have become apparent in the comparison of preclinical and clinical data is that preclinical studies stress effector cells from sites different from those that are readily obtainable during clinical studies. For example, either peritoneal exudate cells (PEC) or spleen cells are examined in preclinical studies, whereas virtually all clinical studies are limited to peripheral blood lymphocytes and monocytes. Furthermore, different assays of immunofunction are utilized in clinical studies as compared to preclinical studies. One example is the use in preclinical studies of macrophage function of PEC and a 72-hr [^{125}I]UdR or [^{3}H]thymidine assay. In contrast, virtually all clinical studies are based on human monocytes whose function is determined by Fc receptor density, HLA-DR expression, and more rarely hydrogen peroxidase assays. Thus, there is a need to utilize parallel assays which correlate with biologic function, such as macrophage tumoricidal activity (to use the same example) and therapeutic response. Furthermore, these analyses should not require the extensive amounts of blood that are presently needed. Continuing with the macrophage example, this analysis could include studies of activated oxygen intermediates such as peroxidase or neopterin levels in the urine or serum as well as Fc and HLA-DR expression which are commonly examined by FACS analysis.

In addition to a lack of correlative studies of monocyte/macrophage function, there are many deficiencies in studies of T cell function. Clinical assays of T cell function have largely been limited to blastogenic response to mitogens or allogeneic mixed lymphocyte responses. Unfortunately, effector cell functions such as cytotoxic T lymphocyte (CTL) tumoricidal activity are limited by the requirement for autochthonous tumor cell lines and are rarely undertaken in clinical trials. In contrast, more relevant T cell function analyses, including CTL analyses, are possible in preclinical studies, since transplantable tumors are commonly utilized. Thus, similar to the monocyte function analyses, new models of T cell activity need to be developed and utilized in both preclinical and clinical studies. Practical considerations suggest that this could take the form of Northern blot analyses of messenger RNA levels for IL-2, TAC, IFN-γ, etc.

BRMs and cytokines can have different kinetics of activation or compartments of activation, which are also influenced by the route of administration. For example, MVE-2 or liposomes incorporating MTP-PE will activate peritoneal macrophages but not alveolar macrophages following i.p. injection (30, 37). Conversely, alveolar macrophages are activated following i.v. injections, while peritoneal macrophages may not be. Thus, it may be useful to examine effector cell activity in various tissues (particularly those commonly affected by metastases, such as the lungs and liver), as well as within the tumors themselves (28) to obtain "true" insight into the microsystem of effector cell augmentation.

Therapeutic activity is dependent on prolonged administration of a BRM (28), but repeated administration of various immunomodulators can also depress the immune response. For example, multiple injections of IFN result in an NK cell hyporesponsive state within the peripheral blood lymphocytes of patients (31, 32) or the spleen cells of mice (33). Studies with animals receiving multiple injections of MVE-2, BCG, OK-432, or C. parvum revealed an NK cell hyporesponsive state within the spleen, and blood; however, there is a sustained increase in the number and function of NK cells in the lungs and liver (30, 34-37). This site selectivity may be attributable to in situ proliferation of effector cells in these organs and/or sequestration-margination of NK cells from the circulation. In contrast, animals receiving multiple injections of rM IFN-γ or rH IFN-α A/D have a systemic NK cell hyporesponsive state. These studies demonstrate the need to examine various tissues for effector activity, as well as the importance of examining effector cell activity following multiple injections. Furthermore, they have important implications for clinical studies where normally only the peripheral blood is sampled. We suggest the potential value, when indicated from preclinical studies, of examining cells from the peritoneal effusions, bronchial alveolar lavage, and/or from the tumor sites themselves for immune function.

In addition to studies of immunoaugmentation, we need to address the protective or reconstitutive effect of a BRM on the myelosuppressive activities of a cytotoxic agent or irradiation. This approach is important not only because of the increased therapeutic activity of combination therapies, but also because of the potential for increased toxicity associated with the CSF induced cell cycling of stem cells, which are at risk to cytoreductive therapies. The combination of chemotherapy and immunotherapy may be active, via a number of mechanisms including: 1) reduction in tumor burden by cytoreductive agents resulting in a tumor burden more ameanable to immunotherapy; 2) decreased suppressor cell activity resulting in an increased T cell adjuvant activity induced by agents such as rH IL-2; 3) the increased therapeutic activity of two cytoreductive - cytostatic modalities such as a chemotherapeutic agent and tumor necrosis factor or interferon; 4) accelerated myeloid restoration due to increased stem cell activity thereby allowing more aggressive doses of chemotherapy and reduced neutropenia; 5) the addition of agents which stimulate stem cell cycling and accelerate myeloid restoration in conjunction with autologous bone marrow following aggressive cytoreductive therapy. Overall, these combination approaches appear to provide the most promising developments for the treatment of neoplasia.

Because metastasis is the major threat to cancer patients, the emphasis in the therapeutic evaluation of BRMs should be on the assessment of therapeutic efficacy against systemic lesions. Two models for the immunotherapy of metastasis are commonly utilized, one directed against tumor cells introduced in the host systemically by i.v. injection (experimental metastases) and the other, potentially more clinically relevant, assessment of activity against metastatic foci which develop after the resection of a primary tumor (spontaneous metastases). In the spontaneous metastasis model, the host has been "conditioned" by tumor growth for some time prior to resection of the primary tumor at which time therapies are initiated. It is expected that the spontaneous metastasis model will be most predictive of systemic therapeutic efficacy in the clinical setting (38).

It is likely that, as one progresses from models of immunoprophylaxis to more stringent and rigorous models of immunotherapy, fewer positive results will be obtained or decreased activity will be observed. Thus, a BRM with prophylactic activity may not have striking therapeutic activity against an ascites tumor or a intradermal-subcutaneous tumor. Similarly, therapeutic activity demonstrated against established systemic disease in a relatively normal host (experimental metastases) or in a host conditioned by the growth of primary tumors with preexistent spontaneous metastases may not be matched in the clinic. This difference may arise because the metastasis models utilize young, relatively healthy hosts bearing transplantable tumors, a situation which is quite different from that of typical cancer patients, who may have been exposed to a carcinogenic insult many years before, and who have gradually developed progressive primary and metastatic tumor lesions over a prolonged period. Thus, one might expect that the most clinically relevant therapeutic model would be metastases from primary autochthonous tumors.

A few metastatic autochthonous tumor models have been developed (38). One utilizes UV light-induced skin tumors, which include a variety of histologic types: squamous cell carcinomas, fibrosarcomas, hemangiosarcomas, rhabdomyosarcomas, and other less well differentiated sarcomas. A second model utilizes the induction of mammary tumors in Sprague-Dawley rats by N-nitrosomethyl urea (NMU). Similar to clinical studies, when palpable tumors are detected in these therapeutic models, the animals are randomly assigned to one of several arms of a protocol, and treatment is initiated. Therapeutic efficacy can be assessed by the survival of animals, rate of tumor growth, response rate, development of metastatic lesions, and multiple primary tumors. These studies are more difficult, tedious, and time-consuming than experiments with transplantable tumors, but they do provide a rigorous and probably more predictive test of therapeutic efficacy. The difficulties associated with such studies are similar to those routinely encountered in clinical trials, including variations and delays in tumor accrual, need for randomization between treatment and control groups, variability of tumor growth and behavior, need for extensive and relevant statistical evaluation, and prolonged study duration.

It is apparent that no one animal tumor model is optimal for studying immunotherapy and that all suffer from various limitations and deviations from the clinical situation. The adequacy of one tumor as compared to another to assess the therapeutic efficacy of a BRM is of secondary importance, compared with the rational development and interpretation of results from the model system and the subsequent development of clinically testable hypotheses. Nonetheless, independent of the model utilized, the level of confidence in the results is increased if positive results can be obtained in more than one tumor or animal model since such an observation is likely to reflect general principles of activity rather than the idiosyncrasies of a single model. Thus, multiple tumor histiotypes involving solid tumors such as melanoma, colon carcinoma, or pulmonary carcinoma might give the greatest reflection of potential clinical efficacy.

One of the potentially most important approaches to the development of BRM is the optimization of the therapeutic protocol. This optimization includes studies not only of dosage, but also route, schedule and duration of administration, as well as sequence and timing of administration for combination therapy. The importance of the route of administration goes beyond its influence on the site of activation as discussed earlier, it also affects the bioavailability and serum half-life of the immunomodulator. This effect is readily apparent when the pharmacokinetics and biodistribution of an immunomodulator are studied. For example, following i.v. injection, rH IL-2 has a short serum half-life requiring high doses to augment NK cell activity. In contrast, the i.p.

injection of IL-2 results in a prolonged serum half-life with the concomitant augmentation of peritoneal NK cells at much lower doses of rH IL-2 than those required to augment splenic NK cells following i.v. administration (39, 40). In addition, greater therapeutic activity is observed following i.p. as opposed to i.v. administration (41).

Applying this concept to recombinant IL-2, the continuous infusion of rH IL-2 by osmotic pumps, in animal models, has therapeutic activity. Indeed, the therapeutic activity is observed at suprisingly low levels of rH IL-2 on an hourly basis. The lymphoid hyperplasia, and therapeutic activity that are observed with continuous administration of high doses of rH IL-2 suggest that *ex vivo* LAK cell induction, cultivation, and infusion technology may be unnecessary (41). Although clinical therapeutic activity and toxicology have been noted when rH IL-2 was administered with an aggressive dosage and schedule, most notably in combination with LAK cells or tumor infiltrating lymphocytes (26, 42-46), the preclinical observation of therapeutic activity at low doses by continuous infusion suggest that rH IL-2 might be therapeutically effective using a less toxic protocol. The hypothesis that moderate doses of rH IL-2 have low toxicity and therapeutic activity when administered by continuous infusion has been studied to some extent in the clinic. Therapeutic activity has been observed when rH IL-2 was administered by continuous infusion in combination with LAK cells (47). In a preliminary report, responses were found in 9 of 16 evaluable melanoma patients (47), and it has been suggested that maintenance administration of rH IL-2 by continuous infusion for 1 week each month following two cycles of IL-2 and LAK therapy may have a more prolonged response as compared to 2 or 3 cycles of rH IL-2 and LAK. In other studies, continuous infusion of rH IL-2 initially produced leukopenia, but later resulted in lymphoid hyperplasia in the peripheral blood and increased LAK cell activity as determined *ex vivo* (48).

Thus, rH IL-2 has been administered clinically, using a wide variety of doses and routes. However, most of these studies have involved a short course of i.v. administration. Both of these parameters may not result in optimal therapeutic or immunomodulatory activity since immunotherapy, at least in rodents, requires chronic administration for significant responses. Thus, the clinical observation of immunomodulation and therapeutic responses following the infusion of 3×10^6 units of rH IL-2 per day/M^2 qd (48) or 3×10^7 units of rH IL-2 for 24 hr/M^2 once a week (49, 50) is very encouraging. These protocols are tolerated better and use lower cummulative doses than those used to date in the aggressive i.v. therapeutic protocols of approximately 100,000 U/kg administered tid for 4 days (26). The latter protocol results in a daily dose of approximately 20 million units per day per patient. While the impaired pharmacokinetics obtained by continuous infusion of rH IL-2 result in increased toxicity on a equidose basis compared to an i.v. bolus, the impression has been that greater immunomodulation with less toxicity is observed in patients receiving a continuous infusion of rH IL-2 at 1 to 3×10^6 units per day qd or a weekly 24 hr infusion of 30×10^7 units as compared to the i.v. administration of 7×10^6 units in 96 hr. Recently, Dr. Sanjo et al. at the Japanese National Cancer Institute compared therapeutic protocols of i.v. bolus, daily continuous infusion, and 5 days continuous infusion with 2 days rest. They observed a significant augmentation of LAK and NK cell activity and lymphoid hyperplasia in the blood with low toxicity, at approximately 1×10^6 units/M^2 per day in cancer patients on a schedule of daily administration, or on a 5 day/2 day rest schedule of administration. In every case, rH IL-2 by continuous infusion resulted in greater immunomodulation and leukocyte counts compared to i.v. push. It remains to be determined in the clinic whether or not *ex vivo* augmentation of LAK cells will be required in combination with rH IL-2 for the greatest therapeutic activity.

In contrast to rH IL-2, which has greater therapeutic activity following i.p. as compared to i.v. administration, recombinant TNF has little therapeutic activity for systemic disease following i.p. administration and modest therapeutic activity following i.v. administration. In the case of rH TNF, studies of serum pharmacokinetics suggest that the increased therapeutic activity following i.v. as opposed to i.p. administration is associated with the high transient levels of TNF that are achieved following i.v. administration. In studies of rM IFN-γ, i.v. administration has greater therapeutic activity on an equidose basis compared to i.p. administration. However, the prolonged serum levels achieved following i.m. administration clinically suggest that this route of administration may be advantageous (51). In preclinical studies, rM IFN-γ, which demonstrates a bell-shaped immunomodulatory and therapeutic response curve, must be administered at specific doses and schedules to produce optimal therapeutic activity. Optimal activity was observed after i.v. administration of 1.5×10^6 - 2.5×10^6 U/kg of rM IFN-γ 3 times per week (51). In other studies, this dose and schedule resulted in optimal pulmonary macrophage tumoricidal and tumoristatic activity and induction of cytotoxic T lymphocyte activity, which correlated significantly with survival of the animals receiving various doses of rM IFN-γ (52). In clinical studies, the dose of approximately 100 μg/M^2 of rH IFN-γ produced maximal macrophage activity. Unfortunately, to date, most clinical trials with rH IFN-γ have been predicated on a "more is better" philosophy, such that they have used doses relatively close to the maximum tolerated dose (4 mg/M^2). In contrast to the significant and reproducible therapeutic activity of rM IFN-γ, treatment protocols with TNF have not shown marked therapeutic activity for tumors at other than intradermal injection sites. Thus, when used as the sole therapeutic agent, rH or rM TNF demonstrates only modest therapeutic activity, and only after i.v. administration, for the treatment of pulmonary metastases or subcutaneous tumors (51). This is in contrast to the studies examining the treatment of Meth A tumors at intradermal sites (53-60), since most tumors are sensitive to the i.v. or intratumoral administration of TNF when implanted at this site, regardless of their response in vitro. In addition, the administration of TNF has also been effective for peritoneal ascites. Nonetheless, this agent is likely to be of limited value for the majority of human cancer patients because most tumors are not readily accessible to intralesional (intratumoral or i.p.) injection and/or are not growing at an intradermal site. Because the therapeutic activity of TNF appears to be route dependent, requiring relatively high serum levels of TNF, it is possible that continuous infusion might be used to advantage in clinical trials. However, this may also increase its toxicity. In contrast to the minimal therapeutic activity of TNF alone, the administration of TNF in combination with rM IFN-γ significantly increases therapeutic activity (51). This was initially noted in vitro for increased cytostasis with tumor cells that were sensitive to TNF and even with cell lines that were refractory to TNF as a single agent. The mechanism of this complementary effect is not yet understood, although recombinant IFN-γ and recombinant TNF are additive for the augmentation of macrophage/mediated tumoricidal activity in vitro. The latter mechanism may involve an upregulation of receptors for TNF, a phenomenon which has been suggested to occur after IFN-γ exposure (61). Unfortunately, the combination of TNF and IFN-γ in both rodents and in the clinic results in increased toxicity (62). This observation points out another important concept in immunotherapy, namely that treatment with cytokines is often associated with life threatening toxicity with very few exceptions. Thus, the therapeutic index is an important component during the development of biological response modifiers. A growing body of evidence suggests that treating patients with rH IFN-γ at a maximum tolerated dose may not always be optimal because therapeutic activity may be increased at lower immunomodulatory doses. One of the more interesting observations with TNF and IFN-γ is that it is possible to decrease the toxicity of these two agents by administering cyclooxygenase/lipoxygenase inhibitors, which appear to have some role in the toxicity of these agents (62).

The colony stimulating factors (CSFs), which are cytokines with "growth factor" like activity for bone marrow stem cells, also have tremendous potential for decreasing the toxicity and increasing the therapeutic index of the chemotherapeutic agents and irradiation used in the treatment of neoplasia. The importance of dosage intensity, both preclinically and clinically in retrospective studies, has supported the idea that the more aggressive use of chemotherapeutic agents also increases their therapeutic activity (63, 64). Increased doses of chemotherapeutic agents can be achieved, in part, by preventing or decreasing the neutropenia associated with their aggressive use. This goal can be achieved by providing autologous bone marrow transplants or administering agents which facilitate the outgrowth of stem cells and thereby reduce or prevent neutropenia. Furthermore, agents which reduce the DNA damage by radio- or chemotherapy such as scavaging agents can be used in a protective-prophylactic manner. Autologous bone marrow transplants and accelerated reconstitution following myelosuppression are both achievable goals, which are facilitated with the use of the CSFs. The successful utilization of the CSFs is dependent on the schedule and timing of their administration. For example, if administered too soon after injection of a chemotherapeutic agent, that is, before the agent(s) has been completely metabolized, increased toxicity or a lack of activity may be observed. In contrast, if the agent is administered after complete metabolism of the agent, either alone or in conjunction with autologous bone marrow transplants, accelerated reconstitution and, in certain cases, a complete lack of neutropenia may be observed, at least in preclinical models. Thus, the CSFs appear to have considerable potential for increasing the therapeutic response due to more aggressive use of chemotherapy and an increased therapeutic index.

The development of successful immunotherapeutic protocols requires new strategies and the development of clinically predictive tumor models and protocols (38). Beyond the determination of a maximum tolerated dose (MTD), it is also important to determine the optimal immunomodulatory dose (OID) in initial clinical trials. The optimal therapeutic dose and protocol, yet to be determined for any BRM in the clinic, likely depends not only on the augmentation of effector cell activity but also on the number, function, and tumor infiltration of effector cells. Furthermore, the determination of an optimal therapeutic protocol (OTP) requires information on therapeutic schedule and duration of treatment. Studies of scheduling parameters have revealed that many BRMs induce host toxicity when delivered frequently at high doses. However, if the BRM is administered at either a lower level or less frequently, there may be a decrease in toxicity and a concomitant increase in therapeutic activity, (see discussion of clinical trials with rH IL-2). Thus, in agreement with studies of *in vitro* and *in vivo* immunomodulation, it has not been unusual to observe a bell-shaped dose-response curve for therapeutic activity. This curve may be due to decreased immunomodulatory potential at higher doses, or it may be associated with toxicity to the host at the higher levels of the agent. In combination with the tumor burden, such toxicity may result in morbidity or mortality as compared to an OTP.

In summary, the OTP appears to be dependent on a complex combination of parameters including dose, frequency of administration, duration of administration, and route of administration. Normally, systemic disease is treated by i.v. injection; however, consideration should also be given to s.c. or i.m. administration, which may be beneficial because of a prolonged half-life of the agent in addition to the more convenient injection site. For agents which can be absorbed through the gastrointestinal tract, oral administration also has clinical practicality since it may allow treatment on an outpatient basis, as does s.c. administration.

Preclinical modeling can also provide information about the effector cell responsible for the therapeutic activity of an immunomodulator (52, 65). Such approaches may identify aspects of effector cell function that could be clinically monitored to guide the development of an OID for subsequent Phase II and III clinical trials. The OID is the minimal dose of a biological response modifier which results in the significant augmentation of effector cell activity associated with the therapeutic response to a BRM. This is a dose of the BRM which can be repeatedly administered to maintain a constant level or repeated increases in an effector cell activity associated with therapeutic activity. This dose may be lower than the MTD, which is defined here as the largest dose of a BRM that can be tolerated either by the clinician or the patient during clinical trials, and as such, may vary not only from patient-to-patient but also among clinical centers depending upon the physician's perception of acceptable toxicity.

Thus, preclinical and clinical studies for the development of a new immunotherapeutic agent have several major goals including:
1) Examine the pharmacokinetics and toxicology.
2) Determine whether a BRM has therapeutic potential.
3) Determine the OTP.
4) Learn which effector cells are responsible for a BRM's therapeutic activity.
5) Determine whether monitoring of immune effector cell function in the blood during Phase Ib trials can provide guidance for subsequent Phase II and III trials.

Studies of the double-stranded polyribonucleotides and recombinant cytokines (IL-2, IFN-γ, and TNF) have helped to focus on these issues. For example, all of these agents (but not all BRMs) have bell-shaped dose-response curves (41, 51). High, toxic doses appear to have less therapeutic activity than intermediate doses of these agents. Studies with poly(I,C)-LC have revealed a direct and significant correlation between the extent of effector cell augmentation within tumors (peritoneal ascites) and host survival (65). Similarly, rM IFN-γ also has significant correlation between the extent of effector cell augmentation within the tumor-bearing organ and host survival (52). In contrast, rH IL-2 appears to have a biphasic dose-response curve for its therapeutic activity, with high dose activity associated with the augmentation of NK or LAK cells, and low dose activity associated with T cell augmentation (27, 39, 40). Recent studies with rH TNF suggest that therapeutic activity results not only from its antiproliferative effect, but also with its immunomodulatory properties (51). Because immunotherapy is only active against limited neoplastic disease, chemoimmunotherapy has in theory, a great therapeutic potential. The mechanism of therapeutic activity may include a reduction in tumor burden by chemotherapy, increased dosage intensity due to myeloid reconstitution by immune modulation and/or direct therapeutic activity by effector cell augmentation which is active on the reduced tumor burden.

An important component of chemoimmunotherapeutic scheduling is the sequence and timing of administration of the various components of chemoimmunotherapy. For example, if one of the goals of a particular approach to chemoimmunotherapy is to accelerate recovery from the myelosuppressive activities of aggressive chemotherapy, it might be counter-productive to administer a myelostimulatory agent, which would increase stem cell cycling, prior to administering the chemotherapeutic agent. It would appear more logical to place the stem cells in the noncycling portion (S phase) of the cell cycle where they are relatively refractory to the myelotoxicity of chemotherapy. Following a short wash-out period, the remaining stem cells should be stimulated to cycle and reconstitute the host effector cells. Finally, one must consider the potential importance of treating systemic disease by administration into multiple compartments. This concept has been developed in studies with mice bearing hepatic and pulmonary metastases from primary renal tumors. In these studies, BRM administration by both the i.p. and i.v. routes resulted in superior therapeutic activity, compared to either route alone (66, 67). Thus, the administration of chemo and biotherapeutic agents by both the i.p. and i.v. routes of administration resulted in highly effective therapy compared to a single route of administration.

Several issues must be considered in the development of BRMs for treating patients with neoplastic disease. The first is the rational design of Phase I, II, and III clinical trials and the use of preclinical findings to guide the development of a BRM in clinical trials. Clinical trials with BRMs, at least until recently, have been designed similarly to those performed with chemotherapeutic agents. Phase I trials of both types of agents have been performed in patients with advanced cancers, for whom other therapies have been ineffective. The major purpose of Phase I trials has been to evaluate toxicity and determine the MTD of a compound. The generally superficial attempts to determine immunological responses in the context of these dose escalation studies have not yielded sufficient information about the immunomodulatory properties of BRMs. In addition to the dose-determining nature of such Phase I trials with BRMs, they should also determine the optimal schedule and route of administration for therapeutic efficacy based on immunomodulation.

One issue in the development of a clinical trial concerns the evaluation of the effect of BRMs on the immune systems of Phase I patients with large tumor burdens, who have been previously treated with aggressive radiotherapy or chemotherapy. Unfortunately, the findings of such studies might not be comparable to those which would be obtained in patients with minimal tumor burdens. Results in animal tumor models predict that BRMs will be most effective in hosts with minimal tumor burdens and minimally effective in individuals with widespread disease. Furthermore, it is probable that the OID will differ in patients with bulky tumors (Phase 1 patients) compared to those with minimal residual disease (Phase III patients and normal clinical utilization). Thus, Phase I and II clinical trials may not accurately assess the immunomodulatory and clinical effects of potentially valuable BRMs as a result of inappropriate study design and/or inclusion of patients with bulky tumor burdens. We must place a greater emphasis in Phase I trials on determining the optimal route, dose, schedule, and duration of a BRM administration that induces a sustained level of effector cell activities. It remains to be seen whether such an approach will facilitate the assessment of therapeutic activity, but it does provide a working hypothesis that the OTP may be associated with the OID rather than the MTD.

A discussion of preclinical and clinical parallels is very subjective. First, because few clinical trials have been undertaken with biologics and BRMs outside of a Phase I toxicity setting. Second, the limited Phase II or III trials to address the therapeutic activity of a BRM have largely been undertaken with agents such as BCG and C. parvum, which are potentially nonreproducible (assay results as well as reagents).

Third, until recently virtually all clinical trials have been undertaken without appropriate immune monitoring to allow a comparison of immunomodulatory activity between preclinical and clinical studies. Fourth, only a limited number of clinical trials have been undertaken with sufficiently pure and refined materials, such as the synthetics and recombinant molecules, to allow a reasonable comparison of preclinical and clinical immunomodulatory activities. In theory, therapeutic principles should hold true for crude BRMs, as well as recombinant or synthetic BRMs, as along as the preparations are adequately standardized and subject to stringent quality control, although in practice this has not occurred. In addition, particularly for culture supernatants, there is the potential of multiple activities within the preparations further obscuring any results. We suggest that stress should be placed on the development of synthetic and recombinant agents as effective immunomodulatory agents. The current emphasis on recombinant cytokines, due to their novelty, may be inappropriate at least as compared to synthetic agents. Synthetic BRM's are reproducible, induce multiple cytokines which may be therapeutically fortuitous as suggested by the current development of "cocktails" of cytokines for clinical and preclinical studies (TNF and IFN-γ, IFN-γ and IFN-α or IFN-β, and TNF and IL-2 etc.) and may be active, via oral administration. In addition, the formulation and cost of development is often superior for such agents as compared to cytokines. This is not to suggest that we should ignore cytokines but that synthetic agents merit serious study and their development should not be ignored.

SUMMARY

It should be apparent in the foregoing discussion that a great deal of information is lost or missed at present in clinical trials and preclinical studies. There is a tremendous need for the coordination of preclinical and clinical studies and among the numerous clinical trials that are ongoing or are about to be initiated. A cohesive and unified approach is needed to conserve our limited resources and, perhaps more importantly, to facilitate the more rapid development of efficacious clinical protocols.

ACKNOWLEDGEMENTS

By acceptance of this article the publisher or recipient acknowledges the right of the U. S. Government to retain a nonexclusive, royalty-free license in and to any copyright covering the acticle.

This research was sponsored by the Biological Resources Branch, Biological Response Modifiers Program, Division of Cancer Treatment, National Cancer Institute, DHHS, under contract No. N01-23910 with Program Resources, Inc. The contents of this publication do not necessarily reflect the views or policies of the Department of Health and Human Services, nor does mention of trade names, commercial products, or organizations imply endorsement by the U. S. Government.

REFERENCES

1. E. Mihich, and A. Fefer (Eds.), Biological response modifiers subcommittee report. In: Natl. Cancer Inst. Monogr. 63, pp. 1-252 (1983).
2. E. Mihich, Biological response modifiers: Their potential and limitations in cancer therapeutics. Cancer Invest. 3, 71-83 (1985).

3. R. V. Smalley, and R. K. Oldham, Biological response modifiers: Preclinical evaluation and clinical activity. CRC Crit. Rev. Oncol. Hematol. 1, 259-294 (1984).
4. R. K. Oldham, Biologicals and biological response modifiers: Design of clinical trials. J. Biol. Resp. Modif. 4, 117-128 (1985).
5. R. B. Herberman, Design of clinical trials with biological response modifiers. Cancer Treat. Rep. 69, 1161-1164 (1985).
6. C. M. Pinsky, Application of phase I trial results in the design of phase II and II biological response modifier trials. Cancer Treat. Rep. 69, 1171-1182 (1985).
7. G. Mathe, J. L. Amiel, L. Schwarzenberg, M. Schneider, A. Cattan, J. R. Schlumberger, M. Hayat, and F. De Vassal, Active immunotherapy for acute lymphoblastic leukemia. Lancet 1, 697-699 (1969).
8. R. Heyn, M. Nesbit, P. Joo, M. Karon, W. Borges, D. Hammond, and N. Breslow, Effect of BCG in the duration of the first remission in children with acute lymphoid leukemia. In Neoplasia Immunity: BCG Vaccination 1974 (R. G. Crispen, ed.) pp. 15-23. Schori Press, Evanston.
9. R. L. Powles, D. Crowther, C. J. T. Bateman, M. E. J. Beard, T. J. McElwain, J. Russel, T. A. Lister, J. M. A. Whitehouse, P. F. M. Wrigley, M. Pike, P. Alexander, and G. Fairley, Immunotherapy for acute myelogenous leukaemia. Br. J. Cancer 28, 365-376 (1973).
10. J. U. Gutterman, E. M. Hersh, V. Rodriguez, K. B. McCredie, G. Mavligit, R. Reed, M. A. Burgess, T. Smith, E. Gehan, G. P. Bodey, and E. J. Freireich, Chemoimmunotherapy of adult acute leukaemia: Prolongation of remission in myeloblastic leukaemia with BCG. Lancet 2, 1405-1409 (1974).
11. W. R. Vogler, and Y. K. Chang, Prolonging remission in myeloblastic leukaemia by Tice-Strain Bacillus Calmette-Guerin. Lancet 2, 128-131 (1974).
12. K. A. Foon, R. V. Smalley, C. W. Riggs, and R. P. Gale, The role of immunotherapy in acute myelogenous leukemia. Arch. Int. Med. 143, 1726-1731 (1983).
13. A. Morales, and A. Ersil, Adjuvant BCG immunotherapy in the prophylaxis and treatment of noninvasive bladder cancer. In Immunotherapy of Human Cancer 1982 (W. D. Terry, and S. A. Rosenberg, Eds.) pp. 301-307. Excerpta Medica, New York.
14. C. M. Pinsky, F. J. Camacho, D. Kerr, D. W. Braun Jr., W. F. Whitmore Jr., and H. Oettgen, Treatment of superficial bladder cancer with intravesical BCG. In Immunotherapy of Human Cancer 1982 (W. D. Terry, and S. A. Rosenberg, Eds.) pp. 309-313. Excerpta Medica, New York.
15. D. L. Lamm, D. E. Thor, S. C. Harris, V. D. Stogdill, and H. M. Radwin, Intravesical and percutaneous BCG immunotherapy of recurrent superficial bladder cancer. In Immunotherapy of Human Cancer 1982 (W. D. Terry, and S. A. Rosenberg, Eds.) pp. 315-322. Excerpta Medica, New York.
16. C. Pinsky, F. Camacko, D. Kerr, H, Herr, H. Geller, W. Whitmore, and H. Oettgen, Treatment of superficial bladder cancer with intravesical BCG. Proc. Amer. Soc. Clin. Oncol. 2, 57 (1983).
17. M. F. McKneally, C. Maver, H. W. Kausel, Regional immunotherapy of lung cancer with intrapleural BCG. Lancet 1, 377-379 (1976).
18. R. C. Bast, J. S. Berek, R. Obrist, C. T. Griffiths, R. S. Berkowitz, N. F. Hacker, L. Parker, L. D. Lagasse, and R. C. Knapp, Intraperitoneal immunotherapy of human ovarian carcinoma with corynebacterium parvum. Cancer Res. 43, 1395-1401 (1983).
19. A. Uchida, M. Micksche, and T. Hoshino, Intrapleural administration of OK-432 in cancer patients: Augmentation of autologous tumor killer actiivty of tumor-associated large granular lymphocytes. Cancer Immunol. Immunother. 18, 5-12 (1984).

20. K. A. Foon, S. A. Sherwin, P. G. Abrams, D. L. Longo, M. F. Fer, H. C. Stevenson, J. J. Ochs, G. C. Bottino, C. S. Schoenberger, J. Zeffren, E. S. Jaffe, and R. K. Oldham, Treatment of advanced non-Hodgkin's lymphoma with recombinant leukocyte A interferon. N. Engl. J. Med. 311, 1148-1152 (1984).
21. J. A. Neidhart, M. M. Gagen, D. Young, R. Tuttle, T. J. Melink, A. Ziccarrelli, and D. Kisner, Interferon-α therapy of renal cancer. Cancer Res. 44, 4140-4143 (1984).
22. P. A. Bunn Jr., K. A. Foon, D. C. Ihde, D. L. Longo, J. Eddy, C. F. Winkler, W. R. Veach, J. Zeffren, S. Sherwin, and R. Oldham, Recombinant leukocyte A interferon: An active agent in advanced cutaneous T-cell lymphomas. Ann. Intern. Med. 101, 484-487 (1984).
23. S. A. Sherwin, K. A. Foon, P. G. Abrams, M. R. Heyman, J. J. Ochs, T. Watson, A. Maluish, and R. K. Oldham, A preliminary phase I trial of partially purified interferon-γ in patients with cancer. J. Biol Resp. Modif. 3, 599-607 (1984).
24. J. M. Kirkwood, and M. S. Ernstoff, Interferons in the treatment of human cancer. J. Clin. Oncol. 2, 336-352 (1984).
25. E. C. Borden, Interferon and cancer: How the promise is being kept. In Interferons, vol. 5. 1984 (I. Gresser, ed.) pp. 43-83. Academic Press, Orlando, Florida.
26. S. A. Rosenberg, M. T. Lotze, L. M. Muul, S. Leitman, A. E. Chang, S. E. Ettinghausen, Y. L. Matory, J. M. Skibber, E. Shiloni, J. T. Vetto, C. A. Seipp, C. Simpson, and C. M. Reichert, Observations on the systemic administration of autologous lymphokine-activated killer cells and recombinant interleukin-2 to patients with metastatic cancer. N. Eng. J. Med. 313, 1485-1492 (1985).
27. J. E. Talmadge, I. J. Fidler, and R. K. Oldham, Screening For Biological Response Modifiers: Methods and Rationale 1985 pp. 1-193. Martinus Nihjoff Publishing, Boston.
28. J. E. Talmadge, and R. B. Herberman, The preclinical screening laboratory: Evaluation of immunomodulatory and therapeutic properties of biological response modifiers. Cancer Treat. Rep. 70, 171-182 (1986).
29. V. T. DeVita, Jr., Principles of Chemotherapy. In Cancer - Principles and Practice of Oncology 1985 (V. T. DeVita, Jr., S. Hellman, and S. A. Rosenberg, eds.) pp. 257-285 Lippincott, Philadelphia.
30. J. E. Talmadge, A. E. Maluish, M. Collins, M. Schneider, R. B. Herberman, R. K. Oldham, and R. H. Wiltrout, Immunomodulation and antitumor effects of MVE-2 in mice. J. Biol. Resp. Mod. 3, 634-652 (1984).
31. C. M. Hawrylowicz, R. C. Rees, B. W. Hancock, and C. W. Potter, Depressed spontaneous natural killing and interferon augmentation in patients with malignant lymphoma. Eur. J. Cancer Clin. Oncol. 18, 1081-1088 (1982).
32. A. E. Maluish, J. R. Ortaldo, J. C. Conlon, S. A. Sherwin, R. Leavitt, D. M. Strong, P. Wiernik, R. K. Oldham, and R. B. Herberman, Depression of natural killer cytotoxicity after in vivo administration of recombinant leukocyte interferon. J. Immunol. 131, 503-507 (1983).
33. M. J. Brunda, D. Rosenbaum, and L. Stern, Inhibition of experimentally-induced murine metastases by recombinant alpha interferon: correlation between the modulatory effect of interferon treatment on natural killer cell activity and inhibition of metastases. Int. J. Cancer 34, 421-426 (1984).
34. J. E. Talmadge, J. Adams, H. Phillips, M. Collins, B. Lenz, M. Schneider, E. Schlick, R. Ruffman, R. H. Wiltrout, and M. A. Chirigos, Immunomodulatory effects in mice of polyinosinic-polycytidylic acid complexed with poly-L-lysine and carboxymethylcellulose. Cancer Res. 45, 1058-1065 (1985).

35. R. H. Wiltrout, B. J. Mathieson, J. E. Talmadge, C. W. Reynolds, S. R. Zhang, R. B. Herberman and J. R. Ortaldo, Augmentation of organ-associated NK activation by biological response modifiers: isolation and characterization of large granular lymphocytes from the liver. J. Exp. Med. 160, 1431-1449 (1984).
36. R. H. Wiltrout, R. B. Herberman, S. Zhang, M. A. Chirigos, J. R. Ortaldo, K. M. Green Jr, and J. E. Talmadge, Role of organ-associated NK cells in decreased formation of experimental metastases in lung and liver. J. Immunol. 134, 4267-4275 (1985).
37. J. E. Talmadge, M. Schneider, M. Collins, H. Phillips, R. B. Herberman, and R. H. Wiltrout, Augmentation of NK cell activity in tissue specific sites by liposomes incorporating MTP-PE. J. Immunol. 135, 1477-1483 (1985).
38. J. E. Talmadge, B. F. Lenz, M. S. Collins, K. A. Uithoven, M. A. Schneider, J. S. Adams, J. W. Pearson, W. J. Agee, R. E. Fox, and R. K. Oldham, Tumor models to investigate the therapeutic efficiency of immunomodulators. Behring Inst. Mitt. 74, 219-229 (1984).
39. J. E. Talmadge, Immunoregulation and immunostimulation of murine lymphocytes by recombinant human interleukin-2. J. Biol. Resp. Modif. 4, 18-34 (1985).
40. J. E. Talmadge, S. Dennis-Tait, M. A. Schneider, A. K. Meeker, J. S. Adams, In vitro and in vivo stimulation of murine lymphocytes by human recombinant interleukin 2 (rIL-2). In Immunity to Cancer 1985 (A. E. Reif, and M. S. Mitchell, Eds.) pp. 531-537. Academic Press, New York.
41. J. E. Talmadge, H. Phillips, J. Schindler, H. R. Tribble, and R. Pennington, A systematic preclinical study on the therapeutic properties of recombinant human interleukin 2 for the treatment of metastatic disease. Cancer Res. in press (1987).
42. P. Kern, J. Toy, and M. Dietrich, Preliminary clinical observations with recombinant interleukin-2 in patients with AIDS or LAS. Blut. 50, 1-6 (1985).
43. M. T. Lotze, R. J. Robb, S. W. Sharrow, L. W. Frana, and S. A. Rosenberg, Systemic administration of interleukin-2 in humans. J. Biol. Resp. Modif. 3, 475-482 (1984).
44. H. C. Lane, J. P. Siegel, A. H. Rook, H. Masur, E. P. Gelmann, G. V. Quinnan, and A. S. Fauci, Use of interleukin-2 in patients with acquired immunodeficiency syndrome. J. Biol. Resp. Modif. 3, 512-516 (1984).
45. G. Pizza, G. Severini, D. Menniti, C. De Vinci, and F. Corrado, Tumour regression after intralesional injection of interleukin 2 (IL-2) in bladder cancer. Preliminary report. Int. J. Cancer 34, 359-567 (1984).
46. J. P. Siegel, H. C. Lane, N. I. Stock, G. V. Quinnan Jr., and A. S. Fauci, Pharmacokinetics of lymphocyte-derived and recombinant DNA-derived interleukin-2 after intravenous administration to patients with the acquired immunodeficiency syndrome. J. Biol. Resp. Modif. 4, 596-601 (1985).
47. W. H. West, K. W. Tauer, J. R. Yarnelli, G. D. Marshall, D. W. Orr, G. B. Thurman, and R. K. Oldham, Constant-infusion recombinant interleukin-2 in adoptive immunotherapy of advanced cancer. J. Biol. Resp. Modif. 6, 221 (1987).
48. J. Hank, N. Rosenthal, P. Kohler, B. Storer, and P. Sondel, Peripheral blood lymphocytes obtained following in vivo IL-2 therapy express lymphokine activated killer activity. Fed. Proc. 46, 1509 (1987).
49. S. P. Creekmore, J. E. Harris, T. M. Ellis, D. P. Braun, J. McMannis, I. I. Cohen, N. Bhoopalam, P. F. Jassak, M. A. Cahill, C. L. Canzoneri, and R. I. Fisher, Phase I/II trial of recombinant interleukin-2 by 24-hr continuous infusion - an Illinois Cancer Council trial. Proc. Amer. Soc. Clin. Oncol. 6, 244 (1987).

50. J. T. Kurnick, R. L. Kradin, L. A. Boyle, and A. Burdeshaw, Assessment of activation of T lymphocytes infiltrating human lung cancers. Fed. Proc. 46, 1341 (1987).
51. J. E. Talmadge, H. R. Tribble, R. W. Pennington, H. Phillips, and R. H. Wiltrout, Immunomodulatory and immunotherapeutic properties of recombinant γ-interferon and recombinant tumor necrosis factor in mice. Cancer Res. 47, 2563-2570 (1987).
52. P. L. Black, H. Phillips, H. R. Tribble, R. Pennington, M. Schneider, and J. E. Talmadge, Correlation of immunomodulatory and therapeutic activities of interferon and interferon inducers in metastatic disease. J. Cell. Biochem. in press (1987).
53. E. A. Carswell, L. J. Old, R. L. Kassel, S. Green, N. Fiore, and B. Williamson, An endotoxin-induced serum factor that causes necrosis of tumors. Proc. Natl. Acad. Sci. U.S.A. 72, 3666-3670 (1975).
54. K. Haranaka, and N. Satomi, Cytotoxic activity of tumor necrosis factor (TNF) on human cancer cells in vitro. Jpn. J. Exp. Med. 51, 191-194 (1981).
55. K. Haranaka, N. Satomi, and A. Sakurai, Antitumor activity of murine tumor necrosis factor (TNF) against transplanted murine tumors and heterotransplanted human tumors in nude mice. Int. J. Cancer 34, 263-267 (1984).
56. B. D. Williamson, E. A. Carswell, B. Y. Rubin, J. S. Predergast, and L. J. Old, Human tumor necrosis factor produced by human B-cell lines: Synergistic cytotoxic interaction with human interferon. Proc. Natl. Acad. Sci. U.S.A. 80, 5397-5401 (1983).
57. Y. Sohmura, K. Nakata, H. Yoshida, S. Kashimoto, Y. Matsui, and H. Furuichi, Recombinant human tumor necrosis factor. II. Antitumor effect on murine and human tumors transplanted in mice. Int. J. Immunopharmacol. 8, 357-368 (1986).
58. N. Watanabe, Y. Niitsu, H. Sone, H. Neda, I. Urushizaki, A. Yamamoto, M. Nagamura, and Y. Sugawara, Therapeutic effect of endogenous tumor necrosis factor on ascites Meth A sarcoma. J. Immunopharmacol. 8, 271-283 (1986).
59. G. G. MacPherson, and R. J. North, Endotoxin-mediated necrosis and regression of established tumours in the mouse. Cancer Immunol. Immunother. 21, 209-216 (1986).
60. F. R. Balkwill, A. Lee, G. Aldam, E. Moodie, J. A. Thomas, J. Tavernier, and W. Fiers, Human tumor xenografts treated with recombinant human tumor necrosis factor alone or in combination with interferons. Cancer Res. 46, 3990-3993 (1986).
61. B. B. Aggarwal, T. E. Eessalu, and P. E. Hass, Characterization of receptors for human tumour necrosis factor and their regulation by γ-interferon. Nature 318, 665-667 (1985).
62. J. E. Talmadge, O. Bowersox, H. Tribble, S. H. Lee, M. Shepard, and D. Liggitt, Toxicity of tumor necrosis factor with interferon-γ and can be reduced with cyclooxygenase inhibitors. Amer. J. Path. in press (1987).
63. W. Hryniuk, and M. N. Levine, Analysis of dose intensity for adjuvant chemotherapy trials in stage II breast cancer. J. Clin. Oncol. 4, 1162-1167 (1986).
64. D. P. Griswold Jr, M. W. Trader, E. Frei III, W. P. Peters, M. K. Wolpert, and W. R. Laster Jr., Response of drug-sensitive and -resistant L1210 leukemias to high-dose chemotherapy. Cancer Res. 47, 2323-2328 (1987).
65. D. Hartmann, J. S. Adams, A. K. Meeker, M. A. Schneider, B. F. Lenz, and J. E. Talmadge, Dissociation of therapeutic and toxic effects of polyinosinic-polycytidylic acid admixed with poly-L-lysine and solubilized with carboxymethyl cellulose in tumor-bearing mice. Cancer Res. 46, 1331-1338 (1985).
66. R. Lafreniere, and S. A. Rosenberg, Adoptive immunotherapy of murine hepatic metastases with lymphokine activated killer (LAK) cells and

recombinant interleukin 2 (rIL 2) can mediate the regression of both immunogenic and nonimmunogenic sarcomas and an adenocarcinoma. J. Immunol. 135, 4273-4280 (1985).
67. R. Salup, T. C. Bock, and R. H. Wiltrout, Successful treatment of advanced murine renal cell carcinoma by biocompartmental adoptive chemoimmunotherapy. J. Immunol. 138, 641-647 (1987).

CIRCUMVENTION OF BIOLOGIC DIVERSITY OF CANCER METASTASIS

Isaiah J. Fidler

Department of Cell Biology-173
The University of Texas
M. D. Anderson Hospital and Tumor Institute
Houston, Texas 77030

INTRODUCTION

The major challenge facing the oncologist is how to treat cancer that has disseminated to and is growing in multiple organs throughout the body. Despite major advances in general patient care, surgical techniques, and adjuvant therapies, most deaths from cancer are caused by the growth of metastases that are resistant to chemotherapy or radiotherapy. A major factor which prevents treatment of metastases is the fact that cancer cells populating both primary and secondary neoplasms are biologically heterogeneous (1,2). By the time of diagnosis, and certainly in clinically advanced lesions, malignant neoplasms contain multiple cell populations exhibiting a wide range of biological characteristics such as cell surface structures, growth rate, sensitivity to various cytotoxic drugs, and the ability to further invade and metastasize. The implication of the heterogeneous responses of tumor cells to conventional anticancer drugs is that the successful therapy of disseminated metastases must circumvent the problems of biologic heterogeneity and be a treatment modality to which resistance is unlikely to develop.

During the last decade, we have accumulated a large body of data demonstrating that a host defense cell - the macrophage - can become activated to lyse tumor cells in a very specific manner and can meet these two demanding criteria (3). In the following pages, I shall review some of these data and concentrate on the methods utilized to achieve systemic activation of macrophages which lead to the destruction of established cancer metastases.

MACROPHAGES IN HOMEOSTASIS AND IN CANCER

Cells of the macrophage-histiocyte series are important components of the host system responsible for homeostasis. A primary function of macrophages in the body is the phagocytosis and disposal of effete cells, such as damaged red blood cells, cellular debris, and serum proteins. In addition to phagocytosis, macrophages are involved in the controlled metabolism of lipids and iron, in host response to injury, and in the defense against microbial infections and parasitic infestations.

Macrophages are also an important component of the afferent and efferent arms of the immune system and are critical to host defense against neoplasms (3).

Normal macrophages do not lyse tumor cells. To become activated to the tumoricidal state, macrophages need to undergo a series of events collectively termed "activation." In vivo, there are two major pathways to achieve macrophage activation: (a) Macrophage interaction with microorganisms or their products, such as endotoxins, bacterial cell wall skeleton, and defined components of the bacterial cell wall, such as muramyl dipeptide (MDP) (4-6); and (b) macrophage interaction with lymphokines - soluble mediators released by antigen- or mitogen-sensitized lymphocytes. The soluble lymphokine that induces macrophage activation is referred to as macrophage-activating factor (MAF) (7) and one such lymphokine with MAF activity is gamma-interferon (IFN-γ) (8-10). To activate macrophages, the lymphokine must first bind to macrophage surface receptors and then be internalized and bound to intracellular receptors (10).

Regardless of the activation signal, tumoricidal mammalian macrophages can distinguish between neoplastic and normal cells, i.e., tumor cells are lysed, whereas, normal cells are left unharmed even under co-cultivation conditions (3). The mechanism for this is not known but appears to be nonimmunologic in nature and to require intimate cell-to-cell contact (3,11-14).

Macrophage lysis of tumor cells appears to be nonselective. Recently, we attempted to select in vitro tumor cell variant lines that exhibit complete resistance to macrophage-mediated lysis. Despite repeated macrophage-tumor interactions, we failed to isolate tumor cells with resistance to macrophage-mediated lysis (15).

ACTIVATION OF MACROPHAGES IN VIVO

In general, past attempts to activate macrophages in vivo to enhance host defense against cancer metastasis have not been successful. This has been due to several reasons: (a) lymphokines injected intravenously have an extremely short half-life, probably because they are rapidly bound to serum proteins or are inactivated (3,16); (b) only a small fraction of the mononuclear phagocyte system may be capable of responding to free lymphokines (17); (c) macrophages can be activated by lymphokines only within a relatively short period of time following their extravasation from the peripheral circulation to the tissues (18); and (d) the tumoricidal properties of lymphokine-activated macrophages are short-lived (2-3 days), and they become refractory to reactivation by free lymphokines (17).

The systemic activation of macrophages with microorganisms or their products also has major limitations. The administration of whole bacteria can activate various effector cells, but it is accompanied by serious toxicity such as allergic reaction and granuloma formation (19). For this reason, little progress had been made until the discovery of N-acetyl-muramyl-L-alanyl-D-isoglutamine (MDP), which has been shown to be the minimal structural unit of Mycobacterium with immune-potentiating activity (4,5) that is capable of activating macrophages (6). Recently, a lipophilic MDP derivative, N-acetyl-muramyl-L-alanyl-D-isoglutamyl-L-alanyl-phosphatidylethanolamine (MTP-PE) has also been synthesized (6). Although MDP influences several macrophage functions in vitro, comparable effects have not been observed in vivo because within minutes after parenteral administration, this molecule is cleared from the body to be excreted in the urine (6). Even when injected at very high doses, MDP fails to induce significant macrophage-mediated antitumor activity (6).

Advances in liposome technology have provided techniques for activating macrophages *in situ* with soluble mediators such as lymphokines and MDP. Liposomes can be used to carry agents to phagocytic cells, since these cells are responsible for the rapid clearance of particulate material from the circulation. The advantages of these carrier vehicles are numerous. Many macrophage-activating agents such as bacterial products or lymphokines can be antigenic, and repeated systemic administrations can lead to adverse reactions. Liposomes, on the other hand, are nonimmunogenic, and elicitation of allergic reactions is uncommon (19-21).

Several criteria must be met for liposomes to be ideal carriers for macrophage activators. Liposomes must: (a) readily bind to and be phagocytosed by microphages; (b) retain the entrapped drug for reasonable periods of time, i.e., be nonleaky; (c) localize to tissues other than those rich in phagocytes, i.e., the lung parenchyma (since the lung is a major site of metastatic disease); and (d) have a relatively long shelf life. In order to better define these characteristics, we have systematically evaluated the role of liposome-lipid composition on macrophage-liposome interactions (20,21).

Through the systematic evaluation of multilamellar liposomes (MLV) phospholipid composition, certain classes of phospholipids that are preferentially recognized by macrophages were identified. The inclusion of negatively charged phospholipids, such as phosphatidylserine (PS) in MLV with phosphatidylcholine (PC) greatly enhances their binding to and phagocytosis by macrophages (17,18,20-22). Indeed, the inclusion of PS within PC MLV leads to recognition of these liposomes by every cell of the reticuloendothelial system. Thus, enhancement of phagocytosis has been shown to occur in mouse peritoneal macrophages, mouse Kupffer cells, rodent alveolar macrophages, human peripheral blood monocytes and human alveolar macrophages (3).

IN SITU ACTIVATION OF MACROPHAGES AND DESTRUCTION OF METASTASES BY IMMUNOMODULATORS ENCAPSULATED IN LIPOSOMES

Studies from our laboratory have shown that immunomodulators entrapped within liposomes can render cells of the monocyte macrophage series tumoricidal much more effectively than free unencapsulated immunomodulators. Efficient *in vivo* activation of mouse macrophages has been achieved by employing liposomes containing entrapped MDP, MTP-PE, IFN-γ, or MAF. In addition, the duration of macrophage activation obtained with liposomes containing entrapped immunomodulators is significantly prolonged, and the required dose of liposome-delivered agents is at least 800 times lower than that required for free macrophage modulators to achieve similar levels of macrophage activation. Moreover, for activators such as MAF, IFN-γ and MDP that are ordinarily degraded or cleared from the body too rapidly for effectiveness, encapsulation in liposomes extends their active half-life within the body and enables these agents to activate macrophages *in situ*.

ERADICATION OF METASTASES BY THE SYSTEMIC ADMINISTRATION OF LIPOSOMES CONTAINING IMMUNOMODULATORS

To test whether systemic administration of liposomes containing immunomodulators could produce destruction of metastases, we injected liposomes containing various immunomodulators into mice bearing spontaneous melanoma metastases. The B16-BL6 melanoma cell line, which is syngeneic to C57BL/6 mice, was used as the model to determine the effectiveness of liposome-encapsulated materials in the treatment of metastases. After implantation in the footpad, this tumor metastasized to lymph nodes and the lungs in more than 90% of the mice (23,24). Mice were given an intrafootpad injection of melanoma cells, and four to five weeks later, when the tumors had reached a size of 10 to 12 mm, the leg bearing the tumor, including the popliteal lymph node, was amputated. Three days after resection of the primary tumor, liposomes containing immunomodulators or placebo preparations were injected intravenously into the mice. Both test and control groups were treated twice weekly for four weeks (8 i.v. injections). We used MLVs consisting of PC/PS since they have been shown to be nontoxic at the doses used and because they are efficiently arrested in the lungs as well as in organs of the reticuloendothelial system following intravenous injection (20,24).

Spontaneous metastases in the lungs and lymph nodes were well established at the time liposome treatment began. Many individual metastases could be seen macroscopically. The mice treated intravenously with saline, with free MAF, with free MDP, or with liposomes containing saline died by day 90 of the experiment. In marked contrast, 66% and 60% of mice injected intravenously with liposome-encapsulated lymphokines or with liposome-encapsulated MDP, respectively, were alive when the experiments were terminated at 200 days. At the start of therapy, we estimate that the metastases contained 10^7 cells. Since the median survival time of mice injected with as few as 10 viable B16 cells (admixed with 10^6 dead cells) is 40 to 50 days (24), the residual tumor burden in the surviving mice must have been reduced to less than 10 viable cells. Similar data on the successful treatment of metastases by the intravenous injections of liposomes containing immunomodulators have been reported for several mouse fibrosarcomas (25,26) and melanomas (27,28), colon carcinomas (25), and spontaneous mouse tumors induced by chronic UV irradiation (29).

IN SITU SYNERGISTIC ACTIVATION OF MOUSE ALVEOLAR MACROPHAGES BY COMBINED IMMUNOMODULATORS ENCAPSULATED WITHIN THE SAME LIPOSOME PREPARATION

Liposomes can deliver more than a single agent to macrophages. Recently, we have shown that the combination of MAF and MDP entrapped in MLV at subthreshold doses leads to significant synergistic activation of macrophages *in vitro* (30) and *in vivo* (31). Moreover, combining suboptimal concentrations of a lymphokine and MDP in the same liposome preparations enhances the efficacy of macrophage-mediated destruction of metastases (31). For the *in vitro* experiments, liposomes containing various dilutions of MAF, various dilutions of MDP, combinations of MAF and MDP, or control liposome preparations were injected intravenously into mice. Alveolar macrophages were harvested 24 hours later, and their antitumor activity was monitored *in vitro*. Liposomes containing combinations of MAF and MDP were most effective in generating cytotoxic properties in lung macrophages. Moreover, this therapeutic regimen was very efficacious in the treatment of relatively large metastatic burdens (31). These therapy experiments described above began 7 days after the surgical removal of the subcutaneous B16 melanoma when lung metastases were large, and some even measured 1 mm in diameter. Liposomes were injected twice weekly for 4 weeks (total of 8 i.v. injections). The data demonstrated that the systemic administration of MLV containing subthreshold amounts of both MAF and MDP was associated with a significant increase in the long-term survival rate of mice (50%, $p = 0.0007$) and demonstrate that the encapsulation of two distinct immunomodulators in the same liposome can lead to marked improvement in treatment of spontaneous metastases.

Synergistic activation of macrophages *in vivo* would be most advantageous if it could be accomplished with well-defined synthetic agents or highly purified lympokines such as IFN-γ. For this reason, we examined whether encapsulation of recombinant IFN-γ and MDP within the same MLV preparations generated synergistic activation of mouse macrophages (8) and human monocytes (9). The delivery of recombinant IFN-γ and MDP to macrophages by liposomes bypassed the necessity of binding unencapsulated lymphokine to the cell surface and was associated with the abrogation of recombinant IFN-γ species specificity, i.e., once encapsulated, mouse IFN-γ or human IFN-γ activated both mouse and human macrophages (10).

EVIDENCE FOR THE PARTICIPATION OF ACTIVATED MACROPHAGES IN TREATMENT OF METASTASES

The regression of lung and lymph node metastases after the systemic administration of liposomes containing immunomodulators required the activation of macrophages to become tumoricidal. Four experiments recommend this conclusion. First, administration of macrophage-activating agents encapsulated within liposomes that are not arrested in the lung capillaries fails to activate lung macrophages(3). Second, the pretreatment of tumor-bearing mice with agents that are toxic for macrophages before systemic therapy with liposome-encapsulated immunomodulators abrogates the response to liposome therapy. Third, intravenous injection of macrophages activated *in vitro* by incubation with liposomes containing MAF effects a reduction in metastatic burden comparable to that achieved by systemic administration of liposome-encapsulated activators (3). Fourth, morphologic and functional analysis of macrophages isolated from pulmonary metastases of mice given intravenous injections of liposomes with MTP-PE showed that macrophages containing phagocytosed liposomes infiltrate and localize within pulmonary metastases. Moreover, once localized, macrophages that have engulfed liposomes containing MDP, but not macrophages that have engulfed liposomes containing a placebo, were tumoricidal against the metastatic tumor cells (3).

CONCLUSIONS

Tumoricidal macrophages can recognize and destroy neoplastic cells without damaging normal cells. Systemic administration of liposomes containing immunomodulators activates rodent macrophages *in situ* and augments host resistance against spontaneous metastases. The stimulation of host antitumor responses by liposome-encapsulated macrophage activators does not require the participation of functional T lymphocytes or natural killer cells. Rather, the destruction of established metastases after multiple systemic administrations of liposome-encapsulated immunomodulators is mediated by activated macrophages. Although these results are encouraging, it is unlikely that the systemic macrophage activation alone could destroy large tumors. Large tumors, however, can be reduced with a variety of other treatment modalities such as radiotherapy or chemotherapy. Tumoricidal macrophages should then be used to destroy those tumor cells that escaped destruction by conventional therapy.

REFERENCES

1. I. J. Fidler and I. R. Hart, Biological diversity in metastatic neoplasms: Origins and implications. Science, 217, 998-1003 (1982).
2. I. J. Fidler and G. Poste, The cellular heterogeneity of malignant neoplasms: Implications for adjuvant chemotherapy. Semin. Oncol. 12, 207-222 (1985).
3. I. J. Fidler, Macrophages and metastases: A biological approach to cancer therapy. Cancer Res. 45, 4714-4726 (1985).
4. L. Chedid, L. Carelli, and F. Audibert, Recent developments concerning muramyl dipeptide, a synthetic immunoregulating molecule. J. Reticuloendothel. Soc. 26, 631-641 (1979).
5. E. Lederer, Synthetic immunostimulants derived from the bacterial cell wall. J. Med. Chem. 23, 819-825 (1980).
6. W. E. Fogler and I. J. Fidler, Modulation of the immune response by muramyl dipeptide. In: Immune Modulation Agents and Their Mechanisms (M.A. Chirigos and R.L. Fenichel, Eds.), pp. 499-512. Marcel Dekker, New York, 1984.
7. I. J. Fidler, The MAF dilemma. Lymphokine Res. 3, 51-54 (1984).
8. I. Saiki and I. J. Fidler, Synergistic activation by recombinant mouse interferon-gamma and muramyl dipeptide of tumoricidal properties in mouse macrophages. J. Immunol. 135, 684-688, (1985).
9. I. Saiki, S. Sone, W. E. Fogler, E. S. Kleinerman, G. Lopez-Berestein, and I. J. Fidler, Synergism between human recombinant gamma-interferon and muramyl dipeptide encapsulated in liposomes for activation of antitumor properties in human blood monocytes. Cancer Res. 45, 6188-6193, (1985).
10. I. J. Fidler, W. E. Fogler, E. S. Kleinerman, and I. Saiki, Abrogation of species specificity for activation of tumoricidal properties in macrophages by recombinant mouse or human gamma interferon encapsulated in liposomes. J. Immunol. 135, 4289-4294, (1985).
11. J. B. Jr. Hibbs. Discrimination between neoplastic and non-neoplastic cells in vitro by activated macrophages. J. Natl. Cancer Inst. 53, 1487-1492, (1974).
12. C. D. Bucana, L. C. Hoyer, A. J. Schroit, E. S. Kleinerman, and I. J. Fidler, Ultrastructural studies of the interaction between liposomes-activated human blood monocytes and allogeneic tumor cells in vitro, Am. J. Pathol. 112, 101-111, (1983).
13. I. J. Fidler, Recognition and destruction of target cells by tumoricidal macrophages. Isr. J. Med. Sci. 14, 177-191, (1978).
14. I. J. Fidler and E. S. Kleinerman, Lymphokine-activated human blood monocytes destroy tumor cells but not normal cells under cocultivation conditions. J. Clin. Oncol. 2, 937-943, (1984).
15. W. E. Fogler and I. J. Fidler, Nonselective destruction of murine neoplastic cells by syngeneic tumoricidal macrophages. Cancer Res. 45, 14-18, (1985).
16. G. Poste, R. Kirsh, W. Fogler, and I. J. Fidler, J. Activation of tumoricidal properties in mouse macrophages by lymphokines encapsulated in liposomes. Cancer Res. 39, 881-892, (1979).
17. G. Poste and R. Kirsh, Rapid decay of tumoricidal activity and loss of responsiveness to lymphokines in inflammatory macrophages. Cancer Res. 39, 2582-2590, (1979).
18. G. Poste, The tumoridal properties of inflammatory tissue microphages and multinucleate giant cells. Am. J. Pathol. 96, 595-606, (1979).
19. A.C. Allison, Mode of action of immunological adjuvants. J. Reticuloendothel. Soc. 26, 619-630, (1979).
20. A. J. Schroit and I. J. Fidler, Effects of liposome structure and lipid composition on the activation of the tumoricidal properties of

macrophages by liposomes containing muramyl dipeptide. Cancer Res. 42, 161-167, (1982).
21. G. Poste, R. Kirsh, and P. Bugelski, Liposomes as a drug delivery system in cancer therapy. In: Novel Approaches to Cancer Chemotherapy (P.S. Sunkara, Ed.), pp. 166-221. Academic Press, New York, 1984.
22. A. J. Schroit, I. R. Hart, J. Madsen, and I. J. Fidler, Selective delivery of drugs encapsulated in liposomes: Natural targeting to macrophages involved in various disease states. J. Biol. Response Modif. 2, 97-100, (1983).
23. I. J. Fidler, Therapy of spontaneous metastases by intravenous injection of liposomes containing lymphokines. Science. 208, 1469-1471, (1980).
24. I. J. Fidler, S. Sone, W. E. Fogler, and Z. L. Barnes, Eradication of spontaneous metastases and activation of alveolor macrophages by intravenous injection of liposomes containing muramyl dipeptide. Proc. Natl. Acad. Sci. U.S.A. 78, 1680-1684, (1981).
25. S. D. Deodhar, B. P. Barna, M. Edinger, and T. Chiang, Inhibition of lung metastases by liposomal immunotherapy in a murine fibrosarcoma model. J. Biol. Response Modif. 1, 27-34, (1982).
26. G. Lopez-Berestein, L. Milas, N. Hunter, K. Mehta, D. Eppstein, M. A. VanderPas, T. R. Mathews, and E. M. Hersh, Prophylaxis and treatment with liposome-encapsulated 6-O-steroyl-N-acetyl muramyl-L-aminobutyryl-D-isoglutamine. Clin. Exp. Metastasis, 2, 366- 367, (1984).
27. N. C. Phillips, M. L. Mora, L. Chedid, P. Lefrancier, and J. M. Bernard, Activation of tumoricidal activity and eradication of experimental metastases by freeze-dried liposomes containing a new lipophilic muramyl derivative. Cancer Res. 45, 128-134, (1985).
28. D. A. Eppstein, M. A. Van Der Pas, E. B. Fraser-Smith, C. G. Kurahara, P. L. Felgner, T. R. Matthews, R. V. Waters, M. C. Venuti, G. H. Jones, R. Metha, G. Lopez-Berestein, Liposome-encapsulated muramyl dipeptide analogue enhances non-specific host immunity. Int. J. Immunother. 2, 115-126, (1986).
29. J. E. Talmadge, B. F. Lenz, R. Klabansky, R. Simon, C. Riggs, S. Guo, R. K. Oldham, and I. J. Fidler, Therapy of autochthonous skin cancers in mice with intravenously injected liposomes containing muramyltripeptide. Cancer Res. 46, 1160-1163, (1986).
30. S. Sone and I. J. Fidler, Synergistic activation by lymphokines and muramyl dipeptide of tumoricidal properties in rat alveolar macrophages. J. Immunol. 125, 2454-2460, (1980).
31. I. J. Fidler and A. J. Schroit, Synergism between lymphokines and muramyl dipeptide encapsulated in liposomes: In situ activation of macrophages and therapy of spontaneous cancer metastasis. J. Immunol. 133, 515-518, (1984).

COMPARISON OF THERAPEUTIC POTENTIAL OF CYTOKINES

James E. Talmadge and Paul L. Black

Preclinical Screening Laboratory, Program Resources, Inc.
National Cancer Institute-Frederick Cancer Research Facility, Frederick, MD 21701[1]

INTRODUCTION

Biological response modifiers (BRMs) are those agents or approaches that influence the relationship between the tumor and host by modifying the hosts response to tumor cells, with resultant therapeutic activity (9). A series of studies with recombinant cytokines designed to better understand their immunomodulatory and therapeutic properties have been undertaken. The present report discusses recombinant murine interferon-gamma (rM IFN-g), recombinant human tumor necrosis factor (rH TNF), rH interleukin-2 (rH IL-2), and rM colony stimulating factor-gm (rM CSF-gm). We discuss the development of Optimal Therapeutic Protocols (OTP) for preclinical studies and initial clinical trials that have been initiated to test the resultant clinical hypotheses.

Despite the appeal of immunotherapy, clinical trials have generally been disappointing, with results inferior to those obtained in a variety of animal models (12). Positive results in experimental systems often appear to be due to the initiation of therapy in normal animals with minimal tumor burden (12). Clinically, metastasis has occurred at the time of diagnosis; therefore, the major problem in cancer treatment, as well as in the development of therapeutic protocols with BRMs, is the control of metastasis.

One approach involves determining the optimal immunomodulatory dose (OID) and protocol (OIP) both in the clinic and in preclinical studies. The OID has been found, at least preclinically, to correlate directly with the optimal therapeutic dose (OTD), both of which differ from the maximum tolerated dose (MTD) (6). Similarly clinical studies with rH IFN-g have revealed that the OID is disparate from the clinical MTD (4,

---

[1]Abbreviations: BRM, biological response modifier; TNF, tumor necrosis factor; IFN, interferon; IL, interleukin; CSF, colony stimulating factor; rM, recombinant murine; rH, recombinant human; OID, optimal immunomodulatory dose; MTD, maximum tolerated dose; OTP, optimal therapeutic protocol; NK, natural killer; LAK, lymphokine-activated killer; BRMP, Biological Response Modifiers Program; CTL, cytotoxic T lymphocyte.

5, 7, 8). The development of an OTP depends upon a complex combination of parameters, including dose, frequency of administration, duration of administration, route of administration, and in the case of combination chemoimmunotherapy, sequence of administration.

RESULTS

Therapeutic activity of rH IL-2. The treatment of metastases with rH IL-2 has a biphasic therapeutic response curve (13). The OTP depends upon the dose and schedule of administration (Table I). The greatest therapeutic activity in mice bearing spontaneous B16 metastases occurs following the daily i.p. administration of 50,000 or 100,000 units of rH IL-2 per animal, as compared with the saline control. In contrast, intermediate doses or less frequent administration of rH IL-2 have less therapeutic activity. The MTD of rH IL-2 is >300,000 units/animal; thus, less toxic, lower levels of IL-2, when given with an OTP, can have significant therapeutic activity. The low dose therapeutic optimum varies from approximately 50 units/animal to 1000 units/animal depending on the length of the hosts' exposure to the tumor and the extent of tumor (i.e., immunologic priming).

Because of the differing pharmacokinetics of IL-2 following i.p. versus i.v. injection (2) and the observation that i.v. administration has significantly less therapeutic activity, the therapeutic activity of rH IL-2 was investigated following continuous infusion. Alzet osmotic pumps (14 day) were placed in the peritoneal cavities of syngeneic tumor-bearing mice. Significant therapeutic activity was observed with the infusion of approximately 14,000 units of rH IL-2 per 24 hour period (approximately 600 units/animal/hour). This level of

Table 1. Treatment of Spontaneous Metastases by i.p. Injection of rH IL-2[a]

Agent	Dose[b] (units/animal)	Median number of metastases (range)	Significance[a]
Saline	-	150 (19->300)	-
rH IL-2	100,000	4 (0-73)	<0.001
rH IL-2	10,000	>300 (8->300)	0.43
rH IL-2	1,000	161 (23->300)	0.72
rH IL-2	100	61.5 (0->300)	0.007
rH IL-2	50	27 (0-200)	<0.001

[a] Syngeneic C57BL/6 mice were injected in a posterior footpad with 5 x 10^4 B16-BL6 cells. Twenty-nine days later, when the tumors were 1 cm in diameter, the tumor-bearing leg was resected. Therapy was initiated 24 hours later. The mice were necropsied 35 days postresection (N = 10).
[b] Mice received saline or rH IL-2 by i.p. injection five times per week for 4 weeks.
[c] Difference in the median number of spontaneous metastases (Mann-Whitney U-test).

activity was significantly greater than that observed at either a log higher or lower dose of rH IL-2. Thus, the chronic infusion of rH IL-2 has therapeutic activity at doses lower than the MTD, as determined by histopathology and clinical toxicology. Furthermore, when given either i.p. or by continuous infusion, rH IL-2 must be chronically administered. Regardless of dose, no therapeutic activity is observed following 1 or 2 weeks of daily administration, whereas 4 weeks of administration resulted in significant therapeutic activity.

One of the most intriguing aspects of the therapeutic activity of rH IL-2 is its biphasic dose response. We suggest that low dose therapeutic activity may be associated with T cells, since the dosage optimum parallels the adjuvant activity of rH IL-2 for effector T cell activity. Low dose therapeutic activity has been observed in four experimental murine tumor models, one guinea pig model (Dr. I. Green), and in an UV-induced autochthonous tumor model, which suggests that this is a highly reproducible observation (10). The hypothesis of an association of low IL-2 dose activity and T cell augmentation is supported by the absence of low dose therapeutic activities in nude mice (13). However, high dose therapeutic activity is present in nude mice at the same doses required to activate natural killer (NK) cells or lymphokine activated killer (LAK) cells in normal mice (13). The observation that intermediate doses suppress the development of cytotoxic T effector cells suggests that dose-responsive helper T cells respond to a rH IL-2 dose that differs from the effective dose for augmentation of suppressor T cells. Although clinical therapeutic activity and toxicity have been noted when rH IL-2 was administered at an aggressive dosage and schedule, most notably in combination with LAK cells or tumor infiltrating lymphocytes (10, 11, 13), the preclinical observation of therapeutic activity at low doses suggests that rH IL-2 might be therapeutically effective using a less toxic protocol. The hypothesis that moderate, nontoxic doses of rH IL-2 have therapeutic activity when administered by continuous infusion has been tested both preclinically (15) and to some extent in the clinic. Clinical studies headed by Dr. William West at Bio Therapeutics, Inc., have shown therapeutic activity when rH IL-2 was administered by continuous infusion in combination with LAK cells (13). Indeed, in a preliminary report, these investigators described responses in 9 of 16 evaluable melanoma patients. Studies by Dr. Paul Sondel at the University of Wisconsin have shown that continuous infusion of rH IL-2 initially produces leukopenia and later results in lymphoid hyperplasia in the peripheral blood and increased LAK cell activity as monitored ex vivo. Thus, with a better understanding of the pharmacokinetics of rH IL-2, it may be possible to increase the agent's therapeutic activity while decreasing its toxicity.

Therapeutic activity of rM IFN-g. An overview of studies to determine the OTP for rM IFN-g against experimental B16 metastases is shown in Table 2. The greatest therapeutic activity was observed when 50,000 units/animal of rM IFN-g was administered i.v. three times per week or daily. Interestingly, i.v. administration of 100,000 units of rM IFN-g per animal produced no therapeutic activity in this experiment, in contrast to the significant activity achieved with injections of 50,000 units/ animal. Moreover, therapeutic activity is absent when IFN-g is injected at 50,000 units/animal twice a week or 10,000 units three times a week. This bell-shaped dose-response curve has been a consistent observation, with the therapeutic optimum observed when rM IFN-g is administered i.v. at 30,000 or 50,000 units/animal tiw (14). Thus, it appears that rM IFN-g has a very narrow window of therapeutic activity that is dose, route, and schedule dependent.

Table 2. Treatment of Experimental Melanoma Metastases with rM IFN-g[a]

Agent	Dose/Animal (U)	Schedule[c]	Route	Median number of metastases (range)	p[b]
Saline	-	tiw	i.v.	>300 (57->300)	-
rM IFN-g	10,000	tiw	i.v.	262 (29->300)	N.S.
rM IFN-g	50,000	biw	i.v.	292 (6->300)	N.S.
rM IFN-g	50,000	tiw	i.v.	46 (8->300)	0.008
rM IFN-g	50,000	qd[c]	i.v.	78 (3->300)	0.008
rM IFN-g	100,000	tiw	i.v.	>300 (8->300)	N.S.

[a]Syngeneic mice (C57BL/6) received 50,000 B16-BL6 tumor cells by i.v. injection, and immunotherapy was initiated 48 hours later. The treatment consisted of i.v. injections using various routes and schedules of rM IFN-g for 4 consecutive weeks. Necropsies were performed on day 35, and the extent of experimental metastasis determined (N = 10).
[b]Probability of no difference in the number of lung nodules compared to mice that received saline control, as determined using the Mann-Whitney U-test. N.S. is not significant.
[c]qd was daily administration 5 days per week.

We have also examined which effector cell is responsible for the therapeutic activity of rM IFN-g (1). In these studies, cohorts of animals were examined at various times, and effector cell activities in the blood, tumor-bearing organ, and spleen were determined. These effector cell activities include cytotoxic T lymphocyte (CTL) activity, NK cell activity, LAK cell activity, and macrophage tumoricidal activity. The effector cell activities with a positive correlation with the median number of pulmonary metastases or host survival included pulmonary (tumor-bearing organ) but not splenic CTL activity and pulmonary but not alveolar macrophage activity. Thus, the therapeutically important mechanisms of immunomodulation with rH IFN-g, at least as it relates to the treatment of metastatic disease, are associated with CTL and macrophage augmentation in the tumor-bearing organ, the lung.

Clinical trials examining immunomodulation with rH IFN-g have revealed a bell-shaped dose-response curve for the activation of macrophages or NK cells following administration, via various routes (4, 5, 8). The immune monitoring in these clinical trials was based on the preclinical observation of a bell-shaped dose-response curve to rM IFN-g (14) as mentioned above. Clinical bell-shaped immunomodulation response curves have been observed with IFN-g or -a in studies directed by Dr. Kirkwood at Yale (5), Dr. Borden at the University of Wisconsin (7), and Dr. Gutterman at M.D. Anderson (47), as well as in studies undertaken by the BRMP in association with Drs. Ron Steis and Annette Maluish (8). Thus, it appears that the absence of clinical therapeutic activity with rH IFN-g may be associated with the administration of doses near the MTD as opposed to the OID. The results of the clinical immune modulating trials previously cited have suggested that rH IFN-g's OID is approximately 0.1 mg/m^2 administered either tiw or daily. Therefore, the BRMP, in conjunction with the Cancer Therapy Evaluation Program is now undertaking a phase III clinical trial to test the therapeutic properties of rH IFN-g at the OID in patients with pulmonary carcinoma.

Therapeutic activity of rH TNF. The results of treating spontaneous metastases with tiw i.v. injections of rH TNF are shown in Table 3. These animals received various doses of rH TNF. When rH TNF was administered to tumor-bearing mice by i.v. injection, 500,000 units/animal was slightly greater than the MTD. However, significant therapeutic activity was observed when 50,000 or 5000 units/animal was administered by i.v. injection (Table 3). In contrast, none of the doses produced significant therapeutic activity when administered i.p. The low level of serum activity observed following i.p. administration (14) may explain the decrease in therapeutic activity when the agent is administered via this route. RH TNF's have T-cell adjuvant activity, as well as macrophage activating properties, suggesting that these may be additional mechanisms of therapeutic activity (14).

When rH TNF is admixed with suboptimal doses of rM IFN-g, a significant increase in therapeutic activity is observed. Unfortunately, when rH TNF and rM IFN-g are injected at doses which individually are nontoxic, an increased level of toxicity is also observed.

Histopathological studies of the toxicity from the combination of rM IFN-g and rH TNF demonstrated systemic multifocal thrombi with secondary ischemic necrosis. This suggests that one mechanism of rH TNF and rM IFN-g toxicity might involve the coagulation system. Preliminary studies revealed that the toxicity of rH TNF and rM IFN-g, as measured by lethality, could be decreased with the injection of aspirin (30 minutes before and 4 hours after BRM administration).

Table 3. Treatment of Spontaneous Melanoma Metastases with rH TNF[a]

Agent	Dose/animal	MST[b]	Median number of metastases (range)	p[c]
Saline	-	>29	11 (16-27)	-
poly(I,C)-LC	10 µg	>29	1 (0-4)	0.000
rH TNF	1,000,000 U	3	0 (0-1)	0.000
rH TNF	500,000 U	17	0 (0-65)	0.027
rH TNF	50,000 U	>29	0 (0-36)	0.034
rH TNF	5,000 U	>29	1 (0-36)	0.020
rH TNF	500 U	>29	25 (0-65)	0.496

[a]B16-BL6 tumor cells were injected into the posterior footpad of each mouse. When the tumors reached a 0.9 cm diameter, the tumor-bearing limb was resected at midfemur to include the popliteal lymph node. Therapy was initiated 1 day later with tiw i.v. injections for 4 weeks. Necropsies were performed on day 29, 1 day after the last injection (N = 10 mice).

[b]MST is median survival time.

[c]Probability of no difference in the number of pulmonary tumor nodules compared to mice that received saline control, as determined using the Mann-Whitney U-test.

Similar results were obtained with corticosteriods and other cyclooxygenase inhibitors.

In summary, rH TNF's therapeutic activity appears to relate mainly to its immunomodulatory and coagulative properties. This hypothesis is supported by the observation that the impressive therapeutic response for the treatment of Meth A sarcomas is related to the i.d. injection of Meth A cells. If Meth A is injected s.c., minimal prolongation of survival is observed. Therefore, we hypothesize that at least some of the therapeutic activity of TNF, especially against i.d. tumors, is associated with the development of ischemic necrosis.

Therapeutic potential of rM CSF and combination chemoimmunotherapy. CSFs are glycoproteins that control the production and function of granulocytes, macrophages, and other hematopoietic cells. Their names relate to their function, i.e., rM CSF-gm acts on a bipotent stem cell, resulting in the production of granulocytes and mononuclear phagocytes. Not too surprisingly, recombinant CSF-gm has not shown any significant therapeutic activity as a single agent. In contrast, the administration of rM CSF to animals that have received lethal doses of cyclophosphamide or irradiation results in significant prolongation of survival and the rescue of a number of animals. The combination therapy with cytotoxic agents and rM CSF has demonstrated therapeutic potential, at least as a myelorestorative measure. However the sequence of administration is critical to the success of the therapeutic protocol since the cell cycling induced by rM CSF places the cells at increased risk to damage by alkylating agents. Thus, it appears that the incorporation of rM CSF into combination chemoimmunotherapy protocols will not be as simple as initially envisioned and that multiple cycles of chemotherapy will have to undertaken very cautiously.

DISCUSSION

At present, cancer immunotherapy appears to have a relatively limited clinical success. This conclusion is in contradistinction to optimistic and premature reports which have resulted in inflated expectations. Nonetheless, several tumor histiotypes have responded to clinical immunotherapy under specific conditions. Our present understanding of immunoregulation and the mechanisms of immunotherapy suggests that we may be able to develop effective clinical therapeutic protocols. Many of the preclinical studies discussed in this report have led to the development of hypotheses that can be tested clinically. This strategy differs from more traditional clinical trials, which rely on an empirical approach centered on the MTD. Although such hypothesis-based clinical trials cannot replace the toxicological aspects of phase I clinical trials, they may take the form of phase "Ib" or phase II trials. We must be aware that immunotherapy, particularly with cytokines, which are not potent cytotoxic-cytostatic agents, will not result in a 99% reduction in tumor cell number. Rather, the therapeutic effects they may have, either directly or indirectly via recovery from myelosuppression or augmentation of effector cell activities, will be slight and require chronic augmentation-activation to produce significant therapeutic activity. Thus, as has been observed with hairy cell leukemia, immunotherapy may require long-term treatment. Because these are potent immunoregulatory products, they will up-regulate effector cell activity, but they may also down- regulate effector cells or produce effects on other parts of the immunological network. Therefore, one must be aware of the possibility of a bell-shaped immunomodulatory and therapeutic response curve, as described here in preclinical studies with rM IFN-g and rH IL-2.

Combination chemoimmunotherapy alone or in combination with autochthonous bone marrow has considerable potential for increased therapeutic application. Surprisingly, in addition to the ability of rM CSF-gm to rescue hosts from lethal or high doses of CTX or irradiation, rH IL-2, rH IL-1, and rM IFN-g can significantly prolong survival and/or protect the host during aggressive chemo- or radio-therapy. In contrast, rH TNF, because of its cytotoxic and cytostatic properties for bone marrow stem cells, has at least an additive toxicity in combination with chemotherapeutic agents. Thus, combination chemoimmunotherapy protocols with BRMs such as the CSFs, the ILs, and the IFNs have considerable clinical potential. This is true not only for the treatment of neoplasia but also for bone marrow transplantation studies.

REFERENCES

1. P. L. Black, H. Phillips, H. Tribble, R. Pennington, M. Schneider, and J. E. Talmadge, Immune responses in tumor-bearing organs and therapeutic activity of rM IFN-G, Fed. Proc. 46:1499 (1987).
2. M. A. Cheever and P. D. Greenberg, In vivo administration of interleukin-2, Contemp. Top. Mol. Immunol. 10:263-282 (1985).
3. M. A. Cheever, J. A. Thompson, D. J. Peace, and P. D. Greenberg, Potential uses of interleukin 2 in cancer therapy, Immunobiology. 172:365-382 (1986).
4. B. S. Edwards, J. A. Merritt, R. C. Fuhlbrigge, and E. C. Borden, Low doses of interferon alpha result in more effective clinical natural killer cell activation, J. Clin. Invest. 75:1908 (1985).
5. M. S. Ernstoff, S. Reich, Y. Nishoda, and J. M. Kirkwood, Immunological assessment of melanoma patients treated with recombinant interferon gamma (rIFN-gamma, Biogen, Inc., Cambridge, MA) in a phase I/II trial, Proc. Am. Assoc. Cancer Res. 26:280 (1985).
6. D. Hartmann, J. S. Adams, A. K. Meeker, M. A. Schneider, B. F. Lenz, and J. E. Talmadge, Dissociation of therapeutic and toxic effects of polyinosinic-polycytidylic acid admixed with poly-L-lysine and solubilized with carboxymethyl cellulose in tumor-bearing mice, Cancer Res. 46:1331-1338 (1986).
7. E. S. Kleinerman, R. Kurzrock, D. Wyatt, J. R. Quesada, J. U. Gutterman, and I. J. Fidler, Activation or suppression of the tumoricidal properties of monocytes from cancer patients following treatment with human recombinant gamma-interferon, Cancer Res. 46:5401-5405 (1986).
8. A. E. Maluish, W. J. Urba, K. Gordon, W. R. Overton, D. Coggin, E. R. Crisp, R. Williams, and S. A. Sherwin, Determination of an optimum biological response modifying (BRM) dose of interferon gamma in melanoma patients, Proc. Am. Soc. Clin. Oncol. in press (1987).
9. E. Mihich and A. Fefer, Biological response modifiers subcommittee report, in: "National Cancer Institute Monograph," 63:1-278 (1983).
10. S. A. Rosenberg, M. T. Lotze, L. M. Muul, S. Leitman, A. E. Chang, S. E. Ettinghausen, Y. L. Matory, and J. M. Skibber, Observations on the systemic administration of autologous lymphokine-activated killer cells and recombinant interleukin-2 to patients with metastatic cancer, N. Eng. J. Med. 313:1485-1492 (1985).
11. S. A. Rosenberg, P. Spiess, and R. Lafreniere, A new approach to

the adoptive immunotherapy of cancer with tumor-infiltrating lymphocytes, Science 233:1318-1322 (1986).
12. J. E. Talmadge and R. B. Herberman, The preclinical screening laboratory: Evaluation of immunomodulatory and therapeutic properties of biological response modifiers, Cancer Treat. Rep. 70:171-182 (1986).
13. J. E. Talmadge, H. Phillips, J. Schindler, H. Tribble, and R. Pennington, A systematic preclinical study on the therapeutic properties of rH IL-2 for the treatment of metastatic disease, Cancer Res. in press (1987).
14. J. E. Talmadge, H. R. Tribble, R. W. Pennington, H. Phillips, and R. H. Wiltrout, Immunomodulatory and immunotherapeutic properties of recombinant gamma interferon and recombinant tumor necrosis factor in mice, Cancer Res. in press (1987).
15. W. H. West, K. W. Tauer, J. R. Yarnelli, G. L. Marshall, D. W. Orr, G. B. Thurman, and R. K. Oldham, Constant infusion recombinant interleukin-2 in adoptive immunotherapy of advanced cancer, Biol. Ther. Cancer A 1986 Update.

ACKNOWLEDGEMENT

By acceptance of this article, the publisher or recipient acknowledges the right of the U. S. Government to retain a nonexclusive, royalty-free license in and to any copyright covering the article.
This research was supported by the National Cancer Institute, DHHS, under contract No. N01-23910 with Program Resources, Inc. The contents of this publication do not necessarily reflect the views or policies of the Department of Health and Human Services, nor does mention of trade names, commercial products, or organizations imply endorsement by the U. S. Government.

CLINICAL TRIALS USING MONOCLONAL ANTIBODIES

John W. Smith and Dan L. Longo

Biological Response Modifiers Program
Division of Cancer Treatment
National Cancer Institute
Frederick, Maryland 21701

Early attempts at passive immunotherapy by administering natural antisera derived from recovered cancer patients to patients with active disease failed to result in significant antitumor responses. Many of the limitations of this serotherapy were remedied by the revolutionary innovation described by Kohler and Milstein (1) of monoclonal antibody production by means of hybridoma technology. Large quantities of homogeneous, highly specific antibody to a defined antigenic determinant became available for clinical trials. Most monoclonal antibodies (mAbs) used to date to treat cancer patients have been murine mAbs. Recently, clinical grade human mAbs have been generated in sufficient quantity to permit investigation.

Monoclonal antibodies injected into the bloodstream are passively carried throughout the body, slowly diffuse into the extracellular space and percolate into the tissues where they specifically bind to cancer cells expressing the appropriate antigen. Gradually, unbound antibody leaves the normal tissues resulting in a relative concencentration of the mAb in the tumor. Studies using biopsies to examine achievable ratios of tumor to normal tissue concentrations of mAb (2-4) indicate that greater than 10:1 ratios are attainable, but not commonly achieved. Furthermore, only 0.005-0.02% of the total injected mAb dose can be detected per gram of tumor.

Monoclonal antibodies can have an antitumor effect by a variety of different mechanisms. Unconjugated antibody can fix complement resulting in cell membrane disruption and lysis. Antibody alone can bind to the Fc receptors expressed by cytolytic effector cells resulting in the concentration of these "armed" cells in the tumor and subsequent cancer cell lysis (antibody dependent cell mediated toxicity, ADCC phenomenon). Cancer cells depend on growth factors to maintain a proliferative advantage and antibodies blocking their receptors can halt the malignant process. Unconjugated murine mAbs have even been theorized (5) to cause immunization of a patient against his own tumor if they provoke a human anti-mouse mAb response against the Fab region and the subsequent development of a second antibody an anti-anti-idiotype antibody (or T cells with the same specificity) which recognizes the tumor antigen. Monoclonal antibodies can be conjugated with a radioisotope, a toxin, a chemotherapeutic agent, or a biological response modifier resulting in selective tumor cell killing if adequate therapeutic ratios are obtained.

Several hundred patients have been safely infused with numerous different murine mAbs. Side effects of fever, chills, diaphoresis, and urticaria are commonly observed, especially in patients with circulating antigen-bearing cells, but are not dose limiting (6). Rare patients have experienced dyspnea associated with rapid antibody infusion thought to be caused by trapping of antigen-coated cells in the lung. More than half of the immunocompetent patients treated with murine mAbs develop human anti-mouse mAb antibodies (HAMA). The HAMA response affects the subsequent clearance of the mAb and can neutralize it in some cases. Patients treated with repetitive courses of mAb have occasionally had allergic reactions including bronchospasm and anaphylaxis which in some studies correlated with the development of the HAMA response (7).

A number of pilot trials using unconjugated mAbs in patients with leukemias and lymphomas have indicated therapeutic efficacy, but responses (except for one) have been partial and transient, limited by the phenomena of antigenic modulation, alterations of the target antigen, and the HAMA response. Administration of the J5 mAb against the CALLA (common acute lymphoblastic leukemia antigen) antigen to four patients with relapsed acute lymphoblastic leukemia resulted in rapid binding of the mAb to the circulating leukemic cells, but the reduction in circulating blasts was short-lived because of the modulation of the CALLA antigen from the cell surface membrane (8). Several patients with chronic lymphocytic leukemia or cutaneous T-cell lymphoma have been treated at different centers with the T101 mAb (directed against the T65 antigen) at a variety of doses and administration schedules (9-15). Tumor cells showed targeting of the mAb and several patients had partial responses of brief duration with antigenic modulation and the HAMA response limiting therapeutic efficacy. At Stanford, eleven patients with B-cell malignancies have been treated with mAbs specific for the idiotype of their malignant cell surface immunoglobulin. This anti-idiotype therapy resulted in a dramatic, long-lived complete response in the first patient (16), but only transient partial responses in five others (17). Blocking antigen in the form of circulating serum idiotype and the presence of HAMA were the major impediments to therapy, however, the emergence of idiotype variants also played a significant role in its failure (18). Recently, five patients with adult T-cell leukemia were treated with the anti-Tac mAb direct against the human interleukin-2 receptor. Two patients had significant, albeit transient, responses and one was successfully retreated (19). Neither side effects nor HAMA response were noted.

Patients with acute lymphoblastic leukemia (ALL) and non-Hodgkin's lymphoma (NHL) have been treated with high-dose chemotherapy and/or irradiation followed by autologous bone marrow transplant with marrow that had been purged by ex vivo treatment with mAb therapy. Purging has not delayed marrow engraftment. The result in ALL patients is similar to that achieved with allogeneic transplants (20). Over one half of the NHL patients have experienced a sustained remission, but this success may be due in part to the requirement that patients be cytoreduced to near-complete remission prior to the treatment program (21). Monoclonal antibodies against contaminating T cells in the donor marrow have been employed to decrease graft vs. host (GVH) disease in patients with acute leukemia undergoing allogeneic bone marrow transplant (22). While marked reductions in GVH disease have been observed, concern has been raised about increased graft failure in these patients (22,23).

Several clinical trials in patients with gastrointestinal carcinoma have utilized the mAb 17-1A, an IgG2a murine immunoglobulin which recognizes a cell surface glycoprotein moiety on adenocarcinoma cells and is postulated to mediate ADCC against tumor cells. Phase I studies showed that single I.V. infusions over a range of doses were well tolerated that the mAb targeted the tumor cells and that no antigenic modulation occurred (24-35).

Repeated infusions occasionally resulted in anaphylaxis (7,28,32). Tumor responses were noted in a minority of patients and with higher frequency in those patients with smaller tumor burden. Additional trials have combined the 17-1A mAb with chemotherapy (33), with interferon gamma (35), with other mAbs (31) or have incubated the mAb in vitro with autologous peripheral blood leukocytes prior to infusions. None of these variations substantially affected response rates. A randomized trial of adjuvant mAb therapy post-surgery for patients with colorectal cancer and high risk of recurrence is currently underway (31).

Several different antibodies have been utilized to treat patients with metastatic malignant melanoma. Twenty patients treated with the 9.2.27 antimelanoma mAb showed tumor localization but none responded (36). Some of the most impressive responses in melanoma patients occurred after treatment with the mAb R24, an IgG3 antibody that detects the disialoganglioside GD3. In vitro this mAb alters tumor cell morphology, exerts a growth inhibitory effect, interferes with cell attachment, activates human complement and mediates ADCC (37). One fourth of the treated patients had major tumor regressions and minor responses occurred in several others (38). Patients experienced pruritus and urticaria around tumor sites or scars from prior resections. Tumor biopsies before and after therapy showed that treatment induced complement deposition and infiltrates of T lymphocytes. Tumor cells showed no evidence of antigenic modulation during therapy and progression of tumor was not related to outgrowth of antigen-negative cells. Lastly, a human IgM mAb L72 that specifically reacts with ganglioside GD2 has been injected intralesionally in patients with cutaneous melanoma resulting in complete or partial responses in most of the lesions (39).

The first clinical studies of a mAb-toxin conjugate in melanoma patients have recently been completed. An IgG2a mAb similar to 9.2.27 conjugated to the ricin A chain was administered in a dose-escalating schema to 22 patients. Dose-limiting toxicities included decreases in total protein, serum albumin, and peripheral edema. In this phase I study where patients received only one 5-day course of therapy, antitumor effects nevertheless occurred with one patient obtaining a complete remission lasting more than 26 months, and four patients showing mixed responses of 2-17 months' duration (40). A follow-up phase II study in previously treated melanoma patients enrolled 46 patients. Patients received 0.4 mg/kg/day for 5 days. Three of 43 evaluable patients had a partial response and ten others had minor responses. All but one patient developed antibodies against both the ricin A chain and the mAb and no patients were retreated (41). An attempt to block the antibody response by pretreatment with cyclophosphamide is currently being investigated.

Radioimmunotherapy - mAb labeled with a radioisotope - has the theoretical advantage of overcoming the problem of antigenic heterogeneity among tumor cells. A phase I trial in melanoma patients using ^{131}I-labeled anti-p97 and anti-chondroitin sulfate proteoglycan Fab fragments demonstrated that multiple doses could be given safely with the major toxicity being marrow suppression. Of the three patients treated at the higher doses, one had a partial response and one a prolonged stabilization (42-44). However, subsequently treated patients have failed to respond (45).

As noted above, attempts to treat cutaneous T-cell lymphomas (CTCL) were frustrated by antigenic modulation. Six patients with previously treated CTCL were given a therapeutic dose of 100-150 mCi of radiolabeled ^{131}I on 10-17 mg of T101 mAb. All patients responded to their initial therapy and two patients have responded to retreatment (46). All patients' chronic pruritus resolved and regression of peripheral adenopathy as well as skin lesions was witnessed. Responses have lasted for 2 weeks to 3 months following one course of therapy. Myelosuppression was noted in the patients treated with 150 mCi ^{131}I.

Because of the difficulty achieving adequate therapeutic ratios and high enough concentrations of mAb in solid tumors treated with I.V. infusions of mAb, some groups have used a regional approach to deliver radioimmunotherapy (47). Six patients with relapsed high grade brain glioma received intra-arterial infusions of ^{131}I-labeled mAb against epidermal growth factor receptor and four patients showed clinical and radiologic improvement (48). Patients with pericardial or pleural effusions were treated with intracavitary ^{131}I-labeled mAb resulting in 12 out of 15 patient responses (49). One ovarian cancer patient who failed chemotherapy and had recurrent malignant ascites was treated with 49 mCi of ^{131}I mAb AUA1 administered intraperitoneally. When the patient's ascites reaccumulated one month later the malignant cells had become negative for the AUA1 mAb and the patient was treated with 39 mCi ^{131}I-labeled HMFG2. This resulted in no ascites reaccumulation for 6+ months (50). Fifteen other patients with ovarian cancer were treated with intraperitoneal injection of ^{131}I-labeled mAb resulting in some complete responses in patients with low volume disease (51).

At the Biological Response Modifiers Program we are evaluating a human anti-colorectal carcinoma mAb in patients with metastatic colorectal cancer. This is one of the first clinical trials of an intravenously injected human mAb. The human IgM mAb 28A32 recognizes an antigen in the cell membrane and cytoplasm of colon cancer cells. The antibody, developed by Bionetics Research, was prepared from a human-mouse heterohybridoma created by fusing NS-1 murine myeloma cells with diploid human B cells from a patient immunized with irradiated autologous colon cancer cells admixed with BCG. The antibody is a completely human antibody and does not fix human complement in vitro.

Patients with measurable metastatic disease whose tumor tissue biopsies stained positively with 28A32 are being treated on a dose-escalating trial designed to assess the pharmacokinetics, toxicity, imaging ability, and therapeutic effects of the mAb. After skin testing and SSKI administration to block thyroidal uptake, all patients received 5 mCi of ^{131}Ilabeled 28A32 intravenously over 2 hours on days 1 and 8. On days 8, 15, 22, and 29 patients received unlabeled antibody at escalating doses. Three patients have been treated at 10 mg and at 50 mg, two patients have received 100 mg. The antibody was labeled using the iodogen method and 97% of the ^{131}I was protein-bound. The immunoreactivity of the radiolabeled mAb against cultured cell lines and in nude mice with a human tumor xenograft was the same as that of the unlabeled mAb. In 6 of 7 patients known sites of metastatic disease were imaged and included sites in liver, lung, and bone. Not all sites of disease were imaged and lesions less than 4 cm were not seen. The one patient who failed to image had high titers of circulating immunoglobulin prior to therapy that competed with 28A32. Tumor to non-tumor contrast improved up to ten days after injection and some lesions were still imaged 23 days after the last dose of radiolabeled mAb. Half of the radiolabeled mAb was cleared from the whole body by 1.5 days; however, there was rapid accumulation of the antibody in the extracellular space. Therapy was very well tolerated and no patient had an allergic reaction. An assay to demonstrate the development of a host IgG anti-28A32 antibody response detected none in any patient four weeks after therapy and none in the two patients that were tested eight weeks after therapy. One patient who did not receive treatment because of a positive skin test reaction had a high titer of IgG anti-28A32 antibody. Only one patient developed a significant rise in the level of immune complexes on treatment as determined by Raji cell and Clq binding assays. No patient had evidence for an anti-tumor effect. The study has demonstrated that the mAb can localize to some sites of metastatic disease. Most importantly it has shown that repeated infusions of human mAb can be safely given without allergic reactions and without provoking host immunoglobulin responses to the mAb. As noted above, the HAMA response limited therapy in several trials because of allergic reactions and the neutralization of the murine mAb by the host immunoglobulin. Human mAbs may overcome this limitation and permit the safe, repeated administration of mAbs.

In summary, a variety of mAbs have been used to treat patients with many different malignancies. In general, therapy has been well-tolerated and produced tumor regressions in some patients. The therapy holds promise but awaits optimization. Future directions include more extensive trials with immunotoxin or radionuclide conjugates, further trials with human mAbs, trials in combination with other mAbs, with other biological response modifiers such as interferon (52) or in combination with effector cells and interleukin-2. Newly developed recombinantly engineered chimeric Abs (53) await testing as well as mAbs against growth factor receptors.

REFERENCES

1. G. Kohler and C. Milstein, Continuous culture of fused cells secreting antibody of predefined specificity. Nature (London) 256: 495-497 (1975).
2. J. Schlom, Basic principles and applications of monoclonal antibodies in the management of carcinomas: The Richard and Hinda Rosenthal Foundation Award Lecture. Cancer Res. 46: 3225-3238 (1986).
3. A.A. Epenetos, D. Snook, H. Durbin, et al., Limitations of radiolabeled monoclonal antibodies for localization of human neoplasms. Cancer Res. 46: 3183-3191 (1986).
4. B.G. Ward, S.J. Mather, J.H. Shepard, et al., Localization of intraperitoneally injected ^{125}I-HMFG2 in patients with ovarian cancer. Lancet ii: 580-581 (1986) (letter).
5. H. Koprowski, D. Herlyn, M. Lubeck, et al., Human anti-idiotype antibodies in cancer patients: Is the modulation of the immune response beneficial for the patient? Proc. Natl. Acad. Sci. U.S.A. 81: 216-219 (1984).
6. R.O. Dillman, J.C. Beauregard, S.E. Halpern, et al., Toxicities and side effects associated with intravenous infusions of murine monoclonal antibodies. J. Biol. Response Mod. 5: 73-84 (1986).
7. A.F. Lobuglio, M. Saleh, L. Peterson, et al., Phase I clinical trial of CO 17-1A monoclonal antibody. Hybridoma 5: 5117-5123 (1986).
8. J. Ritz, J.M. Pesando, S.E. Sallan, et al., Serotherapy of acute lymphoblastic leukemia with monoclonal antibody. Blood 58: 141-152 (1981).
9. R.O. Dillman, D.L. Shawler, R.E. Sobol, et al., Murine monoclonal antibody therapy in two patients with chronic lymphocytic leukemia. Blood 59: 1036 1045 (1982).
10. R.A. Miller and R. Levy, Response of cutaneous T-cell lymphoma to therapy with hybridoma monoclonal antibody. Lancet ii: 226-230 (1981).
11. R.A. Miller, A.R. Oseroft, P.T. Stratte, et al., Monoclonal antibody therapeutic trials in seven patients with T-cell lymphoma. Blood 62: 988-995 (1983).
12. R.O. Dillman, D.L. Shawler, J.B. Dillman, et al., Therapy of chronic lymphocytic leukemia and cutaneous T-cell lymphoma with T101 monoclonal antibody. J. Clin. Oncol. 2: 881-891 (1984).
13. K.A. Foon, R.W. Schroff, P.A. Bunn, et al., Effects of monoclonal antibody therapy in patients with chronic lymphocytic leukemia. Blood 64: 1085-1093 (1984).
14. R.O. Dillman, J. Beauregard, D.L. Shawler, et al., Continuous infusion of T101 monoclonal antibody in chronic lymphocytic leukemia and cutaneous T-cell lymphoma. J. Biol. Response Mod. 5: 394-410 (1986).
15. J.H. Bertram, P.S. Gill, A.M. Levine, et al., Monoclonal antibody T101 in T-cell malignancies: A clinical, pharmacokinetic, and immunologic correlation. Blood 68: 752-761 (1986).
16. R.A. Miller, D.G. Maloney, R. Warnke, et al., Treatment of B-cell lymphoma with monoclonal anti-idiotype antibody. N. Engl. J. Med. 306: 517-522 (1982).

17. T.C. Meeker, J. Lowder, D.G. Maloney, et al., A clinical trial of anti-idiotype therapy for B-cell malignancy. Blood 65: 1349-1363 (1985).
18. J.N. Lowder, T.C. Meeker, M. Campbell, et al., Studies on B lymphoid tumors treated with monoclonal anti-idiotype antibodies: Correlation with clinical responses. Blood 69: 199-210 (1987).
19. T.A. Waldmann, D.L. Longo, W.J. Leonard, et al., Interleukin 2 receptor (Tac antigen) expression in HTLV-1-associated adult T-cell leukemia. Cancer Res. (Suppl.) 45: 4559S-4562S (1985).
20. R.A. Riesfield and M.I. Greeve, Fifth Annual Sapporo Cancer Seminar-Meeting Report. Cancer Res. 46: 2193-2196 (1986).
21. L.M. Nadler, L. Botnick, R. Finberg, et al., Anti-B1 monoclonal antibody and complement treatment in autologous bone-marrow transplantation for relapsed B-cell non-Hodgkin's lymphoma. Lancet ii: 427-430 (1984).
22. W. Heit, D. Bumjes, M. Wiesneth, et al., Ex vivo T-Cell depletion with the monoclonal antibody Campath-1 plus human complement effectively prevents acute graft-vs-host disease in allogeneic bone marrow transplantation. Br. J. Haematol. 64: 479-486 (1986).
23. P.J. Martin, J.A. Hansen, C.D. Buckner, et al., Effects of in vitro depletion of T cells in HLA - Identical allogeneic marrow grafts. Blood 66: 664-672 (1985).
24. H.F. Sears, J. Mattis, D. Herlyn, et al., Phase I clinical trial of monoclonal antibody in treatment of gastrointestinal tumors. Lancet i: 762-765 (1982).
25. H.F. Sears, D. Herlyn, Z. Steplewski, et al., Effects of monoclonal antibody immunotherapy on patients with gastrointestinal adenocarcinoma. J. Biol. Resp. Mod. 3: 138-150 (1984).
26. J.W. Shen, B. Atkinson, H. Koprowski, et al., Binding of murine immunoglobulin to human tissues after immunotherapy with anticolorectal carcinoma monoclonal antibody. Int. J. Cancer 33: 465-468 (1984).
27. H.F. Sears, D. Herlyn, Z. Steplewski, et al., Phase II clinical trial of a murine monoclonal antibody cytotoxic for gastrointestinal adenocarcinoma. Cancer Res. 45: 5910-5913 (1985).
28. H.F. Sears, D. Herlyn, Z. Spteplewski, et al., Initial trial use of murine monoclonal antibodies as immunotherapeutic agents for gastrointestinal adenocarcinoma. Hybridoma 5: 5109-5115 (1986).
29. W.F. Sindelar, M.M. Maher, D. Herlyn, et al., Trial of therapy with monoclonal antibody 17-1A in pancreatic carcinoma: Preliminary results. Hybridoma 5: 5125-5132 (1986).
30. M.A. Tempero, P.M. Pour, E. Uchida, et al., Monoclonal antibody CO17-1A and leukopheresis in immunotherapy of pancreatic cancer. Hybridoma 5: 5133-5138 (1986).
31. J.Y. Douillard, P.A. Lehur, J. Vignoud, et al., Monoclonal antibodies specific immuntherapy of gastrointestinal tumors. Hybridoma 5: 5139-5149 (1986).
32. J. Frodin, P. Biberfield, B. Christensson, et al., Treatment of patients with metastasizing colorectal carcinoma with mouse monoclonal antibodies (Moab 17-1A): A progress report. Hybridoma 5: 5151-5161 (1986).
33. A.R. Paul, P.F. Engstrom, L.M. Weiner, et al., Treatment of advanced measurable or evaluable pancreatic carcinoma with 17-1A murine monoclonal antibody alone or in combination with 5-fluorouracil, adriamycin, and mitomycin (FAM). Hybridoma 5: 5171-5174 (1986).
34. H. Verrill, M. Goldberg, R. Rosenbaum, et al., Clinical trial of Wistar Institute 17-1A monoclonal antibody in patients with advanced gastrointestinal adenocarcinoma: A preliminary report. Hybridoma 5: 5175-5183 (1986).
35. L.M. Weiner, Z. Steplewski, H. Koprowski, et al., Biologic effects of gamma interferon pretreatment followed by monoclonal antibody 17-1A administration in patients with gastrointestinal carcinoma. Hybridoma 5: 565-577 (1986).

36. R.K. Oldham, K.A. Foon, A.C. Morgan, et al., Monoclonal antibody therapy of malignant melanoma: In vivo localization in cutaneous metastasis after intravenous administration. J. Clin. Oncol. 2: 1235-1244 (1984).
37. A. Knuth, Response to monoclonal antibodies in melanoma: Specific or nonspecific? Eur. J. Cancer Clin. Oncol. 22: 1285-1288 (1986).
38. A.N. Houghton, D. Mintzer, C. Cordon-Cardo, et al., Mouse monoclonal IgG3 antibody detecting GD3 ganglioside: A phase I trial in patients with malignant melanoma. Proc. Natl. Acad. Sci. 82: 1242-1246 (1985).
39. R.F. Irie and D.L. Morton, Regression of cutaneous metastatic melanoma by intralesional injection with human monoclonal antibody to ganglioside GD2. Proc. Natl. Acad. Sci. 83: 8694-8698 (1986).
40. L. Spitler, Phase I trial of monoclonal anti-melanoma antibody ricin A chain immunotoxin. Clin. Cancer Lett. 9: 1-2 (1986).
41. L. Spitler, personal communication.
42. S.M. Larson, J.A. Carrasquillo, K.A. Krohn, et al., Localization of ^{131}I-labeled p 97-specific Fab fragments in human melanoma as a basis for radiotherapy. J. Clin. Invest. 72: 2101-2114 (1983).
43. J.A. Carrasquillo, K.A. Krohn, P. Beaumier, et al., Diagnosis of and therapy for solid tumors with radiolabeled antibodies and immune fragments. Cancer Treat. Rep. 68: 317-328 (1984).
44. S.M. Larson, Radiolabeled monoclonal anti-tumor antibodies in diagnosis and therapy. J. Nucl. Med. 26: 538-545 (1985).
45. J.A. Carrasquillo, personal communication.
46. S.T. Rosen, A.M. Zimmer, R.G. Leikin, et al., Radioimmunodetection and radioimmunotherapy of cutaneous T-cell lymphomas utilizing an ^{131}I-labeled T101 monoclonal antibody: An Illinois Cancer Council Study. Blood 68: 241a (1986) (Abstract).
47. N. Courtenay-Luck, A.A. Epenetos, K.E. Halnan, et al., The Hammersmith Oncology Group and the Imperial Cancer Research Fund. Antibody-guided irradiation of malignant lesions: Three cases illustrating a new method of treatment. Lancet i: 1441-1443 (1984).
48. A.A. Epenetos, N. Courtenay-Luck, D. Pickering, et al., Antibody guided irradiation of brain glioma by arterial infusion of radioactive monoclonal antibody against epidermal growth factor receptor and blood group A antigen. Br. Med. J. 290: 1463-1466 (1985).
49. D. Pectasides, S. Stewart, N. Courtenay, et al., Antibody guided irradiation of malignant pleural and pericardial effusions. Br. J. Cancer 53: 727-732 (1986).
50. A.A. Epenetos, G. Hooker,T. Krausy, et al., Antibody guided irradiation of malignant ascites in ovarian cancer: A new therapeutic method possessing specificity against cancer cells. Obstet. Gynecol. 68: 715-745 (1986).
51. A.A. Epenetos, Clinical results with regional antibody-guided irradiation. Cancer Drug Deliv. 2: 233 (1985).
52. J.W. Greiner, P.H. Hand, P. Noguchi, et al., Enhanced expression of surface tumor associated antigens on human breast and colon tumor cells after recombinant human leukocyte gamma-interferon treatment. Cancer Res. 44: 3208-3214 (1984).
53. L.K. Sun, P. Curtis, E. Radkowicz-Szukzynska, et al., Chimeric antibody with human constant regions and mouse variable regions directed against carcinoma-associated antigen 17-1A. Proc. Natl. Acad. Sci. (U.S.A.) 84: 214-218 (1987).

CHEMOPROPHYLAXIS BY INTERFERONS OR INDUCERS AGAINST CHEMICAL CARCINOGENESIS

Ernest C. Borden, Younan A. Sidky, and George T. Bryan

Department of Human Oncology
University of Wisconsin Clinical Cancer Center
Madison, WI 53792

Human malignancies arise commonly after exposure to environmental carcinogens. Bladder cancer is a well-studied and excellent preclinical model for this process. Transitional cell carcinoma (TCC) of the lower urinary tract accounts for approximately 4% of human cancers. Most of the TCC arise in the urinary bladder, with a male:female ratio in the U.S. of about 3:1. Chemical carcinogen-induced mouse and rat tumors in situ and transplantable bladder cancer models, which mimic the pathogenesis of the human disease, have been developed (1-4). Neoplasms can be histologically graded and staged in a manner similar to that for human bladder carcinoma.

Principally used carcinogens have been N-[4-(5-nitro-2-furyl)-2-thioazolyl]formamide (FANFT), N-butyl-N-(4-hydroxybutyl)nitrosamine, and bracken fern (BF). In susceptible species the incidence of carcinoma in a group of exposed animals approaches 90-100% but is a manifestation of carcinogen dose (5). Mucosal hyperplasia occurs within three weeks following administration of FANFT or BF to mice or rats and is followed by superficial mucosal papillary or sessile carcinoma by nine to thirteen weeks (5). These then progress to frank neoplasia. Hyperplasia, though first appearing at three weeks, is biologically reversible despite continued carcinogen administration for an additional three to four weeks. Beyond this time, though carcinogen administration may be discontinued, hyperplastic lesions progress to TCC indicating biological commitment to neoplasia six to eight weeks after carcinogen exposure (5,6). Though light microscopic appearance of the epithelial surface is similar at six and eight weeks, scanning electron microscopy reveals the presence of pleomorphic microvilli on epithelial cells at eight weeks (5,6). These may serve as a morphologic marker of irreversible, and possibly progressive, epithelial proliferative change (6).

Additionally, FANFT-induced neoplasia of the bladder is transplantable subcutaneously or intravesically (2,7). Thus, it is possible to obtain mice or rats bearing in situ, carcinogen-induced preneoplastic or neoplastic bladder epithelial lesions at a defined stage of disease evolution (5), or carrying transplanted intravesical or subcutaneous bladder epithelial neoplasms of defined tumor burden and biological characteristics (2,7,8). These characteristics have permitted the use of FANFT-induced urothelial lesions and TCC in mice or rats for studies of prevention or treatment of bladder cancer. Investigation of a series of cytotoxic antineoplastic agents identified cis-diaminedichloroplatinum as the most effective (8). This drug has been the most effective single agent studied to date for the therapy of human metastatic TCC (9). Other cytotoxic drugs such as adriamycin, cyclophosphamide (CY), 5-fluorouracil, methotrexate or VP-16 have demonstrated some activity against mouse bladder tumor (MBT-2) and FANFT-induced murine TCC, and also have variable, but transient activity in patients with TCC (7,8,10,11).

We recently demonstrated (12,13) that potent interferon (IFN) inducers, polyriboinosinic-polyribocytidylic acid (poly I:C) or pyrimidinones, significantly inhibited MBT-2 growth in mice. In this system, CY combined with poly I:C had an additive antineoplastic effect and at optimal doses 8/14 treated mice were tumor free on Day 60 (vs 0/14 untreated MBT-2 bearing mice) (12). We also reported (14,15) that naturally produced (NP) IFN-β exhibited more antiproliferative activity against 5 human TCC cell lines than did NP-IFN-α. Antiproliferative effects of IFNs were augmented at elevated temperatures against the same human TCC cell lines (16). Of considerable interest is the recent report (17) that IFN-α_2 exhibited activity in superficial TCC following intravesical administration. Minimal local and systemic toxicities were reported in these studies in patients (17). With evidence of _in vitro_ (14,16) and in vivo (17) effects of IFNs on human TCC, we believe that the murine transplantable MBT-2 and FANFT-induced bladder TCC model systems can accelerate data development that would lead to improved design of human clinical trials of IFNs in TCC.

IFNs have antiproliferative effects in a variety of other transplantable or carcinogen-induced rodent tumors (reviewed in 18-21). IFNs or inducers were also protective against spontaneous (23), irradiation-induced (24), virus-induced (25), as well as chemically-induced (26,27) carcinogenesis. An example of inhibition of chemical carcinogenesis by IFNs was reported in mice inoculated with 3-methylcholanthrene (MCA). In this system, none of the 26 mice treated with IFN developed lung adenomas compared to 8/18 for the mice treated with PBS (26). Repeated i.p. administration of poly I:C inhibited two-stage skin carcinogenesis initiated by MCA and promoted by Tween 60. The treatment significantly reduced tumor yield, prolonged latency period, and increased survival time (27).

In addition to TCC, human tumors reported as responsive to IFNs in phase II clinical trials included breast, lung, colorectal, renal, and prostatic carcinomas; myeloma, favorable lymphomas, and leukemias; and melanoma or Kaposi sarcoma (22). Some preclinical studies (21) suggest that IFNs exhibit augmented effects when combined with antineoplastic drugs. Most of the studies in humans or with experimental combination therapies, have used IFN and/or drug schedules that have been semi-empirically derived (17,21,22). Critical questions yet to be addressed include the following: What IFNs will prove most useful for what? What are optimal routes, doses, and schedules of IFN administration? What IFNs and other modalities of therapy can be combined to provide additive or synergistic responses? What are the mechanism(s) of IFNs actions as antineoplastic agents?

Strong rationale exists to further define parameters which enhance effectiveness of IFN treatment and/or prophylaxis using reproducible models of chemical carcinogenesis. As a component of the host response to the malignant process (28), it is not unreasonable to speculate that interferons or other biological response modifiers could eventually be effectively used as chemoprophylactic compounds for human malignancy.

ACKNOWLEDGMENT

Research studies in this area have been supported in our laboratories by the American Cancer Society and the Upjohn Company.

REFERENCES

1. E. Ertürk, J.M. Price, J.E. Morris, S.M. Cohen, R.S. Leith, A.M. VonEsch, and A.J. Crovetti, The production of carcinoma of the urinary bladder in rats by feeding N-[4-(5-nitro-2-furyl)-2-thiazolyl]formamide. Cancer Res. 27: 1998-2002 (1967).
2. E. Ertürk, S.M. Cohen, J.M. Price, and G.T. Bryan, Pathogenesis, histology and transplantability of urinary bladder carcinomas induced by albino rats by oral administration of N-[4-(5-nitro-2-furyl)-2-thiazolyl]formamide. Cancer Res. 29: 2219-2228 (1969).
3. E. Ertürk, S.M. Cohen, and G.T. Bryan, Urinary bladder carcinogenicity of N-[4-(5-nitro-2-furyl)-2-thiazolyl]formamide in female Swiss mice. Cancer Res. 30: 1309-1311 (1970).
4. S.M. Cohen, G.M. Lower, Jr., E. Ertürk, and G.T. Bryan, Comparative carcinogenicity in Swiss mice of N-[4-(5-nitro-2-furyl)-2-thiazolyl]-acetamide and structurally related 5-nitrofurans and 4-nitrobenzenes. Cancer Res. 33: 1593-1597 (1973).
5. G. T. Bryan, The pathogenesis of experimental bladder cancer. Cancer Res. 37: 2813-2816 (1977).
6. J.B. Jacobs, M. Arai, S.M. Cohen, and G.H. Friedell, A long-term study of reversible and progressive urinary bladder cancer lesions in rats fed N-[4-(5-nitro-2-furyl)-2-thiazolyl]formamide. Cancer Res. 37: 2817-2821 (1977).
7. M.S. Soloway, Intravesical and systemic chemotherapy of murine bladder cancer. Cancer Res. 37: 2918-2929 (1977).
8. M.S. Soloway and W.M. Murphy, Experimental chemotherapy of bladder cancer -- Systemic and intravesical. Seminars Oncol. 6: 166-183 (1979).
9. A. Yagoda, Phase II trials with cis-dichlorodiammineplatinum (II) in the treatment of urothelial cancer. Cancer Treat. Rep. 63: 1565-1572 (1979).
10. W.H. Harker, F.S. Freiha, L.D. Shortliffe, F.J. Meyers, J.F. Hannigan, Jr., M.S. Flam, and F.M. Torti, Cisplatin, methotrexate and vinblastine (CMV) chemotherapy for metastatic transitional cell carcinoma of the urinary tract (TCC): evaluation of complete response by site. Proc. Amer. Assoc. Clin. Oncol. 3: 160 (1984).
11. R. Hahn, L. Kvols, S. Frytak, and W. Nichols, Phase II trial of combination VP16 and cis-platinum in advanced measurable bladder cancer. Proc. Amer. Assoc. Clin. Oncol. 3: 162 (1984).
12. E.C. Borden, Y.A. Sidky, D.S. Groveman, and G.T. Bryan, Antitumor effects of polyribonucleotides for mouse transitional cell carcinoma enhanced by cyclophosphamide. Cancer Res. 45: 45-50, 1985.
13. Y.A. Sidky, E.C. Borden, W. Wierenga, and G.T. Bryan, Inhibitory effects of interferon-inducing pyrimidinones on the growth of transplantable mouse bladder tumors. Cancer Res. 46: 3798-3802, (1986).
14. E.C. Borden, N. Yamamoto, T.F. Hogan, B.S. Edwards, and G.T. Bryan, Interferons: preclinical rationale for trials in human bladder carcinoma. In Interferon. Properties, Mode of Action, Production, Clinical Application. Proc. 3rd Intern. Expert Meeting of the Deutsche Stiftung fur Krebsforschung, 1982 (K. Munk and H. Kirchner, Eds.) pp. 42-52. S. Karger-Verlag, Basel.
15. E.C. Borden, D.S. Groveman, T. Nasu, C. Reznikoff, and G.T. Bryan, Antiproliferative activities of interferons against human bladder carcinoma cell lines in vitro. J. Urol. 132: 800-803 (1984).
16. D.S. Groveman, E.C. Borden, J.A. Merritt, H.I. Robins, R. Steeves, and G.T. Bryan, Augmented antiproliferative effects of interferons at elevated temperatures against human bladder carcinoma cell lines. Cancer Res. 44: 5517-5521 (1984).
17. F.M. Torti, L.D. Shortliffe, R.D. Williams, J.T. Spaulding, J.F. Hannigan, Jr., J. Palmer, F.J. Meyers, M. Higgins, and F.S. Freiha, Superficial bladder cancers are responsive to $alpha_2$ interferon

administered intravesically. Proc. Amer. Soc. Clin. Oncol. 3: 160 (1984).
18. E.C. Borden, Interferons: rationale for clinical trials in neoplastic disease. Ann. Int. Med. 91: 472-479 (1979).
19. M. Krim, Towards tumor therapy with interferons, Part I. Interferons: production and properties. Blood 55: 711-721 (1980).
20. M. Krim, Towards tumor therapy with interferons, Part II. Interferons: in vivo effects. Blood 55: 875-884 (1980).
21. E.C. Borden, Interferons and cancer: How the promise is being kept. In: Interferons, Vol. 5, 1984 (I. Gresser, Ed.) pp. 43-83. Academic Press, London.
22. E.C. Borden, Progress toward therapeutic application of interferons, 1979-1983. Cancer 54: 2770-2776 (1984).
23. P.E. Came and D.H. Moore, Effect of exogenous interferon treatment on mouse mammary tumors. J. Natl. Cancer Inst. 48: 1151-1154 (1972).
24. J.K. Ball and J.A. McCarter, Effect of polyinosinic-polycytidylic acid on induction of primary or transplanted tumors by chemical carcinogen or irradiation. J. Natl. Cancer Inst. 46: 1009-1014 (1971).
25. F. Lacour, G. Delage, and C. Chianale, Reduced incidence of spontaneous mammary tumors in C3H/He mice after treatment with polyadenylate-polyuridylate. Science 187: 256-257 (1975).
26. R.A. Salerno, C.E. Whitmire, I.M. Garcia, and R.J. Huebner, Chemical carcinogenesis in mice inhibited by interferon. Nature (London) New Biol. 239: 31-32 (1972).
27. K. Elgjo and M. Degré, Polyinosinic-polycytidylic acid in two-stage skin carcinogenesis. Effect on epidermal growth parameters and interferon induction in treated mice. J. Natl. Cancer Inst. 51: 171-177 (1973).
28. I. Gresser, F. Belardelli, C. Maury, M.T. Maunoury, and M.G. Tovey, Injection of mice with antibody to interferon enhances the growth of transplantable murine tumors. J. Exp.Med. 158: 2095-2107 (1983).

INDEX
(All page numbers refer to the first page of each contribution)

MIX
Papier aus verantwortungsvollen Quellen
Paper from responsible sources
FSC® C105338

If you have any concerns about our products,
you can contact us on
ProductSafety@springernature.com

In case Publisher is established outside the EU,
the EU authorized representative is:
Springer Nature Customer Service Center GmbH
Europaplatz 3, 69115 Heidelberg, Germany

Printed by Libri Plureos GmbH
in Hamburg, Germany